FINITE ELASTICITY

Proceedings of the IUTAM Symposium on

FINITE ELASTICITY

Held at Lehigh University, Bethlehem, PA, USA
August 10–15, 1980

Editors

D. E. Carlson

Department of Theoretical and Applied Mechanics
University of Illinois, Urbana, Illinois, USA

R. T. Shield

Department of Theoretical and Applied Mechanics
University of Illinois, Urbana, Illinois, USA

1982
Martinus Nijhoff Publishers
The Hague/Boston/London

Distributors:

for the United States and Canada
Kluwer Boston, Inc.
190 Old Derby Street
Hingham, MA 02043
USA

for all other countries
Kluwer Academic Publishers Group
Distribution Center
P.O.Box 322
3300 AH Dordrecht
The Netherlands

Library of Congress Cataloging in Publication Data

IUTAM Symposium on Finite Elasticity (1980 : Lehigh University)
Proceedings of the IUTAM Symposium on Finite Elasticity.

1. Elasticity--Congresses. I. Carlson, D. E. (David Emil), 1942- . II. Shield, R. T. III. International Union of Theoretical and Applied Mechanics. IV. Title.
TA418.I87 1980 620.1'1232 81-22540
ISBN 90-247-2629-8 AACR2

ISBN 90-247-2629-8

PRINTED IN THE NETHERLANDS

CONTENTS

FOREWORD

Although finite elasticity theory has its roots in the nineteenth century, its development was largely neglected until the end of the Second World War. Since then it has attracted a substantial amount of attention and considerable progress has been made both in our understanding of the basis of the subject and in its applications.

It occurred to me about three years ago that finite elasticity had reached a level of development at which an international symposium on the subject was overdue. Accordingly, with strong encouragement from Professor P. M. Naghdi and numerous other colleagues, I submitted to the International Union of Theoretical and Applied Mechanics a proposal for their support of such a symposium to be held at Lehigh University during the period August 10–15, 1980.

The proposal received enthusiastic support from the International Union and an international scientific committee under my chairmanship, consisting of Professors G. Fichera (Rome), W. T. Koiter (Delft), L. I. Sedov (Moscow), and A. J. M. Spencer (Nottingham), was assigned responsibility for the scientific program. In constructing the program we aimed at as broad a coverage as possible of the many aspects of the subject on which significant progress is currently being made. These range from theoretical studies of existence and uniqueness of solutions of the governing equations of finite elasticity theory to experimental studies of its application to such problems as tear resistance and friction in vulcanized rubbers.

In addition to the financial support accorded the symposium by the International Union of Theoretical and Applied Mechanics, we were fortunate in obtaining further financial support from the United States Army Research Office, the Office of Naval Research, and the National Science Foundation. I am deeply grateful to all these organizations for their support.

I am grateful also to the members of the Scientific Committee for their wise counsel and friendly cooperation and, above all, to the speakers whose excellent contributions, published here, were the major factor in the success of the Symposium. My thanks go also to Professors D. E. Carlson and R. T. Shield for undertaking the laborious task of editing these Proceedings.

R. S. Rivlin

EXISTENCE OF SOLUTIONS IN FINITE ELASTICITY

J. M. BALL

Department of Mathematics, Heriot-Watt University, Riccarton, Currie, Edinburgh EH14 4AS, Scotland

(Received September 25, 1980)

Very few exact solutions are known to static and dynamic problems of finite elasticity, particularly in the case when the material is compressible. General theorems on existence of solutions provide reassurance that the theory is mathematically sound; for example it is important to understand whether or not solutions of the basic equations have singularities consistent with assumptions used in deriving the equations. But there are several other, equally important, reasons for studying questions of existence of solutions. One such is the establishment of convergence properties for numerical methods in elasticity (in this connection it should be noted that finite-difference schemes for certain partial differential equations may converge to solutions of *different* equations). Experience with other partial differential equations has also taught us that existence theorems are an essential prerequisite for the study of various qualitative properties of solutions (for example, bifurcation, stability and asymptotic behaviour). In a broader context, we today face problems in elasticity similar to unsolved questions in other branches of mechanics and physics, and the unifying nature of the theory of partial differential equations can thus lead us to hope, as has been the case in the past, that advances in elasticity will lead to corresponding progress in other fields. Here, however, we concentrate on a more specific reason for proving, or attempting to prove, existence theorems in elasticity, namely that it leads to information concerning the relationship between constitutive hypotheses (i.e. assumptions on the stored-energy function, or stress–strain law of the material) and smoothness properties of solutions. We will, as much as possible, avoid introducing

Proceedings of the IUTAM Symposium on Finite Elasticity, Lehigh University, August 10–15, 1980. General lecture.

Carlson, D.E. and Shield, R.T. (eds.)
Proceedings of the IUTAM Symposium on Finite Elasticity
ISBN 90 247 2629 8

technicalities from analysis, referring for these to various articles which it is the main purpose of this paper to summarize.

We confine attention mainly to static problems. In general the problem of proving the existence and uniqueness of solutions for initial-boundary value problems of elastodynamics is much harder, and the theory is far from complete even in one dimension. Part of the difficulty in the dynamic case is that solutions which are initially smooth may develop shocks (cf. MacCamy and Mizel [1], Klainerman and Majda [2], John [3]). Short-time existence of smooth solutions for smooth initial data for the case of an elastic body occupying all space has been established by Hughes, Kato and Marsden [4] under a strong ellipticity assumption (see (13) below) on the stored-energy function, but little, if anything, is known about global existence of weak solutions in three dimensions. For the static case we consider only conservative problems, which can be discussed within the framework of the calculus of variations. We do not discuss local existence theorems (cf. Stoppelli [5], van Buren [6], Marsden [7], and Valent [8]) in which solutions nearby a known solution are proved to exist when the data for the known solution is changed slightly. The applicability of such local theorems is severely restricted because there are few trivial solutions around which to perturb, and also by the fact that so far the methods have not been applied successfully to mixed boundary conditions.

We begin by giving two one-dimensional examples in which no absolute minimum for the total energy exists. The total energy $I(x)$ is given by

$$I(x)=\int_0^1 [W(X, x'(X))+\Phi(X, x(X))]dX, \tag{1}$$

where $x(X)$ denotes the deformed position of a particle with position X, $0 \leqq X \leqq 1$, in a reference configuration, $W(X,p)$ is the stored-energy function, $\Phi(X,x)$ the body-force potential, and where $x'(X)=dx(X)/dX$. [Throughout this article we consider deformations $x(\cdot)$ whose derivatives may possess discontinuities. All derivatives of $x(\cdot)$ are to be understood in the sense of distributions (cf. Adams [9]), but while this technicality is important for the statements of the results to be correct, it can be safely ignored at a first reading.] Consider the problem of minimizing (1) subject to the boundary conditions

$$x(0)=0, \qquad x(1)=\alpha>0. \tag{2}$$

Formally, a minimizing function $x(\cdot)$ for (1) subject to (2) will satisfy the Euler–Lagrange, or equilibrium equation,

$$\frac{d}{dX} W_p(X, x'(X)) = \Phi_x(X, x(X)), \qquad 0 < X < 1, \tag{3}$$

where subscripts denote partial derivatives. $W_p(X, \cdot)$ is the stress–strain function of the material at X.

EXAMPLE 1 (cf. [10]).

$$W(X, p) = \frac{(p-1)^2(p-2)^2}{p} + \frac{3}{2} X^2 p,$$

$$\Phi(X, x) = x^2 - \frac{3}{2}, \qquad \alpha = \frac{3}{2}.$$

In this case, an integration by parts shows that

$$I(x) = \int_0^1 \left[\frac{(x'-1)^2(x'-2)^2}{x'} + \left(x - \frac{3X}{2}\right)^2 \right] dX. \tag{4}$$

We are interested only in deformations $x(\cdot)$ for which $x'(X) > 0$, so that interpenetration of matter does not occur and the orientation of the specimen is preserved; the behaviour $W(X, p) \to \infty$ as $p \to 0+$, corresponding to the requirement that infinite energy is required to compress a finite length of the specimen to zero length, will hopefully ensure this. For such functions $x(\cdot)$ the integrand in (4) is strictly positive, and so $I(x) > 0$. But $I(x)$ may be made arbitrarily small by choosing for x a zig-zag line with slopes alternately 1 and 2, approximating closely the line $x = 3X/2$. In fact, by rounding off the corners of such a zig-zag line one can show that *smooth* functions $x(\cdot)$ satisfying (2) exist which make I arbitrarily small. Thus *I does not attain its minimum* subject to (2). Examples of this type are classical in the calculus of variations, seem first to have been discussed by Weierstrass, and are described in the books of Bolza [11] and Young [12]. They occur whenever $W(X, p)$ is not a strictly convex function of p, that is, whenever stress does not increase with strain. When $W(X, p) \to \infty$ as $p \to 0+$ the construction of such examples is possible only when either W or Φ depends explicitly on X (cf. [10]). Although a minimizer for I does not exist, one can think of there being a 'generalized function' (cf. Young [12]), which takes its place. This object consists of the line $x = \frac{3}{2}X$ with a superimposed 'infinitesimal zig-zag' of slopes 1 and 2.

EXAMPLE 2.

$$I(x) = \int_0^1 \left[\frac{1}{x'} + x' + x \right] dX. \tag{5}$$

In this case the strictly convex function $W(x')$ behaves linearly for large x'.

We can think of the problem of minimizing I as corresponding to a vertical bar acted on by gravity, whose lower end $X = 0$ is fixed, and whose upper end is raised to a height $x = \alpha$. Elementary calculations show that if $\alpha \leqq 2$ there is a unique minimizer $x_\alpha(\cdot)$ for I which is smooth for $\alpha > 2$ ($x_2'(\cdot)$ has a singularity at $X = 1$). But if $\alpha > 2$ no minimizer exists. However, in this case minimizing sequences for I tend to a 'generalized function' which consists of $x_2(X)$ for $0 \leqq X < 2$ and the vertical line $\{X = 1,\ 2 \leqq x \leqq \alpha\}$. (W. Noll has remarked to me that an exactly analogous calculation holds for isentropic equilibrium of a polytropic gas under gravity, this model predicting a finite height for the atmosphere.)

In both Examples 1 and 2, no minimizer that is a bona fide function exists, though in each case there is a corresponding 'generalized minimizer'. By assuming that $W(X,\cdot)$ is strictly convex and that $W(X,p)/p \to \infty$ as $p \to \infty$ one can prove the existence of a bona fide minimizer and show that under certain conditions it is smooth. For the details and precise hypotheses of such results see Antman [13] and the references therein, and [10]. Antman treats more general models of rods. For studies of one-dimensional elasticity and viscoelasticity when $W(X,\cdot)$ is not convex (as in Example 1) see Ericksen [14], James [15, 16], Slemrod [17], Andrews [18], and Andrews and Ball [19].

Turning now to three-dimensional elastostatics, we make the simplifying assumptions, for ease of exposition, that the material is homogeneous and that there is no body force.

The problem we consider is to minimize

$$I(x) = \int_\Omega W(\nabla x(X))dX \tag{6}$$

subject to

$$x(X) = \bar{x}(X) \qquad \text{for } X \in \partial\Omega_1, \tag{7}$$

where $\bar{x}$ is a given function. In (6), (7), $x(X)$ denotes the deformed position of a particle occupying the position X in a reference configuration Ω, which we assume to be a bounded, open subset of three-dimensional space with sufficiently smooth boundary $\partial\Omega$. We suppose that $\partial\Omega$ is composed of two disjoint portions $\partial\Omega_1$ and $\partial\Omega_2$, where $\partial\Omega_1$ has positive surface area. $W(F)$ is the stored-energy function of the material. If $x(\cdot)$ minimizes (6) subject to (7) then, formally,

$$\frac{d}{d\varepsilon} I(x + \varepsilon v)\bigg|_{\varepsilon=0} = 0$$

for any smooth function v which vanishes on $\partial\Omega_1$; that is

$$(8) \qquad \int_\Omega \frac{\partial W}{\partial F^i_\alpha}(\nabla x(X))\frac{\partial v^i}{\partial X^\alpha}(X)dX = 0.$$

If (8) holds for all such v we say that x is a *weak solution* of the mixed boundary value problem in which the displacement of $\partial\Omega_1$ is specified by (7) and $\partial\Omega_2$ is traction-free. The reason for this terminology is that if x is a smooth weak solution then integrating (8) by parts and using the arbitrariness of v leads to the equilibrium equations

$$(9) \qquad \frac{\partial}{\partial X^\alpha}\frac{\partial W}{\partial F^i_\alpha}(\nabla x(X)) = 0, \qquad X \in \Omega,$$

and the natural boundary condition

$$(10) \qquad \frac{\partial W}{\partial F^i_\alpha} N_\alpha = 0, \qquad X \in \partial\Omega_2,$$

where $N = N(X)$ denotes the outward normal to $\partial\Omega$ at X. Let $M^{3\times 3}$ denote the set of real 3×3 matrices, and $M^{3\times 3}_+$ consist of those $F \in M^{3\times 3}$ with $\det F > 0$. We introduce the following hypotheses on W:

(H1) $W: M^{3\times 3}_+ \to \mathbb{R}$ is *polyconvex*, i.e. there exists a convex function $g: M^{3\times 3}\times M^{3\times 3}\times(0,\infty)\to\mathbb{R}$ such that

$$W(F) = g(F, \operatorname{adj} F, \det F) \qquad \text{for all } F \in M^{3\times 3}_+,$$

where $\operatorname{adj} F$ denotes the (transposed) matrix of cofactors of F.

(H2) There exist constants $C > 0$, $p \geqq 2$, $q \geqq p/(p-1)$, $r > 1$, $s > 0$ such that

$$W(F) \geqq C(1 + |F|^p + |\operatorname{adj} F|^q + (\det F)^r + (\det F)^{-s})$$

for all $F \in M^{3\times 3}_+$.

THEOREM 1. *Let* (H1), (H2) *hold, and suppose that*

$$\mathcal{A} \stackrel{\text{def}}{=} \{x: \bar{\Omega} \to \mathbb{R}^3 : \det \nabla x(X) > 0 \text{ a.e.}, I(x) < \infty, x_{|\partial\Omega_1} = \bar{x}_{|\partial\Omega_1}\}$$

is nonempty. Then I attains its minimum on $\mathcal{A}$.

(The abbreviation 'a.e.' stands for 'almost everywhere,' meaning 'except possibly on a set of zero volume.')

The theorem is a slight refinement due to Ball, Currie and Olver [20] of a result proved in [21, 22]. These papers can be consulted for a slightly more precise definition of $\mathcal{A}$ and for the detailed proofs. Before discussing the meaning of (H1) and (H2), we show that, at least, these hypotheses are satisfied by some reasonable models of natural rubber. To this end we consider an isotropic material, i.e. one for which

$$W(F) = \Phi(v_1, v_2, v_3)$$

is a symmetric function Φ of the principal stretches $v_i > 0$. We suppose that

$$\Phi = \sum_{i=1}^{M} a_i (v_1^{\alpha_i} + v_2^{\alpha_i} + v_3^{\alpha_i} - 3)$$

$$+ \sum_{j=1}^{N} b_j ((v_2 v_3)^{\beta_j} + (v_3 v_1)^{\beta_j} + (v_1 v_2)^{\beta_j} - 3) + h(v_1 v_2 v_3).$$

Then (H1), (H2) are satisfied (cf. [21, 22]) if

$$\alpha_1 > \cdots > \alpha_M \geqq 1, \quad \beta_1 > \cdots > \beta_N \geqq 1, \quad a_i > 0, \quad b_i > 0, \quad \alpha_1 \geqq 2,$$

$$\beta_1 \geqq \frac{\alpha_1}{\alpha_1 - 1}, \quad h : (0, \infty) \to \mathbb{R} \text{ is convex},$$

$$h(\delta) \geqq k(\delta^r + \delta^{-s}), \qquad \text{for all } \delta > 0,$$

and

$$k > 0, \quad r > 1, \quad s > 0.$$

This class of stored-energy functions is a slight modification of a class introduced by Ogden [23]. We remark that the verification of (H1), and other such convexity conditions, is somewhat simpler when $W(F)$ is expressed in terms of principal stretches than when expressed in terms of the principal invariants of $B = FF^T$.

For incompressible materials a corresponding version of Theorem 1 holds (cf. Ball [21, 22]) and is satisfied by the above material with $h \equiv 0$. Thus we obtain existence for a Mooney–Rivlin material ($M = N = 1$, $\alpha_1 = \beta_1 = 2$), but for a single term stored-energy function

$$\Phi = \mu(v_1^\alpha + v_2^\alpha + v_3^\alpha - 3),$$

$\mu > 0$, we obtain existence only if $\alpha \geqq 3$. The Neo-Hookean material ($\alpha = 2$) is not covered by the theorem.

The polyconvexity condition (H1) is a kind of convexity condition, but it *does not imply that* $W(F)$ *is convex* (as the example $W(F) = \det F$ shows). It is well known that convexity of $W(F)$ is unacceptable physically. There is no direct physical interpretation of polyconvexity known to me, but it is a natural mathematical condition, and, as we shall see below, implies other inequalities having an interpretation in terms of material stability. The mathematical significance of the arguments of the function g in the definition of polyconvexity can be described in several equivalent ways, one of which is the following.

PROPOSITION. *The Euler–Lagrange equations for*

$$\int_\Omega \psi(\nabla x(X))dX$$

are satisfied identically for all smooth functions $x(X)$ if and only if

$$\psi(F) = A + A_i^\alpha F_\alpha^i + B_\alpha^i(\operatorname{adj} F)_i^\alpha + C \det F,$$

where the coefficients are constants.

The earliest proof of (a more general version of) this proposition known to me is that of Landers [24]. For other references and equivalent properties of ψ see [20–22].

To describe the relationship of polyconvexity to other inequalities we make the definitions

DEFINITION 1 (Morrey [25]). W is *quasiconvex* if

$$\int_D W(F + \nabla\phi(Y))dY \geqq \int_D W(F)dY = (\operatorname{vol} D) \times W(F)$$

for all bounded, open sets $D \subset \mathbb{R}^3$, all $F \in M_+^{3\times3}$, and all smooth functions ϕ vanishing on the boundary of D and such that $F + \nabla\phi(Y) \in M_+^{3\times3}$ for all $Y \in D$.

DEFINITION 2. W is *rank* 1 *convex* if

$$W(tF + (1-t)G) \leqq tW(F) + (1-t)W(G) \tag{11}$$

whenever $0 < t < 1$ and $F - G$ has rank 1 (i.e. $F - G = \lambda \otimes \mu \neq 0$; where $(\lambda \otimes \mu)_\alpha^i = \lambda^i \mu_\alpha$ and $a, b \in \mathbb{R}^3$). W is *strictly rank* 1 *convex* if the inequality in (11) is strict.

(Note that the properties of the determinant imply that if $F, G \in M_+^{3\times3}$ with $\operatorname{rank}(F - G) = 1$ then $tF + (1-t)G \in M_+^{3\times3}$ whenever $0 \leqq t \leqq 1$, so that $W(tF + (1-t)G)$ is defined.)

By differentiating twice with respect to t it is easily shown that when W is twice continuously differentiable rank 1 convexity is equivalent to the *Legendre–Hadamard condition*

$$\frac{\partial^2 W}{\partial F_\alpha^i \partial F_\beta^j}(F)\lambda^i\mu_\alpha\lambda^j\mu_\beta \geqq 0 \qquad \text{for all } \lambda, \mu \in \mathbb{R}^3, \tag{12}$$

and that the *strong-ellipticity condition*

$$(13)\qquad \frac{\partial^2 W}{\partial F^i_\alpha \partial F^j_\beta}(F)\lambda^i\mu_\alpha\lambda^j\mu_\beta > 0 \qquad \text{whenever } \lambda, \mu \in \mathbb{R}^3 \text{ are nonzero,}$$

implies that W is strictly rank 1 convex.

We can now state the implications

$$W \text{ polyconvex} \Rightarrow W \text{ quasiconvex} \Rightarrow W \text{ rank 1 convex.}$$

The proofs of these implications and further discussion can be found in [20–22].

To understand the relevance of matrices of rank 1 in elasticity, consider a plane in the reference configuration with normal μ. Then a function x exists such that $\nabla x = F = \text{const.}$ above the plane, $\nabla x = G = \text{const.}$ below the plane, and x is continuous across the plane if and only if $F - G = \lambda \otimes \mu$ for some $\lambda \in \mathbb{R}^3$. Furthermore, such a function x is a weak equilibrium solution if and only if

$$\frac{\partial W}{\partial F^i_\alpha}(F)\mu_\alpha = \frac{\partial W}{\partial F^i_\alpha}(G)\mu_\alpha, \qquad i = 1, 2, 3,$$

which expresses the balance of forces acting on the plane. Suppose now that there exists a natural state $F_0 \in M^{3\times 3}_+$, so that

$$W(F) \geqq W(F_0) \qquad \text{for all } F \in M^{3\times 3}_+.$$

Then we have the following result.

THEOREM 2 ([26]). *Let W be continuously differentiable. Then W is strictly rank* 1 *convex if and only if all weak equilibrium solutions of the above type are trivial* (*i.e.* $F = G$).

Thus strict rank 1 convexity is *necessary* for the smoothness of all weak solutions. Nontrivial piecewise affine weak solutions can occur in the twinning of elastic crystals; and, as has been discussed by Ericksen [27], the stored-energy functions for such crystals are not rank 1 convex. For other work on finite elasticity where rank 1 convexity is not assumed see Knowles and Sternberg [28].

A natural question is now whether strict rank 1 convexity is sufficient for the smoothness of weak equilibrium solutions (presupposing, of course, that W and the other data in the problem are smooth). The answer is negative, and this brings us to a discussion of the growth hypothesis (H2). Note first that (H2) implies that $W(F) \to \infty$ as $\det F \to 0+$. (We could equally well suppose that $W(F) \to \infty$ as $\det F \to b > 0$, and a modified version of Theorem 1 holds.) If F is confined to a bounded region of $M^{3\times 3}_+$ having positive distance from the surface $\{\det F = 0\}$, then by adding a suitable constant to W we see that the inequality in (H2) imposes no restriction on W at all. Indeed if we add to the properties of the set $\mathscr{A}$ of admissible deformations the constraints

$$0 < k \leqq \det \nabla x(X), \qquad \bar{W}(\nabla x(X)) \leqq l, \qquad \text{a.e.},$$

where k, l are constants and $\bar{W}$ is polyconvex, then Theorem 1 holds without assuming (H2) and with $W(F)$ defined only for F satisfying the above constraints. (However, in this case a minimizer x cannot be expected to be an equilibrium solution since in general, for some values of X, $\nabla x(X)$ will lie on the boundary of the constraint set; i.e. $\det \nabla x(X) = k$ or $\bar{W}(\nabla x(X)) = l$.) Thus the growth condition (H2) restricts W only for arbitrarily small $\det F$ or arbitrarily large $|F|$, whereas for real materials one expects such F to be outside the range where elasticity is a good model. Nevertheless the behaviour of $W(F)$ for small $\det F$ and large $|F|$ has important consequences for the existence and smoothness of solutions within the context of pure elasticity theory; we have already seen that this is the case in one dimension. Quantitative estimates in terms of the growth behaviour of W and the magnitude of the boundary data of, for example, the size of the set of points X where $\nabla x(X)$ lies outside a given range might lead to a better understanding of this situation.

Note that (H1) allows, but (H2) excludes, the case of an elastic fluid, for which $W(F) = h(\det F)$; in this case W is rank 1 convex if and only if h is convex. A study of the equilibrium problem for elastic fluids, formulated in material coordinates, and concentrating on the case when h is not convex, can be found in Dacorogna [30].

A sufficient condition for (H2) to hold is that

$$W(F) \geqq C_1(1 + |F|^p + (\det F)^{-s}) \qquad \text{for all } F \in M_+^{3\times 3} \tag{14}$$

where $C_1 > 0$, $p > 3$ and $s > 0$; this follows from the facts that adj F and $\det F$ are quadratic and cubic functions of F respectively. A simple calculation (cf. [22, 29]) shows that a cube of infinitesimally small side ε can be deformed by means of a homogeneous deformation into a parallelepiped having a given finite diameter, with total stored-energy bounded above independently of ε, if and only if

$$\frac{W(F)}{|F|^3} \nrightarrow \infty \qquad \text{as } |F| \to \infty. \tag{15}$$

(Note that (14) and (15) cannot hold simultaneously.) When (15) holds the above calculation suggests the possibility of the existence of weak solutions in which voids or cavities form where none is present in the reference configuration. Such weak solutions exist, and a detailed study of some of them in both the case of compressible and incompressible materials will appear in [31]. Here we give an informal argument exhibiting such solutions in the incompressible case.

We consider radial deformations

$$x(X) = \frac{r(R)}{R} X, \qquad |X| = R, \quad |x| = r$$

of an incompressible, isotropic material which occupies in the reference configuration the unit ball $\{|X|<1\}$. We suppose the boundary of the deformed body to be subjected to uniform radially outward dead load tractions of constant magnitude P. The only kinematically possible deformations have the form

$$r^3 = R^3 + A^3, \tag{16}$$

A representing the radius of the cavity formed in the interior of the body. The total energy I can be expressed in terms of A using (16). Thus

$$E(A) \stackrel{\text{def}}{=} \frac{I(A)}{4\pi} = \int_0^1 R^2 \Phi\left(r', \frac{r}{R}, \frac{r}{R}\right) dR - P(1+A^3)^{1/3}, \tag{17}$$

where the last term represents the potential energy due to the surface tractions. Substituting $v = r/R$ we obtain

$$E(A) = A^3 \int_{(1+A^3)^{1/3}}^{\infty} \frac{v^2}{(v^3-1)^2} \Phi(v^{-2}, v, v)\, dv - P(1+A^3)^{1/3},$$

and thus

$$E'(A) = A^2 \left[\int_{(1+A^3)^{1/3}}^{\infty} \frac{1}{v^3-1} \frac{d\Phi}{dv}\, dv - \frac{P}{(1+A^3)^{2/3}} \right]. \tag{18}$$

The condition for equilibrium is that $E'(A) = 0$. Thus there are two possibilities; either $A = 0$ and $r = R$, or

$$P = (1+A^3)^{2/3} \int_{(1+A^3)^{1/3}}^{\infty} \frac{1}{v^3-1} \frac{d\Phi}{dv}\, dv. \tag{19}$$

The integrals in (17), (18) may or may not converge, depending on the growth properties of Φ. In the special case

$$\Phi(v_1, v_2, v_3) = \mu\,(v_1^\alpha + v_2^\alpha + v_3^\alpha - 3), \tag{20}$$

where $\mu > 0$, these integrals converge provided $-\frac{3}{2} < \alpha < 3$. Note that by (19) the solution with the cavity bifurcates from the trivial solution at the critical traction

$$P_{\text{crit}} = \int_1^{\infty} \frac{1}{v^3-1} \frac{d\Phi}{dv}\, dv. \tag{21}$$

In the case of the Neo-Hookean material, given by (20) with $\alpha = 2$, we obtain $P_{\text{crit}} = 5\mu$ and in this case it is not hard to show that the solution with the cavity minimizes $I(A)$ and is stable, while the trivial solution $r = R$ is unstable for $P > P_{\text{crit}}$. The calculation leading to (21) gives the same result as that of Gent and Lindley [32] in their study of internal rupture of rubber under tension. The reason for this is discussed in [31].

We next turn to the question of whether the energy minimizers whose existence are guaranteed by Theorem 1 are invertible, so that interpenetration of matter does not occur. The following result for a pure displacement boundary-value problem is proved in [33].

THEOREM 3. *Let $\partial\Omega_2$ be empty. Let* (H1), (H2) *hold and suppose further that $p > 3$, $q > 3$, $s > 2q/(q-3)$. Let $\bar{x} \in \mathcal{A}$ be one-to-one in Ω and suppose that $\bar{x}(\Omega)$ satisfies the cone condition. Then the minimizer x is a homeomorphism (continuous with a continuous inverse) of Ω onto $\bar{x}(\Omega)$.*

The cone condition is a mild restriction on the irregularity of the boundary of $\bar{x}(\Omega)$ (cf. [32] for details).

We conclude with a list of three open problems related to the subject matter discussed here.

1. Are the minimizers whose existence are established in Theorem 1 weak solutions of the equilibrium equations?
2. Are they smooth (provided W and the boundary data are smooth)?
3. Does a version of Theorem 1 hold with a weakened version of (H2) allowing for the possibility of cavitation?

REFERENCES

1. R. C. MacCamy and V. J. Mizel, *Existence and non-existence in the large of solutions of quasilinear wave equations*, Arch. Rat. Mech. Anal. **25** (1967), 299–320.
2. S. Klainerman and A. Majda, *Formation of singularities for wave equations including the nonlinear vibrating string*, Comm. Pure and Appl. Math. **33** (1980), 241–263.
3. F. John, article in this volume.
4. T. J. R. Hughes, T. Kato and J. E. Marsden, *Well-posed quasi-linear second order hyperbolic systems with applications to nonlinear elasticity and general relativity*, Arch. Rat. Mech. Anal. **63** (1977), 273–294.
5. F. Stoppelli, *Un teorema di esistenza e di unicità relativo allé equazioni dell' elastostatica isoterma per deformazioni finite*, Ricerche Matematica **3** (1954), 247–267.
6. W. van Buren, *On the existence and uniqueness of solutions to boundary value problems in finite elasticity*, Thesis, Department of Mathematics, Carnegie–Mellon University, 1968. Research Report 68-ID7-MEKMA-RI, Westinghouse Research Laboratories, Pittsburgh, Pa. 1968.
7. J. E. Marsden and T. J. R. Hughes, *Topics in the mathematical foundations of elasticity*, in *Nonlinear Analysis and Mechanics*, Heriot–Watt Symposium, Vol. II (R. J. Knops, ed.), Pitman, London, 1978.
8. T. Valent, article in this volume.

9. R. A. Adams, *Sobolev Spaces*, Academic Press, New York, 1975.

10. J. M. Ball, *Remarques sur l'existence et la régularité des solutions d'élastostatique nonlinéaire*, in *Recent contributions to nonlinear partial differential equations* (H. Berestycki and H. Brezis, ed.), Pitman, Boston, 1981.

11. O. Bolza, *Lectures on the Calculus of Variations*, 3rd ed., reprinted Chelsea, New York, 1973.

12. L. C. Young, *Lectures on the Calculus of Variations and Optimal Control Theory*, W. B. Saunders, Philadelphia, 1969.

13. S. S. Antman, *Ordinary differential equations of nonlinear elasticity II: Existence and regularity theory for conservative boundary value problems*, Arch. Rat. Mech. Anal. **61** (1976), 353–393.

14. J. L. Ericksen, *Equilibrium of bars*, J. of Elasticity **5** (1975), 191–201.

15. R. D. James, *Co-existent phases in the one-dimensional static theory of elastic bars*, Arch. Rat. Mech. Anal. **72** (1979), 99–140.

16. R. D. James, *The propagation of phase boundaries in elastic bars*, Arch. Rat. Mech. Anal. **73** (1980), 125–158.

17. G. Andrews, *On the existence of solutions to the equation* $u_{tt} = u_{xxt} + \sigma(u_x)_x$, J. Differential Equations **35** (1980), 200–231.

18. M. Slemrod, *Inadmissibility of propagating phase boundaries according to certain viscosity criteria*, preprint.

19. G. Andrews and J. M. Ball, *Asymptotic behaviour and changes of phase in one-dimensional nonlinear viscoelasticity*, J. Differential Equations, to appear.

20. J. M. Ball, J. C. Currie and P. J. Olver, *Null Lagrangians, weak continuity, and variational problems of arbitrary order*, J. Functional Analysis, **41** (1981), 135–174.

21. J. M. Ball. *Convexity conditions and existence theorems in nonlinear elasticity*, Arch. Rat. Mech. Anal. **63** (1977), 337–403.

22. J. M. Ball, *Constitutive inequalities and existence theorems in nonlinear elastostatics*, in *Nonlinear Analysis and Mechanics*, Heriot–Watt Symposium Vol. I (R. J. Knops, ed.), Pitman, London, 1977.

23. R. W. Ogden, *Large deformation isotropic elasticity: On the correlation of theory and experiment for compressible rubberlike solids*, Proc. Roy. Soc. London **A328** (1972), 567–583.

24. A. W. Landers, *Invariant multiple integrals in the calculus of variations*, in *Contributions to the Calculus of Variations, 1938–1941*, Univ. Chicago Press, Chicago, 1942, pp. 184–189.

25. C. B. Morrey, *Quasi-convexity and the lower semicontinuity of multiple integrals*, Pacific J. Math. **2** (1952), 25–53.

26. J. M. Ball, *Strict convexity, strong ellipticity, and regularity in the calculus of variations*, Math. Proc. Camb. Phil. Soc. **87** (1980), 501–513.

27. J. L. Ericksen, *Special topics in elastostatics*, in *Advances in Applied Mechanics*, Vol. 17, Academic Press, New York, 1977.

28. J. K. Knowles and E. Sternberg, *On the failure of ellipticity of the equations for finite elastic plane strain*, Arch. Rat. Mech. Anal. **63** (1977), 321–326.

29. J. M. Ball, *Finite time blow-up in nonlinear problems*, in *Nonlinear Evolution Equations* (M. G. Crandall, ed.), Academic Press, New York, 1978.

30. B. Dacorogna, *A relaxation theorem and its application to the equilibrium of gases*, Arch. Rat. Mech. Anal., to appear.

31. J. M. Ball, *Discontinuous equilibrium solutions and cavitation in nonlinear elasticity*, to appear.

32. A. N. Gent and P. B. Lindley, *Internal rupture of bonded rubber cylinders in tension*, Proc. Roy. Soc. London, **A249** (1958), 195–205.

33. J. M. Ball, *Global invertibility of Sobolev functions and the interpenetration of matter*, Proc. Royal Soc. Edinburgh **88A** (1981), 315–328.

ELASTIC STABILITY, BUCKLING AND POST-BUCKLING BEHAVIOUR

W. T. KOITER

Technical University of Delft, Delft, The Netherlands

(Received September 11, 1980)

ABSTRACT

Elastic stability is perhaps the oldest topic in finite elasticity theory. Stability in the sense of Lyapunov is essentially a dynamic concept. Dynamics of continuous media find their proper place under the wings of thermodynamics. The thermodynamic foundation of the theory of elastic stability is now fairly secure, thanks to Duhem and Ericksen. Some subtle mathematical difficulties are still connected with questions of Fréchet differentiability of the elastic energy functional. Applications to specific problems of buckling of structures have preceded the complete development of the theory of elastic stability. This situation is indeed quite common in engineering science. The significance of post-buckling behaviour has been recognized much later. This essentially nonlinear aspect, so characteristic for finite elasticity theory, will be reviewed in some detail.

1. Introduction

We are not aware of earlier work in nonlinear elasticity than Euler's theory of the elastica (1744), and our claim seems therefore justified that elastic stability is the oldest topic in finite elasticity theory. This topic, however, is the subject of many controversies. The most disparaging remarks by Truesdell and Noll [1] have been refuted in our more or less polemic paper [2], and they have also been superseded by the extensive, more balanced and thoughtful review by Knops and Wilkes in the same series [3]. It is now widely recognized that the concept of stability is of a dynamic nature, even if the most popular criteria of stability are quasi-static in character. Their justification, however, depends on a dynamic analysis. There is also fairly wide agreement on the need of an appeal to continuum thermodynamics in the dynamic discussion of elastic stability theory. On the other hand, there seems

Proceedings of the IUTAM Symposium on Finite Elasticity, Lehigh University, August 10–15, 1980. General lecture.

Carlson, D.E. and Shield, R.T. (eds.)
Proceedings of the IUTAM Symposium on Finite Elasticity
ISBN 90 247 2629 8

to be still a divergence of views on the result of the thermodynamic analysis. Some writers find it difficult to accept the isothermal character of the energy criterion. The present survey is offered as an introduction to the subdomain of elastic stability as an essential part of the IUTAM symposium on finite elasticity.

2. Thermodynamics of elastic stability [24, 25]

We identify a particle of the body by the set $\boldsymbol{x}$ of Cartesian coordinates x_i $(i = 1, 2, 3)$ in the fundamental state I of equilibrium and rest which is to be investigated as to its stability. Following upon an initial disturbance of the pre-buckling equilibrium at time $t = 0$ a motion develops, described by a time-dependent displacement field $\boldsymbol{u}(\boldsymbol{x}, t)$ for $t > 0$. The (additional) deformation associated with this displacement field is specified by the (additional) strain tensor $\boldsymbol{\gamma}(\boldsymbol{x}, t)$ with components

$$\gamma_{ij} = \frac{1}{2}[u_{i,j} + u_{j,i} + u_{h,i}u_{h,j}], \tag{2.1}$$

where partial differentiation with respect to a coordinate x_i is denoted by an additional subscript i preceded by a comma. We also employ the summation convention with respect to a repeated subscript.

The first basic property of an elastic material according to our definition is that the thermodynamic state is specified completely by the internal energy density U and by the deformation described by the (additional) strain tensor $\boldsymbol{\gamma}$. The second basic property of an elastic material is expressed by the assumption that any pair of states described by admissible state variables U and $\boldsymbol{\gamma}$ may be connected by a reversible (infinitely slow) process. These properties imply the existence of an entropy density s, and the elastic material may then be characterized by the internal energy density as a function of the specific entropy and the deformation $U(s, \boldsymbol{\gamma})$. Since both the internal energy and the entropy contain an arbitrary additive constant, we may take both U and s equal to zero in the fundamental pre-buckling state I.

We emphasize that our specification of an elastic material by an internal energy density $U(s, \boldsymbol{\gamma})$ does not exclude the occurrence of irreversible internal processes in the motion of an elastic body with non-vanishing strain rates. Our basic assumptions only imply that a reversible (infinitely slow) process always exists for the transition of every particle to a new state of specific entropy and deformation. The internal stresses in such a reversible process are given by the partial derivatives of the internal energy density with respect to the strain components. For finite values of the strain rates, however, additional stresses arise, Ziegler's so-called irreversible stresses [4]. In equilibrium situations the additional viscous stresses vanish, and in such cases there is no difference between the more conventional specification of elasticity and our description. There is ample evidence, however, that viscous stresses play an essential role in elastodynamics. Since elastic stability is essentially a dynamic concept it seems therefore no more than reasonable, if not imperative, to allow for such irreversible phenomena in the analysis. Moreover, the inclusion of such dissipative stresses facilitates the discussion of stability and instability, and their exclusion by a rigid adherence to a more restricted definition of elasticity [e.g. 3, p. 173] can thus hardly be recommended in our opinion.

The absolute temperature is a scalar field $T(\boldsymbol{x}, t)$, specified by

$$T = \frac{\partial U}{\partial s} > 0. \tag{2.2}$$

Essential in our later argument is that the specific heat at constant deformation

$$C_\gamma = T\left(\frac{\partial^2 U}{\partial s^2}\right)^{-1} > 0 \tag{2.3}$$

is positive for all admissible states. This property of material stability enables us to solve equation (2.2) for the specific entropy in terms of the absolute temperature T and the deformation $\boldsymbol{\gamma}$. The free energy per unit mass, defined by

$$U - s\frac{\partial U}{\partial s} = U - Ts = F(T, \boldsymbol{\gamma}) \tag{2.4}$$

may thus also be considered as a function of temperature T and deformation $\boldsymbol{\gamma}$.

No heat flow occurs in the pre-buckling fundamental state of equilibrium and rest, and the temperature has therefore a time-independent and uniform value T_I in this state I. This assumption of equilibrium and rest in state I evidently implies that the surroundings of the body also has the constant and uniform temperature T_I. We now assume that any increase (decrease) in the surface temperature of the elastic body in its motion following upon an initial disturbance results in an outward (inward) heat flow. Let $\boldsymbol{q}(\boldsymbol{x}, t)$ denote the heat flow vector in the body. Our assumption is now expressed by the inequality holding at every point of the surface

$$(T - T_I)\boldsymbol{q} \,.\, \boldsymbol{n} \geqq 0, \tag{2.5}$$

where $\boldsymbol{n}$ is the outward unit normal vector. Associated with the heat flow field $\boldsymbol{q}(\boldsymbol{x}, t)$ is an entropy flow described by a vector field $\boldsymbol{h}(\boldsymbol{x}, t) = T^{-1}\boldsymbol{q}(\boldsymbol{x}, t)$. Let

$\sigma(\boldsymbol{x}, t)$ denote the rate of entropy production per unit mass in the body due to all irreversible internal processes. In the absence of internal heat sources the equation of entropy balance is now expressed by

$$\int \rho \dot{s} dv = -\int \boldsymbol{h} . \boldsymbol{n} dA + \int \rho \sigma dv, \tag{2.6}$$

where $\dot{s}(\boldsymbol{x}, t)$ is the time rate of the specific entropy, $\rho(\boldsymbol{x})$ is the mass density in state I, and the integrals refer to the body and its surface in this fundamental state. The rate of entropy production $\sigma(\boldsymbol{x}, t)$ is always non-negative by the inequality of Clausius and Duhem. We shall assume that the strict inequality $\sigma(\boldsymbol{x}, t) > 0$ always holds whenever the heat flow and/or the strain rates are non-zero.

In our formulation of the energy balance in the motion following upon the initial disturbance we have to consider, in addition to the internal energy, the kinetic energy, the work of the external loads and the heat supply through the surface. We assume that the external loads are conservative. This assumption implies that the external loads have a potential energy, a functional $P_I[\boldsymbol{u}(\boldsymbol{x}, t)]$ of the time-dependent displacement field. Here again the zero level of the potential energy, as in the case of the internal energy and entropy densities, is identified with the fundamental state I. The kinetic energy is expressed in terms of the velocity field $\dot{\boldsymbol{u}}(\boldsymbol{x}, t)$

$$K[\dot{\boldsymbol{u}}(\boldsymbol{x}, t)] = \int \frac{1}{2} \rho \dot{\boldsymbol{u}} . \dot{\boldsymbol{u}} dv. \tag{2.7}$$

The first law of thermodynamics is now expressed by the equation of energy balance

$$\frac{d}{dt}\left\{\int \rho U(s, \boldsymbol{\gamma}) dv + P_I[\boldsymbol{u}] + K[\dot{\boldsymbol{u}}]\right\} = -\int \boldsymbol{q} . \boldsymbol{n} dA. \tag{2.8}$$

Following Duhem [5] we multiply both members of (2.6) by T_I and subtract the result from (2.8). We obtain the modified balance equation

$$\begin{aligned} &\frac{d}{dt}\left\{\int \rho[U(s, \boldsymbol{\gamma}) - T_I s] dv + P_I[\boldsymbol{u}] + K[\dot{\boldsymbol{u}}]\right\} \\ &\quad = \int \left(\frac{T_I}{T} - 1\right) \boldsymbol{q} . \boldsymbol{n} dA - T_I \int \rho \sigma dv \leqq 0, \end{aligned} \tag{2.9}$$

where the strict inequality holds for the right-hand member whenever the heat flow and/or the strain rates are non-zero. In view of the positive definite character of the kinetic energy, inequality (2.9) enabled Duhem to conclude by a Lyapunov-type argument to stability of equilibrium in the fundamental state I whenever the quantity

$$\int \rho[U(s,\gamma) - T_I s]dv + P_I[\boldsymbol{u}] \tag{2.10}$$

is positive definite as a functional of the specific entropy $s(\boldsymbol{x})$ and the displacement field $\boldsymbol{u}(\boldsymbol{x})$.

A drawback of Duhem's stability criterion based on the functional (2.10) is that it is thermomechanical in character, involving both thermal and mechanical quantities. The next and decisive advance was achieved by Ericksen [6, 7] who succeeded in separating the thermal and mechanical quantities in the stability criterion. By means of (2.4) and the implied relation

$$s(T,\gamma) = -\frac{\partial}{\partial T} F(T,\gamma). \tag{2.11}$$

Ericksen writes

$$U(s,\gamma) - T_I s = F(T,\gamma) + (T_I - T)\frac{\partial F}{\partial T}, \tag{2.12}$$

where the derivative is to be evaluated at the current temperature T and deformation γ. This suggests a Taylor-expansion of the free energy around the current temperature

$$F(T_I,\gamma) = F(T,\gamma) + (T_I - T)\frac{\partial F}{\partial T} + \frac{1}{2}(T_I - T)^2\left(\frac{\partial^2 F}{\partial T^2}\right)^*, \tag{2.13}$$

where the second, starred derivative is to be evaluated at an intermediate temperature T^* and the current deformation γ. By differentiation of (2.11) we obtain

$$\left(\frac{\partial^2 F}{\partial T^2}\right)^* = -\frac{C\gamma^*}{T^*} \tag{2.14}$$

in terms of the specific heat $C\gamma^*$ at the constant deformation γ and the intermediate temperature T^*. Substituting from (2.13) and (2.14) into (2.12), we may rewrite (2.10) in the form

$$\begin{aligned} &\int \rho[U(s,\gamma) - T_I s]dv + P_I[\boldsymbol{u}] \\ &= \int \rho\left[F(T_I,\gamma) + \frac{1}{2}\frac{C\gamma^*}{T^*}(T_I - T)^2\right]dv + P_I[u]. \end{aligned} \tag{2.15}$$

We introduce the elastic energy density for isothermal deformations, defined per unit volume in the fundamental state I

$$W(\gamma) = \rho F(T_I, \gamma), \tag{2.16}$$

and the potential energy functional

$$P[u(x)] = \int W(\gamma) dv + P_l[u(x)], \tag{2.17}$$

which is indeed in view of (2.1) a functional of the displacement field $u(x)$ alone. We also introduce the functional

$$\begin{aligned} &V[u(x,t), \dot{u}(x,t), T(x,t)] \\ &\qquad = P[u] + K[\dot{u}] + \int \frac{1}{2} \rho \frac{C\gamma^*}{T^*} (T_I - T)^2 dv, \end{aligned} \tag{2.18}$$

with the following inequality for its time rate

$$\frac{d}{dt} V[u, \dot{u}, T] = \int \left(\frac{T_I}{T} - 1\right) q \,.\, n dA - T_I \int \rho \sigma dv \leqq 0. \tag{2.19}$$

The positive definite character of the second and third terms in (2.18) now permits us to infer the stability of equilibrium in the fundamental state whenever the potential energy functional (2.17) is also positive definite. A more precise formulation of this result will be discussed in the next section.

3. Formulation and application of the energy criterion

We introduce a convenient norm $\|u\|$ in the space of admissible displacement fields $u(x)$, for example the L_2-norm. The potential energy functional is now called positive definite, if its greatest lower bound on a sphere in the space of displacement fields

$$\inf_{\|u\|=c} P[u] = d(c) \tag{3.1}$$

is a strictly increasing function in a range $0 < c < c_1$. The potential energy functional is called indefinite whenever it has negative values within any ball $\|u\| < c$ no matter how small the positive number c is chosen.

The stability theorem in the case of a positive definite potential energy functional (2.17) is now expressed by the statement that the norm $\|u\|_t$ of the displacement field $u(x, t)$ for all positive values of time is bounded above by c whenever the initial conditions satisfy a similar bound $\|u\|_0 < c$ in addition to the energy bound

$$V_0 = V[u(x,0), \dot{u}(x,0), T(x,0)] < d(c). \tag{3.2}$$

In the absence of any alternative configuration of equilibrium in a ball

$\|u\| < c_1$ around the fundamental state the positive definite potential energy functional even ensures asymptotic stability.

The converse of the stability theorem, i.e. the theorem on instability, is proved on the basis of two additional assumptions. First of all it is assumed, similar to the assumption made in the theorem on asymptotic stability, that no alternative equilibrium configuration, here with a negative level of potential energy, exists within a ball $\|u\| < c_1$ around the fundamental state. Secondly it is assumed that the body is properly supported such that the energy dissipation is positive definite not only in the strain rates but also in the particle velocities. Under these assumptions the equilibrium in the fundamental state I is unstable in the case of an indefinite potential energy functional (2.17).

A graphic description of the stability theorem consists of the statement that the existence of a potential energy barrier around the fundamental state ensures stability. Conversely, the existence of any loop-hole in such a barrier, containing a descending energy path, implies instability.

In order to apply the energy criterion of stability we have to expand the potential energy functional (2.17) in the vicinity of the fundamental state I. The expansion of the elastic energy density (2.16) reads

$$\begin{aligned} W(\boldsymbol{\gamma}) &= \frac{\partial W}{\partial \gamma_{ij}}\gamma_{ij} + \frac{1}{2}\frac{\partial^2 W}{\partial \gamma_{ij}\partial \gamma_{kl}}\gamma_{ij}\gamma_{kl} + \cdots \\ &= S_{ij}\gamma_{ij} + \frac{1}{2}E^{I}_{ijkl}\gamma_{ij}\gamma_{kl} + \cdots, \end{aligned} \tag{3.3}$$

where S_{ij} is the Cauchy stress tensor in the fundamental state and E^{I}_{ijkl} is the tensor of elastic moduli for strain increments in that state. The potential energy of dead external loads is linear in the displacement field $\boldsymbol{u}(\boldsymbol{x})$, and it is cancelled by the volume integral of the linear term in (3.3). The potential energy functional thus takes the form for dead loading conditions

$$P[\boldsymbol{u}(\boldsymbol{x})] = \int \left(\frac{1}{2}S_{ij}u_{h,i}u_{h,j} + \frac{1}{2}E^{I}_{ijkl}\gamma_{ij}\gamma_{kl} + \cdots\right) dv. \tag{3.4}$$

In the case of other conservative loads we have still to add the non-linear part of the potential energy of the external loads.

It will be convenient to write

$$\gamma_{ij} = \theta_{ij} + \frac{1}{2}u_{h,i}u_{h,j}, \tag{3.5}$$

where $\boldsymbol{\theta}$ is the linear part of the (additional) strain tensor $\boldsymbol{\gamma}$. A necessary condition for stability in the case of dead loads is now that the second variation

$$P_2[\boldsymbol{u}(\boldsymbol{x})] = \int \left(\frac{1}{2} S_{ij} u_{h,i} u_{h,j} + \frac{1}{2} E^{I}_{ijkl} \theta_{ij} \theta_{kl}\right) dv \geqq 0 \tag{3.6}$$

is non-negative for all admissible displacement fields $\boldsymbol{u}(\boldsymbol{x})$. Unfortunately there is no hope in the classical three-dimensional theory of elasticity, where the elastic energy density W is a function of the displacement gradient $\nabla \boldsymbol{u}$, to establish further conditions under which a positive definite second variation (3.6) would represent a sufficient condition for a positive definite potential energy functional (3.4). This would indeed require that the second variation (3.6) is a continuous second derivative of (3.4) in the sense of Fréchet, and Martini has established unequivocally that this cannot be the case [8]. In other words, a positive definite second variation of the energy is inadequate to ensure an energy barrier around the fundamental state.

Even if it is impossible to prove the existence of an energy barrier around the fundamental state of equilibrium from the criterion of a positive definite second variation of the energy, the latter criterion may still be adequate for stability. In fact as we suggested in [14], the occurrence of irreversible viscous stresses in elastodynamics may possibly be employed to vindicate the energy criterion. Potier–Ferry has recently announced such results to be published shortly [10, 11].

It is also worthwhile to note that the situation is much more satisfactory for one-dimensional and two-dimensional structures such as thin rods or thin plates and shallow shells. The simplified potential energy functionals of these structures contain positive definite quadratic terms in the second derivative of the normal deflection. These additional terms enable us to establish a positive definite second variation directly as a sufficient condition for stability [12], or indirectly by showing first that the energy functional is at least twice continuously differentiable in the Fréchet sense [13]. A similar result has been achieved in the three-dimensional theory by the introduction of additional positive definite terms in the elastic potential involving the strain gradients $\nabla \boldsymbol{\gamma}$ [9]. Combined with the unquestionable success of the energy criterion of stability in all applications in engineering we shall not hesitate to continue our reliance on the second variation of the potential energy as a crucial test of stability [2, 14].

4. The critical case of neutral equilibrium [15, 16]

Having accepted the criterion of the second variation, the equilibrium in the fundamental state is stable or unstable accordingly as the second variation is positive definite or indefinite. The equilibrium is at the stability limit, if the second variation is still non-negative, but equal to zero for one or more special displacement fields, the buckling modes. Such a so-called critical case of neutral equilibrium is thus characterized by a zero minimum of the second variation (3.6) attained for a non-vanishing displacement field. The variational equation for such neutral equilibrium reads

$$\int (S_{ij}u_{h,i}\zeta_{h,j} + E^{I}_{ijkl}u_{i,j}\zeta_{k,l})dv = 0, \tag{4.1}$$

holding for all admissible virtual displacement fields $\boldsymbol{\zeta}(\boldsymbol{x})$. In most cases of neutral equilibrium equation (4.1) has a unique solution $\boldsymbol{u}_1(\boldsymbol{x})$, apart from an arbitrary constant factor, but it may happen that several linearly independent solutions $\boldsymbol{u}_h(\boldsymbol{x})$, $h = 1, 2, \cdots, m$ exist. In that case we may orthogonalize and normalize these buckling modes, and the general solution of (4.1) is then a linear combination of buckling modes, $a_h\boldsymbol{u}_h(\boldsymbol{x})$, if we employ the summation convention here from $h = 1$ to $h = m$.

In a critical case of neutral equilibrium the second variation is clearly unable to give a verdict on stability or instability. We have to continue the expansion of the complete energy functional (3.4) to higher order terms

$$P[\boldsymbol{u}(\boldsymbol{x})] = P_2[\boldsymbol{u}(\boldsymbol{x})] + P_3[\boldsymbol{u}(\boldsymbol{x})] + P_4[\boldsymbol{u}(\boldsymbol{x})] + \cdots. \tag{4.2}$$

Since $P_2[\boldsymbol{u}(\boldsymbol{x})]$ vanishes for the linear combination of buckling modes $a_h\boldsymbol{u}_h(\boldsymbol{x})$ we have as obvious further necessary conditions for stability

$$P_3[a_h\boldsymbol{u}_h(\boldsymbol{x})] \equiv A_{hij}a_ha_ia_j = 0, \qquad P_4[a_h\boldsymbol{u}_h(\boldsymbol{x})] \geqq 0. \tag{4.3}$$

In order to derive an additional necessary, and now also sufficient condition for stability we write

$$\boldsymbol{u}(\boldsymbol{x}) = a_h\boldsymbol{u}_h(\boldsymbol{x}) + \boldsymbol{v}(\boldsymbol{x}), \tag{4.4}$$

where $\boldsymbol{v}(\boldsymbol{x})$ is orthogonal with respect to all buckling modes. Since $P_2[\boldsymbol{v}(\boldsymbol{x})]$ is now positive definite we may minimize the potential energy (4.2) with respect to $\boldsymbol{v}(\boldsymbol{x})$ for fixed (small) values of the amplitudes of the buckling modes. Omitting all details of the analysis the minimizing field $\boldsymbol{v}_{\min}(\boldsymbol{x})$ takes the form

$$\boldsymbol{v}_{\min}(\boldsymbol{x}) = a_ha_k\boldsymbol{v}_{hk}(\boldsymbol{x}) + O(a^3), \tag{4.5}$$

where the fields $v_{hk}(\boldsymbol{x})$ are uniquely defined, and the error is of third order in the amplitudes of the buckling modes. The necessary condition for stability in the critical case of neutral equilibrium, in addition to (4.3), is now

$$A_{hijk}a_ha_ia_ja_k \equiv P_4[a_h\boldsymbol{u}_h(\boldsymbol{x})] - P_2[a_ha_k\boldsymbol{v}_{hk}(\boldsymbol{x})] \geqq 0, \tag{4.6}$$

and it is also sufficient for the strict inequality sign in (4.6).

The foregoing analysis of stability in a critical case of neutral equilibrium is formal in the same sense as the criterion of the second variation. Condition

(4.6) is necessary and sufficient for the existence of a proper weak minimum in the sense of the calculus of variations. In the case of one- and two-dimensional structures (thin rods, thin plates and shallow shells) these conditions may also be shown to imply the presence of an energy barrier around the fundamental state, but in the three-dimensional case we face the same difficulties as in the criterion of the second variation.

5. Post-buckling behaviour [15, 16, 17]

We shall now consider our elastic body under the action of conservative external loads described by the product of a unit load system and a scalar load factor λ. In the case of a properly supported body the fundamental state I of equilibrium, obtained from the undeformed body by a displacement field $\boldsymbol{U}(\lambda)$, is unique and stable for sufficiently small load factors. We assume that a critical case of neutral equilibrium occurs at the critical load factor λ_1 and we also assume that the fundamental state $\boldsymbol{U}(\lambda)$ is continued continuously differentiable with respect to λ beyond λ_1. As a consequence of these assumptions we have a bifurcation of equilibrium at the critical load factor λ_1.

The second variation of the energy in the neighbourhood of the critical load factor and evaluated for the linear combination of buckling modes $a_h\boldsymbol{u}_h$ is approximated by

$$P_2[a_h\boldsymbol{u}_h\,;\lambda] = (\lambda_1 - \lambda)A_{hi}a_ha_i\,[1 + O(|\lambda_1 - \lambda|)], \tag{5.1}$$

where $A_{hi}a_ha_i$ is a positive definite quadratic form. In the vicinity of the critical state the additional displacement from the fundamental state $\boldsymbol{U}(\lambda)$ is again written in the form (4.4). Equilibrium configurations require first of all a stationary value of the potential energy with respect to $\boldsymbol{v}(\boldsymbol{x})$, and these stationary values are for load factors near the critical load factor λ_1 proper minima. If the necessary condition for stability that the cubic form in (4.3) vanishes identically is violated, the potential energy is approximated by

$$F(a_h\,;\lambda) = (\lambda_1 - \lambda)A_{hi}a_ha_i + A_{hij}a_ha_ia_j, \tag{5.2}$$

and there is no need to evaluate the minimizing field (4.5). On the other hand, if the first necessary condition for stability (4.3) is actually satisfied, expression (5.2) is modified by the replacement of the cubic form (4.3) by the quartic form (4.6)

$$F(a_h\,;\lambda) = (\lambda_1 - \lambda)A_{hi}a_ha_i + A_{hijk}a_ha_ia_ja_k. \tag{5.3}$$

The conditions of equilibrium for configurations in the vicinity of the critical case of neutral equilibrium are now obtained by putting the partial derivatives of expression (5.2) or (5.3) with respect to the amplitudes a_h equal to zero. We unite the resulting equations

$$2(\lambda_1 - \lambda)A_{hi}a_i + \begin{matrix} 3A_{hij}a_ia_j \\ 4A_{hijk}a_ia_ja_k \end{matrix} = 0. \tag{5.4}$$

The associated equilibrium configurations are stable or unstable accordingly as the second variation

$$\left[2(\lambda_1 - \lambda)A_{hi} + \begin{matrix} 6A_{hij}a_j \\ 12A_{hijk}a_ja_k \end{matrix}\right] \delta a_h \delta a_i \tag{5.5}$$

is positive definite or indefinite. It is now an immediate consequence of (5.4) by taking $\delta a_i = a_i$ in (5.5) that all branched equilibrium configurations at loads less than the critical load, $\lambda < \lambda_1$, are unstable.

Similar to the investigation of the stability of a critical case of neutral equilibrium in the previous section the foregoing analysis is formal in character. In the case of thin rods, plates and shallow shells, however, a complete justification has been obtained by the application of methods of functional analysis [18–22] under the condition that the matrix occurring between the braces in (5.5) is non-singular. The latter restriction also applies in the general three-dimensional theory but a complete mathematical justification is equally elusive as in the theory of the second variation.

6. Concluding remarks

The practical importance of elastic stability is derived primarily from its many applications to buckling of structures. Since the elastic limit of by far the majority of engineering materials is only a very small fraction of the shear modulus and similar elastic moduli, buckling constitutes a problem mostly for slender structures. It is no wonder therefore that most investigations on buckling are based on simplified one- or two-dimensional theories of elasticity similar to Euler's pioneering contribution. Investigations of neutral equilibrium by means of the three-dimensional theory of elasticity are quite limited in number, and we are not aware of any investigation on the stability in a critical case of neutral equilibrium based on the three-dimensional theory of elasticity, with the exception of the paper to be presented by Sawyers at the present symposium [23].

References

1. C. Truesdell and W. Noll, *The nonlinear field theories of mechanics*, in *Encyclopedia of Physics*, Vol. III/3, in particular sections 68 bis and 89, Springer-Verlag, Berlin, 1965.

2. W. T. Koiter, *Purpose and achievements of research in elastic stability*, in *Recent Advances in Engineering Science*, Gordon and Breach, London, 1969, pp. 197–218.

3. R. J. Knops and E. W. Wilkes, *Theory of elastic stability*, in *Encyclopedia of Physics*, Vol. VIa/3, *Mechanics of Solids* III (C. Truesdell, ed.), Springer-Verlag, Berlin, 1973, pp. 125–302.

4. H. Ziegler and D. McVean, *On the notion of an elastic solid*, in *Recent Progress in Applied Mechanics* (Folke Odqvist volume), Almqvist and Wiksell, Stockholm, 1967, pp. 561–572.

5. P. Duhem, *Traité d'énergetique ou Thermodynamique Générale*, 2 Vols., Paris, 1911.

6. J. L. Ericksen, *A thermo-kinetic view of elastic stability theory*, Intern. Journ. Solids Struct. **2** (1966), 573–580.

7. J. L. Ericksen, *Thermoelastic stability*, Proc. 5th U.S. Nat. Congr. Appl. Mech. (1966), 187–193.

8. R. Martini, *On the Fréchet differentiability of certain energy functionals*, Proc. Kon. Ned. Ak. Wet. **A79** (1976), 326–330.

9. W. T. Koiter, *The energy criterion of stability for continuous elastic bodies*, Proc. Kon. Ned. Ak. Wet. **B68** (1965), 178–202.

10. M. Potier-Ferry, *Critères de l'énergie en élasticité et viscoélasticité*, Séminaire *Flambement des Structures*, St-Rémy-lès-Chevreuse, Mai, 1980.

11. M. Potier-Ferry, *On the mathematical foundations of elastic stability theory I*, Arch. Rat. Mech. Anal., to appear.

12. W. T. Koiter, *A sufficient condition for the stability of shallow shells*, Proc. Kon. Ned. Ak. Wet. **B70** (1967), 367–375.

13. M. Como and A. Grimaldi, *Stability, buckling and post-buckling of elastic structures*, Meccanica, Vol. 10, No. 4 (1975), 254–268; Vol. 12 (1977), 236–248.

14. W. T. Koiter, *A basic open problem in the theory of elastic stability*, Proc. IUTAM/IMU Symposium on Applications of Methods of Functional Analysis to Problems in Mechanics (P. Germain and B. Nayroles, eds.), Springer-Verlag, Berlin, 1976, pp. 366–373.

15. W. T. Koiter, *Over de stabiliteit van het elastisch evenwicht*, Proefschrift Delft, 1945. English translations: *On the stability of elastic equilibrium* (a) NASA TT F 10,833 (1967); (b) Air Force Flight Dynamics Laboratory, TR 70–25 (1970).

16. W. T. Koiter, *Current trends in the theory of buckling*, Proc. IUTAM Symposium on Buckling of Structures (B. Budiansky, ed.), Springer-Verlag, Berlin, 1976, pp. 1–16.

17. B. Budiansky, *Theory of buckling and post-buckling behavior of elastic structures*, Adv. Appl. Mech. **14** (1974), 1–65.

18. G. H. Knightly and D. Sather, *Nonlinear buckled states of rectangular plates*, Arch. Rat. Mech. Anal. **54** (1974), 356–372.

19. D. Sather, *Branching and stability for nonlinear shells*, Proc. IUTAM/IMU Symposium on Applications of Methods of Functional Analysis to Problems in Mechanics (P. Germain and B. Nayroles, eds.), Springer-Verlag, Berlin 1976, pp. 462–473.

20. G. H. Knightly and D. Sather, *Existence and stability of axisymmetric buckled states of spherical shells*, Arch. Rat. Mech. Anal. **63** (1977), 305–319.

21. G. H. Knightly and D. Sather, *Nonlinear buckling and stability of cylindrical panels*, SIAM Journ. Math. Anal. **10** (1979), 389–403.

22. G. H. Knightly and D. Sather. *Buckled states of a spherical shell under uniform external pressure*, Arch. Rat. Mech. Anal. **72** (1980), 315–380.

23. K. N. Sawyers, *Stability of a thick neo-Hookean plate*, article in this volume.

24. W. T. Koiter, *On the thermodynamic background of elastic stability theory*, in *Problems of Hydrodynamics and Continuum Mechanics* (Sedov Anniversary Volume), SIAM, Philadelphia, 1969, pp. 423–433.

25. W. T. Koiter, *Thermodynamics of elastic stability*, Proc. 3rd Can. Congr. Appl. Mech. Calgary (1971), 29–37.

MECHANICS OF FRACTURE OF RUBBER-LIKE MATERIALS

G. J. LAKE AND A. G. THOMAS

The Malaysian Rubber Producers' Research Association, Hertford, England

(Received September 11, 1980)

ABSTRACT

The paper reviews work on the strength properties of rubber based on a fracture mechanics approach. The high extensibility of rubber requires that the theory is developed without restriction to small strains. Because of this, the most appropriate formulation is in terms of the strain energy release rate, which has been termed the "tearing energy" for rubbers since the original work was carried out on tearing. The subsequent application of the approach to enable a variety of failure phenomena to be interrelated — small-scale crack growth and fatigue, tensile failure, ozone attack, abrasion, cutting by sharp objects — and to treat practical problems is discussed. It has been found that there is a limiting tearing energy below which no crack propagation occurs; a theory which takes into account the molecular structure and chemical bond strength of vulcanized rubber successfully predicts the magnitude of this limiting energy.

Introduction

The application of an energetics approach to the fracture of rubber began with a study of tear behaviour [1]. This work was stimulated by the observation that different test pieces, when used in conventional tear tests, rate a series of vulcanizates in different orders. Thus it is clear that such tests do not measure the same 'tear resistance' property. The inability to measure a basic strength property characteristic of the material is a considerable impediment both to fundamental work and to efforts to relate laboratory measurements to service performance. An attempt was made to overcome these difficulties by applying the strain energy release rate concept originally advanced by Griffith [2] to explain the strength of glass. When its application to the tearing of rubber is considered, it is evident from visual observation of

Proceedings of the IUTAM Symposium on Finite Elasticity, Lehigh University, August 10–15, 1980. General lecture presented by A. G. Thomas.

Carlson, D.E. and Shield, R.T. (eds.)
Proceedings of the IUTAM Symposium on Finite Elasticity
ISBN 90 247 2629 8

the tip of a growing tear that energy is being dissipated irreversibly, ultimately as heat, and not appearing exclusively as surface free energy as Griffith's theory requires. Thus some modification of the theory is necessary.

If the energy dissipation is considered more closely, it is apparent that it occurs primarily in a highly localized region around the tip of the growing tear. Thus its magnitude would be expected to be governed by the physical properties of the rubber in this region and by the geometrical form of the tip rather than by the overall shape of the test piece or the detailed way the forces are applied in regions remote from the tip. Thus the energy required to cause tearing may still be characteristic of the material and independent of test piece shape even though it greatly exceeds the surface energy.

The tear criterion and tear behaviour

For rubbers, the tearing energy T has been defined mathematically by

(1) $$T = -(\partial U/\partial A)_l$$

where U is the total strain energy stored in a specimen containing a crack, A the area of one fracture surface of the crack and the partial derivative indicates that the specimen is considered to be held at constant length, l, so that the external forces do no work. A check on the validity of the approach is possible since T can be calculated in terms of the applied forces or deformations for various differently-shaped test pieces, some examples of which are shown in Figure 1. In each of these cases the calculation is possible because the region of complicated strain around the crack tip advances with the crack, as it propagates, without changing. The results of the calculations are as follows [1, 3]

(i) 'trousers' test piece (Figure 1a)

(2) $$T = \frac{2F\lambda}{t} - wW$$

where F is the force applied to each leg, t and w are the test piece thickness and width respectively, and λ is the extension ratio and W the strain energy density in the legs;

(ii) pure shear test piece (Figure 1b)

(3) $$T = Wh_0$$

where h_0 is the unstrained value of the height h and W is the strain energy density in the central region, away from the crack, which is in pure shear;

(iii) 'angled' test piece (Figure 1c), in which the angle α is kept constant

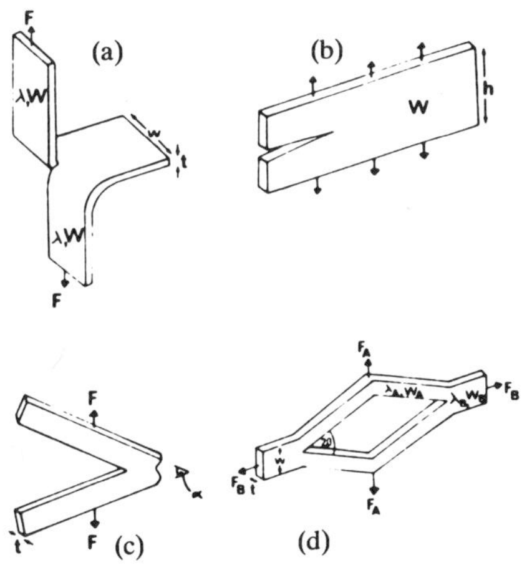

Figure 1. Various tear test pieces: (a) 'trousers'; (b) pure shear; (c) 'angled'; (d) 'split'.

$$T = \frac{2F}{t} \sin \frac{\alpha}{2} \tag{4}$$

where F and t are again the applied force and test piece thickness;
(iv) the 'split' test piece (Figure 1d)

$$T = [F_A \lambda_A \sin \beta + F_B (\lambda_A \cos \beta - \lambda_B)]/t - w(W_A - W_B) \tag{5}$$

where F_A, F_B are the forces applied to the respective pairs of legs, λ_A, λ_B and W_A, W_B the corresponding extension ratios and strain energy densities, 2β the angle of opening of the legs ($\tan \beta = F_A / F_B$) and w and t are as before.

Thus from the measured tearing forces, the values of T at which tearing occurs can be calculated. A non-crystallizing styrene-butadiene rubber (SBR) vulcanizate gives values of T that depend on the rate of tearing [4]. Thus the tear results for each test piece give a relation between T and the rate of tearing r. This relation should, according to theory, be the same for all the test pieces and the results in Figure 2 show that this is the case. Similar agreement is found for other rubbers although the detailed nature of the tearing and the level of the tear strength may vary. It appears therefore that the tearing energy T is a measure of the tear strength of a rubber that is independent of the form of the test piece and is thus a characteristic property of the material. The magnitude found for T is of the order of 10^6 erg/cm^2 and is very much greater than any true surface energy, even when allowance is made for the effect of the long-chain molecular structure of an elastomer [5].

The approach indicates why different conventional tear test pieces rate rubbers in different orders. The tearing force, as can be seen from the theory

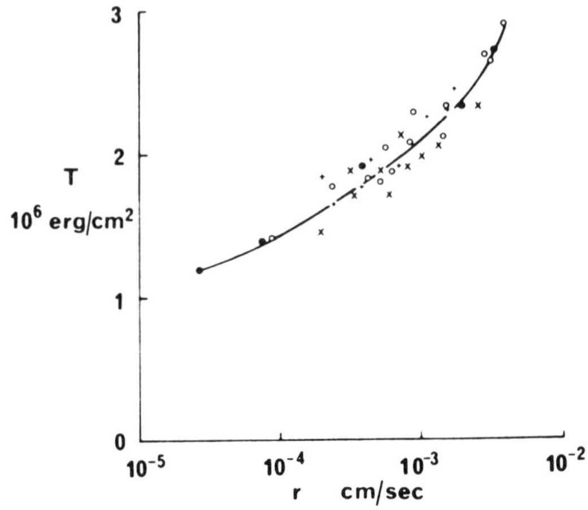

Figure 2. Tearing energy T versus rate of tearing r for an SBR vulcanizate using the test pieces shown in Figure 1: × 'trousers'; + pure shear; ○ split; ● 'angled'.

for the test pieces considered above, is governed by a function of T and the modulus of the rubber. This function varies from one test piece to another, giving different sensitivities to modulus and tearing energy changes. As both the modulus and tearing energy vary from one material to another, different ratings are obtained with different test pieces.

The strain energy concentration at an incision

Insight into why an overall energetics approach should work can be obtained by consideration of the location of the energy loss when a crack propagates. If a test piece which has a force-free edge of arbitrary shape is considered, then it can readily be shown [6] that the energy loss associated with the excision of a strip of material from the edge (Figure 3a) is independent of the detailed shape of the edge, provided this is maintained, and is proportional to the volume of material removed. If now a special force-free edge containing a model crack is considered and a small element of material is excised from the crack tip only (Figure 3b) — to simulate a propagation step — then there is again an energy loss proportional to the volume of excised material and also a further loss resulting from relaxation of the normal stresses at the new tip to zero. However, it can be shown quite generally, for finite strains and independent of any assumptions about the nature of the stress–strain behaviour of the material, that the latter term is of second order compared with the first. Thus the energy lost through a small

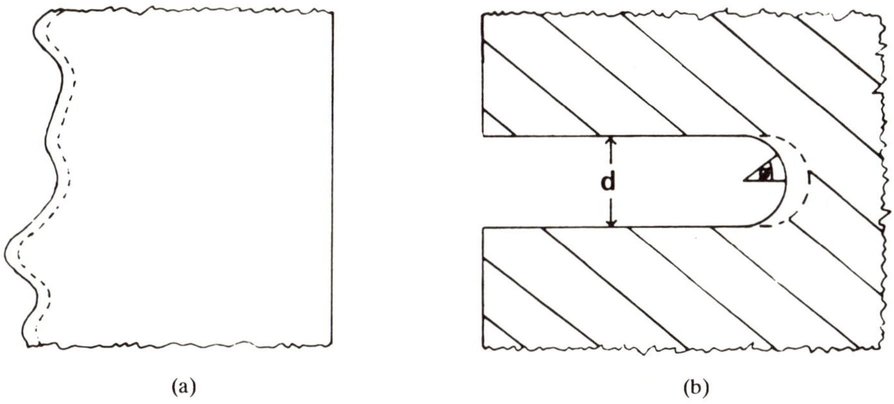

Figure 3. Schematic diagrams illustrating: (a) the excision of a strip of material from a test piece edge of arbitrary shape; (b) the excision of material from a model crack tip to simulate a propagation step.

'propagation' of a model crack in a specimen held at constant deformation is mainly associated with relaxation of material in the immediate vicinity of the tip of the crack. This suggests that the tearing energy is closely related to the energy lost in the region of the tip, as mentioned earlier. In particular, for a model semi-circular tip of diameter d, the energy loss is given by

$$T = d \int_0^{\pi/2} W_\phi \cos \phi d\phi \tag{6}$$

where W_ϕ is the strain energy density at the tip surface at an angle ϕ from the pole (Figure 3b). As an approximation, equation (6) can be simplified to

$$T \simeq W_t d \tag{7}$$

where W_t is the average strain energy density at the tip. Strictly speaking, the term T should not be used directly in equations (6) and (7) since when a real crack propagates, material is not excised from the tip region. However, effects of such a difference would be expected to be small. Direct measurements of the strain, and hence strain energy, distribution around a model crack tip have confirmed the validity of equation (6), good agreement being found between the tearing energies determined in this way and those calculated from the applied forces [6]. Further, tear experiments have shown T to increase with d in the predicted way for model incisions with tip diameters from 1–3 mm and W_t to be of the same order as the value obtained from a tensile test to break [6].

Crack growth and fatigue

The cyclic crack growth behaviour of various rubbers has been studied using the tearing energy concept [7–9]. If the amount of growth per cycle (dc/dn) is plotted against the maximum T value attained during each cycle it is found that the results are independent of the form of the test piece used. This parallels the situation for tear measurements and shows that the tearing energy concept is also valid for small-scale crack growth. For many natural rubber vulcanizates it has been found that the crack growth behaviour, at least under conditions where complete relaxation occurs for part of each cycle, can be approximated at moderate to high T values by the relation

$$\frac{dc}{dn} = BT^2 \tag{8}$$

where B is a constant, characteristic of the material. Similar measurements have been made on other elastomers. For SBR it is found that a relation of the form

$$\frac{dc}{dn} = BT^4 \tag{9}$$

where B is again a constant but with a different value (and dimensions) adequately represents the results in the same region and other elastomers often have exponents of T intermediate between these values [10]. Figure 4 shows results for natural rubber and SBR and the straight lines drawn through them at higher tearing energies have slopes of 2 and 4 respectively and can be seen to represent the results quite well in these regions.

At lower tearing energies the power law relationships break down and for many elastomers there is a linear region with an intercept on the T-axis [11] (cf. inset to Figure 4). This represents the energy (T_0) required for the initiation of mechanical crack growth. Below T_0 only very slow growth occurs at a rate which is essentially independent of tearing energy and this growth is entirely attributable to the effects of ozone. Thus in this region the crack growth characteristics are generally of the form

$$\frac{dc}{dn} = A(T - T_0) + r_z \tag{10}$$

where the first term on the right hand side represents the mechanical growth, A being a constant, and r_z represents the rate of ozone crack growth.

Under circumstances where effects of heat build-up are negligible, fatigue failure can occur in rubber articles subjected to repeated deformations, because of cyclic crack growth. Using the fracture mechanics approach

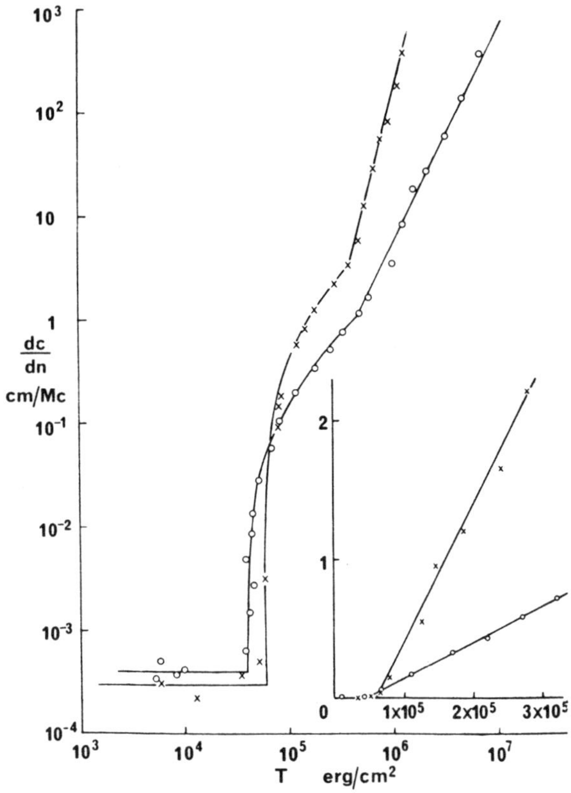

Figure 4. Cyclic crack growth rate dc/dn versus maximum tearing energy of the cycle T for vulcanizates of natural rubber [NR] (○) and SBR (×) [logarithmic scales]; inset: region near T_0 plotted on linear scales.

described above and the observed characteristics of crack growth, an analysis of this fatigue failure can be made. For example, for a parallel-sided strip containing a small crack of length c in one edge (Figure 5) and subjected to repeated tensile stressing, the tearing energy is given by [1]

$$T = 2KWc \tag{11}$$

where W is the strain energy density in the bulk of the test piece (i.e. remote from the crack) and K is a slowly varying function of extension that has been independently determined, both experimentally and by finite element analysis [12–14]. If the amount of crack growth that occurs during one deformation cycle is considered, equation (8) for natural rubber gives, on substituting for T from equation (11)

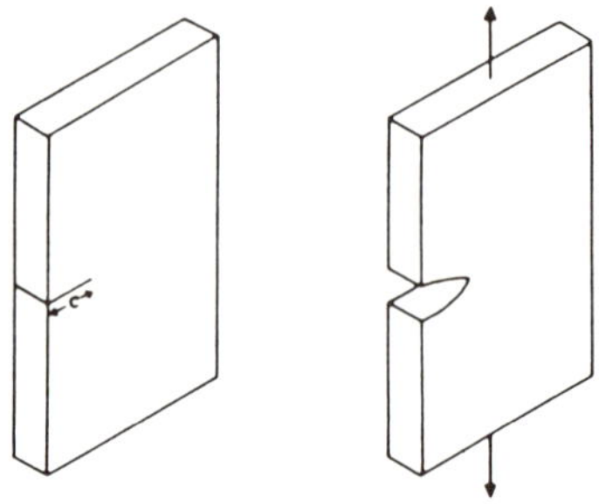

Figure 5. Tensile strip test piece containing an edge crack of length c.

$$dc/dn = B(2KW)^2c^2. \tag{12}$$

In the case of failure from a small initial crack of length c_0, integration of equation (12) gives the number of cycles to failure N as

$$N = \frac{1}{B(2KW)^2}\left(\frac{1}{c_0} - \frac{1}{c_c}\right) \simeq \frac{1}{B(2KW)^2c_0} \tag{13}$$

where the approximation holds provided the crack length at failure, c_c, is much greater than c_0, which is generally the case.

Equation (13) has been checked by model experiments using test pieces with initial razor cuts of known size, good agreement being found [8]. The fatigue of dumbbell test pieces containing no intentionally introduced flaws has also been studied. The change of fatigue life with maximum strain should be given by equation (13) if the mechanism of failure is crack growth from small flaws of effective length c_0. Figure 6 shows results for the moderate to

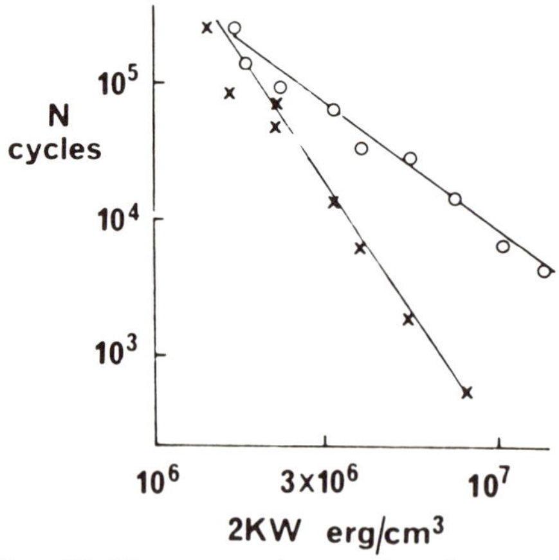

Figure 6. Tensile fatigue life N versus maximum value of the strain energy density factor $2KW$ (minimum value zero) for vulcanizates of NR (○) and SBR (×).

high strain region for which equation (8) would be expected to be appropriate, the life being plotted against $2KW$ on logarithmic scales. The line drawn has the predicted slope of -2 and can be seen to give a good fit to the experimental results.

The fatigue behaviour of the non-crystallizing SBR vulcanizate can be predicted in a similar manner to that used above for natural rubber (NR), but using equation (9) to describe the crack growth characteristics. The result is

$$N = \frac{1}{3B(2KW)^4 c_0^3} \,. \tag{14}$$

Comparing this with equation (13) for NR, it is apparent that the predicted dependence of fatigue life on strain is much greater for SBR and this is confirmed by the results shown in Figure 6, the line through the SBR results having the theoretical slope of -4. A much stronger dependence on flaw size is also predicted for SBR (c_0^{-3} as compared with c_0^{-1}). Thus an increase in flaw size will have a much more harmful effect on the life of SBR than on that of NR, which is in accord with general observation.

In equations (13) and (14) all the quantities are measurable except c_0, so that if these equations are assumed to be applicable, the effective flaw size can be calculated for each rubber. The values obtained are a few thousandths of a centimetre, which is consistent with the size of observable flaws such as moulding and die-stamping imperfections or particulate impurities.

The theory can be extended to lower deformations, where different forms of crack growth characteristic apply, and can be used to assess the contributions of oxygen and ozone to crack growth and fatigue, as well as purely mechanical effects. Figure 7 compares theory with experimental results obtained in the laboratory atmosphere for natural rubber over a wide range of strains. A marked increase in life occurs at a $2KW$ value of about 10^7 erg/cm^3 (which corresponds to a maximum strain of about 75%). This deformation constitutes a mechanical fatigue limit for the rubber and is the lowest strain at which the tearing energy attains the value (T_0) required to initiate mechanical growth from flaws of the naturally-occurring size c_0. According to equation (11), the fatigue limit should occur when

$$2KW = T_0/c_0 \tag{15}$$

and this relation gives good agreement with experiment when the T_0 value obtained from crack growth measurements and the flaw size estimated from high strain fatigue tests are used. Below the fatigue limit, cracks must at first grow by ozone and the fatigue life is correspondingly sensitive to the ozone concentration in the test atmosphere. The effect of ozone can be taken into

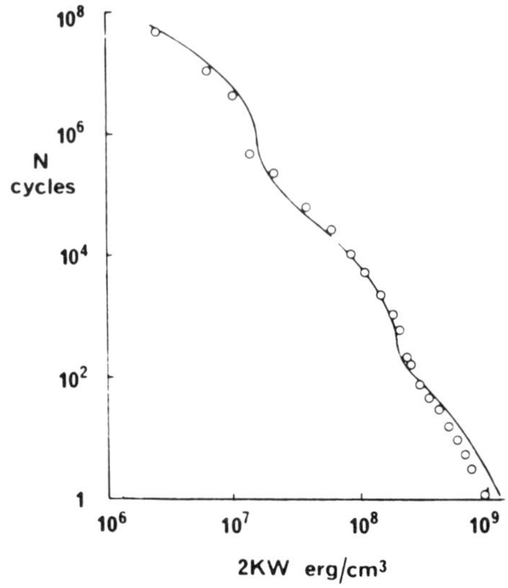

Figure 7. Number of cycles to failure N versus maximum value of the strain energy density factor $2KW$ for a natural rubber vulcanizate for maximum strains ranging from about 30% to 700% (minimum strain zero throughout). The line drawn is based on the theoretical relationships discussed in the text assuming a naturally-occurring flaw size of 2.5×10^{-3} cm.

account by means of equation (10) using an ozone crack growth constant obtained independently from static ozone tests to estimate r_z. Figure 7 also shows results at very high deformations right up to the tensile failure limit and these will be discussed in a subsequent section.

Theoretical limiting strength for a highly-elastic material

The magnitude of the energy T_0 required to initiate mechanical crack growth is of similar order — ca. 5×10^4 erg/cm^2 — for a range of elastomers including natural rubber, SBR, polychloroprene and Butyl rubber, which differ widely in other strength properties such as tear and tensile strengths [11]. The similarity of the T_0 values for different elastomers under various conditions suggests that this property may be governed fairly directly by the primary carbon–carbon bond strength. An estimate of the minimum tearing energy required for bond rupture can be made using equation (7) if it is assumed that the tip diameter has its smallest possible value, which for an elastomer is of the order of the distance between crosslinks in the unstrained state; this gives $d \simeq \xi u^{1/2}$ where ξ is the length of a monomer unit and μ the number of monomer units between crosslinks. Assuming forces to be transmitted primarily via the crosslinks, the energy stored by the carbon–

carbon double bonds will be small at the breaking force of the single bonds, so that the maximum possible energy density will be of the order of bH, where b is the number of single bonds per unit volume and H is the energy stored by each single bond at its rupture point. Substitution in equation (7) yields

$$T_0 \simeq bH\xi\mu^{1/2}. \tag{16}$$

For a typical natural rubber vulcanizate the following approximate values hold: $\xi = 5 \times 10^{-8}$ cm, $b = 2.4 \times 10^{22}$ cm^{-3}, $\mu = 100$; putting $H = 3.3 \times 10^{-12}$ erg (the dissociation energy of the weakest C–C bond in the isoprene unit) equation (16) gives $T_0 \simeq 4 \times 10^4$ erg/cm^2. More precise calculation [5] yields about half this value which is nevertheless in good agreement with the experimental results in view of various uncertainties. Variations between different elastomers would be expected to be fairly small on this basis, as is observed. Thus T_0 can be related approximately to the primary bond strength and molecular structure of a vulcanizate.

The above calculation is similar to a normal surface energy calculation but is modified to take into account the long chain molecular structure of a rubber. Because of this, in order to break a bond crossing the fracture plane, it is necessary to take many other bonds (in the same chain between crosslinks) up to essentially the breaking point; thus the energy required is correspondingly magnified. The question now arises: why does failure not occur in bulk when T_0 is reached? Indeed, for certain rubbers, such as very-highly crosslinked or very-highly swollen ones, bulk failure does occur close to T_0 [11, 29, 30]. However, for most rubbers this is not the case and the tear strength is much greater than T_0. The reason for this is believed to be the irreversible energy dissipation referred to earlier. When T_0 is exceeded, the occurrence of crack growth means that there are regions in the vicinity of a crack tip where retraction of material occurs. Thus the mechanical hysteresis of the material will influence the stress concentration at the tip and hence contribute to the complex picture that applies to crack growth at higher severities. In particular, tear strength and tensile strength (to be discussed in the next section) which represent the upper bounds of the macroscopic failure spectrum, do not appear likely to be amenable to straightforward, if any, fundamental interpretation.

Tensile failure and very high strain fatigue

The fracture mechanics methods used for rubber have found application to several other phenomena in addition to those mentioned so far. The most obvious extension is to tensile strength. In essence, tensile failure can be regarded as a one-cycle fatigue test in which, for non-crystallizing rubbers, failure occurs when the tear strength is reached, although there are complications attributable to scale effects in the crack growth [15, 16]. For crystallizing rubbers there are further complications when the bulk of the test piece becomes crystalline (as is usually the case prior to tensile failure) since then failure does not occur at the normal tear strength and stable crack growth is observed at higher tearing energies. A modified theory appears to take account of this fairly successfully [17, 18]. In fatigue failure the growth per cycle is relatively small but in tensile failure this may not be true and if the growth is comparable with the initial crack length c_1, allowance must be made for it in applying equation (11). If the crack has grown by an amount Δc, then

$$T = 2KW(c_1 + \Delta c). \tag{17}$$

For natural rubber at high tearing energies, the growth from a sharp incision is as shown in Figure 8a; initially there is a period of rapid, 'smooth' growth

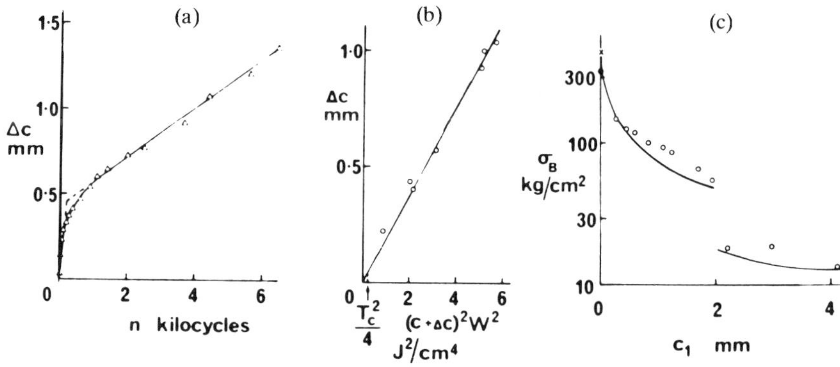

Figure 8. Tensile failure of natural rubber: (a) initial cyclic crack growth Δc from a sharp incision at a high tearing energy; (b) amount of growth during a single extension to very high strains plotted as indicated by equation (19) ($T_c^2/4$ indicates the point at which failure would occur by tearing if the bulk were not crystalline); (c) tensile strength σ_B versus initial flaw size c_1 for inserted cuts of various lengths (○); the lines represent the theoretical relationships based on equations (20) and (21); the experimental and theoretical values with no inserted cut are represented by (●) and (×) respectively.

after which the tip roughens and the rate settles down to the normal, steady value (typically after a few hundred cycles). The smooth growth obeys a relation of the form of equation (8) but a different constant (B_s) applies. This growth is believed to be relevant to tensile failure so that

$$\Delta c = B_s T^2. \tag{18}$$

Elimination of T between equations (17) and (18) gives

$$(c_1 + \Delta c)^2 W^2 = \Delta c / 4K^2 B_s. \tag{19}$$

This quadratic equation has the property that real solutions for Δc only exist up to a value of W denoted by W_1 and given by

$$W_1 = \frac{1}{4K(B_s c_1)^{1/2}}. \tag{20}$$

For W greater than this, Δc will increase indefinitely, that is, rupture will occur.

This theory has been checked by measuring Δc at high strains, where the bulk of the rubber is crystalline, for various values of W. Results for an NR vulcanizate are shown in Figure 8b where Δc is plotted against $(c + \Delta c)^2 W^2$ in accordance with equation (19). The predicted proportionality is found and the slope of the line gives the value of $4K^2B_s$. Also shown on the figure is the value of $T_c^2/4$ for this rubber, that is, the point at which catastrophic tearing would occur according to the simple tear criterion described earlier (K has a value of about 1 at these high strains). It is obvious that very substantially greater T values than this can be sustained in the present case. Thus T_c is not a relevant strength parameter at these high strains when the rubber is crystalline in bulk and failure from a small crack will proceed by the mechanism leading to equation (20).

Measurements were also made of the apparent tensile strength as a function of the initial length c_1 of an inserted cut (Figure 8c). There is an abrupt break in the experimental relation at about $c_1 = 0.2$ cm. The full curve to the left of this point has been calculated from equation (20), using the value of $4KB_s^{1/2}$ obtained from the crack growth measurements. It is seen to be in good agreement with experiment, particularly as there are no adjustable parameters. For cut lengths greater than 0.2 cm rupture occurs before the specimen becomes crystalline in bulk (there is, of course, local crystallization around the tip). The rupture process is now governed by the T_c value for the material, so that from equation (11)

$$W_1 = \frac{T_c}{2Kc_1}. \tag{21}$$

The full curve on the right of Figure 8c is derived from this equation using a T_c value obtained from a trousers test piece (Figure 1a) and compares well with the experimental results.

Tensile failure of a test piece of a crystallizing rubber containing a small cut therefore appears to obey the following pattern. If, on extension, the T_c value for catastrophic tearing is attained before the bulk of the specimen becomes crystalline, failure takes place by a tearing process. However, if bulk crystallization occurs first, failure does not occur until the cut growth mechanism leading to equation (20) is operative. The significance of bulk crystallization is confirmed by the fact that non-crystallizing rubbers show no break in the tensile strength vs. cut length relation.

Equation (20) may be applied when no artificial cuts are present, that is, to the case of a normal tensile test. A naturally-occurring flaw size of 2.5×10^{-3} cm satisfactorily explains the fatigue behaviour of the NR vulcanizate in terms of the relevant crack growth characteristics. Using this value in equation (20) predicts a work to break of 72 Jcm^{-3} compared with an experimentally observed value for this vulcanizate of 52 Jcm^{-3}. Using the measured stress–strain curve, the tensile stress equivalent to this theoretical value can be estimated and is shown in Figure 8c for comparison with the observed tensile strength.

The agreement between theory and experiment is satisfactory considering that no arbitrary parameters are involved and suggests that tensile failure is essentially a crack growth process from naturally-occurring flaws, although the complications discussed mean that it is a difficult property to interpret. These complications can also cause departures from the simple power law relationships for fatigue life, such as equation (13), at very high strains. Using a similar approach to that adopted for tensile failure, allowance can be made for the effects of 'smooth' growth and relatively large growth steps. The high-strain end of the theoretical curve in Figure 7 is based upon this procedure assuming that smooth growth persists for the first 200 cycles of a fatigue test (cf. Figure 8a; values of the other constants used are as before). As can be seen, this gives a reasonably good approximation to the results right down to the tensile failure limit. Thus the modifications successfully bridge the gap between the relatively simple fatigue theories that apply for long lives and the ultimate single-cycle failure.

Ozone cracking

Chemical attack by ozone has already been mentioned as a contributory factor in fatigue at low strains. In addition, ozone can cause cracking in rubber held at constant deformation. Indeed, for rubbers that do not contain added protective agents such cracking commonly has a threshold strain as low as a few percent extension. Experiments on test pieces containing inserted cuts indicate that the cracking process has an energy requirement of about 100 erg/cm^2 — of the order expected for ozone-degraded rubber [19]. Using this energy and the observed threshold strain, a flaw size of some 10^{-3} cm is required to satisfy equation (11); as discussed above, this is plausible for rubber and furthermore agrees well with the independent estimate from fatigue data. Thus the fracture mechanics approach enables the threshold condition for chemically-induced cracking by ozone to be satisfactorily accounted for.

Abrasion

Abrasion is another phenomenon that has been successfully treated by the energetics approach. Patterns consisting of small tongues of rubber often develop during abrasion and the production of these tongues is attributable to crack growth. For an idealized, uniform pattern formed by an abrading force F per unit width acting on a line abrader (Figure 9a) it can be shown that the tearing energy is given by [20]

$$T = F(1 + \cos\theta) \tag{22}$$

where θ is the angle a crack makes with the surface. It is assumed that abrasion occurs because the tongues break off; since the crack length does not appear in equation (22), the detailed way in which this happens — i.e. whether in a single cycle or many cycles — does not matter. Thus if the cracks advance a distance v for each pass of the abrader (i.e. each cycle) the average abraded volume per unit area of surface per cycle will be

$$V \doteq v \sin\theta. \tag{23}$$

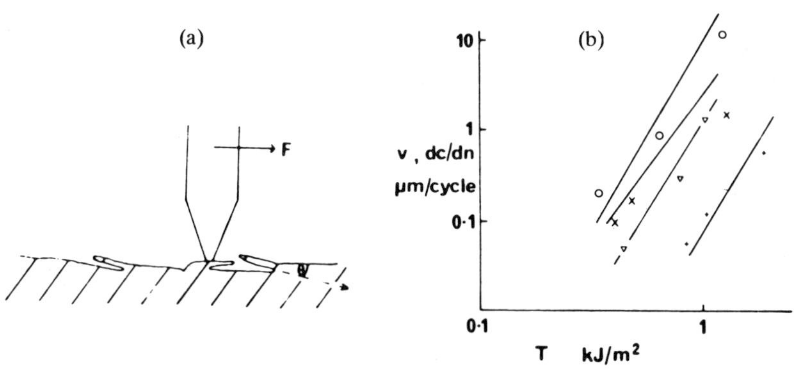

Figure 9. Abrasion: (a) sketch of the profile of part of an abrasion pattern formed by a line abrader acted on by a force F per unit width (the arrow indicates the direction of crack growth at angle θ to the surface); (b) comparison of crack growth rates (v) estimated from abrasion measurements [cis-trans isomerized natural rubber (○), butadiene rubber (×), SBR (▽), acrylonitrile-butadiene rubber (+)] with the independently-measured crack growth characteristics of the rubbers (full lines).

Hence v can be calculated from the rate of abrasion V and the angle θ (which can be determined experimentally from measurements of the movement of the pattern over the surface during abrasion). By using equation (22) to estimate T, v can be compared with the independently measured cyclic crack growth characteristics [10]. Good agreement is obtained for various non-crystallizing rubbers (Figure 9b); agreement for strain-crystallizing natural rubber is less good, the abrasion being faster than predicted, suggesting that crystallization either does not occur or is ineffective during abrasion.

Cutting by sharp objects

A fracture mechanics approach has also proved useful in connection with studies of the cutting of rubber by sharp objects [21]. Cutting is of importance in service, for example for tyres or conveyor belting, primarily under wet conditions when friction between the rubber and the cutting object is much reduced [otherwise the frictional forces tend to be very large and cutting correspondingly difficult]. In order to study the process under conditions where frictional effects were largely absent, a method was developed in which the cutting implement — a razor blade — was applied to the tip of a crack in a stretched tear test piece. An energetics approach again enables results from differently shaped test pieces and different deformations to be superimposed. Furthermore in a low tearing energy region, rapid 'catastrophic' cutting is found to occur when the total energy available, from both the deformation of the rubber and the force applied to the blade, reaches a constant value (Table 1). This applies for all vulcanizates examined. The magnitude of S_c varies for different vulcanizates, being approximately proportional to the corresponding T_0 value (Table 1). The ratio of T_0/S_c is of the order of the ratio of the

Table 1. Comparison of the total energy required for cutting (S_c) with the threshold energy for mechanical crack growth (T_0)

Rubber[a]		S_c kJ/m^2	T_0[b] kJ/m^2
Natural rubber (NR): A	0.30	0.04	0.13
Natural rubber (NR): B	0.26	0.03	0.12
Styrene-butadiene rubber (SBR)	0.33	0.06	0.18
Butadiene rubber (BR)	0.66	ca 0.1	ca 0.15
Isomerized natural rubber[c]	0.47	ca 0.07	ca 0.15
70 NR/30 BR blend	0.45	0.065	0.14

a. unfilled conventional accelerated-sulphur vulcanizates throughout, except for natural rubber vulcanizate B which was vulcanized with di-cymyl peroxide; b. measured at atmospheric pressure for rubber containing antioxidant; c. cis-trans isomerization — trans content about 60%.

minimum tip diameter for a crack in rubber [i.e. $\xi\mu^{1/2}$ — cf. equation (16)] to the tip diameter believed to be representative for the blades used.

Friction and adhesion

Similar energetics methods have proved useful in tackling sliding friction under conditions where waves of detachment (Schallamach waves) are formed and also various adhesion problems (see, for example, the reviews by Gent and Hamed [22] and Roberts [23]).

Practical problems

A major problem in applying the fracture mechanics approach in practice is to determine the relation between the tearing energy, crack size, and overall forces or deformations for a rubber article of complex shape and structure. Calculation of this relation is prohibitively difficult except for certain simple cases, while direct measurement of the change in strain energy produced by a crack is of limited utility. Two alternative methods have been developed. One involves numerical computation using finite element analysis and has been used for assessing stress concentrations and tearing energies in model components [14].

The second method is an empirical one for estimating the tearing energy from measurements of the amount by which a crack opens under stress. By considering the work required to close a crack, it can be shown that for a crack in a sheet the tearing energy is given approximately by [13]

$$T = (\sigma/2)(\partial S/\partial c')_l \tag{24}$$

where σ is the stress (referred to the unstrained state) acting normally to the crack plane, S the area of the crack opening, c' the crack length in the strained state (Figure 10a) and the partial derivative indicates, as before, that the specimen is held at constant deformation. Hence the tearing energy can be found by measuring S as a function of crack length, provided that the bulk stress σ is known. For certain forms of deformation the strain in the rubber is simply related to the maximum opening of the crack (b) so that the stress can also be estimated from the measurements. This method has been checked by experiments on model test pieces, deformed in simple extension or pure shear, and found to work well [13]. It has also been applied successfully to the practical problem of groove cracking in tyres [24], as shown in Figure 10b which compares the observed and predicted rates of growth.

Rubber-cord laminated structures are often used in practice — for example, in tyres, conveyor belting and hose. In common with other laminates,

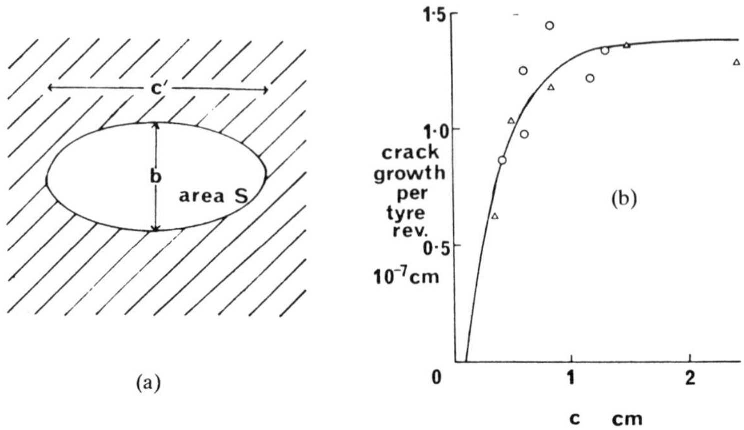

Figure 10. (a) Parameters involved in the crack opening method for estimating the tearing energy; (b) comparison of observed and predicted (——) groove cracking behaviour; experimental results: from service tests (on buses running in Bath, Somerset) (○), from rig tests (△).

failure can occur in these structures and can be complex in nature; possible causes include loss of adhesion between the cords and the rubber and various forms of cracking. Two forms of cracking in rubber-cord laminates have been treated by the fracture mechanics approach [25, 26]: (i) growth of cracks between the ply layers and (ii) growth of roughly cylindrical cracks around individual cords — 'socketing'.

The type of structure considered consists of a sheet of rubber containing two layers of symmetrically disposed cords that are strongly bonded to the rubber (Figure 11a). In the case of inter-ply cracking, an idealised crack is considered which is of uniform length along the length of the test piece and which propagates transversely (Figure 11b). For this case the analysis gives

$$T = W_c t \tag{25}$$

where W_c is the strain energy density in the central region and t the laminate thickness.

If, alternatively, failure results from cracking around the individual cords then again an ideal case in which this 'socketing' is uniform along the test piece is considered. For socketing in one of the ply layers, the analysis yields

$$T = \frac{W_c t}{\pi m D} \tag{26}$$

where m is the number of cords per unit length (perpendicular to the cord direction) and D the diameter of a crack around a cord.

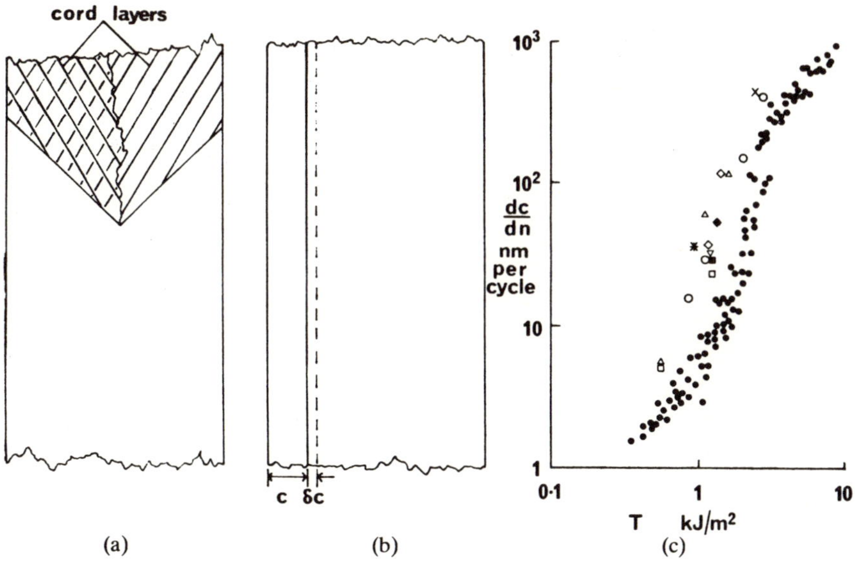

Figure 11. Failure in laminated structures: (a) schematic diagram illustrating the structure of a symmetrical 2-ply rubber-cord laminate; (b) idealised inter-ply crack showing a propagation step δc; (c) comparison of the independently-measured crack growth characteristics (●) with results from laminates of various constructions:

laminate thickness, mm	6.9	3.4	7.0	9.7	6.9	3.5	3.5	6.2	6.5
ply separation, mm	1.1	2.0	2.3	2.0	3.9	1.3	1.3	1.3	1.3
cord angle, deg	22	22	22	22	22	30	15	22	22
cords per inch	10	10	10	10	10	10	10	5	$3\frac{1}{3}$
symbol	○	△	▽	□	◇	■	◆	×	*

In the last two cases failure occurred by socketing.

Comparison of equations (25) and (26) indicates that the energy available for crack growth will be the same for the two mechanisms when

$$m = 1/\pi D. \tag{27}$$

Consistent with the independence of crack length indicated by equations (25) and (26), crack growth rates in laminates subjected to repeated deformations of fixed maximum amplitude were found to be constant once an initiation stage ($c > \sim t/2$) had occurred and before a final failure stage was reached. Figure 11c compares laminate results with the independently measured crack growth characteristics of the rubber (using tensile strip test pieces — cf. Figure 5). The different tearing energies for the laminates were obtained by varying the maximum strain and also the construction parameters. As can be seen, there is quite good agreement, particularly bearing in mind the simplicity of the theory; also a transition from interply cracking to socketing occurs between 10 and 5 cords per inch, which compares with a prediction from equation (27) of 8 cords per inch. Thus the fracture mechanics approach also appears to be applicable to failure in rubber-cord laminates.

General discussion

Insight into reasons for the success of the fracture mechanics approach is provided by the general analysis relating the strain energy release rate to the strain energy density at a crack tip; thus events in the fracture zone are directly related to those occurring macroscopically. Use of this relation enables the energy required for the onset of mechanical crack growth in a rubber to be calculated, to within about a factor of two, from the primary bond strength and (long-chain) molecular structure. The latter theory assumes the tip diameter to have its smallest possible value for a vulcanized rubber (typically rather less than 10^{-6} cm) while the strain energy density at the tip approaches 10^4 J/cm^3. This energy density contrasts with results from experiments on model tips of diameter of the order of a mm, for which the energy density at break is estimated to be less than 10^2 J/cm^3 (similar to the value obtained in standard tensile failure tests). The reason for the difference is believed to be simply that in the latter case large flaws, comparable in size with those to be found in bulk specimens, are present in the volume of material that is ruptured at the tip. As this volume decreases, so will its strength increase from the bulk value until the molecular upper limit is reached. Fracture surfaces generally tend to become rougher at higher tearing energies and there is evidence which suggests that the crack growth characteristics are influenced by progressive changes in the natural tip diameter [27, 16]. The transition to 'smooth' growth at very high severities is in contrast with the general tendency and it has been suggested that this is associated with a change in the nature of failure [28]; according to this picture, the normal, rough surfaces are produced by cavitation which is suppressed at high rates of crack propagation because the material approaches the glassy state and the consequent increase in modulus leads to bond rupture prior to cavitation.

The work described has shown that a fracture mechanics approach based on the strain energy release rate has been very successful in enabling various properties of rubber to be correlated and understood in both the laboratory and in service. The properties include not only those that are commonly regarded as strength properties but also others such as ozone attack, abrasion and cutting by sharp objects. It is noteworthy that with the exception of the naturally-occurring flaw size, on which there are several independent checks as well as agreement with direct observation as to the order of magnitude, there are no arbitrarily adjustable parameters in the theories that have been developed.

References

1. R. S. Rivlin and A. G. Thomas, J. Polym. Sci. **10** (1953), 291.
2. A. A. Griffith, Phil. Trans. Roy. Soc. **A221** (1920), 163.
3. G. J. Lake, P. B. Lindley and A. G. Thomas, Proc. 2nd International Conf. on Fracture, Brighton, Chapman and Hall, London, 1969, p. 493.
4. A. G. Thomas, J. Appl. Polym. Sci. **3** (1960), 168.
5. G. J. Lake and A. G. Thomas, Proc. Roy. Soc. **A300** (1967), 108.
6. A. G. Thomas, J. Polym. Sci. **18** (1955), 177.
7. A. G. Thomas, J. Polym. Sci. **31** (1958), 467.
8. A. N. Gent, P. B. Lindley and A. G. Thomas, J. Appl. Polym. Sci. **8** (1964), 455.
9. G. J. Lake and P. B. Lindley, J. Appl. Polym. Sci. **8** (1964), 707.
10. G. J. Lake and P. B. Lindley, Rubb. J. **146** (10) (1964), 24 and (11) (1964), 30.
11. G. J. Lake and P. B. Lindley, J. Appl. Polym. Sci. **9** (1965), 1233.
12. H. W. Greensmith, J. Appl. Polym. Sci. **7** (1963), 993.
13. G. J. Lake, Proc. International Conf. on the Yield, Deformation and Fracture of Polymers, 5.3/1, Institute of Physics, London, 1970.
14. P. B. Lindley, J. Strain Anal. **7** (1972), 132.
15. H. W. Greensmith, J. Appl. Polym. Sci. **8** (1964), 1113.
16. G. J. Lake and O. H. Yeoh, in preparation.
17. A. G. Thomas, *Physical Basis of Yield and Fracture*, Conf. Proc. Oxford, Institute of Physics, London, 1966, p. 134.
18. A. G. Thomas and J. M. Whittle, Rubb. Chem. Technol. **43** (1970), 222.
19. M. Braden and A. N. Gent, J. Appl. Polym. Sci. **3** (1960), 100.
20. E. Southern and A. G. Thomas, *Plast. Rubb.*, Mat. Appl. **3** (1978), 133.
21. G. J. Lake and O. H. Yeoh, Int. J. Fracture **14** (1978), 509.
22. A. N. Gent and G. R. Hamed, Rubb. Chem. Technol. **51** (1978), 354.
23. A. D. Roberts, Rubb. Chem. Technol. **52** (1979), 23.
24. B. E. Clapson and G. J. Lake, Rubb. J. **152** (12) (1970), 36.
25. R. F. Breidenbach and G. J. Lake, Rubb. Chem. Technol. **52** (1979), 96.
26. R. F. Breidenbach and G. J. Lake, *Fracture Mechanics in Design and Service: Living with Defects*, Discussion Meeting, The Royal Society, London, 1979; Phil. Trans. **A299** (1981), 189.
27. A. N. Gent and A. W. Henry, Proc. International Rubb. Conf., Brighton, Maclaren and Sons, London, 1967, p. 193.
28. A. Kadir and A. G. Thomas, *Elastomers: Criteria for Engineering Design* (C. Hepburn and R. J. W. Reynolds, eds.), Applied Science Publishers, Barking, 1979, p. 67.
29. A. Ahagon and A. N. Gent, J. Polym. Sci. (Polym. Phys. Ed.) **13** (1975), 1903.
30. H. K. Mueller and W. G. Knauss, Trans. Soc. Rheol. **15** (1971), 217.

FINITE DEFORMATION OF ELASTIC RODS AND SHELLS

P. M. NAGHDI

Department of Mechanical Engineering, University of California, Berkeley, CA 94720, USA

(Received November 17, 1980)

ABSTRACT

The objective of this paper is to present an account of recent developments in the direct formation of theories of rods and shells based on 1- and 2-dimensional continuum *models* originating in the works of Duhem and E. and F. Cosserat. Following some preliminaries and description of (3-dimensional) shell-like and rod-like bodies, the rest of the paper is arranged in two parts namely Part A (for shells) and Part B (for rods) and can be read independently of each other. In each part, after providing the main ingredients of the direct model and a statement of the conservation laws, a rapid outline is given of the derivation of the basic equations and nonlinear constitutive equations for elastic materials. Each part also includes a discussion of constrained theories and an account of recent developments on the subject.

1. Introduction

Rods and shells are a class of 3-dimensional bodies whose boundary surfaces have special characteristic features. In general, two entirely different approaches may be adopted for the construction of 1-dimensional and 2-dimensional mechanical theories of rods and shells and similarly these two approaches may be used in the construction of theories of fluid jets and fluid sheets. One approach starts with the 3-dimensional equations of classical continuum mechanics and by applying approximation procedures strives to obtain 1-dimensional (in the case of rods and jets) and 2-dimensional (in the case of shells and sheets) field equations and constitutive equations for the medium under consideration. In the other approach, the medium response is *modelled* as a 1-dimensional and a 2-dimensional *directed* continuum, called a Cosserat curve and a Cosserat surface, respectively; and one then proceeds to

Proceedings of the IUTAM Symposium on Finite Elasticity, Lehigh University, August 10–15, 1980. General lecture.

Carlson, D.E. and Shield, R.T. (eds.)
Proceedings of the IUTAM Symposium on Finite Elasticity
ISBN 90 247 2629 8

the development of the field equations and the appropriate constitutive equations.[†] If *full* information is desired regarding the motion and deformation of the continuum under study in the context of the classical 3-dimensional theory, then there would be no need to develop a particular 1-dimensional and a 2-dimensional theory. In fact, the aim of 1-dimensional and 2-dimensional theories of the type mentioned above is to provide only partial information in some sense: for example, in the case of shells, information concerning quantities which can be regarded as representing the medium response confined to a surface or its neighborhood as a consequence of the (3-dimensional) motion of the body, or the determination of certain weighted averages of quantities resulting from the (3-dimensional) motion of the body.

The nature of the difficulties in the development of both the theory of shells and the theory of rods from the full 3-dimensional equations is well known and has been elaborated upon in various contexts by Green, Laws and Naghdi (1968), Green and Naghdi (1970), Naghdi (1972, Sections 1, 4; 1974; 1979a) and Ericksen (1979). In view of these it is reasonable to attempt to formulate 1-dimensional and 2-dimensional theories of the types described above by replacing the continuum characterizing the (3-dimensional) medium in question with an alternative model which would reflect the main features of the response of the 3-dimensional medium and which would then permit the formulation of appropriate 1-dimensional and 2-dimensional theories by a *direct* approach and without appeal to the special assumptions or approximations generally employed in the derivation from the 3-dimensional equations. It should be emphasized that a Cosserat surface and a Cosserat curve are not, respectively, just a 2-dimensional surface and a 1-dimensional curve; but are, in fact, endowed with some structure in the form of additional primitive kinematical vector fields.

The concept of 'directed' or 'oriented' media originated in the work of Duhem (1893) and a first systematic development of theories of oriented media in one, two and three dimensions was carried out by E. and F. Cosserat (1909). In their work, the Cosserats represented the orientation of each point of their continuum by a set of mutually perpendicular rigid vectors. The pruely kinematical aspects of oriented bodies characterized by ordinary displacement and the independent deformation of N deformable vectors in N-dimensional space has been discussed by Ericksen and Truesdell (1958), who also introduced the terminology of *directors*.

A complete general theory of a Cosserat surface with a single deformable

[†] Other 2-dimensional and 1-dimensional models may also be used to construct direct theories of shells and rods but we postpone further remarks on this until later in this section.

director given by Green, Naghdi and Wainwright (1965) was developed within the framework of thermomechanics. This derivation (Green et al. 1965) is carried out mainly from an appropriate energy equation, together with invariance requirements under superposed rigid body motions. A related development utilizing three directors at each point of the surface, in the context of a purely mechanical theory and with the use of a virtual work principle, is given by Cohen and DeSilva (1966b). A further development of the basic theory of a Cosserat surface along with certain general considerations regarding the construction of nonlinear constitutive equations for elastic shells is given by Naghdi (1972, Sec. 8), which also contains additional historical remarks relevant to oriented continua and to the theory of thin elastic shells. A hierarchical theory of Cosserat surfaces, namely that comprising a material surface with K ($\geqq 1$) directors, is contained in a paper by Green and Naghdi (1976a) which deals with fluid sheets and its application to water waves.

A parallel development in the theory of a Cosserat curve with two deformable directors begins with a paper of Green and Laws (1966) whose derivation is carried out mainly from an appropriate energy equation, together with invariance requirements under superposed rigid body motions. A related development of a directed curve with three deformable directors at each point of the curve, in the context of a purely mechanical theory and with the use of a virtual work principle, is given by Cohen (1966). A further development of the basic theory of a Cosserat curve along with certain general developments regarding the construction of nonlinear constitutive equations for elastic rods is given by Green, Naghdi and Wenner (1974b). A hierarchical theory of Cosserat curves, namely that comprising a material curve with L ($\geqq 2$) directors, is contained in a paper by Naghdi (1979b) which is concerned with applications to Newtonian and non-Newtonian flows in pipes.

Of course, the introduction of an alternative model and formulation of 1-dimensional and 2-dimensional theories by the direct approach does not mean that one ignores the nature of the field equations in the 3-dimensional theory. In fact, some of the developments of the field equations by direct procedure are materially aided or influenced by available information which can be obtained from the 3-dimensional theory. For example, the integrated equations of motion from the 3-dimensional equations provide guidelines for a statement of 1- and 2-dimensional conservation laws in conjunction with the 1- and 2-dimensional models, and also provide some insight into the nature of inertia terms and the kinetic energy in the direct formulation of the 1-dimensional and 2-dimensional theories.

Inasmuch as most of the difficulties associated with the derivation of the 1-dimensional and 2-dimensional theories from the 3-dimensional equations occur in the construction of the constitutive equations, it is in fact here that the direct approach offers a great deal of appeal. These constructions, as well as the entire development by the direct approach, are exact in the sense that they rest on (1-dimensional and 2-dimensional) postulates valid for nonlinear behavior of materials but clearly they cannot be expected to represent all the features that could only be predicted by the relevant full 3-demensional equations. Theories constructed via a direct approach necessarily satisfy the requirements of invariance under superposed rigid body motions that arise from physical considerations and, of course, they are also consistent and fully invariant in the mathematical sense. Moreover, the development by the direct approach is conceptually simple and does not have the difficulties associated with approximations usually made in the development of the theory of thin shells or the theories of slender rods from their corresponding 3-dimensional equations.

Although the direct approach to shells and rods employed in this paper is based on the 2-dimensional and 1-dimensional *directed* continuum models, respectively, other direct 2-dimensional and 1-dimensional models may also be used to construct theories of shells and rods. For example in the case of shells, instead of developing a theory based on a Cosserat surface, we may consider only a material surface and construct a direct theory in which the basic kinematical ingredients are the position vector of the surface together with its first and second gradients. A theory of this kind has been discussed by Balaban, Green and Naghdi (1967) and a somewhat less general theory has been considered by Cohen and DeSilva (1966a, 1968). Although these developments have some overlapping features with corresponding results in the theory of Cosserat surfaces, they are more restrictive. Additional related remarks are made in Section 6 of this paper.

Following some general background information and definitions of shell-like and rod-like bodies in Section 2, the remainder of the paper is arranged in two parts which can be read independently of each other: one part (Part A) is concerned with the theory of shells and the other (Part B) is devoted to the theory of rods. In Part A (Sections 3–8), first a concise development of the basic theory of a Cosserat surface with a single director followed by its generalization is presented. For a Cosserat surface with a single director, constitutive equations are discussed in the context of finite deformation of elastic shells and a procedure is indicated for identification of the assigned fields and the inertia coefficients which occur in the basic theory. Next, a fairly detailed account of constrained theories of shells is presented which includes the construction of an interesting nonlinear constrained theory not discussed previously in the literature. This is followed by an account of recent developments pertaining to elastic shells and a representation of the basic equations of a Cosserat surface in direct (coordinate-free) notation. A table of contents for Part A is listed in the introductory paragraph of Section 3.

Similarly, in Part B (Sections 9–13), first a concise development of the Cosserat curve with two directors and its generalization is presented. Next, with reference to a Cosserat curve with two directors, constitutive equations are discussed for finite deformation of elastic rods and a procedure is indicated for identification of the assigned fields and the inertia coefficients which occur in the basic theory. This is followed by some additional remarks pertaining to elastic rods, together with a brief discussion of the constrained theories of rods, and a representation of the basic equations for a Cosserat curve in direct (coordinate-free) notation. A table of contents for Part B is listed in the introductory paragraph of Section 9.

2. General background

In this section, we provide appropriate definitions for *shell-like* and *rod-like* bodies. To this end, consider a finite three-dimensional body $\mathscr{B}$ in a Euclidean 3-space, and let convected (or Lagrangian) coordinates θ^i $(i = 1,2,3)$, be assigned to each particle (or material point) of $\mathscr{B}$. Further, let[†] $\boldsymbol{r}^*$ be the position vector, from a fixed origin, of a typical particle of $\mathscr{B}$ in the present configuration at time t. Then, a motion of the (three-dimensional) body is defined by a vector-valued function $\hat{\boldsymbol{r}}^*$ which assigns position $\boldsymbol{r}^*$ to each particle of $\mathscr{B}$ at each instant of time, i.e.,[‡]

$$\boldsymbol{r}^* = \hat{\boldsymbol{r}}^*(\theta^1, \theta^2, \theta^3, t). \tag{2.1}$$

We assume that the vector function $\hat{\boldsymbol{r}}^*$ — a 1-parameter family of configurations with t as the real parameter — is sufficiently smooth in the sense that it is differentiable with respect to θ^i and t as many times as required. In some developments, it is convenient to set $\theta^3 = \xi$ and adopt the notation

$$\theta^i = (\theta^\alpha, \xi), \qquad \theta^3 = \xi. \tag{2.2}$$

We recall the formulas

[†] The use of an asterisk attached to various symbols is for later convenience. The corresponding symbols without the asterisks are reserved for different definitions or designations to be introduced later.

[‡] Recall that when the particles of a continuum are referred to a convected coordinate system, the numerical values of the coordinates associated with each particle remain the same for all time.

$$\boldsymbol{g}_i = \frac{\partial \hat{\boldsymbol{r}}^*}{\partial \theta^i}, \qquad g_{ij} = \boldsymbol{g}_i \cdot \boldsymbol{g}_j, \qquad g = \det(g_{ij}),$$

(2.3)
$$\boldsymbol{g}^i \cdot \boldsymbol{g}_j = \delta^i_j, \qquad \boldsymbol{g}^i = g^{ij}\boldsymbol{g}_j, \qquad \boldsymbol{g}^i \cdot \boldsymbol{g}^j = g^{ij},$$

(2.4)
$$dv = g^{1/2} d\theta^1 d\theta^2 d\theta^3$$

and further assume that†

(2.5)
$$g^{1/2} = [\boldsymbol{g}_1 \boldsymbol{g}_2 \boldsymbol{g}_3] > 0.$$

In (2.4), $\boldsymbol{g}_i$ and $\boldsymbol{g}^i$ are the covariant and the contravariant base vectors at time t, respectively, g_{ij} is the metric tensor, g^{ij} is its conjugate, δ^i_j is the Kronecker symbol in 3-space and dv the volume element in the present configuration.

The velocity vector $\boldsymbol{v}^*$ of a particle of the three-dimensional body in the present configuration is defined by

(2.6)
$$\boldsymbol{v}^* = \dot{\boldsymbol{r}}^*,$$

where a superposed dot denotes material time differentiation with respect to t holding θ^i fixed. The stress vector $\boldsymbol{t}$ across a surface in the present configuration with outward unit normal $\boldsymbol{\nu}^*$ is given by

(2.7)
$$\boldsymbol{t} = \nu^*_i \frac{\boldsymbol{T}^i}{g^{1/2}} = \nu^*_i \tau^{ik} \boldsymbol{g}_k, \qquad \boldsymbol{T} = \boldsymbol{g}_i \otimes g^{-1/2} \boldsymbol{T}^i = \tau^{ik} \boldsymbol{g}_i \otimes \boldsymbol{g}_k,$$

where

(2.8)
$$\boldsymbol{T}^i = g^{1/2} \tau^{ij} \boldsymbol{g}_j = g^{1/2} \tau^i_j \boldsymbol{g}^j, \qquad \boldsymbol{\nu}^* = \nu^*_i \boldsymbol{g}^i = \nu^{*i} \boldsymbol{g}_i, \qquad \tau^{ij} = \boldsymbol{g}^i \cdot \boldsymbol{T} \boldsymbol{g}^j,$$

where $\boldsymbol{T}$ is the symmetric Cauchy stress tensor, τ^{ik} its contravariant components and $\otimes$ denotes the tensor product of two vectors. In terms of the quantities defined in (2.5)–(2.8), the local field equations which follow from the integral forms of the three-dimensional conservation laws for mass, linear momentum and moment of momentum, respectively, are

$$\dot{\overline{\rho^* g^{1/2}}} = 0,$$

(2.9)
$$\boldsymbol{T}^i{}_{,i} + \rho^* \boldsymbol{f}^* g^{1/2} = \rho^* g^{1/2} \dot{\boldsymbol{v}}^*, \qquad \boldsymbol{g}_i \times \boldsymbol{T}^i = 0,$$

where ρ^* is the 3-dimensional mass density, $\boldsymbol{f}^*$ is the body force field per unit mass and a comma denotes partial differentiation with respect to θ^i. For later

† The choice of positive sign in (2.5) is for definiteness. Alternatively, for physically possible motions we only need to assume that $g^{1/2} \neq 0$ with the understanding that in any given motion $[\boldsymbol{g}_1 \boldsymbol{g}_2 \boldsymbol{g}_3]$ is either >0 or <0. The condition (2.5) also requires that θ^1 be a right-handed coordinate system.

reference, we note that for an incompressible medium, the condition of incompressibility may be expressed as

$$\overline{g^{1/2}}^{\,\cdot} = 0 \qquad \text{or} \qquad \operatorname{div} \boldsymbol{v}^* = 0. \tag{2.10}$$

A material surface in $\mathscr{B}$ can be defined by the equation $\xi = \xi(\theta^\alpha)$; the equation resulting from (2.1) with $\xi = \xi(\theta^\alpha)$ represents the parametric form of this material surface in the current configuration and defines a 1-parameter family of surfaces in space, each of which we assume to be smooth and non-intersecting. We refer to the surface $\xi = 0$ in the current configuration as s. Any point of the surface s is specified by the position vector $\boldsymbol{r}$, relative to the same fixed origin to which $\boldsymbol{r}^*$ is referred, where

$$\boldsymbol{r} = \hat{\boldsymbol{r}}(\theta^\alpha, t) = \hat{\boldsymbol{r}}^*(\theta^\alpha, 0, t). \tag{2.11}$$

Let $\boldsymbol{a}_\alpha$ denote the base vectors along the θ^α-curves on the surface s. By (2.11) and $(2.3)_1$,

$$\boldsymbol{a}_\alpha = \boldsymbol{a}_\alpha(\theta^\gamma, t) = \frac{\partial \hat{\boldsymbol{r}}}{\partial \theta^\alpha} = \boldsymbol{g}_\alpha(\theta^\gamma, 0, t), \tag{2.12}$$

and the unit normal $\boldsymbol{a}_3 = \boldsymbol{a}_3(\theta^\gamma, t)$ to s may be defined by[†]

$$\boldsymbol{a}_\alpha \cdot \boldsymbol{a}_3 = 0, \qquad \boldsymbol{a}_3 \cdot \boldsymbol{a}_3 = 1, \qquad \boldsymbol{a}_3 = \boldsymbol{a}^3, \qquad [\boldsymbol{a}_1 \boldsymbol{a}_2 \boldsymbol{a}_3] > 0. \tag{2.13}$$

We also recall the formulas

$$\begin{gathered} a_{\alpha\beta} = \boldsymbol{a}_\alpha \cdot \boldsymbol{a}_\beta, \qquad a = \det(a_{\alpha\beta}), \\ \boldsymbol{a}^\alpha = a^{\alpha\beta} \boldsymbol{a}_\beta, \quad \boldsymbol{a}^\alpha \cdot \boldsymbol{a}^\beta = a^{\alpha\beta}, \qquad a^{\alpha\gamma} a_{\gamma\beta} = \delta^\alpha_\beta \end{gathered} \tag{2.14}$$

and

$$\begin{gathered} b_{\alpha\beta} = b_{\beta\alpha} = -\boldsymbol{a}_\alpha \cdot \boldsymbol{a}_{3,\beta} = \boldsymbol{a}_3 \cdot \boldsymbol{a}_{\alpha,\beta}. \\ \boldsymbol{a}_{\alpha\,|\,\beta} = b_{\alpha\beta} \boldsymbol{a}_3, \qquad \boldsymbol{a}_{3,\alpha} = -b^\gamma_\alpha \boldsymbol{a}_\gamma, \qquad b_{\alpha\beta\,|\,\gamma} = b_{\alpha\gamma\,|\,\beta}, \end{gathered} \tag{2.15}$$

where $\boldsymbol{a}^\alpha$ denote the reciprocal base vectors of the surface s, $a_{\alpha\beta}$ and $b_{\alpha\beta}$ are the components of its first and second fundamental forms, a comma denotes partial differentiation with respect to the surface coordinates θ^γ, a vertical bar stands for covariant differentiation with respect to $a_{\alpha\beta}$, and δ^α_β is the Kronecker symbol in 2-space.

A material line (not necessarily a straight line) in $\mathscr{B}$ can be defined by the equations $\theta^\alpha = \theta^\alpha(\xi)$; the equation resulting from (2.1) with $\theta^\alpha = \theta^\alpha(\xi)$

[†] The use of the same symbols for base vectors of a surface in (2.12)–(2.13) and for the triad of a space curve in (2.17)–(2.18) should not give rise to confusion. The main developments for shells and rods are dealt with separately in the rest of the paper; this permits the use of the same symbol for different quantities in the case of shells and rods without confusion.

represents the parametric form of this material line in the current configuration and defines a 1-parameter family of curves in space, each of which we assume to be smooth and nonintersecting. We refer to the space curve $\theta^\alpha = 0$ in the current configuration as c. Any point of this curve is specified by the position vector $\boldsymbol{r}$, relative to the same fixed origin to which $\boldsymbol{r}^*$ is referred, where

$$\boldsymbol{r} = \hat{\boldsymbol{r}}(\xi, t) = \hat{\boldsymbol{r}}^*(0, 0, \xi, t). \tag{2.16}$$

Let $\boldsymbol{a}_3$ denote the tangent vector along the ξ-curve.[†] By (2.16) and $(2.3)_1$,

$$\boldsymbol{a}_3 = \boldsymbol{a}_3(\xi, t) = \frac{\partial \hat{\boldsymbol{r}}}{\partial \xi} = \boldsymbol{g}_3(0, 0, \xi, t) \tag{2.17}$$

and the unit principal normal $\boldsymbol{a}_1$ and the unit binormal vector $\boldsymbol{a}_2$ to c may be introduced as

$$\begin{gathered} \boldsymbol{a}_1 = \boldsymbol{a}_1(\xi, t) = \frac{\partial \boldsymbol{a}_3/\partial \xi}{|\partial \boldsymbol{a}_3/\partial \xi|}, \qquad \boldsymbol{a}_2 = \boldsymbol{a}_2(\xi, t) = \frac{\boldsymbol{a}_3}{|\boldsymbol{a}_3|} \times \boldsymbol{a}_1, \\ |\boldsymbol{a}_3| = (a_{33})^{1/2}, \qquad a_{33} = \boldsymbol{a}_3 \cdot \boldsymbol{a}_3, \qquad [\boldsymbol{a}_1 \boldsymbol{a}_2 \boldsymbol{a}_3] > 0, \end{gathered} \tag{2.18}$$

where the notation $|\boldsymbol{a}_3|$ stands for the magnitude of $\boldsymbol{a}_3$. The system of base vectors $\boldsymbol{a}_i$ are oriented along the Serret–Frenet triad and satisfy the differential equations

$$\begin{gathered} \frac{\partial \boldsymbol{a}_1}{\partial \xi} = \tau (a_{33})^{1/2} \boldsymbol{a}_2 - \kappa \boldsymbol{a}_3, \qquad \frac{\partial \boldsymbol{a}_2}{\partial \xi} = -\tau (a_{33})^{1/2} \boldsymbol{a}_1, \\ \frac{\partial \boldsymbol{a}_3}{\partial \xi} = a_{33} \kappa \boldsymbol{a}_1 + \frac{1}{2a_{33}} \frac{\partial a_{33}}{\partial \xi} \boldsymbol{a}_3, \end{gathered} \tag{2.19}$$

where κ and τ denote, respectively, the curvature and the torsion of c. In the special case that c is a plane curve, we may choose $\boldsymbol{a}_1$ as the unit normal to the curve and then $\boldsymbol{a}_2$ will be perpendicular to the plane of $\boldsymbol{a}_1$ and $\boldsymbol{a}_3$. If c is a straight curve, then there is no unique Serret–Frenet triad and $\boldsymbol{a}_i$ may be chosen as any orthogonal triad with $\boldsymbol{a}_1$, $\boldsymbol{a}_2$ as unit vectors. Equations (2.19) are not identical to the formulas of Frenet because the parameter ξ is not necessarily the arc length of c. It may be noted here that the convected coordinate ξ may be chosen to coincide with the arc length in any one configuration of the material curve, e.g., in the present configuration. However, in a general motion (involving different configurations) the arc length between any pair of particles changes while the convected coordinates

[†] The designation of the tangent vector to a curve by $\boldsymbol{a}_3$ should not be confused with the use of the same symbol for a different purpose in (2.13). In this connection, see the preceding footnote.

of each particle must remain the same. Therefore, arc length would not qualify as a convected coordinate.

In the next four paragraphs (identified as subsections 2A and 2B) we provide appropriate definitions for shell-like and rod-like bodies in fairly precise terms.

2A. *Definition of a shell-like body. A representation for the motion of a thin shell*

Consider a two-dimensional surface s defined by the parametric equation $\xi = 0$, over a finite coordinate patch $\alpha' \leqq \theta^1 < \alpha''$, $\beta' \leqq \theta^2 \leqq \beta''$. Let $\boldsymbol{r}$ and $\boldsymbol{a}_3$ denote, respectively, the position vector and the unit normal to s. At each point of s, imagine material filaments projecting normally above and below the surface s. The surface formed by the material filaments constructed at the points of the closed boundary curve of s is called the *lateral surface*. Such a 3-dimensional body (depicted in Figure 1) is called a shell if the dimension of the body along the normals, called the height and denoted by h, is *small*. A shell is said to be *thin* if its thickness is much smaller than a certain characteristic length $L(s)$ of the surface s, for example, the local minimum radius of curvature of the surface, or the smallest dimension of s in the case of a plane. If h is constant, the shell is said to be of uniform thickness, otherwise

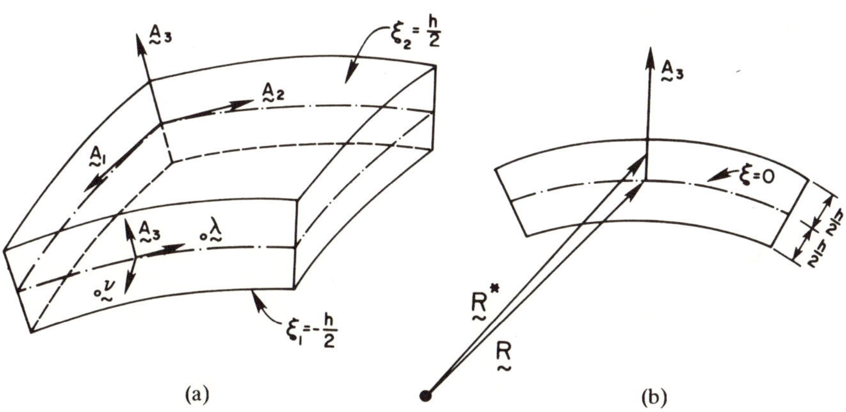

Figure 1. (a) Element of a shell-like body in a reference configuration showing the middle surface $\xi = 0$ and the major surfaces $\xi = \pm(h/2)$, h being the shell thickness; and (b) the position vector $\boldsymbol{R}$ from a fixed origin to a point on the middle surface $\xi = 0$ and the position vector $\boldsymbol{R}^*$ to any point of the region of space occupied by the shell in the reference configuration. Also shown are the base vectors $\boldsymbol{A}_\alpha$ on the middle surface, the unit normal $\boldsymbol{A}_3$ to the middle surface and the right-handed triad $\{{}_0\boldsymbol{\nu}, {}_0\boldsymbol{\lambda}, \boldsymbol{A}_3\}$ with ${}_0\boldsymbol{\nu}$ and ${}_0\boldsymbol{\lambda}$ being, respectively, the unit normal to a curve on the middle surface and the unit tangent to a curve on the middle surface.

of variable thickness. Since a material surface in the three-dimensional body can be defined by the equation $\xi = \xi(\theta^\alpha)$, it follows that the equation resulting from (2.1) with $\xi = \xi(\theta^\alpha)$ represents the parametric form of the material surface in the present configuration. In particular, the equation $\xi = 0$ defines a surface in space at time t, which we assume to be smooth and nonintersecting. Every point of this surface has a position vector $\boldsymbol{r}$ specified by (2.11). Let the boundary of the three-dimensional continuum be specified by the material surfaces

$$\xi = \xi_1(\theta^1, \theta^2), \qquad \xi = \xi_2(\theta^1, \theta^2), \qquad \xi_1 < \xi_2, \tag{2.20}$$

with the surface $\xi = 0$ lying either on one of the two surfaces $(2.20)_{1,2}$ or between them (see, for example, Figure 1), and a material surface

$$f(\theta^1, \theta^2) = 0, \tag{2.21}$$

which is chosen such that $\xi =$ const. form closed smooth curves on the surface (2.21). As pointed out previously by Naghdi (1975a), in the development of a general theory, it is preferable to leave unspecified the choice of the relation of the surface s ($\xi = 0$) to the major surfaces s^+ and s^-. In special cases of the general theory or in specific applications, however, it is necessary to fix the relation of s to the surfaces $(2.20)_{1,2}$.

We now suppose that $\hat{\boldsymbol{r}}^*$ in (2.1) can be represented by the Taylor expansion in the bounded region $\xi_1 \leqq \xi \leqq \xi_2$ with coefficients which are continuous functions of θ^α, t and have continuous space and time derivatives of order 2. Thus, for shell-like bodies, we write

$$\hat{\boldsymbol{r}}^* = \boldsymbol{r} + \sum_{N=1}^{K} \xi^N \boldsymbol{d}_N, \qquad \boldsymbol{d}_N = \boldsymbol{d}_N(\theta^\alpha, t) \tag{2.22}$$

and by $(2.3)_1$ and (2.6) we also have

$$\boldsymbol{g}_\alpha = \boldsymbol{a}_\alpha + \sum_{N=1}^{K} \xi^N \frac{\partial \boldsymbol{d}_N}{\partial \theta^\alpha}, \qquad \boldsymbol{g}_3 = \sum_{N=1}^{K} N \xi^{N-1} \boldsymbol{d}_N, \tag{2.23}$$

$$\boldsymbol{v}^* = \boldsymbol{v} + \sum_{N=1}^{K} \xi^N \boldsymbol{w}_N, \qquad \boldsymbol{v} = \dot{\boldsymbol{r}}, \qquad \boldsymbol{w}_N = \dot{\boldsymbol{d}}_N, \tag{2.24}$$

where $\boldsymbol{r}$ is defined by (2.1) and a superposed dot in (2.22) denotes material time differentiation with respect to t holding θ^α fixed. A special case of (2.22) which is of particular interest in subsequent developments is when $N = 1$, namely

$$\hat{\boldsymbol{r}}^* = \boldsymbol{r} + \xi \boldsymbol{d}, \tag{2.25}$$

where we have set $\boldsymbol{d}_1 = \boldsymbol{d}$.

2B. *Definition of a rod-like body. A representation for the motion of a slender rod*

Consider a space curve c defined by the parametric equations $\theta^\alpha = 0$, over a finite interval $\xi_1 \leqq \xi \leqq \xi_2$. Let $\boldsymbol{r}$ be the position vector of any point of c and let $\boldsymbol{a}_1$, $\boldsymbol{a}_2$ and $\boldsymbol{a}_3$ denote its unit principal normal, unit binormal and the tangent vector, respectively. At each point of c, imagine material filaments lying in the normal plane, i.e., the plane perpendicular to $\boldsymbol{a}_3$, and forming the normal cross-section $\mathscr{A}_n$. The surface swept out by the closed boundary curve $\partial\mathscr{A}_n$ of $\mathscr{A}_n$ is called the lateral surface. Such a 3-dimensional body (depicted in Figure 2) is called rod-like if the dimensions in the plane of the normal cross-section are *small* compared to some characteristic dimension $L(c)$ of c (see Figure 2), e.g., its local radius of curvature $1/\kappa$, or the length of c in the case of a straight curve. A rod-like body is said to be *slender* if the largest dimension of $\mathscr{A}_n$ is much smaller than $L(c)$. If $\mathscr{A}_n$ is independent of ξ, the body is said to be of uniform cross-section, otherwise of variable cross-section. Since a material curve in the three-dimensional body $\mathscr{B}$ can be defined by the equations $\theta^\alpha = \theta^\alpha(\xi)$, it follows that the equation resulting from (2.1) with $\theta^\alpha = \theta^\alpha(\xi)$ represents the parametric form of the material curve in the present configuration and defines a curve c in space at time t, which we assume to be sufficiently smooth and nonintersecting. Every point

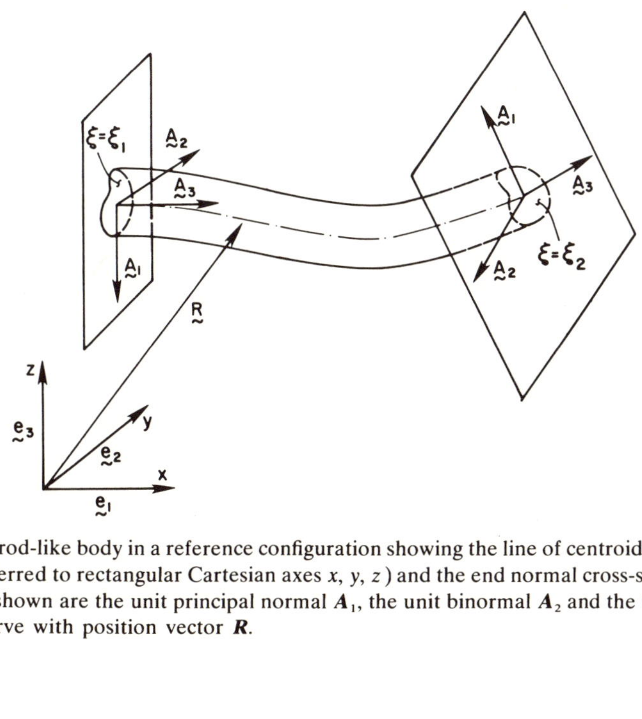

Figure 2. A rod-like body in a reference configuration showing the line of centroids with position vector $\boldsymbol{R}$ (referred to rectangular Cartesian axes x, y, z) and the end normal cross-sections $\xi = \xi_1$, $\xi = \xi_2$. Also shown are the unit principal normal $\boldsymbol{A}_1$, the unit binormal $\boldsymbol{A}_2$ and the tangent vector $\boldsymbol{A}_3$ to the curve with position vector $\boldsymbol{R}$.

of this curve has a position vector specified by (2.14). Let the (3-dimensional) rod-like body in some neighborhood of c be bounded by material surfaces $\xi = \xi_1$, $\xi = \xi_2$ (indicated in Figure 2) and a material surface of the form

$$F(\theta^1, \theta^2, \xi) = 0, \tag{2.26}$$

which is chosen such that $\xi =$ constant are curved sections of the body bounded by closed curves on this surface with c lying on or within (2.26). In the development of a general theory, it is preferable to leave unspecified the choice of the relation of the curve c to one on the boundary surface (2.26). In special cases or in specific applications, however, it is necessary to fix the relation of c to the surface (2.26).

We now suppose that $\hat{\boldsymbol{r}}^*$ in (2.1) can be represented by the Taylor expansion in the bounded region lying inside the surface (2.26) and between $\xi = \xi_1$, $\xi = \xi_2$, with coefficients which are continuous functions of ξ, t and have continuous space and time derivatives of order 2. Thus, for rod-like bodies, we write

$$\hat{\boldsymbol{r}}^* = \boldsymbol{r} + \sum_{N=1}^{K} \theta^{\alpha_1}\theta^{\alpha_2} \cdots \theta^{\alpha_N}\boldsymbol{d}_{\alpha_1\cdots\alpha_N}, \qquad \boldsymbol{d}_{\alpha_1\cdots\alpha_N} = \boldsymbol{d}_{\alpha_1\cdots\alpha_N}(\xi, t) \tag{2.27}$$

and by $(2.3)_1$ and (2.6) we also have

$$\boldsymbol{g}_\beta = \boldsymbol{d}_\beta + \sum_{N=1}^{K} N\theta^{\alpha_2} \cdots \theta^{\alpha_N}\boldsymbol{d}_{\beta\alpha_2\cdots\alpha_N}, \quad \boldsymbol{g}_3 = \boldsymbol{a}_3 + \sum_{N=1}^{K} \theta^{\alpha_1} \cdots \theta^{\alpha_N}(\partial \boldsymbol{d}_{\alpha_1\cdots\alpha_N}/\partial\xi), \tag{2.28}$$

$$\boldsymbol{v}^* = \boldsymbol{v} + \sum_{N=1}^{K} \theta^{\alpha_1} \cdots \theta^{\alpha_N}\boldsymbol{w}_{\alpha_1\cdots\alpha_N}, \qquad \boldsymbol{w}_{\alpha_1\cdots\alpha_N} = \dot{\boldsymbol{d}}_{\alpha_1\cdots\alpha_N}, \tag{2.29}$$

where $\boldsymbol{r}$ in (2.27) is defined by (2.16), $\boldsymbol{d}_{\alpha_1\cdots\alpha_N}$ is symmetric with respect to indices $\alpha_1 \cdots \alpha_N$ and a superposed dot in (2.29) denotes material time differentiation with respect to t holding ξ fixed. A special case of (2.27) which is of particular interest in subsequent developments is when $N = 1$, namely

$$\hat{\boldsymbol{r}}^* = \boldsymbol{r} + \theta^\alpha \boldsymbol{d}_\alpha, \tag{2.30}$$

where we have put $\alpha_1 = \alpha$.

Part A

Elastic shells: A direct formulation

In Part A (Sections 3–8), we summarize the main kinematics and the basic principles of the theory of Cosserat (or directed) surfaces and then discuss the constitutive equations for elastic shells, as well as several related aspects of the basic theory and recent developments on the subject. Although we are concerned here mainly with the purely mechanical theory involving appropriate forms of the conservation laws for mass, linear momentum, director momentum and moment of momentum, we also include a statement of the conservation of energy. The latter provides motivation in the development of certain constitutive equations, such as those for an elastic material, and in the discussion of aspects of some special solutions involving jump in energy. The contents of Part A are as follows:

3. The basic theory of Cosserat surfaces

Having introduced the notion of a (three-dimensional) shell-like body in Section 2, we now formally define a direct *model* for such a body. Thus, deformable media which are *modelled* by a material surface $\mathscr{S}$ embedded in a Euclidean 3-space, together with K ($K = 1, 2, \cdots, N$) deformable vector fields — called directors — attached to every point of the material surface are called *Cosserat surfaces* or *directed surfaces* and may be conveniently referred to as $\mathscr{C}_K$. The directors which are not necessarily along the unit normals to the surface have, in particular, the property that they remain unaltered in length under superposed rigid body motions.

In the absence of the directors, we merely have a 2-dimensional material surface $\mathscr{S}$ which can serve as a model for the construction by direct approach of the membrane theory of shells. With $K = 1$, the directed medium is a body $\mathscr{C}_1 = \mathscr{C}$ comprising a material surface and a single deformable director attached to every point of the material surface of $\mathscr{C}$. The latter is the simplest model for the construction of a general bending theory of thin shells; and, for simplicity, we restrict attention to this particular model in most of the development[†] of Section 3.

3.1. *Kinematics of a Cosserat surface* $\mathscr{C}$

Let the particles of the material surface $\mathscr{S}$ of $\mathscr{C}$ be identified by means of a system of convected coordinates θ^α $(\alpha = 1, 2)$ and let the 2-dimensional region occupied by the material surface $\mathscr{S}$ in the present configuration of at time t be referred to as $\mathscr{s}$. Let $\boldsymbol{r}$ and $\boldsymbol{d}$ denote the position vector of a typical point of $\mathscr{s}$ and the director at the same point, respectively. Also, let $\boldsymbol{a}_\alpha$, $\boldsymbol{a}_3$ designate, respectively, the base vectors along the θ^α-curves on $\mathscr{s}$ and the outward unit normal to $\mathscr{s}$. Then, a motion of the Cosserat surface is defined by vector-valued functions which assign position $\boldsymbol{r}$ and director $\boldsymbol{d}$ to each particle of $\mathscr{C}$ at each instant of time, i.e.[‡]

$$\boldsymbol{r} = \hat{\boldsymbol{r}}(\theta^\alpha, t), \qquad \boldsymbol{d} = \hat{\boldsymbol{d}}(\theta^\alpha, t), \qquad [\boldsymbol{a}_1\boldsymbol{a}_2\boldsymbol{d}] > 0, \tag{3.1}$$

where

$$\boldsymbol{a}_\alpha = \boldsymbol{a}_\alpha(\theta^\alpha, t) = \frac{\partial \hat{\boldsymbol{r}}}{\partial \theta^\alpha} \tag{3.2}$$

and the condition $(3.1)_3$ ensures that the director $\boldsymbol{d}$ is nowhere tangent to $\mathscr{s}$. The velocity and the director velocity vectors are defined by

$$\boldsymbol{v} = \dot{\boldsymbol{r}}, \qquad \boldsymbol{w} = \dot{\boldsymbol{d}}, \tag{3.3}$$

and since the coordinate curves on $\mathscr{s}$ are convected from (3.2), we have

$$\dot{\boldsymbol{a}}_\alpha = \boldsymbol{v}_{,\alpha}, \tag{3.4}$$

where a superposed dot denotes differentiation with respect to t holding θ^α fixed.

It is convenient to introduce here a slightly different notation than that adopted in Naghdi (1972) and a number of earlier papers on the subject. Thus, we put

$$\boldsymbol{d}_\alpha = \boldsymbol{a}_\alpha, \qquad \boldsymbol{d}_3 = \boldsymbol{d} \tag{3.5}$$

and observe that, in view of $(3.1)_3$ and (3.5), $\boldsymbol{d}_1$, $\boldsymbol{d}_2$, $\boldsymbol{d}_3$ are linearly independent vectors. Hence, we may introduce a set of reciprocal vectors $\boldsymbol{d}^i$ such that

[†] A brief account of the more general theory for Cosserat surfaces $\mathscr{C}_K$ is indicated at the end of this section.

[‡] For convenience, we adopt the notation for $\boldsymbol{r}$ in (2.11) and (2.25) also for the surface $(3.1)_1$. This permits an easy identification of the two surfaces, if desired. The choice of positive sign in $(3.1)_3$ is for definiteness. Alternatively, it will suffice to assume that $[\boldsymbol{a}_1\boldsymbol{a}_2\boldsymbol{d}] \neq 0$ with the understanding that in any given motion the scalar triple product $[\boldsymbol{a}_1\boldsymbol{a}_2\boldsymbol{d}]$ is either > 0 or < 0.

$$d_i \cdot d^j = \delta_i^j, \tag{3.6}$$

where δ_i^j is the Kronecker symbol in 3-space. Whenever desirable, the notations $\boldsymbol{d}_i = (\boldsymbol{d}_1, \boldsymbol{d}_2, \boldsymbol{d}_3)$ and $(\boldsymbol{a}_\alpha, \boldsymbol{d})$ will be used interchangeably throughout Part A depending on the particular context. Consider now a reference configuration, not necessarily the initial configuration, of the Cosserat surface $\mathscr{C}$. In the reference configuration, let the material surface of $\mathscr{C}$ be referred to by $\mathscr{S}_R$ with $\boldsymbol{R}$ as its position vector; let $\boldsymbol{D}$ be the director at $\boldsymbol{R}$; and let $\boldsymbol{A}_\alpha$, $\boldsymbol{A}_3$ denote, respectively, the base vectors along the θ^α-curves on $\mathscr{S}_R$ and the unit normal to $\mathscr{S}_R$. Then, in the reference configuration we have

$$\boldsymbol{R} = \boldsymbol{R}(\theta^\alpha), \qquad \boldsymbol{D} = \boldsymbol{D}(\theta^\alpha), \qquad [\boldsymbol{A}_1\boldsymbol{A}_2\boldsymbol{D}] > 0, \tag{3.7}$$

where

$$\boldsymbol{A}_\alpha = \boldsymbol{A}_\alpha(\theta^\gamma) = \frac{\partial \boldsymbol{R}}{\partial \theta^\alpha} \tag{3.8}$$

and $(3.7)_3$ ensures that $\boldsymbol{D}$ is nowhere tangent to the surface $\mathscr{S}_R$. If the reference configuration of $\mathscr{C}$ is specified to be the initial configuration, say at time $t = 0$, then the vector-valued functions on the right-hand sides of $(3.7)_{1,2}$ can be identified with $\hat{\boldsymbol{r}}(\theta^\alpha, 0)$ and $\hat{\boldsymbol{d}}(\theta^\alpha, 0)$, respectively. Analogously to (3.5), we set

$$\boldsymbol{D}_\alpha = \boldsymbol{A}_\alpha, \qquad \boldsymbol{D}_3 = \boldsymbol{D} \tag{3.9}$$

and note that the dual of (3.6) is given by

$$\boldsymbol{D}_i \cdot \boldsymbol{D}^j = \delta_i^j. \tag{3.10}$$

3.2. *Basic principles of a Cosserat surface* $\mathscr{C}$

In the development of this subsection, we follow the mode of derivation of the basic theory of a Cosserat surface employed by Naghdi (1972, Section 8). Let $\mathscr{P}$, bounded by a closed curve $\partial\mathscr{P}$, be a part of $\mathscr{s}$ occupied by an arbitrary material region of $\mathscr{S}$ in the present configuration at time t and let

$$\boldsymbol{\nu} = \nu^\alpha \boldsymbol{a}_\alpha = \nu_\alpha \boldsymbol{a}^\alpha \tag{3.11}$$

be the outward unit normal to $\partial\mathscr{P}$. It is convenient at this point to define certain additional quantities as follows: The mass denisty $\rho = \rho(\theta^\gamma, t)$ of the surface $\mathscr{s}$ in the present configuration; the contact force† $\boldsymbol{n} = \boldsymbol{n}(\theta^\gamma, t; \boldsymbol{\nu})$ and

† The notations for the contact force $\boldsymbol{n}$, the contact director force $\boldsymbol{m}$ and the surface director force $\boldsymbol{k}$ are the same as those in Naghdi (1977), but differ from Naghdi (1972) and most of the previous papers on the subject. In fact, the vector fields $\boldsymbol{n}$, $\boldsymbol{m}$, $\boldsymbol{k}$ of Part A of the present paper correspond, respectively, to $\boldsymbol{N}$, $\boldsymbol{M}$, $\boldsymbol{m}$ in Naghdi (1972) and most of the previous papers on the

the contact director force $\boldsymbol{m} = \boldsymbol{m}(\theta^\gamma, t; \boldsymbol{\nu})$, each per unit length of a curve in the present configuration; the assigned force $\boldsymbol{f} = \boldsymbol{f}(\theta^\gamma, t)$ and the assigned director force $\boldsymbol{l} = \boldsymbol{l}(\theta^\gamma, t)$, each per unit mass of the surface $\mathscr{s}$; the intrinsic director force $\boldsymbol{k}$ per unit area of $\mathscr{s}$; the inertia coefficients $y^1 = y^1(\theta^\gamma)$ and $y^2 = y^2(\theta^\gamma)$ which are independent of time; the specific internal energy $\varepsilon = \varepsilon(\theta^\gamma, t)$; the heat flux $h = h(\theta^\gamma, t; \boldsymbol{\nu})$ per unit time and per unit length of a curve $\partial\mathscr{P}$; the specific heat supply $r = r(\theta^\gamma, t)$ per unit time; and the element of area $d\sigma$ of the surface $\mathscr{P}$, and the line element ds of the curve $\partial\mathscr{P}$. The assigned field $\boldsymbol{f}$ may be regarded as representing the combined effect of (i) the stress vector on the major surfaces of the shell-like body denoted by $\boldsymbol{f}_c$, e.g., that due to the ambient pressure of the surrounding medium, and (ii) an integrated contribution arising from the three-dimensional body force denoted by $\boldsymbol{f}_b$, e.g., that due to gravity. A parallel statement holds for the assigned field $\boldsymbol{l}$. Similarly, the assigned heat supply r may be regarded as representing the combined effect of (i) heat supply entering the major surfaces of the shell-like body from the surrounding environment, denoted by r_c, and (ii) a contribution arising from the three-dimensional heat supply, denoted by r_b. Thus, we may write

$$\boldsymbol{f} = \boldsymbol{f}_b + \boldsymbol{f}_c, \qquad \boldsymbol{l} = \boldsymbol{l}_b + \boldsymbol{l}_c, \qquad r = r_b + r_c. \tag{3.12}$$

We assume that the kinetic energy of the Cosserat surface $\mathscr{C}$ per unit area of $\mathscr{s}$ in the present configuration is given by

$$\kappa = \tfrac{1}{2}\rho(\boldsymbol{v}\cdot\boldsymbol{v} + 2y^1\boldsymbol{v}\cdot\boldsymbol{w} + y^2\boldsymbol{w}\cdot\boldsymbol{w}). \tag{3.13}$$

We further define the momentum corresponding to the velocity $\boldsymbol{v}$ and the director momentum corresponding to the director velocity $\boldsymbol{w}$ by

$$\frac{\partial\kappa}{\partial\boldsymbol{v}} = \rho(\boldsymbol{v} + y^2\boldsymbol{w}), \qquad \frac{\partial\kappa}{\partial\boldsymbol{w}} = \rho(y^1\boldsymbol{v} + y^2\boldsymbol{w}). \tag{3.14}$$

Also, the physical dimensions of ρ, $\boldsymbol{n}$, $\boldsymbol{f}$ are

$$\begin{gathered}\text{phys. dim. } \rho = [\mathrm{ML}^{-2}],\\ \text{phys. dim. } \boldsymbol{n} = [\mathrm{MT}^{-2}], \qquad \text{phys. dim. } \boldsymbol{f} = [\mathrm{LT}^{-2}],\end{gathered} \tag{3.15}$$

where the symbols [L], [M] and [T] stand for the physical dimensions of length, mass and time. The dimensions of the vector fields $\boldsymbol{m}$, $\boldsymbol{l}$ and $\boldsymbol{k}$ depend

---

subject. Also the notations for the inertia coefficients y^1 and y^2, which occur in (3.13)–(3.14), differ from the corresponding notations in previous papers. In most of the previous papers (for example, Green and Naghdi, 1976a, Naghdi, 1975a or Naghdi, 1979a) the notations k^1, k^2 or α_1, α_2 were used in place of y^1, y^2.

upon the physical dimension of[†] $\boldsymbol{d}$. Here we choose $\boldsymbol{d}$ to have the dimension of length and then $\boldsymbol{m}$, $\boldsymbol{l}$ will have the same physical dimensions as $\boldsymbol{n}$, $\boldsymbol{f}$ in (3.15) while $\boldsymbol{k}$ will have the physical dimension of $[\mathrm{ML}^{-1}\mathrm{T}^{-2}]$.

In terms of the above definitions of the various field quantities and with reference to the present configuration, the conservation laws in the purely mechanical theory of a Cosserat surface $\mathscr{C}$ are[‡],

$$
\begin{aligned}
&\frac{d}{dt}\int_{\mathscr{P}} \rho d\sigma = 0,\\
&\frac{d}{dt}\int_{\mathscr{P}} \rho(\boldsymbol{v} + y^1\boldsymbol{w})d\sigma = \int_{\mathscr{P}} \rho \boldsymbol{f} d\sigma + \int_{\partial\mathscr{P}} \boldsymbol{n} ds,\\
&\frac{d}{dt}\int_{\mathscr{P}} \rho(y^1\boldsymbol{v} + y^2\boldsymbol{w})d\sigma = \int_{\mathscr{P}} (\rho\boldsymbol{l} - \boldsymbol{k})d\sigma + \int_{\partial\mathscr{P}} \boldsymbol{m} ds,\\
&\frac{d}{dt}\int_{\mathscr{P}} \rho[\boldsymbol{r}\times(\boldsymbol{v} + y^1\boldsymbol{w}) + \boldsymbol{d}\times(y^1\boldsymbol{v} + y^2\boldsymbol{w})]d\sigma\\
&\qquad = \int_{\mathscr{P}} \rho(\boldsymbol{r}\times\boldsymbol{f} + \boldsymbol{d}\times\boldsymbol{l})d\sigma + \int_{\partial\mathscr{P}} (\boldsymbol{r}\times\boldsymbol{n} + \boldsymbol{d}\times\boldsymbol{m})ds.
\end{aligned}
\tag{3.16}
$$

The first of (3.16) is a mathematical statement of the conservation of mass, the second that of the linear momentum, the third that of the director momentum and the fourth is the conservation of moment of momentum. We also record the law of conservation of energy in the form

$$
\begin{aligned}
&\frac{d}{dt}\int_{\mathscr{P}} \rho[\varepsilon + \kappa]d\sigma\\
&\qquad = \int_{\mathscr{P}} \rho(\boldsymbol{f}\cdot\boldsymbol{v} + \boldsymbol{l}\cdot\boldsymbol{w} + r)d\sigma + \int_{\partial\mathscr{P}} (\boldsymbol{n}\cdot\boldsymbol{v} + \boldsymbol{m}\cdot\boldsymbol{w} - h)ds.
\end{aligned}
\tag{3.17}
$$

The basic structures of $(3.16)_{1,2}$ and (3.17) and their forms are analogous to the corresponding conservation laws of the classical 3-dimensional continuum field theory. The structures of $(3.16)_3$ and $(3.16)_4$ are less obvious, but a motivation for their forms is provided by a derivation of the basic field equations for shell-like bodies obtained from the 3-dimensional equations of continuum mechanics in which the position vector $\boldsymbol{r}^*$ in 3-space is approxi-

[†] Depending on the choice of the physical dimension of $\boldsymbol{d}$ and with reference to $\boldsymbol{m}$, $\boldsymbol{l}$ and $\boldsymbol{k}$ the terminologies of the contact director couple, the assigned director couple and the intrinsic director couple, respectively, are also used in the literature. In particular, the latter terminologies are employed in Naghdi (1972), where $\boldsymbol{d}$ is taken to be dimensionless.

[‡] As the integrals on the left-hand sides of $(3.16)_{2,3,4}$ allow for coupling in inertia terms, they are slightly more general than the corresponding expressions in Naghdi (1972). The conservation laws (3.16) with coefficients $y^1 = 0$ and $y^2 = \alpha \neq 0$ reduce to those given by Equations (8.11) in Naghdi (1972).

mated by an expression of the form (2.19). It should be noted here that the conservation laws (3.16)–(3.17) are consistent with the invariance conditions under superposed rigid body motions, which ordinarily have wide acceptance in continuum mechanics. Moreover, as shown in Naghdi (1972, Section 8), the conservation laws $(3.16)_1$, $(3.16)_2$ and $(3.16)_4$ are equivalent to, and can be derived from the conservation of energy (3.17) and the invariance conditions under superposed rigid body motions. The conservation law $(3.16)_3$ for the director momentum must be postulated separately.

Returning to the conservation laws (3.16) and (3.17), we note that under suitable continuity assumptions the contact force $\boldsymbol{n}$, the contact director force $\boldsymbol{m}$ and the heat flux h can be expressed in the forms (for details see Naghdi, 1972, Section 8):

$$\boldsymbol{n} = \boldsymbol{N}^{\alpha}\nu_{\alpha}, \qquad \boldsymbol{m} = \boldsymbol{M}^{\alpha}\nu_{\alpha}, \qquad h = q^{\alpha}\nu_{\alpha} \tag{3.18}$$

where $\boldsymbol{N}^{\alpha}$, $\boldsymbol{M}^{\alpha}$ transform as contravariant surface vectors and q^{α} are the contravariant components of the heat flux vector

$$\boldsymbol{q} = q^{\alpha}\boldsymbol{a}_{\alpha}. \tag{3.19}$$

With the use of (3.18) and by usual procedures, from the conservation laws (3.15) and (3.16) follow the local field equations

$$\begin{gathered}
\rho a^{1/2} = \lambda \quad \text{or} \quad \dot{\rho} + \rho \boldsymbol{a}^{\alpha} \cdot \boldsymbol{v}_{,\alpha} = 0, \\
\boldsymbol{N}^{\alpha}{}_{|\alpha} + \rho \boldsymbol{f} = \rho(\dot{\boldsymbol{v}} + y^{1}\dot{\boldsymbol{w}}), \\
\boldsymbol{M}^{\alpha}{}_{|\alpha} + \rho \boldsymbol{l} - \boldsymbol{k} = \rho(y^{1}\dot{\boldsymbol{v}} + y^{2}\dot{\boldsymbol{w}}), \\
\boldsymbol{a}_{\alpha} \times \boldsymbol{N}^{\alpha} + \boldsymbol{d} \times \boldsymbol{k} + \boldsymbol{d}_{,\alpha} \times \boldsymbol{M}^{\alpha} = \boldsymbol{0}
\end{gathered} \tag{3.20}$$

and

$$\rho r - q^{\alpha}{}_{|\alpha} - \rho\dot{\varepsilon} + P = 0, \tag{3.21}$$

where

$$P = \boldsymbol{N}^{\alpha} \cdot \boldsymbol{v}_{,\alpha} + \boldsymbol{k} \cdot \boldsymbol{w} + \boldsymbol{M}^{\alpha} \cdot \boldsymbol{w}_{,\alpha} \tag{3.22}$$

is the mechanical power, λ in $(3.19)_1$ is a function of θ^{α} only, a comma denotes partial differentiation with respect to θ^{α}, a vertical line stands for covariant differentiation with respect to the metric tensor of the surface $\mathscr{s}$ and

$$a^{1/2} = [\boldsymbol{a}_1\boldsymbol{a}_2\boldsymbol{a}_3]. \tag{3.23}$$

3.3. *Hierarchical theories of Cosserat surfaces*

Although the theory outlined in Subsection 3.2 is sufficiently general for many applications, on occasion it becomes necessary to consider Cosserat surfaces with more than one director. Therefore, we now briefly discuss the kinematics and the balance laws of Cosserat surfaces $\mathscr{C}_K$ having K ($K = 1, 2, \cdots$) directors attached to every point of a material surface $\mathscr{S}$. Thus, we admit K directors at $\boldsymbol{r}$ denoted by $\boldsymbol{d}_M$ ($M = 1, 2, \cdots, K$); and, instead of $(3.1)_{1,2}$ specify a motion of $\mathscr{C}_K$ by

$$\boldsymbol{r} = \hat{\boldsymbol{r}}(\theta^\alpha, t), \qquad \boldsymbol{d}_M = \hat{\boldsymbol{d}}_M(\theta^\alpha, t). \tag{3.24}$$

The velocity vector is still given by $(3.3)_1$ but corresponding to $(3.3)_2$ we now define the director velocities

$$\boldsymbol{w}_M = \dot{\boldsymbol{d}}_M. \tag{3.25}$$

We recall for $K = 1$ ($\mathscr{C}_1 = \mathscr{C}$), that the kinetical quantities introduced in Subsection 3.2 consisted of $\boldsymbol{n}$, $\boldsymbol{k}$, $\boldsymbol{m}$ and the assigned fields $\boldsymbol{f}$, $\boldsymbol{l}$. Keeping this in mind, for a body $\mathscr{C}_K$ we admit more general kinetical quantities and assigned fields

$$\begin{gathered} \boldsymbol{n}, \qquad \overset{N}{\boldsymbol{k}}, \qquad \overset{N}{\boldsymbol{m}}, \\ \boldsymbol{f}, \qquad \overset{N}{\boldsymbol{l}} \end{gathered} \tag{3.26}$$

for $N = 1, 2, \cdots, K$, and corresponding to (3.13) and $(3.14)_{1,2}$ write the more general expressions for kinetic energy of $\mathscr{C}_K$ and associated momentum and director momentum, namely

$$\begin{aligned} \kappa &= \tfrac{1}{2}\rho \sum_{M,N=0}^{K} y^{M+N} \boldsymbol{w}_M \cdot \boldsymbol{w}_N, \qquad \boldsymbol{w}_0 = \boldsymbol{v}, \\ \frac{\partial \kappa}{\partial \boldsymbol{v}} &= \rho\left(\boldsymbol{v} + \sum_{M=1}^{K} y^M \boldsymbol{w}_M\right) = \rho \sum_{M=0}^{K} y^M \boldsymbol{w}_M, \\ \frac{\partial \kappa}{\partial \boldsymbol{w}_N} &= \rho\left(y^N \boldsymbol{v} + \sum_{M=1}^{K} y^{M+N} \boldsymbol{w}_M\right) = \rho \sum_{M=0}^{K} y^{M+N} \boldsymbol{w}_M, \end{aligned} \tag{3.27}$$

each per unit area of the surface $\mathscr{P}$. The inertia coefficients y^{M+N} are functions of θ^α only and satisfy the conditions

$$y^{M+N} = y^{N+M}, \qquad y^{M+0} = y^{0+M} = y^M, \qquad y^0 = 1. \tag{3.28}$$

In the special case of $\mathscr{C}_1$ ($= \mathscr{C}$) we may use the notations

$$\boldsymbol{d}_1 = \boldsymbol{d}, \qquad \boldsymbol{w}_1 = \boldsymbol{w}. \tag{3.29}$$

For a detailed statement of conservation laws appropriate for Cosserat surfaces $\mathscr{C}_K$ we refer the reader to Green and Naghdi (1976a, Section 2) but indicate here the structure of the corresponding local field equations. In this connection, we first note that for a purely mechanical theory by usual procedure in addition to $(3.18)_1$ we now obtain $\boldsymbol{m}^N = \boldsymbol{M}^{N\alpha}\nu_\alpha$. Then, the local field equations for Cosserat surface $\mathscr{C}_K$ are

$$\overline{\lambda y^M}^{\,\cdot} = 0, \qquad \dot{\lambda} = 0, \qquad \lambda = \rho a^{1/2},$$

$$\boldsymbol{N}^\alpha{}_{|\alpha} + \rho \boldsymbol{f} = \rho \sum_{M=0}^{K} y^M \dot{\boldsymbol{w}}_M,$$

(3.30)
$$\boldsymbol{M}^{N\alpha}{}_{|\alpha} + \rho \boldsymbol{l}^N - \boldsymbol{k}^N = \rho \sum_{M=0}^{K} y^{M+N} \dot{\boldsymbol{w}}_M, \qquad (N = 1, 2, \cdots, K),$$

$$\boldsymbol{a}_\alpha \times \boldsymbol{N}^\alpha + \sum_{N=1}^{K} \boldsymbol{d}_N \times \boldsymbol{k}^N + \sum_{N=1}^{K} \boldsymbol{d}_{N,\alpha} \times \boldsymbol{M}^{N\alpha} = 0.$$

Also, for Cosserat surfaces $\mathscr{C}_K$, the expression for mechanical power corresponding to (3.22) is

(3.31)
$$P = \boldsymbol{N}^\alpha \cdot \boldsymbol{v}_{,\alpha} + \sum_{N=1}^{K} \boldsymbol{k}^N \cdot \boldsymbol{w}_N + \sum_{N=1}^{K} \boldsymbol{M}^{N\alpha} \cdot \boldsymbol{w}_{N,\alpha}.$$

The general development for Cosserat surfaces $\mathscr{C}_K$ outlined above is contained in a paper by Green and Naghdi (1976a, Section 2) which deals with application of the theory to fluid sheets and to propagation of water waves. When $K = 1$, the results in Subsection 3.3 reduce to those of Subsection 3.2 for a Cosserat surface $\mathscr{C}$.

4. Elastic shells

Within the scope of the theory of a Cosserat surface $\mathscr{C}$ outlined in Section 3, we discuss briefly the constitutive equations for elastic shells in the presence of finite deformation. Preliminary to the discussion that follows, we assume the existence of a strain energy or stored energy per unit mass $\psi = \psi(\theta^\alpha, t)$ such that $\rho\dot{\psi}$ is equal to the mechanical power defined by (3.22), i.e.,

(4.1)
$$P = \rho \dot{\psi}.$$

In the development of nonlinear constitutive equations for elastic shells, we assume that the strain energy density ψ at each material point of $\mathscr{C}$ and for all t is specified by a response function which depends on $\boldsymbol{r}$, $\boldsymbol{d}$ and their partial derivatives with respect to θ^α. But since the response function must remain unaltered under superposed rigid body translational displacement, the dependence on $\boldsymbol{r}$ must be excluded. Thus, the constitutive assumption for the strain energy density can be written as

(4.2)
$$\psi = \psi'(\boldsymbol{r}_{,\alpha}, \boldsymbol{d}, \boldsymbol{d}_{,\alpha}; X),$$

and we also make similar constitutive assumptions for $\boldsymbol{N}^\alpha$, $\boldsymbol{k}$, $\boldsymbol{M}^\alpha$. In these constitutive equations, which represent the mechanical response of the medium, the dependence of the response functions on the local geometrical properties of a reference state and material inhomogeneity is indicated through the argument X.

A general development of various aspects of constitutive theory of elastic shells based on assumptions of the type (4.2) or variants thereof is given in Naghdi (1972, Section 13). In the rest of this section, we limit the discussion to an elastic shell which is homogeneous in its reference configuration and suppose also that the dependence of the response functions on the properties of the reference state occurs through the values of the kinematical variables in the reference state (Carroll and Naghdi, 1972). Then, in place of (4.2), we have

$$\psi = \bar{\psi}(\boldsymbol{r}_{,\alpha}, \boldsymbol{d}, \boldsymbol{d}_{,\alpha}\,; \boldsymbol{A}_\alpha, \boldsymbol{D}, \boldsymbol{D}_{,\alpha}), \tag{4.3}$$

with similar assumptions for $\boldsymbol{N}^\alpha$, $\boldsymbol{k}$, $\boldsymbol{M}^\alpha$. After substituting (4.3) into (4.1), by usual techniques we obtain the following forms for the constitutive equations:

$$\boldsymbol{N}^\alpha = \rho\frac{\partial\bar{\psi}}{\partial\boldsymbol{r}_{,\alpha}}, \qquad \boldsymbol{k} = \rho\frac{\partial\bar{\psi}}{\partial\boldsymbol{d}}, \qquad \boldsymbol{M}^\alpha = \rho\frac{\partial\bar{\psi}}{\partial\boldsymbol{d}_{,\alpha}}, \tag{4.4}$$

along with the restriction

$$\boldsymbol{r}_{,\alpha}\times\frac{\partial\bar{\psi}}{\partial\boldsymbol{r}_{,\alpha}} + \boldsymbol{d}\times\frac{\partial\bar{\psi}}{\partial\boldsymbol{d}} + \boldsymbol{d}_{,\alpha}\times\frac{\partial\bar{\psi}}{\partial\boldsymbol{d}_{,\alpha}} = 0, \tag{4.5}$$

which is obtained from the conservation of moment of momentum and which must be satisfied by the response function $\bar{\psi}$ (Naghdi, 1972, Section 8).

We do not discuss here the reduced forms of the above constitutive equations resulting from invariance requirements under superposed rigid body motions, but for such reductions refer the reader to Naghdi (1972, Section 13). Just as with the equations of motion, it is necessary in applications to specific problems to obtain alternative forms of the above constitutive equations or their reduced forms in terms of tensor components. Such component forms may be expressed with respect to bases $\boldsymbol{a}_i$, or $\boldsymbol{d}_i$, or corresponding bases in the reference configuration. Reduced forms of (4.4) have been utilized extensively in Chapters D and E of Naghdi (1972).

5. Identification of the assigned fields and the inertia coefficients

The local field equations (3.20) in the mechanical theory of a Cosserat surface have the same forms as those that can be derived from the three-dimensional field equations $(2.9)_{1,2,3}$ by suitable integration between the limits

ξ_1, ξ_2 [recall (2.17) and the definition of a shell-like body in Section 2] and in terms of certain definitions for integrated mass density and resultants of stress (for details, see Naghdi, 1972, Section 11–12 or Naghdi (1974)). Similarly, the energy equation (3.21) has the same form as the one that can be derived from the energy equation in the three-dimensional theory by suitable integration between the limits ξ_1, ξ_2 and in terms of certain definitions for integrated internal energy density and heat flux in the three-dimensional theory as given in (Naghdi, 1972). To elaborate further, we confine attention to the purely mechanical theory and recall the definitions

$$(5.1)\qquad \rho a^{1/2} = \lambda = \int_{\xi_1}^{\xi_2} \lambda^* d\xi, \qquad \lambda^* = \rho^* g^{1/2},$$

$$(5.2)\qquad \rho a^{1/2} k^M = \lambda k^M = \int_{\xi_1}^{\xi_2} \lambda^* \xi^M d\xi, \qquad (M = 1,2),$$

and the expressions

$$(5.3)\qquad \lambda \boldsymbol{f} = \rho a^{1/2} \boldsymbol{f} = \int_{\xi_1}^{\xi_2} \lambda^* \boldsymbol{f}^* d\xi + [\boldsymbol{t} g^{1/2} \mathfrak{f}_{(1)}(\xi)]_{\xi_1} + [\boldsymbol{t} g^{1/2} \mathfrak{f}_{(2)}(\xi)]_{\xi=\xi_2},$$

$$(5.4)\qquad \lambda \boldsymbol{l} = \rho a^{1/2} \boldsymbol{l} = \int_{\xi_1}^{\xi_2} \lambda^* \boldsymbol{f}^* \xi d\xi + [\boldsymbol{t} g^{1/2} \xi \mathfrak{f}_{(1)}(\xi)]_{\xi=\xi_1} + [\boldsymbol{t} g^{1/2} \xi \mathfrak{f}_{(2)}(\xi)]_{\xi=\xi_2},$$

where ρ^*, $\boldsymbol{t}$, $\boldsymbol{f}^*$ which occur in (5.1)–(5.4) are defined in Section 2 [following (2.9)] and in order to indicate the nature of the functions $\mathfrak{f}_{(\alpha)}$, $(\alpha = 1,2)$ in (5.3)–(5.4), it will suffice to record

$$(5.5)\qquad \mathfrak{f}_{(1)} = [(\xi_{1,1})^2 g^{11} (\xi_{1,2})^2 g^{22} + g^{33} + 2(\xi_{1,1}\xi_{1,2} g^{12} - \xi_{1,1} g^{13} - \xi_{1,2} g^{23})]^{1/2},$$

which involves the partial derivatives of $\xi_1(\theta^\alpha)$ and the components of the metric tensor in (2.3). The expression for $\mathfrak{f}_{(2)}$ can be stated analogously.

If we now adopt the approximation (2.25), then there is a 1–1 correspondence between the two-dimensional field equations that follow from the conservation laws of a Cosserat surface and those that can be derived from $(2.9)_{1,2,3}$ provided we identify $\boldsymbol{r}$ and $\boldsymbol{d}$ in (2.25) with $(3.1)_1$ and $(3.1)_2$, respectively, and adopt the definitions (5.1)–(5.4), as well as the definitions of the resultants mentioned above. A similar 1–1 correspondence can be shown to hold between the two-dimensional energy equation in the theory of a Cosserat surface and an integrated energy equation derived from the three-dimensional energy equation.

The various quantities in (3.12) are free to be specified in a manner which depends on the particular application in mind and, in the context of the theory of a Cosserat surface, the inertia coefficients y^1, y^2 and the mass density ρ

require constitutive equations. Indeed, $\boldsymbol{f}_c$, $\boldsymbol{l}_c$ and r_c, as well as $\boldsymbol{f}_b$, $\boldsymbol{l}_b$ and r_b, can be identified with corresponding expressions in a derivation from the three-dimensional equations (for details see Naghdi, 1972, 1979a). Likewise, ρ and the coefficients y^1, y^2 may be identified with easily accessible results from the three-dimensional theory.

In what follows, we assume that the above identifications have been made and that the quantities ρ, y^1, y^2, $\boldsymbol{f}_b$, $\boldsymbol{l}_b$ are known or specified. The knowledge of $\boldsymbol{f}_c$, $\boldsymbol{l}_c$ depends on the nature of the boundary conditions on the major surfaces of the particular shell-like body under consideration: they may be specified as known quantities on the surfaces $(2.20)_{1,2}$ or they are unknown (possibly on one of the two surfaces $(2.20)_{1,2}$ only) and must be determined as part of the solution of the problem.

6. Constrained theories of shells

A development of a constrained directed medium in the 3-dimensional theory, with particular reference to an incompressible liquid crystal having a single director of constant length, is contained in a paper of Green, Naghdi and Trapp (1970, Section 6). For a Cosserat surface with a single director, a number of constrained theories have been discussed previously. These pertain to a class of shell-like bodies for which the director is constrained to be of constant length (Green and Naghdi, 1974), an incompressible Cosserat surface (Green, Laws and Naghdi, 1974, Green and Naghdi, 1976a) and a class of fluid sheets in which the director is constrained to remain always parallel to a fixed direction (Green and Naghdi, 1977).

A special case of the constrained theory of elastic shells discussed by Green and Naghdi (1974) includes that for which the director is coincident with the unit normal[†] $\boldsymbol{a}_3$ to the surface $\mathscr{s}$. This special form of the theory can be brought into 1–1 correspondence with that of a restricted theory of elastic shells given by Naghdi (1972, Sections 10, 15), where the director is not admitted and the basic kinematical ingredients that occur in the argument of the strain energy response function are $\boldsymbol{a}_\alpha$ and $\boldsymbol{a}_{3,\alpha}$ (compare with (4.3)). Related developments include the construction of a theory of a deformable surface with simple force multipoles by Balaban, Green and Naghdi (1967), where the position vector $\boldsymbol{r}$ and its first and second gradient $(\boldsymbol{r}_{,\alpha}, \boldsymbol{r}_{,\alpha\beta})$ are taken as the basic kinematic variables. A similar theory, but less general than that of Balaban et al. (1967),

[†] The equations resulting from such a constrained theory of elastic shells in which $\boldsymbol{d} = \boldsymbol{a}_3$ correspond to those which can be obtained from a derivation of shell theory under the so-called Kirchhoff–Love assumption (see Naghdi and Nordgren, 1963).

is given by Cohen and DeSilva (1966a, 1968). For additional related comments see Naghdi (1972, Section 10).

In this section, we begin by considering a class of constraints which are linear relations between the kinematic variables

$$(6.1) \qquad \boldsymbol{v}_{,\alpha}, \qquad \boldsymbol{w}_N, \qquad \boldsymbol{w}_{N,\alpha} \qquad (N = 1, 2, \cdots, K)$$

in the form (Green and Naghdi, 1976a)

$$(6.2) \quad \boldsymbol{A}^{M\alpha} \cdot \boldsymbol{v}_{,\alpha} + \sum_{N=1}^{K} \boldsymbol{B}^{MN} \cdot \boldsymbol{w}_N + \sum_{N=1}^{K} \boldsymbol{C}^{MN\alpha} \cdot \boldsymbol{w}_{N,\alpha} = 0 \quad (M = 0, 1, 2, \cdots, Q),$$

where $\boldsymbol{A}^{M\alpha}$, $\boldsymbol{B}^{MN}$, $\boldsymbol{C}^{MN\alpha}$ are vector functions of $\boldsymbol{a}_\alpha$, $\boldsymbol{d}_N$, $\boldsymbol{d}_{N,\alpha}$ only and do not depend explicitly on the variables (6.1). We assume that each of the functions $\boldsymbol{N}^\alpha$, $\boldsymbol{k}^N$, $\boldsymbol{M}^{N\alpha}$ are determined to within an additive constraint response so that[†]

$$(6.3) \qquad \boldsymbol{N}^\alpha = \bar{\boldsymbol{N}}^\alpha + \hat{\boldsymbol{N}}^\alpha, \qquad \boldsymbol{k}^N = \bar{\boldsymbol{k}}^N + \hat{\boldsymbol{k}}^N, \qquad \boldsymbol{M}^{N\alpha} = \bar{\boldsymbol{M}}^{N\alpha} + \hat{\boldsymbol{M}}^{N\alpha},$$

where $\hat{\boldsymbol{N}}^\alpha$, $\hat{\boldsymbol{k}}^N$, $\hat{\boldsymbol{M}}^{N\alpha}$ are specified by constitutive equations and

$$(6.4) \qquad \bar{\boldsymbol{N}}^\alpha, \qquad \bar{\boldsymbol{k}}^N, \qquad \bar{\boldsymbol{M}}^{N\alpha},$$

which represent the response due to constraints are arbitrary functions of θ^α, t and are workless. Thus, recalling the expression (3.30) for mechanical power, we set

$$(6.5) \qquad \bar{\boldsymbol{N}}^\alpha \cdot \boldsymbol{v}_{,\alpha} + \sum_{N=1}^{K} \bar{\boldsymbol{k}}^N \cdot \boldsymbol{w}_N + \sum_{N=1}^{K} \bar{\boldsymbol{M}}^{N\alpha} \cdot \boldsymbol{w}_{N,\alpha} = 0$$

for all values of the variables (6.1) subject to the constraint conditions (6.2). It then follows that

$$(6.6) \quad \bar{\boldsymbol{N}}^\alpha = -\sum_{M=0}^{Q} \boldsymbol{A}^{M\alpha} p_M, \quad \bar{\boldsymbol{k}}^N = -\sum_{M=0}^{Q} \boldsymbol{B}^{MN} p_M, \quad \bar{\boldsymbol{M}}^{N\alpha} = -\sum_{M=0}^{Q} \boldsymbol{C}^{MN\alpha} p_M,$$

where $p_M = p_M(\theta^\alpha, t)$, $(M = 0, 1, \cdots, Q)$ are arbitrary functions which play the role of Lagrange multipliers.

In the rest of this section, we illustrate the nature of constrained theories with reference to two particular kinematical constraints. One of these constraints is that appropriate for incompressible media and the other pertains to a restriction on the director in the theory of a Cosserat surface $\mathscr{C}$ with a single director.

[†] The development between (6.2)–(6.6) is similar to that for mechanical constraints in the 3-dimensional theory (see section 30 of Truesdell and Noll 1965). For a corresponding thermodynamical theory of a continuum in the presence of thermo-mechanical constraints see Green, Naghdi and Trapp (1970) and Green and Naghdi (1977).

6.1. *Incompressible Cosserat surfaces*

The conditions representing approximately the (3-dimensional) incompressibility condition (2.10) may be derived with the use of the approximation $(2.22)_1$. However, in the interest of brevity, we confine attention to a special case of $(2.22)_1$ for $N = 1$ given by (2.25). Under the approximation (2.25), the base vectors are given by $\boldsymbol{g}_\alpha = \boldsymbol{a}_\alpha + \xi\boldsymbol{d}_{,\alpha}$, $\boldsymbol{g}_3 = \boldsymbol{d}$, where $\boldsymbol{a}_\alpha$ are the base vectors of the surface $\xi = 0$ calculated from (2.25). Then, the incompressibility condition (2.10) may be expressed approximately in the form

$$\frac{d}{dt}[\boldsymbol{a}_1\boldsymbol{a}_2\boldsymbol{d}] + \xi\frac{d}{dt}\left\{\left[\frac{\partial \boldsymbol{d}}{\partial\theta^1}\,\boldsymbol{a}_2\boldsymbol{d}\right] + \left[\boldsymbol{a}_1\,\frac{\partial \boldsymbol{d}}{\partial\theta^2}\,\boldsymbol{d}\right]\right\} + \xi^2\frac{d}{dt}\left[\frac{\partial \boldsymbol{d}}{\partial\theta^1}\,\frac{\partial \boldsymbol{d}}{\partial\theta^2}\,\boldsymbol{d}\right] = 0 \tag{6.7}$$

or equivalently as

$$\begin{aligned}&[(\boldsymbol{d}\cdot\boldsymbol{a}_3)\boldsymbol{a}^\alpha - (\boldsymbol{d}\cdot\boldsymbol{a}^\alpha)\boldsymbol{a}_3 + \xi(\varepsilon^{\alpha\beta}\boldsymbol{d}_{,\beta}\times\boldsymbol{d})]\cdot\boldsymbol{v}_{,\alpha}\\ &\quad+[\boldsymbol{a}_3 + \xi\varepsilon^{\alpha\beta}\boldsymbol{a}_\alpha\times\boldsymbol{d}_{,\beta} + \tfrac{1}{2}\xi^2\varepsilon^{\alpha\beta}\boldsymbol{d}_{,\alpha}\times\boldsymbol{d}_{,\beta}]\cdot\boldsymbol{w}\\ &\quad+[\xi\varepsilon^{\alpha\beta}\boldsymbol{a}_\beta\times\boldsymbol{d} + \xi^2\varepsilon^{\alpha\beta}\boldsymbol{d}\times\boldsymbol{d}_{,\beta}]\cdot\boldsymbol{w}_{,\alpha} = 0\end{aligned} \tag{6.8}$$

where in (6.7) and (6.8) use is made of the notation $(3.29)_{1,2}$ and $\varepsilon^{\alpha\beta}$, $\varepsilon_{\alpha\beta}$ denote the components of the ε-system in 2-space defined by

$$\begin{aligned}\varepsilon^{\alpha\beta} &= a^{-1/2}e^{\alpha\beta}, & e^{11} &= e^{22} = 0, & e^{12} &= -e^{21} = 1,\\ \varepsilon_{\alpha\beta} &= a^{1/2}e_{\alpha\beta}, & e_{11} &= e_{22} = 0, & e_{12} &= -e_{21} = 1.\end{aligned} \tag{6.9}$$

We now generate two conditions representing incompressibility: One of these is obtained from integration of (6.8) with respect to ξ between the limits ξ_1, ξ_2 and another by first multiplying (6.8) by ξ, neglecting terms involving ξ^3 and then integrating the resulting equation with respect to ξ between the limits ξ_1, ξ_2. The resulting two conditions are

$$\begin{aligned}&\{\gamma^0[(\boldsymbol{d}\cdot\boldsymbol{a}_3)\boldsymbol{a}^\alpha - (\boldsymbol{d}\cdot\boldsymbol{a}^\alpha)\boldsymbol{a}_3] + \gamma^1\varepsilon^{\alpha\beta}(\boldsymbol{d}_{,\beta}\times\boldsymbol{d})\}\cdot\boldsymbol{v}_{,\alpha}\\ &\quad+\{\gamma^0\boldsymbol{a}_3 + \varepsilon^{\alpha\beta}[\gamma^1(\boldsymbol{a}_\alpha\times\boldsymbol{d}_{,\beta}) + \tfrac{1}{2}\gamma^2(\boldsymbol{d}_{,\alpha}\times\boldsymbol{d}_{,\beta})\}\cdot\boldsymbol{w}\\ &\quad+\{\varepsilon^{\alpha\beta}[\gamma^1(\boldsymbol{a}_\beta\times\boldsymbol{d}) + \gamma^2(\boldsymbol{d}_{,\beta}\times\boldsymbol{d})]\}\cdot\boldsymbol{w}_{,\alpha} = 0,\\ &\{\gamma^1[(\boldsymbol{d}\cdot\boldsymbol{a}_3)\boldsymbol{a}^\alpha - (\boldsymbol{d}\cdot\boldsymbol{a}^\alpha)\boldsymbol{a}_3] + \gamma^2\varepsilon^{\alpha\beta}(\boldsymbol{d}_{,\beta}\times\boldsymbol{d})\}\cdot\boldsymbol{v}_{,\alpha}\\ &\quad+\{\gamma^1\boldsymbol{a}_3 + \gamma^2\varepsilon^{\alpha\beta}(\boldsymbol{a}_\alpha\times\boldsymbol{d}_{,\beta})\}\cdot\boldsymbol{w} + \gamma^2\varepsilon^{\alpha\beta}(\boldsymbol{a}_\beta\times\boldsymbol{d})\cdot\boldsymbol{w}_{,\alpha} = 0,\end{aligned} \tag{6.10}$$

where the coefficients γ^K are defined by

$$\gamma^K = \int_{\xi_1}^{\xi_2}\xi^K d\xi, \qquad (K = 0, 1, 2). \tag{6.11}$$

It is perhaps interesting to observe that in special circumstances in which the quantity λ^* in (5.1)–(5.2) is or can be approximated to be independent of ξ, then the coefficients γ^1 and γ^2 in (6.10) will have the same numerical values as the inertia coefficients y^1 and y^2, respectively.

For an incompressible Cosserat surface under discussion, from (6.2) the constraint conditions are

$$\begin{aligned} \boldsymbol{A}^{0\alpha} \cdot \boldsymbol{v}_{,\alpha} + \boldsymbol{B}^{01} \cdot \boldsymbol{w} + \boldsymbol{C}^{01\alpha} \cdot \boldsymbol{w}_{,\alpha} &= 0, \\ \boldsymbol{A}^{1\alpha} \cdot \boldsymbol{v}_{,\alpha} + \boldsymbol{B}^{11} \cdot \boldsymbol{w} + \boldsymbol{C}^{11\alpha} \cdot \boldsymbol{w}_{,\alpha} &= 0, \end{aligned} \tag{6.12}$$

and the corresponding constrained response obtained from (6.6) has the form

$$\begin{aligned} \bar{\boldsymbol{N}}^{\alpha} &= -(p_0 \boldsymbol{A}^{0\alpha} + p_1 \boldsymbol{A}^{1\alpha}), \\ \bar{\boldsymbol{k}} &= -(p_0 \boldsymbol{B}^{0\alpha} + p_1 \boldsymbol{B}^{11}), \\ \bar{\boldsymbol{M}}^{\alpha} &= -(p_0 \boldsymbol{C}^{01\alpha} + p_1 \boldsymbol{C}^{11\alpha}), \end{aligned} \tag{6.13}$$

where p_0, p_1 are the Lagrange multipliers. Guided by the two conditions which follow from (6.10), we select the vectors $\boldsymbol{A}^{0\alpha}, \boldsymbol{B}^{01}, \boldsymbol{C}^{01\alpha}, \cdots$, which occur in (6.12) to have the special values

$$\left.\begin{aligned} \boldsymbol{A}^{K\alpha} &= \text{coeff. of } \boldsymbol{v}_{,\alpha} \text{ in } (6.10) \\ \boldsymbol{B}^{K1} &= \text{coeff. of } \boldsymbol{w} \text{ in } (6.10) \\ \boldsymbol{C}^{K1\alpha} &= \text{coeff. of } \boldsymbol{w}_{,\alpha} \text{ in } (6.10) \end{aligned}\right\} \quad (K = 0, 1). \tag{6.14}$$

Then, it follows from (6.13) and (6.14) that the expressions for the constraint response are given by[†]

$$\begin{aligned} \bar{\boldsymbol{N}}^{\alpha} &= -P_0[(\boldsymbol{d} \cdot \boldsymbol{a}_3)\boldsymbol{a}^{\alpha} - (\boldsymbol{d} \cdot \boldsymbol{a}^{\alpha})\boldsymbol{a}_3] - P_1 \varepsilon^{\alpha\beta} \boldsymbol{d}_{,\beta} \times \boldsymbol{d}, \\ \bar{\boldsymbol{k}} &= -P_0 \boldsymbol{a}_3 - P_1 \varepsilon^{\alpha\beta} \boldsymbol{a}_{\alpha} \times \boldsymbol{d}_{,\beta} - \tfrac{1}{2}\gamma^2 p_0 \varepsilon^{\alpha\beta} \boldsymbol{d}_{,\alpha} \times \boldsymbol{d}_{,\beta}, \\ \bar{\boldsymbol{M}}^{\alpha} &= -P_1 \varepsilon^{\alpha\beta} \boldsymbol{a}_{\beta} \times \boldsymbol{d} - \gamma^2 p_0 \varepsilon^{\beta\alpha} \boldsymbol{d} \times \boldsymbol{d}_{,\beta}. \end{aligned} \tag{6.15}$$

The arbitrary coefficient functions P_0, P_1 are related to the Lagrange multipliers p_0, p_1 and $\gamma^2 p_0$ can be expressed in terms of P_0, P_1 as follows:

$$P_0 = \gamma^0 p_0 + \gamma^1 p_1, \qquad P_1 = \gamma^1 p_0 + \gamma^2 p_1, \qquad \gamma^2 p_0 = \frac{(\gamma^2 P_0 - \gamma^1 P_1)\gamma^2}{\gamma^0 \gamma^2 - (\gamma^1)^2}. \tag{6.16}$$

In obtaining the results (6.10) and (6.15), no identification has been made between the surfaces $(3.1)_1$ in the theory of a Cosserat surface $\mathscr{C}$ and an appropriate reference surface in the (3-dimensional) shell-like body. Indeed,

[†] Although the expressions (6.15) have the same form as those given previously (Green and Naghdi, 1976a, Equations (4.2)), they are not the same in view of the relations (6.16).

different values for the coefficients γ^K in (6.10) will result depending on the choice of the identification with the surface $\xi = 0$. For example, if this surface is chosen between the major surfaces of the shell-like body in such a way that $\xi_1 = -\xi_2 = -\frac{1}{2}$, then the coefficients γ^K in (5.8) and P_0, P_1 in (6.15) become

$$\begin{gathered}\gamma^0 = 1, \qquad \gamma^1 = 0, \qquad \gamma^2 = \frac{1}{12}, \\ P_0 = p_0, \qquad P_1 = \frac{1}{12} p_1\end{gathered} \tag{6.17}$$

and the incompressibility conditions (6.10) reduce to those used by Green and Naghdi (1976a, Equations (4.3)) for a directed fluid sheet with a single director. On the other hand, if we identify $(3.1)_1$ with the bottom surface of the shell-like body so that $\xi_1 = 0$, $\xi_2 = 1$, then the coefficients γ^K and P_0, P_1 become

$$\begin{gathered}\gamma^0 = 1, \qquad \gamma^1 = \tfrac{1}{2}, \qquad \gamma^2 = \tfrac{1}{3}, \\ P_0 = p_0 + \tfrac{1}{2} p_1, \qquad P_1 = \tfrac{1}{2} p_0 + \tfrac{1}{3} p_1.\end{gathered} \tag{6.18}$$

For a complete theory of an incompressible Cosserat surface, constitutive equations are required for the quantities $\hat{\boldsymbol{N}}^\alpha$, $\hat{\boldsymbol{k}}$ and $\hat{\boldsymbol{M}}^\alpha$ but a discussion of these can be carried out as in Section 4.

6.2. *A constrained theory with director along the normal to the surface of $\mathscr{C}$*

We turn now to the development of a constrained theory of a Cosserat surface in which the director is always along the normal to the material surface so that

$$\boldsymbol{d} \cdot \boldsymbol{a}_\alpha = 0, \qquad \boldsymbol{d} = \phi \boldsymbol{a}_3, \qquad \phi = \phi(\theta^\gamma, t). \tag{6.19}$$

Differentiating the constraint condition $(6.19)_1$ with respect to time and using $(3.3)_2$ and (3.4) we obtain

$$\boldsymbol{d} \cdot \boldsymbol{v}_{,\alpha} + \boldsymbol{a}_\alpha \cdot \boldsymbol{w} = 0, \tag{6.20}$$

which represents two constraint conditions. From $(6.19)_2$, along with the use of $(2.13)_1$ and $(2.15)_3$, follows the expression

$$\boldsymbol{d}_{,\alpha} \cdot \boldsymbol{a}_\beta = -\phi b_{\alpha\beta}, \tag{6.21}$$

which is symmetric in α, β. Hence

$$\varepsilon^{\alpha\beta}(\boldsymbol{d}_{,\alpha} \cdot \boldsymbol{a}_\beta) = 0, \tag{6.22}$$

where $\varepsilon^{\alpha\beta}$ is defined in (6.9). Differentiating (6.22) with respect to t and observing that $\dot{\varepsilon}^{\alpha\beta}(\boldsymbol{d}_{,\alpha}\cdot\boldsymbol{a}_\beta)=0$ in view of (6.22) and the fact that $\dot{\varepsilon}^{\alpha\beta}=\dot{a}^{-1/2}e^{\alpha\beta}$, we also have

$$\varepsilon^{\alpha\beta}(\boldsymbol{d}_{,\alpha}\cdot\boldsymbol{v}_{,\beta}+\boldsymbol{a}_\beta\cdot\boldsymbol{w}_{,\alpha})=0, \tag{6.23}$$

as a third constraint condition.

The two conditions $(6.20)_{1,2}$ can be regarded as a special case (6.2) for $K=1$ and with the coefficient of $\boldsymbol{w}_{,\alpha}$ equal to zero. Similarly, (6.23) is a special case of (6.2) for $M=0$, $K=1$ and with the coefficient of $\boldsymbol{w}$ equal to zero. Thus, each of the three constraint conditions $(6.20)_{1,2}$ and (6.23) may be viewed as a special case of (6.2) with coefficient functions $\boldsymbol{A}^0$, $\boldsymbol{B}^{01}$, $\boldsymbol{C}^{01\alpha}$, $\boldsymbol{A}^{M\alpha}$, $\boldsymbol{B}^{M1}$, $\boldsymbol{C}^{M1\alpha}$ conveniently identified as

$$\begin{aligned}
&\boldsymbol{A}^{0\alpha}=\varepsilon^{\beta\alpha}\boldsymbol{d}_{,\beta}, \qquad \boldsymbol{B}^0=\boldsymbol{0}, \qquad \boldsymbol{C}^{01\alpha}=\varepsilon^{\alpha\beta}\boldsymbol{a}_\beta,\\
&\boldsymbol{A}^{M\alpha}=\delta^{\alpha}_{M}\boldsymbol{d}, \qquad \boldsymbol{B}^{M1}=\boldsymbol{a}_M, \qquad \boldsymbol{C}^{M1\alpha}=\boldsymbol{0}, \qquad (M=1,2).
\end{aligned} \tag{6.24}$$

Now according to (6.6) and with the help of (6.24), the expressions for the constraint response are

$$\begin{aligned}
&\bar{\boldsymbol{N}}^\alpha=-\left[p^0\varepsilon^{\beta\alpha}\boldsymbol{d}_{,\beta}+\sum_{M=1}^{2}p^M\delta^{\alpha}_{M}\boldsymbol{d}\right]=-[p^0\varepsilon^{\beta\alpha}\boldsymbol{d}_{,\beta}+p^\alpha\boldsymbol{d}],\\
&\bar{\boldsymbol{k}}=-\sum_{M=1}^{2}p^M\boldsymbol{a}_M=-p^\alpha\boldsymbol{a}_\alpha, \qquad \bar{\boldsymbol{M}}^\alpha=-p^0\varepsilon^{\alpha\beta}\boldsymbol{a}_\beta,
\end{aligned} \tag{6.25}$$

where p^0, p^α are the Lagrangian multipliers, and in line with the notation of $(3.29)_{1,2}$ and that of Subsection 3.2 we have set $\boldsymbol{k}^1=\boldsymbol{k}$, $\boldsymbol{M}^{1\alpha}=\boldsymbol{M}^\alpha$.

In anticipation of the final form of the equations of the constrained theory, we could set the skew-symmetric parts of both $M^{\alpha\gamma}$ and $\hat{M}^{\alpha\gamma}$ equal to zero and thus require also the vanishing of the skew-symmetric part of $\bar{M}^{\alpha\gamma}$ which is equivalent to setting $p^0=0$. However, we postpone such stipulations until later in this section, and retain in (6.25) the Lagrange multiplier p^0 which arises from the constraint condition (6.23).

Before recording the modified equations of motion appropriate for the constrained theory under discussion, we introduce the functions $S^\alpha=S^\alpha(\theta^\gamma,t)$ defined by

$$S^\alpha=-[p^\alpha+\varepsilon^{\alpha\beta}p^0_{,\beta}] \tag{6.26}$$

and note that

$$S^\alpha{}_{|\alpha}=p^\alpha{}_{|\alpha}. \tag{6.27}$$

Also, for convenience, we introduce the abbreviation

$$\begin{aligned}
&\hat{\boldsymbol{f}}=\boldsymbol{f}-(\dot{\boldsymbol{v}}+y^1\dot{\boldsymbol{w}}),\\
&\hat{\boldsymbol{l}}=\boldsymbol{l}-(y^1\dot{\boldsymbol{v}}+y^2\dot{\boldsymbol{w}}).
\end{aligned} \tag{6.28}$$

Then, after substituting (6.3) and (6.25) into $(3.20)_{3,4,5}$ and making use of (6.26) and (6.27), we obtain

$$\hat{N}^{\alpha}_{\ |\alpha} + \rho \hat{f} = -(\phi S^{\alpha} a_3)_{|\alpha},$$

$$(6.29) \qquad \hat{M}^{\alpha}_{\ |\alpha} - \hat{k} + \rho \hat{l} = S^{\alpha} a_{\alpha},$$

$$a_{\alpha} \times \hat{N}^{\alpha} + d \times \hat{k} + d_{,\alpha} \times \hat{M}^{\alpha} = 0,$$

as the equations of motion of the constrained theory. It should be noted that the above equations involve only two arbitrary functions of position and time related to the three multipliers p^0, p^{α} by (6.26). Moreover, $(6.29)_3$ and the normal component of $(6.29)_2$ are free from S^{α}.

A further reduction of the system of equations (6.29) may be effected by eliminating S^{α} between $(6.29)_{1,2}$. For this purpose it is convenient to refer the various vector quantities in (6.29) to the base vectors a_i and write the equations of motion in tensor components. Thus, we write†

$$(6.30) \qquad N^{\alpha} = N^{i\alpha} a_i, \qquad k = k^i a_i, \qquad M^{\alpha} = M^{i\alpha} a_i,$$

$$(6.31) \qquad f = f^i a_i, \qquad l = l^i a_i,$$

with similar expressions for $\hat{N}^{\alpha}$, $\hat{k}$, $\hat{M}^{\alpha}$, $\hat{f}$, $\hat{l}$. Recalling $(6.19)_1$ and making use of formulas of the type (6.30), from the scalar product of $(6.29)_3$ with a^{β} and again with a^3 we deduce

$$\varepsilon_{\alpha\beta}(\hat{N}^{\alpha\beta} - \phi b^{\beta}_{\gamma} \hat{M}^{\alpha\gamma}) = 0,$$

$$(6.32) \qquad \hat{N}^{3\alpha} - \phi \hat{k}^{\alpha} - \phi b^{\alpha}_{\gamma} \hat{M}^{3\gamma} - \phi_{,\gamma} \hat{M}^{\alpha\gamma} = 0,$$

where $\varepsilon^{\alpha\beta}$ is defined in (6.9).

It is instructive at this point to express the mechanical power (3.22) in terms of the tensor components (6.30). To this end, we first note from $(6.19)_1$ and $(2.13)_{1,2}$ that the tensor components of w and $w_{,\alpha}$ referred to a_i are

$$w \cdot a_{\alpha} = -\phi(v_{,\alpha} \cdot a_3), \qquad w \cdot a_3 = \dot{\phi},$$

$$(6.33) \qquad w_{,\alpha} \cdot a_{\beta} = -\overline{(\phi b_{\beta\alpha})}^{\boldsymbol{\cdot}} + \phi b_{\alpha\gamma}(v_{,\beta} \cdot a^{\gamma}) - \phi_{,\alpha}(v_{,\beta} \cdot a_3),$$

$$w_{,\alpha} \cdot a_3 = \dot{\phi}_{,\alpha} - \phi b_{\alpha\beta}(v_{,\gamma} \cdot a_3) a^{\gamma\beta}.$$

Then, remembering (6.4), (6.5) for $N = 1$ and the notation $(3.29)_{1,2}$, we may write (3.22) as

$$P = (\hat{N}^{\beta\alpha} - \phi b^{\beta}_{\gamma} \hat{M}^{\alpha\gamma})(v_{,\alpha} \cdot a_{\beta}) + \hat{k}^3 \dot{\phi} - \hat{M}^{\beta\alpha} \overline{(\phi b_{\alpha\beta})}^{\boldsymbol{\cdot}} + \hat{M}^{3\alpha} \dot{\phi}_{,\alpha}$$
$$+ [\hat{N}^{3\alpha} - \phi \hat{k}^{\alpha} - \hat{M}^{\alpha\beta} \phi_{,\beta} - \phi b^{\alpha}_{\beta} \hat{M}^{3\beta}](v_{,\alpha} \cdot a_3).$$

† As noted earlier, the order of indices in $(6.30)_{1,3}$ are opposite to those used in Naghdi (1972) and most of the earlier papers on the subject.

Since the coefficient of $(\boldsymbol{v}_{,\alpha}\cdot\boldsymbol{a}_3)$ vanishes identically in view of $(6.32)_2$, the last expression reduces to

$$P=(\hat{N}^{\beta\alpha}-\phi b^{\beta}_{\gamma}\hat{M}^{\alpha\gamma})(\boldsymbol{v}_{,\alpha}\cdot\boldsymbol{a}_{\beta})+\hat{k}^3\dot{\phi}-\hat{M}^{\beta\alpha}\overline{(\phi b_{\alpha\beta})}^{\cdot}+\hat{M}^{3\alpha}\dot{\phi}_{,\alpha} \tag{6.34}$$

and does not involve the components $\hat{N}^{3\alpha}$ and $\hat{k}^{\alpha}$. Next, with the help of $\phi\hat{k}^{\alpha}+\phi\hat{M}^{3\gamma}b^{\alpha}_{\gamma}=\hat{N}^{3\alpha}-\hat{M}^{\alpha\gamma}\phi_{,\gamma}$ which follows from $(6.32)_2$, the component form of the equations of motion $(6.29)_{1,2}$ referred to $\boldsymbol{a}^i$ can be written as

$$\hat{N}^{\beta\alpha}{}_{|\alpha}-b^{\beta}_{\alpha}N^{3\alpha}+\rho\hat{f}^{\beta}=0,\qquad N^{3\alpha}{}_{|\alpha}+b_{\beta\alpha}\hat{N}^{\alpha\beta}+\rho\hat{f}^3=0, \tag{6.35}$$

$$(\phi M^{\beta\alpha})_{|\alpha}+\rho\phi\hat{l}^{\beta}=N^{3\alpha},\qquad M^{3\alpha}{}_{|\alpha}+b_{\beta\alpha}\hat{M}^{\beta\alpha}-\hat{k}^3+\rho\hat{l}^3=0, \tag{6.36}$$

where in recording (6.35) and (6.36) we have also substituted $N^{3\alpha}$ for the quantity $(\hat{N}^{3\alpha}+\phi S^{\alpha})$. By substitution from $(6.36)_1$, we can now eliminate $N^{3\alpha}$ from $(6.35)_{1,2}$. In this way, the resulting two equations may be put in the form

$$\begin{aligned}&\{(\hat{\boldsymbol{N}}^{\alpha}\cdot\boldsymbol{a}^{\beta})_{|\alpha}-b^{\beta}_{\alpha}(\phi\boldsymbol{M}^{\gamma}\cdot\boldsymbol{a}^{\alpha})_{|\gamma}\}\boldsymbol{a}_{\beta}+\{(\phi\boldsymbol{M}^{\beta}\cdot\boldsymbol{a}^{\alpha})_{|\beta\alpha}+b_{\alpha\beta}(\hat{\boldsymbol{N}}^{\alpha}\cdot\boldsymbol{a}^{\beta})\}\boldsymbol{a}_3\\&+\{\rho\hat{\boldsymbol{f}}+(\rho\phi\hat{l}^{\alpha}\boldsymbol{a}_3)_{|\alpha}\}=0.\end{aligned} \tag{6.37}$$

In a general theory of an elastic Cosserat surface (Section 4), constitutive equations for both the symmetric and the skew-symmetric parts of $N^{\alpha\beta}$, $M^{\alpha\beta}$ can be provided through the expression for mechanical power. Here, however, since $b_{\alpha\beta}$ is symmetric, the term $-\hat{M}^{\beta\alpha}\overline{(\phi b_{\alpha\beta})}^{\cdot}$ in (6.34) provides constitutive equations for only the symmetric part of $\hat{M}^{\alpha\beta}$. Moreover, the quantity $(\hat{N}^{\alpha\beta}-\phi b^{\beta}_{\gamma}\hat{M}^{\alpha\gamma})$ is symmetric by virtue of $(6.32)_1$ and the two differential equations resulting from (6.37) involve only the symmetric parts of $\hat{N}^{\alpha\beta}$ and $\hat{M}^{\alpha\beta}$. Thus, in line with classical results in shell theory, in order to obtain a determinate theory we now put

$$\hat{M}^{[\alpha\beta]}=0,\qquad \hat{M}^{[\alpha\beta]}=\tfrac{1}{2}(\hat{M}^{\alpha\beta}-\hat{M}^{\beta\alpha}). \tag{6.38}$$

In summary, the relevant system of equations of the constrained theory under discussion are given by (6.37), the normal component of $(6.29)_2$, i.e., $(6.36)_2$ and the skew-symmetric part $N^{\alpha\beta}$ is determined from $(6.32)_1$. This completes the development of the constrained theory in which the director is constrained to have the form $(6.19)_1$.

If in addition to (6.38) we also set the multiplier $p^0=0$, then $\bar{M}^{[\alpha\beta]}=0$ and hence the skew-symmetric $M^{[\alpha\beta]}=0$. It then follows that

$$S^{\alpha}=-p^{\alpha}, \tag{6.39}$$

$$N^{3\alpha}=\hat{N}^{3\alpha}-\phi S^{\alpha},\qquad k^{\alpha}=\hat{k}^{\alpha}-S^{\alpha} \tag{6.40}$$

and the relevant equations of motion of the determinate constrained theory remain as before. It is of interest to examine the reduction of the forgoing development when $\phi = 1$. In this case, we have $\boldsymbol{d} = \boldsymbol{a}_3$ instead of $(6.19)_1$ and the resulting equations are identical with those of a restricted theory discussed by Naghdi (1972, Sections 10 and 15). The results with $\phi = 1$ can also be brought into correspondence with a special case of the constrained theory discussed by Green and Naghdi (1974) or those contained in the paper of Naghdi and Nordgren (1963).

The nature of the boundary conditions in the theory of a Cosserat surface $\mathscr{C}$ discussed in Subsection 3.2 is clear from the expression for the rate of work R_c of contact force and contact director force over the closed boundary curve $\partial\mathscr{P}$, namely

$$R_c(\mathscr{P}) = \int_{\partial\mathscr{P}} (\boldsymbol{n} \cdot \boldsymbol{v} + \boldsymbol{m} \cdot \boldsymbol{w}) ds. \tag{6.41}$$

However, in a constrained theory of the type discussed in Subsection 6.2, the question of the boundary conditions must be reconsidered in view of the reduction in the number of differential equations.[†] Since the development of the reduced boundary conditions is similar to that of a restricted theory (Naghdi, 1972, p. 552), our discussion will be brief.

Recalling $(6.30)_{1,2,3}$ and (6.33), from (6.41) we obtain

$$R_c(\mathscr{P}) = \int_{\partial\mathscr{P}} \nu_\alpha \{(N^{\gamma\alpha} - \phi b^\gamma_\beta M^{\beta\alpha}) v_\gamma + N^{3\alpha} v_3 + M^{3\alpha}\dot{\phi} - \phi M^{\gamma\alpha} v_{3,\gamma}\} ds. \tag{6.42}$$

Let $\partial/\partial\nu$ stand for the directional derivative along the unit normal $\boldsymbol{\nu}$ to the boundary curve $\partial\mathscr{P}$ and let $\partial/\partial s$ denote the directional derivative along the tangent to $\partial\mathscr{P}$. Then, provided the quantities in (6.42) are single-valued on a (sufficiently smooth) closed curve $\partial\mathscr{P}$, with the use of

$$v_{3,\gamma} = \frac{\partial v_3}{\partial \nu} \nu_\gamma - \frac{\partial v_3}{\partial s} \varepsilon_{\gamma\beta} \nu^\beta \tag{6.43}$$

and an integration by parts, (6.42) can be reduced to

$$R_c(\mathscr{P}) = \int_{\partial\mathscr{P}} \left\{ P^\beta v_\beta + P^3 v_3 - G \frac{\partial v_3}{\partial \nu} + H\dot{\phi} \right\} ds, \tag{6.44}$$

where

$$\begin{aligned} P^\beta &= (N^{\beta\alpha} - \phi b^\beta_\gamma M^{\gamma\alpha}) \nu_\alpha, \qquad & G &= \phi M^{\gamma\alpha} \nu_\gamma \nu_\alpha = \phi M^{(\gamma\alpha)} \nu_\gamma \nu_\alpha, \\ P^3 &= N^{3\alpha} \nu_\alpha - \frac{\partial}{\partial s} (\phi M^{\gamma\alpha} \nu_\alpha \varepsilon_{\gamma\beta} \nu^\beta), \qquad & H &= M^{3\alpha} \nu_\alpha. \end{aligned} \tag{6.45}$$

[†] The number of the relevant scalar differential equations of the constrained theory is five as compared to the nine scalar equations in the theory of Subsection 3.2.

The nature of the reduced boundary conditions of the constrained theory is now clear from (6.44) and (6.45).

7. Additional remarks on shells

The theory of Cosserat surfaces can easily allow for the effect of surface tension (Naghdi, 1972, p. 547 and 1974) and can accommodate the specification of either tractions or displacements on major surfaces of the shell-like (or sheet-like) bodies for application to various interfacial and contact problems. Even the theory of a Cosserat surface with a single director can be used to formulate a fairly broad class of contact problems of elastic shells and plates, as discussed by Naghdi (1975a). The relevance and applicability of the basic theory of a Cosserat surface to problems of an incompressible, inviscid fluid sheet is discussed by Green, Laws and Naghdi (1974) and by Green and Naghdi (1975, 1976a). The nonlinear differential equations derived in these papers include the effects of gravity and surface tension and are also valid for propagation of fairly long water waves in a stream of initial variable depth. A discussion of an incompressible viscous fluid sheet, along with further recent developments on the subject, can be found in the papers of Green and Naghdi (1976a, b; 1977a; 1979c). The basic theory is also applicable to problems of cell membranes, as has been emphasized by Ericksen (1979).

In the remainder of this section we briefly comment on some special cases of the general theory and also mention some recent researches which bear on the various aspects of elastic shells. Although these developments will be described mainly in the context of a mechanical theory, some recent results pertaining to thermal effects in shells are also discussed.

The well-known membrane theory of shells can be obtained as a special case of the general theory by essentially suppressing the effect of the director and corresponding kinetical variables and this is discussed briefly in Naghdi (1972, Section 14). A development of another special theory, known as the inextensional theory, wherein the length of each element of the surface of $\mathfrak{s}$ is assumed to remain constant throughout all motions is also contained in Naghdi (1972, Section 14). Similarly, a nonlinear *restricted* theory of shells by direct approach, motivated mainly by the classical theory corresponding to Kirchhoff–Love theory of shells and plates, is given by Naghdi (1972, Sections 10 and 15). Related constrained theories of an elastic Cosserat surface are already mentioned in Section 6 and need not be repeated here.

The nonlinear constitutive equations in Section 4 are valid for an elastic Cosserat surface which may be anisotropic with reference to preferred directions associated with material points of $\mathscr{S}$. A general discussion of material symmetries for shells is given by Naghdi (1972, Section 13). Carroll and Naghdi (1972) have subsequently examined the influence of the reference geometry on the response of elastic shells by assuming the existence of a local *preferred* state of the body and then stipulating that the influence of the reference geometry, as in (4.3), occurs through the values of the constitutive variables in the preferred state. Material symmetry restrictions for elastic shells have been discussed also, from a different point of view, by Ericksen (1972a, 1973b) who has also indicated (Ericksen, 1973) a comparison with the results contained in the paper of Carroll and Naghdi (1972).

Some general aspects of wave propagation in elastic shells, based on the theory of a Cosserat surface have been discussed by Ericksen (1971). A related study on the subject, limited only to wave propagation in a surface not endowed with a director, was given earlier by Cohen and Suh (1970). The theory of small deformation superposed on a large deformation of an elastic Cosserat surface, along with related problems of stability and vibrations of initially deformed plates, is discussed by Green and Naghdi (1971). Related developments concerning plane waves and stability of elastic plates are given by Ericksen (1973c, 1974). For a system of linear equations characterizing the initial mixed boundary-value problem of elastic shells, Naghdi and Trapp (1972) have obtained a uniqueness theorem without the use of a definiteness assumption for the strain energy density. This result (Naghdi and Trapp, 1972) holds for nonhomogeneous and anisotropic shells undergoing small motions superposed on a large deformation.

In still another study, the theory of a Cosserat surface has been employed by Naghdi (1975a) to formulate contact problems of shells and plates mentioned above. In the derivation of shell theory from the 3-dimensional equations, equations of motion in terms of resultants and detailed consideration of constitutive equations for shells are usually obtained relative to an interior surface, rather than one of the major surfaces of the shell-like body which may be the contacting surface; the interior surface ordinarily is identified with the middle surface of the shell or plate in the reference configuration. In the development of shell theory by the direct approach, although the material surface of $\mathscr{S}$ may be identified with any surface of the (3-dimensional) shell-like body, nevertheless the complete discussion of constitutive equations and the identification of the inertia coefficients and the assigned fields may again require explicit use of a reference surface in the shell-like body. For certain problems it is more natural and conceptually more appealing to select one of the two major surfaces as the reference surface but then the detailed available development of the constitutive equations, as well as identification of such quantities as the inertia coefficients, have to be reconsidered relative to the new surface. This problem can be resolved by deriving appropriate transformation relations (Nagdhdi, 1975a), which relate the kinetic variables $\boldsymbol{n}$, $\boldsymbol{k}$, $\boldsymbol{m}$ (and hence the response functions) in the two formulations. The results (Naghdi, 1975a) are applicable to any shell-like medium and their validity is not limited to elastic shells alone.

Controllable solutions in the theory of a Cosserat surface have been studied by Crochet and Naghdi (1969), Ericksen (1972b) and Naghdi (1975b). In a more recent study, Naghdi and Tang (1977) have discussed controllable deformations that can be maintained, in the absence of body force, in every isotropic elastic membrane by the application of edge loads and/or uniform normal surface loads on the major surfaces of the thin shell-like body. The static solutions of finitely deformed membranes, which are valid for both compressible and incompressible materials, are obtained with the use of a strain energy response function which depends on the metric tensor of the membrane in its deformed configuration. The main results are summarized by several theorems and their corollaries in accordance with three mutually exclusive cases for which the initial undeformed surface of the membrane (which may be a sector of a complete or closed surface) is, respectively, developable, spherical and a surface of variable Gaussian curvature satisfying certain differential criteria. The corresponding deformed surfaces are, respectively, a plane or a right circular cyclinder, a sphere and a surface of constant mean curvature. These results are exhaustive in that they represent *all* finite deformation solutions possible in every isotropic elastic material characterized by the strain energy response mentioned above. Also discussed in the paper of Naghdi and Tang (1977) are some special cases of the general results and several families of solutions in terms of an alternative description which should be useful in application and which permit easy interpretation.

The development of the theory of Cosserat surfaces in Section 3 is carried out within the scope of the purely mechanical theory. In earlier work on a thermo-mechanical theory of shells by the direct approach (Green and Naghdi, 1970, Naghdi, 1972), only one temperature field was admitted and this allowed for the characterization of temperature changes along some reference surface, such as the middle surface, of the (3-dimensional) shell-like body, but not for temperature changes along the shell thickness. The latter effect has been incorporated recently by Green and Naghdi (1979a) into the thermo-mechanical theory of Cosserat surfaces, together with appropriate thermodynamical restrictions arising from the second law of thermodynamics for shells.

8. The basic equations in direct notation

For some purposes it is convenient to have available the basic equations for a Cosserat surface in a direct (coordinate-free) notation and this is the main purpose of the present section. As will be evident presently, the forms of the basic equations in coordinate-free notation are very similar to those of the corresponding equations in the classical 3-dimensional theory and thus may be more suitable in the discussion of general theorems or in developments which parallel those in the 3-dimensional theory.

As in the papers of Carroll and Naghdi (1972) and Naghdi (1977), we introduce the notations grad and Grad to denote the right spatial and material gradient operators, respectively, with respect to the position on the surface $\mathscr{s}$ in the current configuration and on the surface $\mathscr{S}_R$ in the reference configuration. The corresponding divergence operators will be denoted by div and Div, respectively. In particular, for a vector-valued function $\boldsymbol{V}(\theta^\alpha, t)$, we write[†]

$$\text{(8.1)} \qquad \begin{aligned} \operatorname{grad} \boldsymbol{V} &= \boldsymbol{V}_{,\alpha} \otimes \boldsymbol{d}^\alpha, \qquad & \operatorname{div} \boldsymbol{V} &= \boldsymbol{V}_{,\alpha} \cdot \boldsymbol{d}^\alpha, \\ \operatorname{Grad} \boldsymbol{V} &= \boldsymbol{V}_{,\alpha} \otimes \boldsymbol{D}^\alpha, \qquad & \operatorname{Div} \boldsymbol{V} &= \boldsymbol{V}_{,\alpha} \cdot \boldsymbol{D}^\alpha, \end{aligned}$$

where the symbol $\otimes$ denotes the tensor product. Also, the spatial surface gradient and the spatial surface divergence operators are defined by

$$\text{(8.2)} \qquad \operatorname{grad}_s V = V_{,\alpha} \boldsymbol{a}^\alpha, \qquad \operatorname{div}_s \boldsymbol{V} = \boldsymbol{V}_{,\alpha} \cdot \boldsymbol{a}^\alpha$$

for all scalar-valued functions V and all vector-valued functions $\boldsymbol{V}$.

We introduce a measure of deformation by the tensor $\boldsymbol{F}$, namely[‡]

$$\text{(8.3)} \qquad \boldsymbol{F} = \boldsymbol{d}_i \otimes \boldsymbol{D}^i = \operatorname{Grad} \boldsymbol{r} + \boldsymbol{d}_3 \otimes \boldsymbol{D}^3,$$

and in view of the notations (3.5) and (3.9) we observe that

$$\text{(8.4)} \qquad \begin{aligned} \boldsymbol{F}\boldsymbol{D}_\alpha &= \boldsymbol{F}\boldsymbol{A}_\alpha = \boldsymbol{a}_\alpha = \boldsymbol{d}_\alpha, \\ \boldsymbol{F}\boldsymbol{D}_3 &= \boldsymbol{F}\boldsymbol{D} = \boldsymbol{d} = \boldsymbol{d}_3. \end{aligned}$$

From the definition of the determinant of a second order tensor $\boldsymbol{T}$ given by

[†] We take this opportunity to correct an error in a previous paper (Naghdi 1977). The definitions $(2.9)_{1,2}$ of Naghdi (1977) should be replaced with those in $(8.1)_{1,2}$ of the present paper with $\boldsymbol{d}^\alpha$ defined through (3.5). Also, the "div" operator in $(3.10)_1$ of Naghdi (1977) should be replaced by "div_s" in $(8.2)_2$. The definitions $(2.9)_{3,4}$ of Naghdi (1977) remain unchanged since previously (Naghdi, 1977) the director in the reference configuration was specified to have the form $\boldsymbol{D} = D\boldsymbol{A}_3$. Except for the modifications noted, all other results in the paper of Naghdi (1977) remain intact.

[‡] This definition of $\boldsymbol{F}$ is the same as that used by Naghdi (1977). The symbol $\boldsymbol{F}$ in the paper of Carroll and Naghdi (1972) stands for a different quantity. The term Grad $\boldsymbol{r}$ in (8.3) corresponds to the deformation gradient tensor $\boldsymbol{F}$ in the paper of Carroll and Naghdi (1972).

$$\det \boldsymbol{T}[\boldsymbol{v}_1\boldsymbol{v}_2\boldsymbol{v}_3] = [\boldsymbol{T}\boldsymbol{v}_1, \boldsymbol{T}\boldsymbol{v}_2, \boldsymbol{T}\boldsymbol{v}_3]$$

for all arbitrary vectors $\boldsymbol{v}_1$, $\boldsymbol{v}_2$, $\boldsymbol{v}_3$, and the conditions $(3.1)_3$ and $(3.7)_3$, we obtain

$$\det \boldsymbol{F} = [\boldsymbol{d}_1\boldsymbol{d}_2\boldsymbol{d}_3]/[\boldsymbol{D}_1\boldsymbol{D}_2\boldsymbol{D}_3] > 0. \tag{8.5}$$

The tensor $\boldsymbol{F}$, a linear operator on vectors in 3-space, is nonsingular; and there exists, therefore, the inverse deformation gradient tensor $\boldsymbol{F}^{-1}$ defined by

$$\boldsymbol{F}^{-1} = \boldsymbol{D}_i \otimes \boldsymbol{d}^i. \tag{8.6}$$

The inverse operator $\boldsymbol{F}^{-1}$ transforms vectors in the present configuration into vectors in the reference configuration, i.e.,

$$\boldsymbol{F}^{-1}\boldsymbol{d}_i = \boldsymbol{D}_i \tag{8.7}$$

and it follows that

$$\boldsymbol{F}^{-1}\boldsymbol{F} = \boldsymbol{F}\boldsymbol{F}^{-1} = \boldsymbol{I} = \boldsymbol{d}_i \otimes \boldsymbol{d}^i = \boldsymbol{D}_i \otimes \boldsymbol{D}^i, \tag{8.8}$$

where $\boldsymbol{I}$ is the unit tensor in 3-space. We also introduce here the director gradient tensor $\boldsymbol{G}$ by

$$\boldsymbol{G} = \operatorname{Grad} \boldsymbol{d} = \boldsymbol{d}_{3,\alpha} \otimes \boldsymbol{D}^\alpha = \boldsymbol{d}_{,\alpha} \otimes \boldsymbol{D}^\alpha. \tag{8.9}$$

Recalling the definitions $(3.3)_{1,2}$ for the velocity and the director velocity and since $\dot{\boldsymbol{d}}_\alpha = \boldsymbol{v}_{,\alpha}$, we have

$$\begin{aligned} \dot{\boldsymbol{F}} &= \dot{\boldsymbol{d}}_i \otimes \boldsymbol{D}^i = \dot{\boldsymbol{d}}_\alpha \otimes \boldsymbol{D}^\alpha + \dot{\boldsymbol{d}}_3 \otimes \boldsymbol{D}^3 = \boldsymbol{v}_{,\alpha} \otimes \boldsymbol{D}^\alpha + \boldsymbol{w} \otimes \boldsymbol{D}^3, \\ \dot{\boldsymbol{G}} &= \dot{\boldsymbol{d}}_{3,\alpha} \otimes \boldsymbol{D}^\alpha = \boldsymbol{w}_{,\alpha} \otimes \boldsymbol{D}^\alpha. \end{aligned} \tag{8.10}$$

Also,

$$\begin{aligned} \dot{\boldsymbol{F}}\boldsymbol{F}^{-1} &= \dot{\boldsymbol{d}}_i \otimes \boldsymbol{d}^i = \operatorname{grad} \boldsymbol{v} + \boldsymbol{w} \otimes \boldsymbol{d}^3, \\ \dot{\boldsymbol{G}}\boldsymbol{F}^{-1} &= \boldsymbol{w}_{,\alpha} \otimes \boldsymbol{d}^\alpha = \operatorname{grad} \boldsymbol{w}. \end{aligned} \tag{8.11}$$

Having disposed of the main kinematical results in terms of the gradient tensors $\boldsymbol{F}$, $\boldsymbol{G}$ and their rates, we now turn to kinetical quantities. The expressions corresponding to $(3.18)_{1,2}$ for the contact force $\boldsymbol{n}$ and the contact director $\boldsymbol{m}$ can now be expressed in the form[†]

[†] The second order tensors $\boldsymbol{N}$, $\boldsymbol{M}$ in (8.12) and their tensor components $N^{i\alpha}$, $M^{i\alpha}$ in (8.13) are the transpose of the corresponding quantities in Naghdi (1977). The components $N^{i\alpha}$, $M^{i\alpha}$ were used in the paper of Green, Naghdi and Wainwright (1965) but subsequently their transpose, namely $N^{\alpha i}$ and $M^{\alpha i}$, were adopted in subsequent papers so that the notation would be in agreement with that of the classical shell theory. It may be noted that in terms of the latter notation, instead of (8.12), one would have $\boldsymbol{n} = \boldsymbol{N}^T\boldsymbol{\nu}$, $\boldsymbol{m} = \boldsymbol{M}^T\boldsymbol{\nu}$, where the superposed T denotes transpose. Compare (3.6) and (3.10) of Naghdi (1977) with (8.12) and (8.15) of the present paper.

$$n = N\nu, \qquad m = M\nu, \tag{8.12}$$

with the second order tensors N, M defined by

$$\begin{aligned} N &= N^\alpha \otimes d_\alpha = N^{i\alpha} d_i \otimes d_\alpha, & N^\alpha &= Md^\alpha, \\ M &= M^\alpha \otimes d_\alpha = M^{i\alpha} d_i \otimes d_\alpha, & M^\alpha &= Md^\alpha, \end{aligned} \tag{8.13}$$

which also relate the tensors N, M to N^α, M^α in (4.8). Also, for convenience, we introduce a second order tensor K through

$$K = k \otimes d_3 = k^i d_i \otimes d_3, \qquad k = Kd^3. \tag{8.14}$$

With the use of (8.13) and by usual procedures, from the conservation laws (3.16) follow the local equations

$$\begin{gathered} \dot{\rho} + \rho \operatorname{div}_s v = 0, \\ \operatorname{div}_s N + \rho f = \rho(\dot{v} + y^1 \dot{w}), \\ \operatorname{div}_s M + \rho l - k = \rho(y^1 \dot{v} + y^2 \dot{w}), \\ [N + K + M(GF^{-1})^T] = [N + K + M(GF^{-1})^T]^T, \end{gathered} \tag{8.15}$$

which are equivalent to (3.20). Also, by the definition of the right divergence of a tensor field, we have

$$\operatorname{div}_s N = N^\alpha_{\ |\alpha}, \qquad \operatorname{div}_s M = M^\alpha_{\ |\alpha}. \tag{8.16}$$

It is interesting that the last statement in (8.15) is analogous to the symmetry of the stress tensor in the 3-dimensional theory. In particular, it may be observed that $a_\alpha \times N^\alpha$, $d \times k$ and $d_{,\alpha} \times M^\alpha$ are, respectively, the axial vectors of $[N - N^T]$, $[K - K^T]$ and $[M(GF^{-1})^T - M(GF^{-1})]$. Furthermore, in terms of the kinetical quantities, N, M, K in (8.12)–(8.14) and the rate quantities $(8.10)_{1,2}$, the mechanical power becomes

$$P = tr\{(\dot{F}^T(N + K) + \dot{G}^T M](F^{-1})^T\}. \tag{8.17}$$

With reference to constitutive equations for elastic shells, instead of the kinematical variables in section 4, we now employ the variables (8.3) and (8.9). Thus, corresponding to constitutive assumption (4.3), we now write

$$\psi = \tilde{\psi}(F, G; {}_R G), \tag{8.18}$$

where

$${}_R G = \operatorname{Grad} D = D_{3,\alpha} \otimes D^\alpha, \tag{8.19}$$

along with similar assumptions for N, K, M. Then, with the use of (4.1) and (8.17), by usual techniques we obtain the following alternative forms of the constitutive equations:

$$\text{(8.20)} \qquad \boldsymbol{N} + \boldsymbol{K} = \rho \frac{\partial \tilde{\psi}}{\partial \boldsymbol{F}} \boldsymbol{F}^T, \qquad \boldsymbol{M} = \rho \frac{\partial \tilde{\psi}}{\partial \boldsymbol{G}} \boldsymbol{F}^T,$$

the first of which can be resolved into

$$\text{(8.21)} \qquad \boldsymbol{N} = \rho \frac{\partial \tilde{\psi}}{\partial \boldsymbol{F}} (\boldsymbol{D}^\alpha \otimes \boldsymbol{d}_\alpha), \qquad \boldsymbol{K} = \rho \frac{\partial \tilde{\psi}}{\partial \boldsymbol{F}} (\boldsymbol{D}^3 \otimes \boldsymbol{d}_3).$$

Also, the response function $\tilde{\psi}$ is restricted by

$$\text{(8.23)} \qquad \begin{gathered} \boldsymbol{S} = \boldsymbol{S}^T, \\ \boldsymbol{S} = \frac{\partial \tilde{\psi}}{\partial \boldsymbol{F}} \boldsymbol{F}^T + \frac{\partial \tilde{\psi}}{\partial \boldsymbol{G}} \boldsymbol{G}^T = \frac{\partial \tilde{\psi}}{\partial \boldsymbol{F}} (\boldsymbol{D}^\alpha \otimes \boldsymbol{d}_\alpha) + \frac{\partial \tilde{\psi}}{\partial \boldsymbol{F}} (\boldsymbol{D}^3 \otimes \boldsymbol{d}_3) + \frac{\partial \tilde{\psi}}{\partial \boldsymbol{G}} \boldsymbol{G}^T. \end{gathered}$$

Part B

Elastic rods: A direct formulation

In Part B (Sections 9–13), we first summarize the main kinematics and the basic principles of the theory of Cosserat (or directed) curves and then discuss the constitutive equations for elastic rods, as well as some related aspects of the basic theory and recent developments on the subject. Although we are concerned here mainly with the purely mechanical theory involving appropriate forms of the conservation laws for mass, linear momentum, director momentum and moment of momentum, we also include a statement of the conservation of energy. The latter provides motivation in the development of certain constitutive equations, such as those for an elastic material, and in the discussion of aspects of some special solutions involving jump in energy. The contents of Part B are as follows:

9. The basic theory of Cosserat curves.
   9.1. Kinematics of a Cosserat curve $\mathscr{R}$.
   9.2. Basic principles of a Cosserat curve $\mathscr{R}$.
   9.3. Hierarchical theories of Cosserat curves.
10. Elastic rods.
11. Identification of the assigned fields and the inertia coefficients.
12. Additional remarks on rods.
13. The basic equations for elastic rods in direct notation.

9. The basic theory of Cosserat curves

Having defined a (three-dimensional) rod-like body in Section 2, we now formally introduce a direct *model* for such a body. Thus, deformable media which are *modelled* by a material curve $\mathscr{L}$ embedded in a Euclidean 3-space, together with L $(L \geqq 2)$ deformable vector fields — called directors — attached to every point of the material curve are called *Cosserat curves* or *directed curves* and may be conveniently referred to as $\mathscr{R}_K$ $(K = 1, 2, \cdots, N)$. The directors which are not necessarily along the unit principal normal and the unit binormal vectors to the curve have, in particular, the property that they remain unaltered in length under superposed rigid body motions.

In the absence of the directors, we merely have a 1-dimensional material curve $\mathscr{L}$ which can serve as a model for the construction by direct approach of a string theory. The relationship between the number of directors L and the number K which identifies the order of the hierarchical theory of Cosserat curves can be shown to be $L = \Sigma_1^K (N+1)$ so that (see Naghdi, 1979b)

$$L = K(K+3)/2. \tag{9.1}$$

With $K = 1$, the directed curve is a body $\mathscr{R}_1 = \mathscr{R}$ comprising a material curve and two deformable directors attached to every point of the material curve of $\mathscr{R}$. The latter is the simplest model for the construction of a general bending theory of slender rods; and, for simplicity, we restrict attention to this particular model in most of the development[†] in Section 9.

We now turn to a brief account of the basic theory of a Cosserat curve.

9.1. *Kinematics of a Cosserat curve* $\mathscr{R}$

Let the particles of the material curve $\mathscr{L}$ of $\mathscr{R}$ be identified by means of the convected coordinate ξ and let the curve occupied by $\mathscr{L}$ in the present configuration of $\mathscr{R}$ at time t be referred to as ℓ. Let $\boldsymbol{r}$ and $\boldsymbol{d}_\alpha$ $(\alpha = 1, 2)$ denote the position vector of a typical point of ℓ and the directors at the same point, respectively, and also designate the tangent vector to the curve ℓ by $\boldsymbol{a}_3$. Then, a motion of the Cosserat curve is defined by vector-valued functions which assign a position $\boldsymbol{r}$ and a pair of directors $\boldsymbol{d}_\alpha$ to each particle of $\mathscr{R}$ at each instant of time, i.e.,[‡]

$$\boldsymbol{r} = \hat{\boldsymbol{r}}(\xi, t), \qquad \boldsymbol{d}_\alpha = \hat{\boldsymbol{d}}_\alpha(\xi, t), \qquad [\boldsymbol{d}_1 \boldsymbol{d}_2 \boldsymbol{a}_3] > 0 \tag{9.2}$$

where

$$\boldsymbol{a}_3 = \boldsymbol{a}_3(\xi, t) = \frac{\partial \hat{\boldsymbol{r}}}{\partial \xi}\,. \tag{9.3}$$

[†] A brief account of the more general theory for Cosserat curves is indicated at the end of this section.

[‡] For convenience, we adopt the notation for $\boldsymbol{r}$ in (2.16) and (2.30) also for the surface $(9.2)_1$. This permits an easy identification of the two curves, if desired. The choice of positive sign in $(9.2)_3$ is for definiteness. Alternatively, it will suffice to assume that $[\boldsymbol{d}_1 \boldsymbol{d}_2 \boldsymbol{a}_3] \neq 0$ with the understanding that in any given motion the scalar triple product $[\boldsymbol{d}_1 \boldsymbol{d}_2 \boldsymbol{a}_3]$ is either > 0 or < 0.

The condition $(9.2)_3$ ensures that the directors $\boldsymbol{d}_\alpha$ are nowhere tangent to ℓ and that $\boldsymbol{d}_1$, $\boldsymbol{d}_2$ never change their relative orientation with respect to each other and $\boldsymbol{a}_3$. The velocity and the director velocities are defined by

(9.4) $$\boldsymbol{v} = \dot{\boldsymbol{r}}, \qquad \boldsymbol{w}_\alpha = \dot{\boldsymbol{d}}_\alpha,$$

and from (9.3) and $(9.4)_1$ we have

(9.5) $$\dot{\boldsymbol{a}}_3 = \frac{\partial \boldsymbol{v}}{\partial \xi},$$

where a superposed dot denotes material time differentiation with respect to t holding ξ fixed.

It is convenient to introduce here a slightly different notation than adopted in a number of previous papers, e.g., Naghdi (1979a). Thus, we put

(9.6) $$\boldsymbol{d}_3 = \boldsymbol{a}_3, \qquad \boldsymbol{d}_i = (\boldsymbol{d}_\alpha, \boldsymbol{a}_3)$$

and observe that in view of $(9.2)_3$ and (9.6), $\boldsymbol{d}_1$, $\boldsymbol{d}_2$, $\boldsymbol{d}_3$ are linearly independent vectors. Hence, we may introduce a set of reciprocal vectors $\boldsymbol{d}^i$ such that

(9.7) $$\boldsymbol{d}_i \cdot \boldsymbol{d}^j = \delta_i^j,$$

where δ_i^j is the Kronecker symbol in 3-space. Whenever desirable, the notations $\boldsymbol{d}_i = (\boldsymbol{d}_1, \boldsymbol{d}_2, \boldsymbol{d}_3)$ and $(\boldsymbol{d}_\alpha, \boldsymbol{a}_3)$ will be used interchangeably throughout Part B depending on the particular context. Consider now a reference configuration, not necessarily the intitial configuration, of the Cosserat curve $\mathcal{R}$. In the reference configuration, let the material curve of $\mathcal{R}$ be referred to as $\mathcal{L}_R$ and designate the unit principal normal, the unit binormal and the tangent vector to $\mathcal{L}_R$ by $\boldsymbol{A}_1$, $\boldsymbol{A}_2$ and $\boldsymbol{A}_3$, respectively. Further, let $\boldsymbol{R}$ and $\boldsymbol{D}_\alpha$ $(\alpha = 1, 2)$ stand for the position of a typical point of $\mathcal{L}_R$ and the directors at the same point, respectively. Then, in the reference configuration we have

(9.8) $$\boldsymbol{R} = \hat{\boldsymbol{R}}(\xi), \qquad \boldsymbol{D}_\alpha = \hat{\boldsymbol{D}}_\alpha(\xi), \qquad [\boldsymbol{D}_1 \boldsymbol{D}_2 \boldsymbol{A}_3] > 0,$$

where

(9.9) $$\boldsymbol{A}_3 = \boldsymbol{A}_3(\xi) = \frac{\partial \hat{\boldsymbol{R}}}{\partial \xi}$$

and $(9.8)_3$ ensures that $\boldsymbol{D}_\alpha$ are nowhere tangent to the curve $\mathcal{L}_R$. If the reference configuration of $\mathcal{R}$ is specified to be the initial configuration, say at time $t = 0$, then the vector-valued functions on the right-hand sides of $(9.8)_{1,2}$ can be identified with $\hat{\boldsymbol{r}}(\xi, 0)$ and $\hat{\boldsymbol{d}}_\alpha(\xi, 0)$, respectively. Analogously to (9.6), we set

(9.10) $$\boldsymbol{D}_3 = \boldsymbol{A}_3, \qquad \boldsymbol{D}_i = (\boldsymbol{D}_\alpha, \boldsymbol{A}_3)$$

so that the dual of (9.7) is given by

$$\boldsymbol{D}_i \cdot \boldsymbol{D}^j = \delta_i^j. \tag{9.11}$$

9.2. *Basic principles of a Cosserat curve* $\mathcal{R}$

Consider an arbitrary part of the material curve $\mathcal{L}$ in the present configuration, i.e., a part of the space curve ℓ bounded by $\xi = \xi_1$ and $\xi = \xi_2$ $(\xi_1 < \xi_2)$, and let

$$ds = (a_{33})^{1/2} d\xi, \qquad a_{33} = \boldsymbol{a}_3 \cdot \boldsymbol{a}_3 \tag{9.12}$$

be the element of the arc length of ℓ. It is convenient at this point to define the following additional quantities: The mass density $\rho = \rho = (\xi, t)$ of the curve ℓ; the contact force[†] $\boldsymbol{n} = \boldsymbol{n}(\xi, t)$ and the contact director forces $\boldsymbol{m}^\alpha = \boldsymbol{m}^\alpha(\xi, t)$, each a 3-dimensional vector field in the present configuration; the assigned force $\boldsymbol{f} = \boldsymbol{f}(\xi, t)$ and the assigned director forces $\boldsymbol{l}^\alpha = \boldsymbol{l}^\alpha(\xi, t)$, each a 3-dimensional vector field and each per unit mass of the curve ℓ; the intrinsic (curve) director forces $\boldsymbol{k}^\alpha = \boldsymbol{k}^\alpha(\xi, t)$ per unit length of ℓ which make no contribution to the supply of moment of momentum; the inertia coefficients $y^\alpha = y^\alpha(\xi)$ and $y^{\alpha\beta} = y^{\alpha\beta}(\xi)$, with $y^{\alpha\beta}$ being components of a symmetric tensor, which are independent of time; the specific internal energy $\varepsilon = \varepsilon(\xi, t)$; the specific heat supply $r = r(\xi, t)$ per unit time; and the heat flux $h = h(\xi, t)$ along ℓ, in the direction of increasing ξ, per unit time. The assigned field $\boldsymbol{f}$ represents the combined effect of (i) the stress vector on the lateral surface (2.26) of the rod-like body denoted by $\boldsymbol{f}_c$, and (ii) an integrated contribution arising from the 3-dimensional body force denoted by $\boldsymbol{f}_b$, e.g., that due to gravity. A parallel statement holds for the assigned fields $\boldsymbol{l}^\alpha$. Similarly, the assigned heat supply r represents the combined effect of (i) heat supply entering the lateral surface (2.26) of the rod-like body from the surrounding environment, denoted by r_c, and (ii) an integrated contribution arising from the 3-dimensional heat supply denoted by r_b. Thus, we may write

$$\boldsymbol{f} = \boldsymbol{f}_b + \boldsymbol{f}_c, \qquad \boldsymbol{l}^\alpha = \boldsymbol{l}_b^\alpha + \boldsymbol{l}_c^\alpha, \qquad r = r_b + r_c. \tag{9.13}$$

We assume that the kinetic energy of a Cosserat curve $\mathcal{R}$ per unit length of the curve ℓ in the present configuration is given by

$$\kappa = \tfrac{1}{2}\rho[\boldsymbol{v} \cdot \boldsymbol{v} + 2y^\alpha \boldsymbol{v} \cdot \boldsymbol{w}_\alpha + y^{\alpha\beta} \boldsymbol{w}_\alpha \cdot \boldsymbol{w}_\beta]. \tag{9.14}$$

[†] The notations for the contact force $\boldsymbol{n}$, the contact director forces $\boldsymbol{m}^\alpha$ and the curve director forces $\boldsymbol{k}^\alpha$ differ from those in Green and Laws (1966), Green, Naghdi and Wenner (1974a, b), Naghdi (1979a, b) and most of the previous papers on the subject. In fact, the vector fields $\boldsymbol{n}$, $\boldsymbol{m}^\alpha$, $\boldsymbol{k}^\alpha$ of Part B of this paper correspond, respectively, to $\boldsymbol{n}$, $\boldsymbol{p}^\alpha$, $\boldsymbol{\pi}^\alpha$ of Green, Naghdi and Wenner (1974a, b), and most of the previous papers on the subject.

We further define the momentum corresponding to the velocity $\boldsymbol{v}$ and the director momentum corresponding to the director velocities $\boldsymbol{w}_\alpha$ by

$$\frac{\partial \kappa}{\partial \boldsymbol{v}} = \rho(\boldsymbol{v} + y^\alpha \boldsymbol{w}_\alpha), \qquad \frac{\partial \kappa}{\partial \boldsymbol{w}_\alpha} = \rho(y^\alpha \boldsymbol{v} + y^{\alpha\beta}\boldsymbol{w}_\beta) \tag{9.15}$$

per unit length of ℓ. Also, the physical dimensions of ρ, $\boldsymbol{n}$, $\boldsymbol{f}$ are

$$\begin{gathered} \text{phys. dim. } \rho = [\mathrm{ML}^{-1}], \\ \text{phys. dim. } \boldsymbol{n} = [\mathrm{MLT}^{-2}], \qquad \text{phys. dim. } \boldsymbol{f} = [\mathrm{LT}^{-2}], \end{gathered} \tag{9.16}$$

where as in Section 3 the symbols [L], [M] and [T] stand for the physical dimensions of length, mass and time. The dimensions of the vector fields $\boldsymbol{m}^\alpha$, $\boldsymbol{l}^\alpha$ and $\boldsymbol{k}^\alpha$ depend upon the physical dimensions of $\boldsymbol{d}_\alpha$. Here we choose $\boldsymbol{d}_\alpha$ to have the dimension of length. Then, $\boldsymbol{m}^\alpha$, $\boldsymbol{l}^\alpha$ will have the same physical dimensions as $\boldsymbol{n}$, $\boldsymbol{f}$ in (9.16) while $\boldsymbol{k}^\alpha$ will have the physical dimension of $[\mathrm{ML}^{-2}\mathrm{T}^{-2}]$.

With the above definitions of the various field quantities and the notion

$$[f(\xi, t)]_{\xi_1}^{\xi_2} = f(\xi_2, t) - f(\xi_1, t), \tag{9.17}$$

with reference to the present configuration, the conservation laws for a Cosserat curve are[†]

$$\begin{gathered} \frac{d}{dt}\int_{\xi_1}^{\xi_2} \rho\, ds = 0, \\ \frac{d}{dt}\int_{\xi_1}^{\xi_2} \rho(\boldsymbol{v} + y^\alpha \boldsymbol{w}_\alpha)\, ds = \int_{\xi_1}^{\xi_2} \rho \boldsymbol{f}\, ds + [\boldsymbol{n}]_{\xi_1}^{\xi_2}, \\ \frac{d}{dt}\int_{\xi_1}^{\xi_2} \rho(y^\alpha \boldsymbol{v} + y^{\alpha\beta}\boldsymbol{w}_\beta)\, ds = \int_{\xi_1}^{\xi_2} (\rho \boldsymbol{l}^\alpha - (a_{33})^{-1/2}\boldsymbol{k}^\alpha)\, ds + [\boldsymbol{m}^\alpha]_{\xi_1}^{\xi_2}, \\ \frac{d}{dt}\int_{\xi_1}^{\xi_2} \rho[\boldsymbol{r}\times\boldsymbol{v} + y^\alpha(\boldsymbol{r}\times\boldsymbol{w}_\alpha + \boldsymbol{d}_\alpha\times\boldsymbol{v}) + \boldsymbol{d}_\alpha \times y^{\alpha\beta}\boldsymbol{w}_\beta]\, ds \\ = \int_{\xi_1}^{\xi_2} \rho(\boldsymbol{r}\times\boldsymbol{f} + \boldsymbol{d}_\alpha\times\boldsymbol{l}^\alpha)\, ds + [\boldsymbol{r}\times\boldsymbol{n} + \boldsymbol{d}_\alpha\times\boldsymbol{m}^\alpha]_{\xi_1}^{\xi_2}. \end{gathered} \tag{9.18}$$

The first of (9.18) is a statement of the conservation of mass, the second is the conservation of linear momentum, the third that of the director momentum and the fourth is the conservation of moment of momentum. We also record the law of conservation of energy in the form

$$\frac{d}{dt}\int_{\xi_1}^{\xi_2} \rho(\varepsilon + \kappa)\, ds = \int_{\xi_1}^{\xi_2} \rho(r + \boldsymbol{f}\cdot\boldsymbol{v} + \boldsymbol{l}^\alpha\cdot\boldsymbol{w}_\alpha)\, ds + [\boldsymbol{n}\cdot\boldsymbol{v} + \boldsymbol{m}^\alpha\cdot\boldsymbol{w}_\alpha - h]_{\xi_1}^{\xi_2}. \tag{9.19}$$

[†] The conservation laws (9.18) correspond to Equation (6.16) in the paper of Green, Naghdi and Wenner (1974b).

The basic structures of $(9.18)_{1,2}$ and (9.19) are analogous to the corresponding conservation laws of the classical 3-dimensional theory. The structures of $(9.18)_3$ and $(9.18)_4$ are less obvious, but a motivation for their forms is provided by a derivation of the basic field equations for rod-like bodies obtained from the 3-dimensional equations of continuum mechanics in which the position vector $\boldsymbol{r}^*$ in 3-space is approximated by an expression of the form (2.30). It should be noted that the conservation laws (9.18) and (9.19) are consistent with the invariance conditions under superposed rigid body motions, which ordinarily have wide acceptance in continuum mechanics. Moreover, the conservation laws $(9.18)_1$, $(9.18)_2$ and $(9.18)_4$ are equivalent to, and can be derived from the conservation of energy (9.19) and the invariance conditions under superposed rigid body motions. The conservation law $(9.18)_3$ for the director momentum must be postulated separately.

Returning to the conservation laws, after making suitable continuity assumptions, we see that by usual procedures from $(9.18)_{1,2,3,4}$ and (9.19) follow the local field equations

$$\lambda = \lambda(\xi) = \rho(a_{33})^{1/2} \qquad \text{or} \qquad \dot{\rho} a_{33} + \rho \boldsymbol{a}_3 \cdot \frac{\partial \boldsymbol{v}}{\partial \xi} = 0, \tag{9.20}$$

$$\frac{\partial \boldsymbol{n}}{\partial \xi} + \lambda \boldsymbol{f} = \lambda(\dot{\boldsymbol{v}} + y^\alpha \dot{\boldsymbol{w}}_\alpha), \tag{9.21}$$

$$\frac{\partial \boldsymbol{m}^\alpha}{\partial \xi} + \lambda \boldsymbol{l}^\alpha = \boldsymbol{k}^\alpha + \lambda(y^\alpha \dot{\boldsymbol{v}} + y^{\alpha\beta} \dot{\boldsymbol{w}}_\beta), \tag{9.22}$$

$$\boldsymbol{a}_3 \times \boldsymbol{n} + \boldsymbol{d}_\alpha \times \boldsymbol{k}^\alpha + \frac{\partial \boldsymbol{d}_\alpha}{\partial \xi} \times \boldsymbol{m}^\alpha = 0, \tag{9.23}$$

and

$$\lambda r - \frac{\partial h}{\partial \xi} - \lambda \dot{\varepsilon} + P = 0, \tag{9.24}$$

where

$$P = \boldsymbol{n} \cdot \frac{\partial \boldsymbol{v}}{\partial \xi} + \boldsymbol{k}^\alpha \cdot \boldsymbol{w}_\alpha + \boldsymbol{m}^\alpha \cdot \frac{\partial \boldsymbol{w}_\alpha}{\partial \xi} \tag{9.25}$$

is the mechanical power.

9.3. *Hierarchical theories of Cosserat curves*

Although the theory outlined in Subsection 9.2 is sufficiently general for many applications, on occasion it becomes necessary to consider a more general theory of Cosserat curves. Therefore, we now briefly discuss the kinematics and the balance laws of Cosserat curves $\mathscr{R}_K$ having L ($\geqq 2$) directors attached to every point of a material line $\mathscr{L}$, the number L being given by (9.1).

Thus, instead of $(9.2)_{1,2}$, we specify a motion of $\mathscr{R}_K$ by

$$r = \hat{r}(\xi, t), \qquad d_{\alpha_1\alpha_2\cdots\alpha_N} = d_{\alpha_1\alpha_2\cdots\alpha_N}(\xi, t) \qquad (N = 1, 2, \cdots, K), \tag{9.26}$$

where the vector functions $d_{\alpha_1\alpha_2\cdots\alpha_N}$ are assumed to be symmetric in the indices $\alpha_1\alpha_2\cdots\alpha_N$. The velocity vector is still given by $(9.4)_1$ but corresponding to $(9.4)_2$ we now define the director velocities

$$w_{\alpha_1\alpha_2\cdots\alpha_N} = \dot{d}_{\alpha_1\alpha_2\cdots\alpha_N}. \tag{9.27}$$

We recall that for $K = 1$ ($\mathscr{R}_1 = \mathscr{R}$), the kinetical quantities and the assigned fields introduced in Subsection 9.2 consist of n, k^α, m^α and f, l^α. Keeping this in mind, for a body $\mathscr{R}_K$ we admit the more general kinetical quantities and assigned fields

$$\begin{aligned} &n, \qquad k^{\alpha_1\alpha_2\cdots\alpha_N}, \qquad m^{\alpha_1\alpha_2\cdots\alpha_N},\\ &f, \qquad l^{\alpha_1\alpha_2\cdots\alpha_N}, \end{aligned} \tag{9.28}$$

and corresponding to (9.14) and $(9.15)_{1,2}$ write the more general expressions for kinetic energy of $\mathscr{R}_K$ and associated momentum and director momenta, namely

$$\begin{aligned} \kappa &= \tfrac{1}{2}\rho\left[v\cdot v + 2\sum_{N=1}^{K} y^{\alpha_1\alpha_2\cdots\alpha_N} v\cdot w_{\alpha_1\alpha_2\cdots\alpha_N} + \sum_{N=1,M=1}^{K} y^{\alpha_1\cdots\alpha_N\beta_1\cdots\beta_M} w_{\alpha_1\cdots\alpha_N}\cdot w_{\beta_1\cdots\beta_M}\right],\\ \frac{\partial\kappa}{\partial v} &= \rho\left[v + \sum_{N=1}^{K} y^{\alpha_1\cdots\alpha_N} w_{\alpha_1\cdots\alpha_N}\right],\\ \frac{\partial\kappa}{\partial w_{\alpha_1\cdots\alpha_N}} &= \rho\left[y^{\alpha_1\cdots\alpha_N} v + \sum y^{\alpha_1\cdots\alpha_N\beta_1\cdots\beta_M} w_{\beta_1\cdots\beta_M}\right], \end{aligned} \tag{9.29}$$

each per unit length of the curve ℓ. The inertia coefficients $y^{\alpha_1\cdots\alpha_N}$, $y^{\alpha_1\cdots\alpha_N\beta_1\cdots\beta_M}$ in (9.29) are functions of ξ only, $y^{\alpha_1\cdots\alpha_N}$ are symmetric with respect to indices $\alpha_1\cdots\alpha_N$, $y^{\alpha_1\cdots\alpha_N\beta_1\cdots\beta_M} = y^{\beta_1\cdots\beta_M\alpha_1\cdots\alpha_N}$ and are also symmetric with respect to $\alpha_1\alpha_2\cdots\alpha_N$ and $\beta_1\beta_2\cdots\beta_M$. In the special case of $K = 1$ ($\mathscr{R}_1 = \mathscr{R}$), $L = 2$, we may use the notations

$$\begin{aligned} &d_{\alpha_1} = d_\alpha, \qquad w_{\alpha_1} = w_\alpha,\\ &y^{\alpha_1} = y^\alpha, \qquad y^{\alpha_1\beta_1} = y^{\alpha\beta}. \end{aligned} \tag{9.30}$$

For a detailed statement of conservation laws appropriate for Cosserat curves $\mathscr{R}_K$ we refer the reader to Naghdi (1979b, Section 2), but indicate below the structure of the corresponding local field equations. Thus, for the purely mechanical theory of Cosserat curves $\mathcal{R}_K$, the local field equations are

$$\overline{\lambda y^{\alpha_1 \cdots \alpha_N}}^{\,\cdot} = 0 \qquad (N = 1, \cdots, 2K), \qquad \dot{\lambda} = 0, \qquad \lambda = \lambda(\xi) = \rho a_{33}^{1/2},$$

(9.31)
$$\frac{\partial \boldsymbol{n}}{\partial \xi} + \lambda \bar{\boldsymbol{f}} = \boldsymbol{0},$$
$$\frac{\partial \boldsymbol{m}^{\alpha_1 \cdots \alpha_N}}{\partial \xi} + \lambda \boldsymbol{q}^{\alpha_1 \cdots \alpha_N} = \boldsymbol{k}^{\alpha_1 \cdots \alpha_N} \qquad (N = 1, \cdots, K),$$
$$\boldsymbol{a}_3 \times \boldsymbol{n} + \sum_{N=1}^{K} \left(\boldsymbol{d}_{\alpha_1 \cdots \alpha_N} \times \boldsymbol{k}^{\alpha_1 \cdots \alpha_N} + \frac{\partial \boldsymbol{d}_{\alpha_1 \cdots \alpha_N}}{\partial \xi} \times \boldsymbol{m}^{\alpha_1 \cdots \alpha_N} \right) = \boldsymbol{0},$$

where

(9.32)
$$\boldsymbol{f} = \boldsymbol{f}_b + \boldsymbol{f}_c, \qquad \boldsymbol{l}^{\alpha_1 \alpha_2 \cdots \alpha_N} = \boldsymbol{l}_b^{\alpha_1 \alpha_2 \cdots \alpha_N} + \boldsymbol{l}_c^{\alpha_1 \alpha_2 \cdots \alpha_N},$$

(9.33)
$$\bar{\boldsymbol{f}} = \boldsymbol{f} - \dot{\boldsymbol{v}} - \sum_{N=1}^{K} y^{\alpha_1 \cdots \alpha_N} \dot{\boldsymbol{w}}_{\alpha_1 \cdots \alpha_N} \qquad (N = 1, \cdots, K),$$
$$\boldsymbol{q}^{\alpha_1 \cdots \alpha_N} = \boldsymbol{l}^{\alpha_1 \cdots \alpha_N} - y^{\alpha_1 \cdots \alpha_N} \dot{\boldsymbol{v}} - \sum_{M=1}^{K} y^{\alpha_1 \cdots \alpha_N \beta_1 \cdots \beta_M} \dot{\boldsymbol{w}}_{\beta_1 \cdots \beta_M} \qquad (N = 1, \cdots, K).$$

Also, for Cosserat curves $\mathcal{R}_K$, the expression for mechanical power corresponding to (9.26) is

(9.34)
$$P = \boldsymbol{n} \cdot \frac{\partial \boldsymbol{v}}{\partial \xi} + \sum_{N=1}^{K} \boldsymbol{k}^{\alpha_1 \cdots \alpha_N} \cdot \boldsymbol{w}_{\alpha_1 \cdots \alpha_N} + \sum_{N=1}^{K} \boldsymbol{m}^{\alpha_1 \cdots \alpha_N} \cdot \frac{\partial \boldsymbol{w}_{\alpha_1 \cdots \alpha_N}}{\partial \xi}.$$

The general development for Cosserat curves $\mathcal{R}_K$ outlined above is contained in a paper by Naghdi (1979, Section 2). When $K = 1$, the results in Subsection 9.3 reduce to those of Subsection 9.2 for a Cosserat curve $\mathcal{R}$.

10. Elastic rods

Within the scope of the theory of a Cosserat curve $\mathcal{R}$ outlined in Section 9, we discuss briefly the constitutive equations for elastic rods in the presence of finite deformation. As in Section 4, we again suppose the existence of a strain energy or stored energy per unit mass $\psi = \psi(\xi, t)$ such that $\rho\dot{\psi}$ is equal to the mechanical power P defined by (9.25), i.e.,

(10.1)
$$P = \lambda \dot{\psi}.$$

In the development of nonlinear constitutive equations for elastic rods, we assume that the strain energy density ψ at each material point of $\mathcal{R}$ and for all t is specified by a response function which depends on $\boldsymbol{r}$, $\boldsymbol{d}_\alpha$ in (9.2) and their partial derivatives with respect to ξ. But since the response function must remain unaltered under superposed rigid body translational displacement, the dependence on $\boldsymbol{r}$ must be excluded. Thus, the constitutive assumption for the strain energy density can be written as

$$\psi = \bar{\bar{\psi}}(\boldsymbol{r}', \boldsymbol{d}_\alpha, \boldsymbol{d}'_\alpha; X), \tag{10.2}$$

where a superposed prime stands for

$$(\ )' \equiv \partial(\ )/\partial\xi, \tag{10.3}$$

and we also make similar constitutive assumptions for $\boldsymbol{n}$, $\boldsymbol{k}^\alpha$, $\boldsymbol{m}^\alpha$ in (9.25). In these constitutive equations, which represent the mechanical response of the medium, the dependence of the response functions on the local geometrical properties of a reference state and material inhomogeneity is indicated through the argument X.

A general development of constitutive theory of elastic rods based on an assumption of the type (10.2) is contained in the paper of Green, Naghdi and Wenner (1974b). In the rest of this section, we limit the discussion to an elastic rod which is homogeneous in its reference configuration and suppose also that the dependence of the response functions on the properties of the reference state occurs through the values of the kinematical variables in the reference state. Then, in place of (10.2), we have

$$\psi = \bar{\psi}(\boldsymbol{r}', \boldsymbol{d}_\alpha, \boldsymbol{d}'_\alpha; \boldsymbol{R}', \boldsymbol{D}, \boldsymbol{D}'_\alpha), \tag{10.4}$$

with similar assumptions for $\boldsymbol{n}$, $\boldsymbol{k}^\alpha$, $\boldsymbol{m}^\alpha$. After substituting (10.4) into (10.1), by usual techniques we obtain the following forms for the constitutive equations:

$$\boldsymbol{n} = \lambda \frac{\partial\bar{\psi}}{\partial \boldsymbol{r}'}, \qquad \boldsymbol{k}^\alpha = \lambda \frac{\partial\bar{\psi}}{\partial \boldsymbol{d}_\alpha}, \qquad \boldsymbol{m}^\alpha = \lambda \frac{\partial\bar{\psi}}{\partial \boldsymbol{d}'_\alpha}, \tag{10.5}$$

along with the restriction

$$\boldsymbol{d}_i \times \left[\frac{\partial\bar{\psi}}{\partial \boldsymbol{d}_i} + (\boldsymbol{d}'_\alpha \cdot \boldsymbol{d}^i)\frac{\partial\bar{\psi}}{\partial \boldsymbol{d}'_\alpha}\right] = \boldsymbol{0}, \tag{10.6}$$

which is obtained from the conservation of moment of momentum and which restricts the response function $\bar{\psi}$.

We do not discuss here the reduced forms of the above constitutive equations resulting from invariance requirements under superposed rigid body motions, but for such reduction refer the reader to Green, Naghdi and Wenner (1974b). Just as with the equations of motion, it is necessary in applications to specific problems to obtain alternative forms of the above constitutive equations or their reduced forms in terms of tensor components.

Such component forms may be expressed with respect to bases $\boldsymbol{a}_i$, or $\boldsymbol{d}_i$, or corresponding bases in the reference configuration. Reduced forms of (10.5) are discussed in Green et al. (1974b, Section 7).

11. Identification of the assigned fields and the inertia coefficients

The local field equations in the mechanical theory of a Cosserat curve $\mathcal{R}$ have the same forms as those that can be derived from the 3-dimensional field equations $(2.9)_{1,2,3}$ by suitable integration over the cross-sectional area of the rod-like body with respect to θ^1 and θ^2 [recall the definition of a rod-like body at the end of Section 2] and in terms of certain definitions for integrated mass density and resultants of stress (for details, see Green, Naghdi and Wenner, 1974a). Similarly, the energy equation (9.24) has the same form as that which can be derived from the energy equation in the 3-dimensional theory by suitable integration over the cross-section area of the rod-like body with respect to θ^1 and θ^2 and in terms of certain definitions for integrated internal energy density and heat flux in the 3-dimensional theory (see Green and Naghdi, 1970). To elaborate further, we confine attention to the purely mechanical theory and recall the definitions

$$\lambda = \rho a_{33}^{1/2} = \int_{\mathcal{A}} \lambda^* d\theta^1 d\theta^2, \qquad \lambda^* = \rho^* g^{1/2}, \tag{11.1}$$

$$\lambda y^\alpha = \int_{\mathcal{A}} \lambda^* \theta^\alpha d\theta^1 d\theta^2, \qquad \lambda y^{\alpha\beta} = \int_{\mathcal{A}} \lambda^* \theta^\alpha \theta^\beta d\theta^1 d\theta^2, \tag{11.2}$$

and the expressions

$$\lambda \boldsymbol{f} = \int_{\mathcal{A}} \lambda^* \boldsymbol{f}^* d\theta^1 d\theta^2 + \int_{\partial\mathcal{A}} [d\theta^2(\boldsymbol{T}^1 - \grave{\lambda}^1 \boldsymbol{T}^3) - d\theta^1(\boldsymbol{T}^2 - \grave{\lambda}^2 \boldsymbol{T}^3)], \tag{11.3}$$

$$\lambda \boldsymbol{l}^\alpha = \int_{\mathcal{A}} \lambda^* \boldsymbol{f}^* \theta^\alpha d\theta^1 d\theta^2 + \int_{\partial\mathcal{A}} \theta^\alpha [d\theta^2(\boldsymbol{T}^1 - \grave{\lambda}^1 \boldsymbol{T}^3) - d\theta^1(\boldsymbol{T}^2 - \grave{\lambda}^2 \boldsymbol{T}^3)], \tag{11.4}$$

where ρ^*, $\boldsymbol{T}^\alpha$, $\boldsymbol{f}^*$ which occur in (11.1)–(11.4) are defined in Section 2 [following (2.9)], the line integrals are taken along the curve $\xi = \text{const.}$ on the material surface (2.26), $\grave{\lambda}^\alpha = \grave{\boldsymbol{\lambda}} \cdot \boldsymbol{g}^\alpha$ and $\grave{\boldsymbol{\lambda}} = \grave{\lambda}^\alpha \boldsymbol{g}_\alpha + \boldsymbol{g}_3$ is a vector tangential to the surface (2.26) so that $\grave{\boldsymbol{\lambda}} \cdot \boldsymbol{\nu}^* = \grave{\lambda}^\alpha \nu^*_\alpha + \nu^*_3 = 0$.

If we now adopt the approximation (2.30), then there is a 1–1 correspondence between the 1-dimensional field equations that follow from the conservation laws of a Cosserat curve and those that can be derived from the 3-dimensional equations provided we identify $\boldsymbol{r}$ and the director $\boldsymbol{d}_\alpha$ in (2.30) with $(9.2)_1$ and $(9.2)_2$, respectively, and adopt the definitions (11.1)–(11.4), as well as the definitions of the resultants mentioned above. A similar 1–1 correspondence can be shown to hold between the 1-dimensional energy equation in the theory of a Cosserat curve and an integrated energy equation derived from the 3-dimensional energy equation.

The various quantities in (9.13) are free to be specified in a manner which depends on the particular application in mind. Also, we remark that in the context of the theory of a Cosserat curve, the inertia coefficients y^{α}, $y^{\alpha\beta}$ and the mass density ρ require constitutive equations. Indeed, $\boldsymbol{f}_c$, $\boldsymbol{l}_c^{\alpha}$ and r_c, as well as $\boldsymbol{f}_b$, $\boldsymbol{l}_b^{\alpha}$ and r_b, can be identified with corresponding expressions in a derivation from the 3-dimensional equations indicated above (for details, see Naghdi, 1979a). Likewise, ρ and the coefficients y^{α}, $y^{\alpha\beta}$ may be identified with easily accessible results from the 3-dimensional theory.

In what follows, we assume that the above identifications have been made and that the quantities ρ, y^{α}, $y^{\alpha\beta}$, $\boldsymbol{f}_b$, $\boldsymbol{l}_b^{\alpha}$ are known or specified. The knowledge of $\boldsymbol{f}_c$, $\boldsymbol{l}_c^{\alpha}$ depends on the nature of the boundary conditions on the lateral surface of the particular rod-like body under consideration; they may be specified as known quantities on the surface (2.26) or they are unknown and must be determined as part of the solution of the problem.

12. Additional remarks on rods

Topics corresponding to those in Sections 6 and 7 have so far received less attention in the case of rods and consequently the discussions that follow are somewhat brief. We first consider a class of constraints, apply the results to an incompressible Cosserat curve and then go an to briefly comment on some recent researches which bear on various aspects of elastic rods.

Consider a class of constraints which are linear relations between the kinematic variables

$$\boldsymbol{v}', \qquad \boldsymbol{w}_{\alpha_1\alpha_2\cdots\alpha_N}, \qquad \boldsymbol{w}_{\alpha_1\alpha_2\cdots\alpha_N} \qquad (N = 1, 2, \cdots, K). \tag{12.1}$$

Similar to the development in Section 6 for shells, we consider $(Q+1)$ constraint equations of the form[†]

$$\begin{aligned} &\boldsymbol{A}^M \cdot \boldsymbol{v}' + \sum_{N=1}^{K} \boldsymbol{B}^{M\alpha_1\alpha_2\cdots\alpha_N} \cdot \boldsymbol{w}_{\alpha_1\alpha_2\cdots\alpha_N} \\ &\quad + \sum_{N=1}^{K} \boldsymbol{C}^{M\alpha_1\alpha_2\cdots\alpha_N} \cdot \boldsymbol{w}'_{\alpha_1\alpha_2\cdots\alpha_N} = \boldsymbol{0} \quad (M = 0, 1, 2, \cdots, Q), \end{aligned} \tag{12.2}$$

where $\boldsymbol{A}^M$, $\boldsymbol{B}^{M\alpha_1\alpha_2\cdots\alpha_N}$, $\boldsymbol{C}^{M\alpha_1\alpha_2\cdots\alpha_N}$ are vector functions of $\boldsymbol{d}_i$, $\boldsymbol{d}'_i$ only and do not depend explicitly on the variables (12.1). We assume that each of the

[†] The development between (12.2)–(12.6) is similar to that for mechanical constraints in the 3-dimensional theory.

functions $\boldsymbol{n}$, $\boldsymbol{k}^{\alpha_1\alpha_2\cdots\alpha_N}$, $\boldsymbol{m}^{\alpha_1\alpha_2\cdots\alpha_N}$ are determined to within an additive constraint response so that

$$\boldsymbol{n} = \bar{\boldsymbol{n}} + \hat{\boldsymbol{n}},$$

(12.3) $$\boldsymbol{k}^{\alpha_1\alpha_2\cdots\alpha_N} = \bar{\boldsymbol{k}}^{\alpha_1\alpha_2\cdots\alpha_N} + \hat{\boldsymbol{k}}^{\alpha_1\alpha_2\cdots\alpha_N}, \qquad \boldsymbol{m}^{\alpha_1\alpha_2\cdots\alpha_N} = \bar{\boldsymbol{m}}^{\alpha_1\alpha_2\cdots\alpha_N} + \hat{\boldsymbol{m}}^{\alpha_1\alpha_2\cdots\alpha_N},$$

where

$$\hat{\boldsymbol{n}}, \qquad \hat{\boldsymbol{k}}^{\alpha_1\alpha_2\cdots\alpha_N}, \qquad \hat{\boldsymbol{m}}^{\alpha_1\alpha_2\cdots\alpha_N}$$

are specified by consitutive equations and

(12.4) $$\bar{\boldsymbol{n}}, \qquad \bar{\boldsymbol{k}}^{\alpha_1\alpha_2\cdots\alpha_N}, \qquad \bar{\boldsymbol{m}}^{\alpha_1\alpha_2\cdots\alpha_N},$$

which represent the response due to constraints are arbitrary functions of ξ, t and are workless. Thus, recalling the expression (9.34), we set

(12.5) $$\bar{\boldsymbol{n}} \cdot \boldsymbol{v}' + \sum_{N=1}^{K} \bar{\boldsymbol{k}}^{\alpha_1\cdots\alpha_N} \cdot \boldsymbol{w}_{\alpha_1\cdots\alpha_N} + \sum_{N=1}^{K} \bar{\boldsymbol{m}}^{\alpha_1\cdots\alpha_N} \cdot \boldsymbol{w}'_{\alpha_1\cdots\alpha_N} = 0$$

for all values of the variables (12.1) subject to the constraint conditions (12.2). It then follows that

$$\bar{\boldsymbol{n}} = -\sum_{M=0}^{Q} \boldsymbol{A}^M p_M, \qquad \bar{\boldsymbol{k}}^{\alpha_1\cdots\alpha_N} = -\sum_{M=1}^{Q} \boldsymbol{B}^{M\alpha_1\cdots\alpha_N} p_M,$$

(12.6) $$\bar{\boldsymbol{m}}^{\alpha_1\cdots\alpha_N} = -\sum_{M=1}^{Q} \boldsymbol{C}^{M\alpha_1\cdots\alpha_N} p_M,$$

where $p_M = p_M(\xi, t)$ are arbitrary functions which play the role of Lagrange multipliers.

We consider now an incompressible Cosserat curve $\mathscr{R}$ with two directors within the scope of the above constrained theory. As in the case of an incompressible shell-like body discussed in Section 6, the conditions representing approximately the (3-dimensional) incompressibility condition (2.10) may be derived with the use of approximation (2.27) for $N = 1$ given by (2.30). Under this approximation (2.30), the base vectors are given by $\boldsymbol{g}_\alpha = \boldsymbol{d}_\alpha$, $\boldsymbol{g}_3 = \boldsymbol{a}_3 + \theta^\alpha \boldsymbol{d}'_\alpha$, where $\boldsymbol{a}_3$ is the tangent vector to the curve $\theta^\alpha = 0$ and a superposed prime is defined by (10.3). Then, from the incompressibility condition $(2.10)_1$ we obtain an approximate expression as a linear function of θ^1, θ^2 in the form

(12.7) $$\frac{d}{dt}[\boldsymbol{d}_1 \boldsymbol{d}_2 \boldsymbol{a}_3] + \theta^\alpha \frac{d}{dt}[\boldsymbol{d}_1 \boldsymbol{d}_2 \boldsymbol{d}'_\alpha] = 0,$$

or equivalently as

$$\boldsymbol{d}^3 \cdot \boldsymbol{v}' + [\boldsymbol{d}^\alpha + \theta^\beta (\boldsymbol{d}'_\beta \cdot \boldsymbol{d}^3)\boldsymbol{d}^\alpha - \theta^\beta (\boldsymbol{d}'_\beta \cdot \boldsymbol{d}^\alpha)\boldsymbol{d}^3] \cdot \boldsymbol{w}_\alpha + \theta^\alpha \boldsymbol{d}^3 \cdot \boldsymbol{w}'_\alpha = 0, \tag{12.8}$$

where in (12.7) and (12.8) use is made of the notation $(9.6)_{1,2}$ and (9.7). We now generate three conditions representing incompressibility: One of these is obtained from integration of (12.8) with respect to θ^1, θ^2 over the cross-section of the rod-like body and the other two are obtained by first multiplying (12.8) by θ^λ $(\lambda = 1,2)$ and then integrating the resulting equation with respect to θ^1, θ^2 over the cross-section of the rod-like body. These three conditions can be written as

$$\gamma^0 \boldsymbol{d}^3 \cdot \boldsymbol{v}' + \{\gamma^0 \boldsymbol{d}^\alpha + \gamma^\beta [(\boldsymbol{d}'_\beta \cdot \boldsymbol{d}^3)\boldsymbol{d}^\alpha - (\boldsymbol{d}'_\beta \cdot \boldsymbol{d}^\alpha)\boldsymbol{d}^3]\} \cdot \boldsymbol{w}_\alpha + \gamma^\alpha \boldsymbol{d}^3 \cdot \boldsymbol{w}'_\alpha = 0, \tag{12.9}$$

$$\gamma^\lambda \boldsymbol{d}^3 \cdot \boldsymbol{v}' + \{\gamma^\lambda \boldsymbol{d}^\alpha + \gamma^{\lambda\beta} [(\boldsymbol{d}'_\beta \cdot \boldsymbol{d}^3)\boldsymbol{d}^\alpha - (\boldsymbol{d}'_\beta \cdot \boldsymbol{d}^\alpha)\boldsymbol{d}^3]\} \cdot \boldsymbol{w}_\alpha + \gamma^{\lambda\alpha} \boldsymbol{d}^3 \cdot \boldsymbol{w}'_\alpha = 0, \tag{12.10}$$

where

$$\gamma^0 = \int_{\mathscr{A}} d\theta^1 d\theta^2, \qquad \gamma^\alpha = \int_{\mathscr{A}} \theta^\alpha d\theta^1 d\theta^2, \qquad \gamma^{\alpha\beta} = \int_{\mathscr{A}} \theta^\alpha \theta^\beta d\theta^1 d\theta^2. \tag{12.11}$$

For the incompressible Cosserat curve under discussion, from (12.2) the constraint conditions are

$$\boldsymbol{A}^M \cdot \boldsymbol{v}' + \boldsymbol{b}^{M\alpha} \cdot \boldsymbol{w}_\alpha + \boldsymbol{C}^{M\alpha} \cdot \boldsymbol{w}'_\alpha = 0 \qquad (M = 0,1,2) \tag{12.12}$$

and the constrained response obtained from (12.6) has the form

$$\begin{aligned} \bar{\boldsymbol{n}} &= -(p_0 \boldsymbol{A}^0 + p_1 \boldsymbol{A}^1 + p_2 \boldsymbol{A}^2), \\ \bar{\boldsymbol{k}}^\alpha &= -p_0 \boldsymbol{B}^{0\alpha} + p_1 \boldsymbol{B}^{1\alpha} + p_2 \boldsymbol{B}^{2\alpha}, \\ \bar{\boldsymbol{m}}^\alpha &= -(p_0 \boldsymbol{C}^{0\alpha} + p_1 \boldsymbol{C}^{1\alpha} + p_2 \boldsymbol{C}^{2\alpha}), \end{aligned} \tag{12.13}$$

where p_0, p_1, p_2 are the Lagrange multipliers. Guided by the three conditions (12.9) and (12.10) for $\lambda = 1,2$, we select the vector-valued functions $\boldsymbol{A}^M$, $\boldsymbol{B}^{M\alpha}$, $\boldsymbol{C}^{M\alpha}$ in (12.12) and (12.13) to have the special values

$$\left.\begin{aligned} \boldsymbol{A}^M &= \text{coeff. of } \boldsymbol{v}' \text{ in (12.9) and (12.10)} \\ \boldsymbol{B}^{M\alpha} &= \text{coeff. of } \boldsymbol{w}_\alpha \text{ in (12.9) and (12.10)} \\ \boldsymbol{C}^{M\alpha} &= \text{coeff. of } \boldsymbol{w}'_\alpha \text{ in (12.9) and (12.10)} \end{aligned}\right\} (M = 0,1,2). \tag{12.14}$$

Then, it follows from (12.13) and (12.4) that the expressions for the constraint response are given by

$$\begin{aligned} \bar{\boldsymbol{n}} &= -P_0 \boldsymbol{d}^3, \\ \bar{\boldsymbol{k}}^\alpha &= -P_0 \boldsymbol{d}^\alpha - P_1^\beta [(\boldsymbol{d}'_\beta \cdot \boldsymbol{d}^3)\boldsymbol{d}^\alpha - (\boldsymbol{d}_\beta \cdot \boldsymbol{d}^\alpha)\boldsymbol{d}^3], \\ \bar{\boldsymbol{m}}^\alpha &= -P_1^\alpha \boldsymbol{d}^3, \end{aligned} \tag{12.15}$$

where the arbitrary coefficients P_0, P_1^α are related to the Lagrange multipliers by

$$P_0 = \gamma^0 p_0 + \gamma^\alpha p_\alpha, \qquad P_1^\alpha = \gamma^\alpha p_0 + \gamma^{\alpha\beta} p_\beta. \tag{12.16}$$

The applicability of the theory of Cosserat curves is not limited to only elastic rods but in fact can be applied also to problems of fluid jets. These developments, which pertain to both inviscid and viscous jets, have been discussed in the papers of Green and Laws (1968), Green (1975, 1976) and Naghdi (1979b).

A constrained theory of a Cosserat curve with two directors is discussed by Green and Laws (1973) and includes as a special case results correponding to those of the Bernoulli–Euler beam theory. The theory of small deformation superposed on a large deformation of an elastic Cosserat curve, together with a discussion of stability problems of rods, is given by Green, Knops and Laws (1968) and some simpler problems in the context of the nonlinear theory of rods are discussed by Ericksen (1970).

The development of the theory of Cosserat curves in Section 9 is carried out within the scope of the purely mechanical theory. In earlier work on the thermo-mechanical theory of rods by direct approach (Green and Naghdi, 1970), only one temperature field was admitted and this allowed for the characterization of the temperature changes along some reference curve such as the central line of a rod in the (3-dimensional) rod-like body, but not for temperature changes in the cross-section of the rod. The latter effect has been incorporated recently by Green and Naghdi (1979b) into the thermo-mechanical theory of Cosserat curves, together with appropriate thermodynamical restrictions arising from the second law of thermodynamics for rods.

13. The basic equations for rods in direct notation

In parallel to the development of Section 8 for shells, for some purposes it is convenient to have available the basic equations of a Cosserat curve in a direct (coordinate-free) notation and this is the main purpose of the present section. Just as in the case of shells, we shall see that the basic equations for rods in coordinate-free notation are very similar to those of the corresponding equations in the 3-dimensional theory and thus may be more suitable in the discussion of general theorems or in the developments which parallel those in the 3-dimensional theory.

We introduce the notations grad and Grad to denote the right spatial and material gradient operators, respectively, with respect to the position on the curve c in the current configuration and on the curve $\mathscr{L}_R$ in the reference configuration. The corresponding divergence operators will be denoted by div and Div, respectively. In particular, for a vector-valued function $\boldsymbol{V}(\xi, t)$ we write[†]

$$\text{(13.1)} \qquad \begin{aligned} \operatorname{grad} \boldsymbol{V} &= \boldsymbol{V}' \otimes \boldsymbol{d}^3, & \operatorname{div} \boldsymbol{V} &= \boldsymbol{V}' \cdot \boldsymbol{d}^3, \\ \operatorname{Grad} \boldsymbol{V} &= \boldsymbol{V}' \otimes \boldsymbol{D}^3, & \operatorname{Div} \boldsymbol{V} &= \boldsymbol{V}' \cdot \boldsymbol{D}^3, \end{aligned}$$

where a prime denotes partial differentiation with respect to ξ and the symbol $\otimes$ denotes tensor product. Also, the spatial curve gradient operator is defined by

$$\text{(13.2)} \qquad \operatorname{grad}_c V = V' \boldsymbol{a}^3, \qquad \operatorname{div}_c \boldsymbol{V} = \boldsymbol{V}' \cdot \boldsymbol{a}^3,$$

for all scalar-valued functions $V(\xi, t)$.

As in Section 8, we introduce a measure of deformation by the tensor $\boldsymbol{F}$, namely

$$\text{(13.3)} \qquad \boldsymbol{F} = \boldsymbol{d}_i \otimes \boldsymbol{D}^i = \operatorname{Grad} \boldsymbol{r} + \boldsymbol{d}_\alpha \otimes \boldsymbol{D}^\alpha,$$

and in view of the notations (9.6) and (9.10) we observe that

$$\text{(13.4)} \qquad \boldsymbol{F}\boldsymbol{D}_3 = \boldsymbol{d}_3 = \boldsymbol{a}_3, \qquad \boldsymbol{F}\boldsymbol{D}_\alpha = \boldsymbol{d}_\alpha .$$

From the definition of the determinant of a second order tensor used in Section 8 [following (8.4)] and the conditions $(9.2)_3$ and $(9.8)_3$, we obtain

$$\text{(13.5)} \qquad \det \boldsymbol{F} = [\boldsymbol{d}_1 \boldsymbol{d}_2 \boldsymbol{d}_3]/[\boldsymbol{D}_1 \boldsymbol{D}_2 \boldsymbol{D}_3] > 0.$$

The tensor $\boldsymbol{F}$, a linear operator on vectors in 3-space, is nonsingular; and there exists, therefore, the inverse deformation gradient $\boldsymbol{F}^{-1}$ defined by

$$\text{(13.6)} \qquad \boldsymbol{F}^{-1} = \boldsymbol{D}_i \otimes \boldsymbol{d}^i.$$

The inverse operator $\boldsymbol{F}^{-1}$ transforms vectors in the present configuration into vectors in the reference configuration, i.e.,

$$\text{(13.7)} \qquad \boldsymbol{F}^{-1} \boldsymbol{d}_i = \boldsymbol{D}_i$$

and it follows that

$$\text{(13.8)} \qquad \boldsymbol{F}^{-1} \boldsymbol{F} = \boldsymbol{F}\boldsymbol{F}^{-1} = \boldsymbol{I} = \boldsymbol{d}_i \otimes \boldsymbol{d}^i = \boldsymbol{D}_i \otimes \boldsymbol{D}^i,$$

where $\boldsymbol{I}$ is the unit tensor in 3-space. We also introduce here the gradient of the directors by

[†] It is clear that the notations grad, Grad, div and Div in this section stand for operators with respect to position on the curve c and need not be confused with the similar notations in Section 8 for surface operators.

$$(13.9)\qquad \boldsymbol{G}_\alpha = \operatorname{Grad} \boldsymbol{d}_\alpha = \boldsymbol{d}'_\alpha \otimes \boldsymbol{D}^3.$$

Recalling the definitions $(9.4)_{1,2}$ for the velocity and the director velocities, as well as (9.5), we have

$$(13.10)\qquad \begin{aligned} \dot{\boldsymbol{F}} &= \dot{\boldsymbol{d}}_i \otimes \boldsymbol{D}^i = \dot{\boldsymbol{d}}_3 \otimes \boldsymbol{D}^3 + \dot{\boldsymbol{d}}_\alpha \otimes \boldsymbol{D}^\alpha = \boldsymbol{v}' \otimes \boldsymbol{D}^3 + \boldsymbol{w}_\alpha \otimes \boldsymbol{D}^\alpha, \\ \dot{\boldsymbol{G}}_\alpha &= \dot{\boldsymbol{d}}'_\alpha \otimes \boldsymbol{D}^3 = \boldsymbol{w}'_\alpha \otimes \boldsymbol{D}^\alpha. \end{aligned}$$

Also

$$(13.11)\qquad \begin{aligned} \dot{\boldsymbol{F}}\boldsymbol{F}^{-1} &= \dot{\boldsymbol{d}}_3 \otimes \boldsymbol{d}^3 + \dot{\boldsymbol{d}}_\alpha \otimes \boldsymbol{d}^\alpha = \operatorname{grad} \boldsymbol{v} + \boldsymbol{w}_\alpha \otimes \boldsymbol{d}^\alpha, \\ \dot{\boldsymbol{G}}_\alpha \boldsymbol{F}^{-1} &= \boldsymbol{w}'_\alpha \otimes \boldsymbol{d}^3 = \operatorname{grad} \boldsymbol{w}_\alpha. \end{aligned}$$

The formulas (13.3)–(13.10) represent the main kinematical results in terms of the gradient tensors $\boldsymbol{F}$, $\boldsymbol{G}_\alpha$ and their rates. We now turn to kinetical quantitites and note that the contact force $\boldsymbol{n}$ and the contact director force $\boldsymbol{m}^\alpha$, as linear functions of $\boldsymbol{d}^3$, can be expressed in the form

$$(13.12)\qquad \boldsymbol{n} = d_{33}^{1/2} \boldsymbol{N} \boldsymbol{d}^3, \qquad \boldsymbol{m}^\alpha = d_{33}^{1/2} \boldsymbol{M}^\alpha \boldsymbol{d}^3,$$

with the second order tensors $\boldsymbol{N}$, $\boldsymbol{M}^\alpha$ defined by

$$(13.13)\qquad \begin{aligned} d_{33}^{1/2} \boldsymbol{N} &= \boldsymbol{n} \otimes \boldsymbol{d}_3 = n^i \boldsymbol{d}_i \otimes \boldsymbol{d}_3, \\ d_{33}^{1/2} \boldsymbol{M}^\alpha &= \boldsymbol{m}^\alpha \otimes \boldsymbol{d}_3 = m^{i\alpha} \boldsymbol{d}_i \otimes \boldsymbol{d}_3, \end{aligned}$$

where

$$(13.14)\qquad n^i = \boldsymbol{n} \cdot \boldsymbol{d}^i, \qquad m^{i\alpha} = \boldsymbol{m}^\alpha \cdot \boldsymbol{d}^i.$$

Also, it is convenient to introduce a tensor $\boldsymbol{K}$ through

$$(13.15)\qquad \begin{aligned} \boldsymbol{k}^\alpha &= d_{33}^{1/2} \boldsymbol{K} \boldsymbol{d}^\alpha, \\ d_{33}^{1/2} \boldsymbol{K} &= \boldsymbol{k}^\alpha \otimes \boldsymbol{d}_\alpha = k^{i\alpha} \boldsymbol{d}_i \otimes \boldsymbol{d}_\alpha, \end{aligned}$$

where

$$(13.16)\qquad k^{i\alpha} = \boldsymbol{k}^\alpha \cdot \boldsymbol{d}^i.$$

Before proceeding further, we recall that the divergence of a second order tensor field $\boldsymbol{T}$ is defined by

$$\boldsymbol{c} \cdot \operatorname{div} \boldsymbol{T} = \operatorname{div}(\boldsymbol{T}^T \boldsymbol{c})$$

for all constant vectors $\boldsymbol{c}$, where superscript T denotes transpose. Applying the above definition to the tensor $\boldsymbol{N}$ in $(13.12)_1$, and recalling $(9.12)_1$, we obtain

$$\begin{aligned} \operatorname{div}_c (\boldsymbol{N}^T \boldsymbol{c}) &= \operatorname{div}_c [(\boldsymbol{n} \cdot \boldsymbol{c}) \boldsymbol{d}_3 / d_{33}^{1/2}] \\ &= [(\boldsymbol{n} \cdot \boldsymbol{c}) \boldsymbol{d}_3 / d_{33}^{1/2}]' \cdot \boldsymbol{a}^3 = \frac{\partial \boldsymbol{n}}{\partial s} \cdot c, \end{aligned}$$

with a similar result for the tensor $\boldsymbol{M}$. Thus, we have

$$\text{div}_c \boldsymbol{N} = \frac{\partial \boldsymbol{n}}{\partial s}, \qquad \text{div}_c \boldsymbol{M}^\alpha = \frac{\partial \boldsymbol{m}^\alpha}{\partial s}. \tag{13.17}$$

With the use of (13.1), the kinematical results (13.8)–(13.9), and (13.17) from the conservation laws (9.18) follows the local field equations

$$\begin{gathered} \dot{\rho} + \rho \, \text{div}_c \boldsymbol{v} = 0, \\ \text{div}_c \boldsymbol{N} + \rho \boldsymbol{f} = \rho (\dot{\boldsymbol{v}} + y^\alpha \dot{\boldsymbol{w}}_\alpha), \\ \text{div}_c \boldsymbol{M}^\alpha + \rho \boldsymbol{l}^\alpha - \boldsymbol{k}^\alpha = \rho (y^\alpha \dot{\boldsymbol{v}} + y^{\alpha\beta} \dot{\boldsymbol{w}}_\beta), \\ [\boldsymbol{N} + \boldsymbol{K} + \boldsymbol{M}^\alpha (\boldsymbol{G}_\alpha \boldsymbol{F}^{-1})^T] = [\boldsymbol{N} + \boldsymbol{K} + \boldsymbol{M}^\alpha (\boldsymbol{G}_\alpha \boldsymbol{F}^{-1})^T]^T. \end{gathered} \tag{13.18}$$

As in the corresponding results in Section 8, the last statement in (13.18) is similar to the symmetry of the stress tensor in the 3-dimensional theory. In particular, it may be observed that $\boldsymbol{a}_3 \times \boldsymbol{n}$, $\boldsymbol{d} \times \boldsymbol{k}$ and $\boldsymbol{d}'_\alpha \times \boldsymbol{m}^\alpha$ are, respectively, the axial vectors of $a_{33}^{1/2}[\boldsymbol{N} - \boldsymbol{N}^T]$, $a_{33}^{1/2}[\boldsymbol{K} - \boldsymbol{K}^T]$ and $a_{33}^{1/2}[\boldsymbol{M}^\alpha (\boldsymbol{G}_\alpha \boldsymbol{F}^{-1})^T - (\boldsymbol{G}_\alpha \boldsymbol{F}^{-1})(\boldsymbol{M}^\alpha)^T]$. Furthermore, in terms of the kinetical quantities $\boldsymbol{N}$, $\boldsymbol{M}^\alpha$, $\boldsymbol{K}$ in (13.12) and $(13.15)_1$ and the rate quantities $(13.10)_{1,2}$, the mechanical power becomes

$$a_{33}^{-1/2} P = \text{tr}\{[\dot{\boldsymbol{F}}^T (\boldsymbol{N} + \boldsymbol{K}) + \dot{\boldsymbol{G}}_\alpha^T \boldsymbol{M}^\alpha](\boldsymbol{F}^{-1})^T\}. \tag{13.19}$$

With reference to constitutive equations for elastic rods, instead of the kinematic variables used in Section 10, we now employ the variables (13.3) and (13.9). Thus, corresponding to the constitutive assumption (10.4), we now write

$$\psi = \tilde{\psi}(\boldsymbol{F}, \boldsymbol{G}_\alpha \,;\, {}_R\boldsymbol{G}_\alpha), \tag{13.20}$$

where

$${}_R\boldsymbol{G}_\alpha = \text{Grad}\, \boldsymbol{D}_\alpha = \boldsymbol{D}'_\alpha \otimes \boldsymbol{D}^3, \tag{13.21}$$

along with similar assumptions for $\boldsymbol{N}$, $\boldsymbol{K}$, $\boldsymbol{M}^\alpha$. Then, with the use of (10.1), (13.19) and (13.20), by usual techniques we obtain the following alternative forms of the constitutive equations:

$$\boldsymbol{N} + \boldsymbol{K} = \rho \frac{\partial \tilde{\psi}}{\partial \boldsymbol{F}} \boldsymbol{F}^T, \qquad \boldsymbol{M}^\alpha = \rho \frac{\partial \tilde{\psi}}{\partial \boldsymbol{G}_\alpha} (\boldsymbol{D}^3 \otimes \boldsymbol{d}_3), \tag{13.22}$$

the first of which can be resolved into

$$\boldsymbol{N} = \rho \frac{\partial \tilde{\psi}}{\partial \boldsymbol{F}} (\boldsymbol{D}^3 \otimes \boldsymbol{d}_3), \qquad \boldsymbol{K} = \rho \frac{\partial \tilde{\psi}}{\partial \boldsymbol{F}} (\boldsymbol{D}^\alpha \otimes \boldsymbol{d}_\alpha). \tag{13.23}$$

ACKNOWLEDGMENT

The results reported here were obtained in the course of research supported by the U.S. Office of Naval Research under Contract N00014–75–C–0148, Project NR 064–436 with the University of California,

REFERENCES

M. M. Balaban, A. E. Green and P. M. Naghdi, *Simple force multiples in the theory of deformable surfaces*, J. Math. Phys. **8**, (1967), 1026–1036.

M. M. Carroll and P. M. Naghdi, *The influence of reference geometry on the response of elastic shells*, Arch. Rational Mech. Analys. **48** (1972), 302–318.

H. Cohen, *A nonlinear theory of elastic directed curves*, Int. J. Engng. Sci. **4** (1966), 511–524.

H. Cohen and C. N. DeSilva, *Nonlinear theory of elastic surfaces*, J. Math. Phys. **7** (1966a), 246–253.

H. Cohen and C. N. DeSilva, *Theory of directed surfaces*, J. Math. Phys. **7** (1966b), 960–966.

H. Cohen and C. N. DeSilva, *On a nonlinear theory of elastic shells*, J. Mécanique **7** (1968), 459–464.

H. Cohen and S. L. Suh, Wave propagation in elastic surfaces, J. Math. and Mech. **19** (1970), 1117–1129.

E. and F. Cosserat, *Theorie des corps deformables*, A. Hermann et Fils, Paris (1909), also *Theory of deformable bodies* [transl. from original 1909 edition], NASA TTF-11, Washington, D.C., 1968, p. 561.

M. J. Crochet and P. M. Naghdi, *Large deformation solutions for an elastic Cosserat surface*, Int. J. Engng. Sci. **7** (1969), 309–335.

P. Duhem, *Le potentiel thermodynamic et la pression hydrostatique*, Ann. Ecole Norm. (3), **10** (1893), 187–230.

J. L. Ericksen and C. Truesdell, *Exact theory of stress and strain in rods and shells*, Arch. Rational Mech. Anal. **1** (1958), 295–323.

J. L. Ericksen, *Simpler static problems in nonlinear theories of rods*, Int. J. Solids Structures **6** (1970), 371–377.

J. L. Ericksen, *Wave propagation in thin elastic shells*, Arch. Rational Mech. Anal. **43** (1971), 167–178.

J. L. Ericksen, *Symmetry transformations for thin elastic shells*, Arch. Rational Mech. Anal. **47** (1972a), 1–14.

J. L. Ericksen, *The simplest problems for elastic Cosserat surfaces*, J. Elasticity **2** (1972b), 101–107.

J. L. Ericksen, *Apparent symmetry of plates of variable height and thickness*, Ist. Lombardo Rend. Sci. A **107** (1973a), 71–82.

J. L. Ericksen, *Apparent symmetry of certain thin elastic shells*, J. Mécanique **12** (1973b), 173–181.

J. L. Ericksen, *Plane infinitesimal waves in homogeneous elastic plates*, J. Elasticity **3** (1973c), 161–167.

J. L. Ericksen, *Plane waves and stability of elastic plates*, Quart. Appl. Math. **32** (1974), 343–345.

J. L. Ericksen, *Theory of Cosserat surfaces and its applications to shells, interfaces and cell membranes*, Proc. International Symp. on Recent Developments in the Theory and Application of Generalized and Oriented Media (P. G. Glockner, N. Epstein and D. J. Malcom, eds.), Calgary, 1979, pp. 27–39.

A. E. Green, *Compressible fluid jets*, Arch. Rational Mech. Anal. **59** (1975), 189–205.

A. E. Green, *On the nonlinear behavior of fluid jets*, Int. J. Engng. Sci. **14** (1976), 49–63.

A. E. Green, R. J. Knops and N. Laws, *Large deformations, superposed small deformations and stability of elastic rods*, Int. J. Solids Structures **4** (1968), 555–557.

A. E. Green and N. Laws, *A general theory of rods*. Proc. Royal Soc. Lond. **A293** (1966), 145–155.

A. E. Green and N. Laws, *Ideal fluid jets*, Int. J. Engng. Sci. **6** (1968), 317–328.

A. E. Green and N. Laws, *Remarks on the theory of rods*, J. Elasticity **3** (1973), 179–184.

A. E. Green, N. Laws and P. M. Naghdi, *Rods, plates and shells*, Proc. Cambridge Phil. Soc. **64** (1968), 895–913.

A. E. Green, N. Laws and P. M. Naghdi, *On the theory of water waves*, Proc. Royal Soc. Lond. **A338** (1974), 43–55.

A. E. Green and P. M. Naghdi, *Non-isothermal theory of rods, plates and shells*, Int. J. Solids and Structures **6** (1970), 209–244.

A. E. Green and P. M. Naghdi, *On superposed small deformations on a large deformation of an elastic Cosserat surface*, J. Elasticity **1** (1971), 1–17.

A. E. Green and P. M. Naghdi, *On the derivation of shell theories by direct approach*, J. Appl. Mech. **41** (1974), 173–176.

A. E. Green and P. M. Naghdi, *Uniqueness and continuous dependence for water waves*, Acta Mechanica **23** (1975), 297–299.

A. E. Green and P. M. Naghdi, *Directed fluid sheets*, Proc. Royal Soc. Lond. **A347** (1976a), 447–473.

A. E. Green and P. M. Naghdi, *A derivation of equations for wave propagation in water of variable depth*, J. Fluid Mech. **78** (1976b), 237–246.

A. E. Green and P. M. Naghdi, *Water waves in a nonhomogeneous incompressible fluid*, J. Appl. Mech. **44** (1977a), 523–528.

A. E. Green and P. M. Naghdi, *A note on thermodynamics of constrained materials*, J. Appl. Mech. **44** (1977b), 787–788.

A. E. Green and P. M. Naghdi, *On thermal effects in the theory of shells*, Proc. Royal. Soc. Lond. **A365** (1979a), 161–190.

A. E. Green and P. M. Naghdi, *On thermal effects in the theory of rods*, Int. J. Solids Structures **15** (1979b), 829–853.

A. E. Green and P. M. Naghdi, *Directed fluid sheets and gravity waves in compressible and incompressible fluids*, Int. J. Engng. Sci. **17** (1979c), 1257–1272.

A. E. Green, P. M. Naghdi and J. A. Trapp, *Thermodynamics of a continuum with internal constraints*, Int. J. Engng. Sci. **8** (1970), 891–908.

A. E. Green, P. M. Naghdi and W. L. Wainwright, *A general theory of a Cosserat surface*, Arch. Rational Mech. Anal. **20** (1965), 287–308.

A. E. Green, P. M. Naghdi and M. L. Wenner, *On the theory of rods. I. Derivations from the three-dimensional equations*, Proc. Royal Soc. Lond. **A337** (1974a), 451–483.

A. E. Green, P. M. Naghdi and M. L. Wenner, *On the theory of rods. II. Developments by direct approach*, Proc. Royal Soc. Lond. **A337** (1974b), 485–507.

P. M. Naghdi, *The theory of shells and plates*, in *S. Flugge's Handbuch der Physik*, Vol. VIa/2 (C. Truesdell, ed.), Springer-Verlag, Berlin, 1972, pp. 425–640.

P. M. Naghdi, *Direct formation of some two-dimensional theories of mechanics*, Proc. 7th U.S. National Congr. Appl. Mech., Amer. Soc. Mechanical Engineers, New York, N.Y., 1974, pp. 3–21.

P. M. Naghdi, *On the formulation of contact problems of shells and plates*, J. Elasticity **5** (1975a), 379–398.

P. M. Naghdi, *A note on finite torsion and expansion of a cylindrical Cosserat surface* (in Russian). *Mechanics of deformable bodies and structures* (volume honoring Iu. N. Rabotnov), Moscow, U.S.S.R., 1975b, pp. 318–326.

P. M. Naghdi, *Shell theory from the standpoint of finite elasticity*, Proc. Symp. on "Finite Elasticity" (R. S. Rivlin, ed.), AMD-Vol. 27, Amer. Soc. Mechanical Engineers, 1977, pp. 77–89.

P. M. Naghdi, *Fluid jets and fluid sheets: A direct formulation*, Proc. 12th Symp. on Naval Hydrodynamics, National Academy of Sciences, Wash., DC., 1979a, pp. 500–515.

P. M. Naghdi, *On the applicability of directed fluid jets to Newtonian and non-Newtonian flows*, J. Non-Newtonian Fluid Mech. **5** (1979b) 233–265.

P. M. Naghdi and R. P. Nordgren, *On the nonlinear theory of elastic shells under the Kirchhoff hypothesis*, Quart. Appl. Math. **21** (1963), 49–59.

P. M. Naghdi and P. Y. Tang, *Large deformation possible in every isotropic elastic membrane*, Phil. Trans. Royal Soc. Lond. **A207** (1977), 145–187.

P. M. Naghdi and J. A. Trapp, *A uniqueness theorem in the theory of Cosserat surface*. J. Elasticity **2** (1972), 9–20.

C. Truesdell and W. Noll, *The nonlinear field theories of mechanics*, in *S. Flugge's Handbuch der Physik*, Vol. III/3, Springer-Verlag, Berlin, 1965, pp. 1–602.

SOME THOUGHTS ON MATERIAL STABILITY

R. S. RIVLIN

Lehigh University, Bethlehem, PA, USA

(Received November 4, 1980)

ABSTRACT

The paper is concerned with some underlying physical considerations which bear on the development of restrictions on the strain-energy function for an elastic material which may undergo finite deformations. Some comments are also made on the possible use of the strain-energy function at deformations for which a necessary condition for material stability is violated.

1. Introduction

The main difficulty which arises in discussing material stability is any clear concept in physical terms of what the term implies. We have only the vague notion that some restrictions should be placed on constitutive equations, other than those which result from invariance under superposed rigid motions and material symmetry, to ensure that the material modelled is well-behaved physically. However, the precise sense in which the material is to be well-behaved is not entirely clear.

Nevertheless, we can recognize certain physical behavior which is clearly unacceptable, i.e. certain situations can be recognized as evidencing instability. This fact enables us to discuss in a meaningful way necessary conditions for material stability even though it may be fruitless to attempt to obtain sufficient conditions.

In classical elasticity theory certain restrictions can be placed on the elastic moduli, the violation of which implies material instability. For an isotropic material, these are the conditions that the shear and compression moduli both be positive. They result from the fundamental consideration that if either of the moduli is negative, the material will undergo a deformation from its assumed initial homogeneous state even though no forces are applied to it.

Proceedings of the IUTAM Symposium on Finite Elasticity, Lehigh University, August 10–15, 1980. General lecture.

Carlson, D.E. and Shield, R.T. (eds.)
Proceedings of the IUTAM Symposium on Finite Elasticity
ISBN 90 247 2629 8

However, even in this case, it is by no means evident that the assumption of positive shear and compression moduli are sufficient to ensure stability. That this is the case is evident from the discussion in Sections 5 and 6 of the role of thermal and other fluctuations in effecting instability.

The search for necessary conditions for material stability of isotropic elastic materials subjected to finite deformations has been greatly influenced by the corresponding problem in classical elasticity theory. However, physical points of departure, which in the classical infinitesimal theory lead to the same conditions on the elastic moduli, when transposed into the context of finite elasticity theory may lead to quite different restrictions on the strain-energy function. We shall not, in this paper, discuss the various rather complicated restrictions on the strain-energy function for isotropic elastic materials which have been obtained. Rather, we will concentrate on some rather general considerations of a physical character which provide the basis for such relations. We shall also make some comments on the possible use of a strain-energy function at deformations for which one or other of the necessary conditions for material stability is violated. It is emphasized that these comments are advanced tentatively and are intended only to draw attention to certain questions which arise when a strain-energy function is used at such deformations.

2. Infinitesimal deformations

If an isotropic elastic material is subjected to infinitesimal deformations, the strain-energy W per unit volume is given by

$$W = \tfrac{1}{2}\lambda\,(\operatorname{tr}\boldsymbol{e})^2 + \mu\,\operatorname{tr}\boldsymbol{e}^2, \tag{2.1}$$

where $\boldsymbol{e}$ is the infinitesimal strain matrix and λ and μ are the Lamé constants for the material. $\boldsymbol{e}$ is defined in terms of the displacement vector $\boldsymbol{u}$ by

$$\boldsymbol{e} = \tfrac{1}{2}\{\nabla\boldsymbol{u} + (\nabla\boldsymbol{u})^{\dagger}\}, \tag{2.2}$$

where the dagger (†) denotes the transpose.

A possible definition of material stability, in this case, is that W is positive definite. That this is a necessary condition is evident from the fact that if W is negative for some $\boldsymbol{e}$, the material will be unstable in its undeformed state. The condition that W be positive definite is easily seen to be equivalent to the conditions

$$\mu > 0, \qquad 2\mu + 3\lambda > 0. \tag{2.3}$$

In physical terms, the conditions (2.3) state that the shear and compression moduli must be positive. Together they imply that

(2.4) $$\lambda + \mu > 0,$$

i.e. the equibiaxial plane strain modulus is positive. The conditions (2.3) and (2.4) together imply that Young's modulus E is positive, i.e.

(2.5) $$E = \frac{\mu(2\mu + 3\lambda)}{\lambda + \mu} > 0.$$

However, (2.4) and (2.5) do not imply (2.3) and hence do not provide sufficient conditions for W to be positive definite. Nor can one of the conditions (2.4) and (2.5) be taken with one of the conditions (2.3) to imply that W is positive definite.

A necessary condition for material stability which is sometimes considered is the Hadamard condition that the speeds of all plane waves, propagated in a body of the material filling three-dimensional space, be positive. The necessary and sufficient conditions for this to be the case are

(2.6) $$\mu > 0, \qquad \lambda + 2\mu > 0.$$

These conditions are equivalent to the condition that the acoustic tensor be strongly elliptic. The second of the conditions (2.6) also has the statical interpretation that the tensile modulus, when the dimensions normal to the direction of stretch are fixed, be positive.

While the conditions (2.3) imply the conditions (2.6), the converse is not true. Accordingly, the Hadamard conditions are weaker than the condition that W be positive definite.

If the deformations are infinitesimal, we do not distinguish between the Cauchy stress $\boldsymbol{\sigma}$ and the Piola–Kirchhoff stress $\mathbf{\Pi}$. They are given by

(2.7) $$\boldsymbol{\sigma} = \mathbf{\Pi} = \partial W / \partial \boldsymbol{e}.$$

It can easily be seen that the condition that W be positive definite is equivalent to the condition that it be a convex function of $\boldsymbol{e}$. The convexity of W can be expressed algebraically as

(2.8) $$W(\boldsymbol{e}_2) - W(\boldsymbol{e}_1) > \operatorname{tr}\{(\boldsymbol{e}_2 - \boldsymbol{e}_1)\mathbf{\Pi}_1\},$$

where $\boldsymbol{e}_1$ and $\boldsymbol{e}_2$ are two strains and $\mathbf{\Pi}_1$ and $\mathbf{\Pi}_2$ are the corresponding stresses. The condition (2.8) is in turn equivalent to the condition

(2.9) $$\operatorname{tr}\{(\mathbf{\Pi}_2 - \mathbf{\Pi}_1)(\boldsymbol{e}_2 - \boldsymbol{e}_1)\} > 0.$$

If the elastic material is anisotropic, it can still be shown quite easily that positive definiteness of W is equivalent to convexity expressed by either of the conditions (2.8) or (2.9).

3. An attempted generalization

It is evident that if we remove the condition that the deformations be infinitesimal, positive semi-definiteness of W, regarded as a function of the deformation gradients, remains a necessary condition for material stability. It is then tempting to try to find equivalent convexity conditions with which to replace the convexity conditions (2.8) and (2.9) which are valid for the case of infinitesimal deformations. Such a generalization was attempted by Coleman and Noll [1] in 1959.

They considered a deformation in which a particle initially in vector position $\boldsymbol{X}$ moves to $\boldsymbol{x}$. Let X_A $(A = 1,2,3)$ and x_i $(i = 1,2,3)$ be the components of $\boldsymbol{X}$ and $\boldsymbol{x}$ respectively in a rectangular cartesian coordinate system. Then the deformation gradient matrix $\boldsymbol{g}$ is defined by

$$\boldsymbol{g} = \| g_{iA} \| = \| \partial x_i / \partial X_A \|. \tag{3.1}$$

The strain-energy W, per unit initial volume, is then a scalar function of $\mathbf{g}$. Coleman and Noll [1] suggested that the convexity condition (2.8) be replaced by the condition

$$W(\boldsymbol{g}_2) - W(\boldsymbol{g}_1) > \operatorname{tr}\{(\boldsymbol{g}_2 - \boldsymbol{g}_1)\boldsymbol{\Pi}_1\}, \tag{3.2}$$

where $\boldsymbol{g}_1$ and $\boldsymbol{g}_2$ are the deformation gradient matrices corresponding to any two deformations related by

$$\boldsymbol{g}_2 = \boldsymbol{s}\boldsymbol{g}_1, \tag{3.3}$$

where $\boldsymbol{s}$ is a symmetric matrix, and $\boldsymbol{\Pi}_1$ and $\boldsymbol{\Pi}_2$ are the corresponding Piola–Kirchhoff stress matrices.† The relation (3.3) implies that the deformation corresponding to $\boldsymbol{g}_2$ may be obtained from that corresponding to $\boldsymbol{g}_1$ by superposing on the latter a pure homogeneous deformation. The condition (3.2), with (3.3), has been called by Truesdell and Toupin [2] the C–N condition, and they showed that it is equivalent to the condition

$$\operatorname{tr}\{(\boldsymbol{g}_2 - \boldsymbol{g}_1)(\boldsymbol{\Pi}_2 - \boldsymbol{\Pi}_1)\} > 0, \tag{3.4}$$

which they called the GCN condition.

That the C–N and GCN conditions are not necessary conditions for material stability is evident from a much earlier result obtained by Rivlin [3] in 1948. He considered the pure homogeneous deformation of a unit cube of incompressible neo-Hookean material by specified forces acting normally to its faces and uniformly distributed over them. If the strain-energy W is given in terms of the principal extension ratios λ_1, λ_2, λ_3 by

† $\boldsymbol{\Pi}_1 = \|\Pi^{(1)}_{Ai}\|$, $\boldsymbol{\Pi}_2 = \|\Pi^{(2)}_{Ai}\|$.

$$W = \tfrac{1}{2}C(\lambda_1^2 + \lambda_2^2 + \lambda_3^2 - 3), \tag{3.5}$$

where C is a positive constant, then the applied forces f_1, f_2, f_3 are given by

$$f_i = C\lambda_i - p/\lambda_i \qquad (i = 1, 2, 3), \tag{3.6}$$

where p is a hydrostatic pressure which is arbitrary if the λ's are specified. Since the material is incompressible, the λ's must satisfy the relation

$$\lambda_1\lambda_2\lambda_3 = 1. \tag{3.7}$$

If the f's are specified, then equations (3.6) and (3.7) provide four simultaneous equations for the determination of the λ's and p. These do not necessarily have a unique solution and consequently if the applied forces are specified the resultant equilibrium states of pure homogeneous deformation may not be uniquely determined. Certain of these equilibrium states may be unstable and which of the stable states is attained will, in general, depend on the manner in which the applied forces are increased from zero to their final values.

For example [4], if the f's are all equal and tensile, i.e. $f_1 = f_2 = f_3 = f$, say, with $f > 0$, there are seven possible pure homogeneous equilibrium states:

(i) $\lambda_1 = \lambda_2 = \lambda_3 = 1$,
(ii) $\lambda_1 = \lambda_2$, $\lambda_3 < \frac{1}{3}f/C$,
(iii) $\lambda_1 = \lambda_2$, $f/C > \lambda_3 > \frac{1}{3}f/C$,

and states obtained from (ii) and (iii) by cyclic permutation of the subscripts on the λ's.

The state (ii) and the two further states obtained from it are stable, while the state (iii) and the two further states obtained from it are unstable. The undeformed state (i) is stable or unstable accordingly as $f/C < 2$ or > 2.

It was shown by Truesdell and Toupin [2] that if the C–N or GCN conditions are satisfied, the relation between the forces and principal extension ratios for pure homogeneous deformation of a cube of the material are uniquely invertible. Accordingly, these conditions are not satisfied by an incompressible neo-Hookean material to which three equal tensile forces are applied, even if these are small. Since materials exist for which the neo-Hookean strain-energy function is fairly accurately valid and which show no evidence of material instability, we conclude that the C–N and GCN conditions are not necessary conditions for material stability. A heroic attempt was made by Truesdell and Noll [5] to salvage these conditions as necessary for material stability. They asserted that the C–N and GCN conditions should be applied only to compressible materials. However, they failed to explain how the unique invertibility of the force-principal extension ratio relations for a slightly compressible material can, in the limiting case when the material is incompressible, yield multiple widely-separated equilibrium states of pure homogeneous deformation for specified forces.

The condition that the strain-energy function, regarded as a function of $\mathbf{g}$, be globally convex may be expressed as

$$W(\mathbf{g}_2) - W(\mathbf{g}_1) > \operatorname{tr}\{(\mathbf{g}_2 - \mathbf{g}_1)\Pi_1\}, \tag{3.8}$$

where $\mathbf{g}_2$ and $\mathbf{g}_1$ are the deformation gradient matrices corresponding to any two states of deformation, not necessarily connected by the relation (3.3).

Since the C–N and GCN conditions are less severe restrictions on the strain-energy function than global convexity, the latter is not a necessary condition for material stability. Again, since violation of global convexity implies violation of local convexity at some point, the latter is not a necessary condition for material stability.

Local convexity at a point may be expressed by the relation

$$\delta^2 W > 0, \tag{3.9}$$

where $\delta^2 W$ denotes the second variation of W at that point. This may, in turn, be written as

$$\frac{\partial^2 W}{\partial x_{i,A} \partial x_{j,B}} \delta x_{i,A} \delta x_{j,B} > 0, \tag{3.10}$$

for all arbitrary variations $\delta x_{i,A}$ of the deformation gradients. The relation (3.10) may be rewritten as

$$\delta \Pi_{Ai} \delta x_{i,A} > 0, \tag{3.11}$$

where $\delta \Pi_{Ai}$ is the variation in the Piola–Kirchhoff stress corresponding to the variation $\delta x_{i,A}$ in the deformation gradients. In matrix notation (3.11) becomes

$$\operatorname{tr}(\delta \mathbf{\Pi} \delta \mathbf{g}) > 0. \tag{3.12}$$

Although the local convexity condition (3.11) is not a necessary condition for material stability, it becomes such if certain restrictions are placed on the variations $\delta x_{i,A}$, $\delta \Pi_{Ai}$ and on the underlying Piola–Kirchhoff stress Π_{Ai}. Illustrations of such cases are given in the next section. In each of them the total work in the incremental deformation is done by only one of the components of $\mathbf{\Pi}$.

We denote this component of $\mathbf{\Pi}$ by f and the corresponding deformation gradient by ε. Then, the condition (3.11) becomes

$$\delta f \delta \varepsilon > 0; \tag{3.13}$$

i.e. the $f-\varepsilon$ relation must have positive slope. This is evidently a necessary condition for material stability. The implications of its violation will be discussed in Section 5. The condition (3.13) is, of course, equivalent to the condition that the incremental modulus μ, defined by

$$\mu = \partial f / \partial \varepsilon, \tag{3.14}$$

be positive.

4. Some restricted convexity conditions

In this section we describe certain situations in which the local convexity condition (3.11) reduces to the form (3.13) and accordingly provides a necessary condition for material stability. In each of these situations a rectangular block of elastic material, not necessarily isotropic, with its edges parallel to the axes of a rectangular cartesian coordinate system $\bar{x}$, is assumed to undergo a pure homogeneous deformation with principal extension ratios λ_1, λ_2, λ_3 and principal directions parallel to the axes of the system $\bar{x}$.

We consider a rectangular block of the material to be cut from the parent block, with edges parallel to the axes of a rectangular cartesian coordinate system x which may or may not coincide with the system $\bar{x}$. This is held in its deformed state by appropriate forces. We now superpose on the deformation existing in this block an infinitesimal static deformation. For our purposes this is most conveniently described in the coordinate system x.

Case 1. The system x has arbitrary orientation with respect to the system $\bar{x}$ and the infinitesimal superposed deformation consists of a simple shear with the x_1-direction as the direction of shear and the x_1x_2-plane as the plane of shear. The incremental shear modulus must then be positive for material stability.

Case 2. The system x has an arbitrary orientation with respect to the system $\bar{x}$ and the infinitesimal superposed deformation is a pure homogeneous deformation with one principal direction parallel to the x_1-, say, axis, while the dimensions of the block parallel to the x_2- and x_3- axes are held fixed. We shall call this *constrained simple extension.* The incremental modulus in the 1-direction must then be positive for material stability. It is evident, however, that this condition becomes meaningless when the material is incompressible.

Case 3. The system x has one axis, say x_2, parallel to one of the axes, say $\bar{x}_2$, of the system $\bar{x}$, but is otherwise of arbitrary orientation. In the underlying pure homogeneous deformation, the faces of the block normal to $\bar{x}_2$ are force-free and remain so in the infinitesimal superposed deformation which is pure homogeneous and has principal directions parallel to the axes of the system x and zero extension in the x_3, say, direction. The incremental tensile modulus must then be positive for material stability.

Case 4. The system x has all of its axes parallel to those of the system $\bar{x}$, while both the initial finite and superposed infinitesimal deformations are simple extensions parallel to one of the axes, say x_1. The incremental tensile modulus must then be positive for material stability.

The restrictions which these conditions impose on the strain-energy function for an isotropic elastic material have been and remain the subject of extensive investigation. The results which have already been obtained are often complicated and in some cases surprisingly difficult to achieve. We shall not pursue this rather intricate matter here.

5. Instability for one-dimensional deformations

In this section we shall discuss simplistically the extension e by a tensile force f of a thin uniform weightless rod of elastic material with initial length L. We assume that f increases monotonically with e for values of e below some value e_m at which f has the maximum value f_m. For higher values of e, f is assumed to decrease monotonically with increase in e. The f–e relation is shown schematically in Figure 1a. The portion of the f–e curve on which the modulus $\mu = df/de$ is positive is denoted I. That on which it is negative is denoted II.

The energy W stored elastically in the rod is given by

$$W = \int_0^e f(e)de, \tag{5.1}$$

and

$$f(e) = \frac{dW(e)}{de}. \tag{5.2}$$

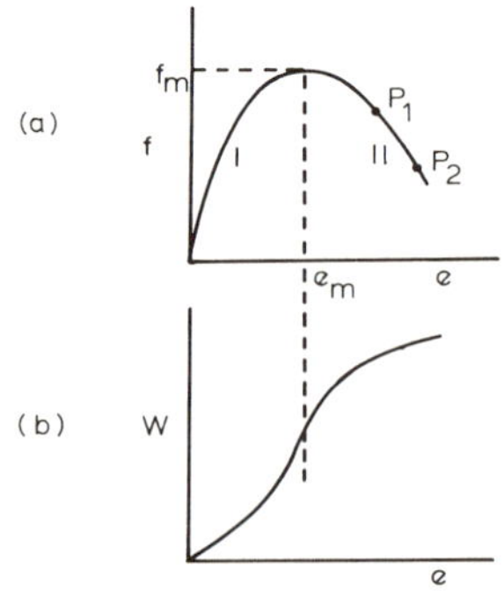

Figure 1.

The dependence of W on e is shown schematically in Figure 1b and we note that W increases monotonically with e, but has a point of inflexion when $e = e_m$.

Suppose the rod is held, with its ends fixed, in the homogeneous state of deformation corresponding to the point P_1 on the portion II of the f–e curve. Then, $f, e = f_1, e_1$, say. Now, suppose the applied force is reduced to f_2. If the extension increases to e_2, as we might expect from Figure 1a, so that the point P_2 is reached, the work done is $(e_2 - e_1)f_2$, while the increase in strain-energy is $\int_{e_1}^{e_2} f(e)de$. Since

$$(e_2 - e_1)f_2 < \int_{e_1}^{e_2} f(e)de, \tag{5.3}$$

the deformation cannot take place. The question arises — what will, in fact, occur? We shall defer discussion of this until later in this section.

Again, suppose that the rod is held, with its ends fixed, in the homogeneous state of deformation corresponding to the point P_1 in Figure 1a and now the constraint on one end of the rod is removed so that a dead-load f_1 acts on the rod. If the extension of the rod increases to e_2, the work done by the dead-load is $(e_2 - e_1)f_2$ and the strain-energy increases to $\int_{e_1}^{e_2} f(e)de$. Since

$$(e_2 - e_1)f_1 > \int_{e_1}^{e_2} f(e)de, \tag{5.4}$$

the deformation can take place and when the extension e_2 is attained the load will have a kinetic energy of amount

$$(e_2 - e_1)f_1 - \int_{e_1}^{e_2} f(e)de. \tag{5.5}$$

If a homogeneous equilibrium state can be attained by the rod, the f–e curve must reverse slope, either as shown in Figure 2a or in Figure 2b. We denote by III the rising portion of the curve beyond the extension at which the minimum value of f occurs.

We now return to the situation in which the rod is held with its ends fixed a distance $L + e$ apart. We shall, however, not assume that the rod is

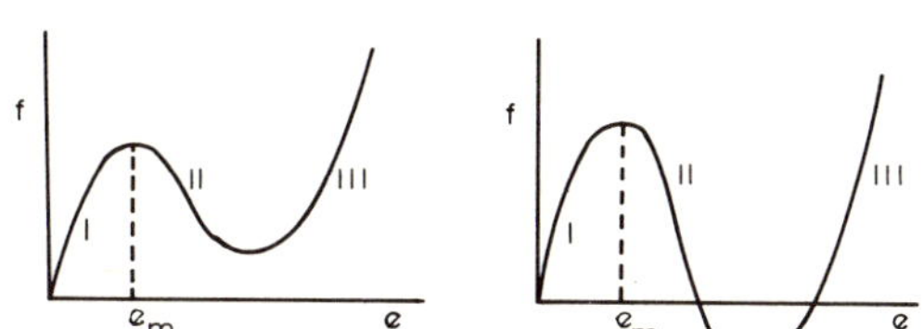

(a) (b)

Figure 2.

homogeneously deformed. Let l be the distance, in the undeformed state, of a generic particle of the rod from one end. Let $\varepsilon(l)$ be the strain at this particle in the deformed state and let $w(\varepsilon)$ be the strain-energy per unit initial length. Then, the total strain-energy W in the rod is given by

$$W = \int_0^L w(\varepsilon)dl. \tag{5.6}$$

The kinematic constraint on the rod is expressed by

$$\int_0^L \varepsilon(l)dl = e. \tag{5.7}$$

For equilibrium, the first variation δW of W must be zero, for all variations $\delta\varepsilon(l)$ of $\varepsilon(l)$ which satisfy the constraint condition (5.7) and hence the condition

$$\int_0^L \delta\varepsilon(l)dl = 0. \tag{5.8}$$

With (5.6), we obtain

$$\delta W = \int_0^L \frac{dw(\varepsilon)}{d\varepsilon}\,\delta\varepsilon(l)dl = 0. \tag{5.9}$$

Taking account of the constraint (5.8) by introducing the (constant) Lagrange multiplier $-f$, we obtain from (5.9)

$$f = dw(\varepsilon)/d\varepsilon. \tag{5.10}$$

f is the tension in the rod. Thus, the equilibrium state of the rod is characterized by the condition that the tension is constant throughout the rod.

From Figure 2 we see that the following possibilities exist:

(i) The rod is homogeneously deformed.

(ii) Part of the rod is in one state and the remainder in another state; these are states corresponding to the constant value of f which lie on two of the three segments I, II, III.

(iii) Parts of the rod are in each of three states. These are the three states corresponding to the constant value of f which lie on the segments I, II, III of the f–e curve.

We now consider the stability of these possible equilibrium states.

An equilibrium state is unstable if the second variation $\delta^2 W$ of W is negative for some kinematically possible variation $\delta\varepsilon(l)$ of $\varepsilon(l)$. From (5.6) and (5.10), we have

$$\delta^2 W = \frac{1}{2}\int_0^L \frac{d^2 w(\varepsilon)}{d\varepsilon^2}(\delta\varepsilon)^2 dl = \frac{1}{2}\int_0^L \frac{df(\varepsilon)}{d\varepsilon}(\delta\varepsilon)^2 dl. \tag{5.11}$$

In case (i) when the rod is homogeneously deformed, $df(\varepsilon)/d\varepsilon$ and hence $\delta^2 w$ is negative if the extension of the rod corresponds to a point on the segment II of the f–e curve. The corresponding equilibrium state is, accordingly, unstable. We observe that this instability certainly obtains as soon as the extension e_m is passed and we shall see later that it may arise earlier.

We now pass to case (ii). We consider an equilibrium state in which a part $\mathscr{L}_1$ of the rod, of total initial length L_1, is subjected to a uniform strain ε_1, while the strain in the remainder $\mathscr{L}_2$, of total initial length L_2, is also uniform but of magnitude ε_2. Let $w(\varepsilon)$ be the strain-energy per unit initial length when the strain is ε. Then, the total strain-energy W in the rod is given by

$$W = L_1 w(\varepsilon_1) + L_2 w(\varepsilon_2). \tag{5.12}$$

We have also

$$L = L_1 + L_2, \qquad e = L_1\varepsilon_1 + L_2\varepsilon_2. \tag{5.13}$$

We obtain from (5.12)

$$\delta W = w(\varepsilon_1)\delta L_1 + w(\varepsilon_2)\delta L_2 + L_1 \frac{dw(\varepsilon_1)}{d\varepsilon_1}\delta\varepsilon_1 + L_2 \frac{dw(\varepsilon_2)}{d\varepsilon_2}\delta\varepsilon_2. \tag{5.14}$$

From (5.13) we obtain the kinematic constraints on δL_1, δL_2, $\delta\varepsilon_1$, $\delta\varepsilon_2$:

$$\delta L_1 + \delta L_2 = 0, \qquad L_1\delta\varepsilon_1 + L_2\delta\varepsilon_2 + \varepsilon_1\delta L_1 + \varepsilon_2\delta L_2 = 0. \tag{5.15}$$

The equilibrium condition $\delta W = 0$ for all kinematically possible variations $\delta L_1, \cdots, \delta\varepsilon_2$ yields

$$\frac{dw(\varepsilon_1)}{d\varepsilon_1} = \frac{dw(\varepsilon_2)}{d\varepsilon_2} = f, \qquad \text{say,} \tag{5.16}$$

and

$$w(\varepsilon_2) - w(\varepsilon_1) = f(\varepsilon_2 - \varepsilon_1). \tag{5.17}$$

The relation (5.16) expresses the fact that at equilibrium the tensions in the parts $\mathscr{L}_1$ and $\mathscr{L}_2$ of the rod are equal. The relation (5.17) is the well-known Maxwell relation. If ε_1 and ε_2 correspond to points B, C lying on the segments I and II of the f–ε curve, as shown in Figure 3, the relation (5.17) cannot be

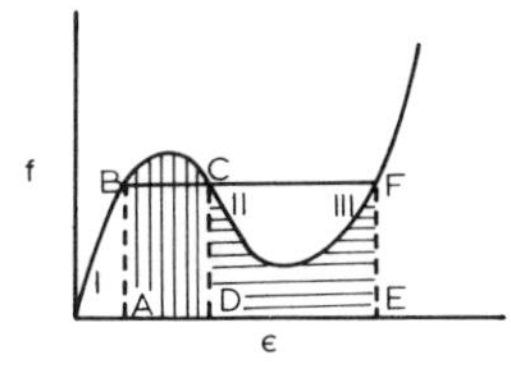

Figure 3.

satisfied for any value of f, since $w(\varepsilon_2) - w(\varepsilon_1)$ is the area of the vertically hatched region in Figure 3, while $f(\varepsilon_2 - \varepsilon_1)$ is the necessarily smaller area of rectangle ABCD.

Similarly, if ε_1 and ε_2 correspond to points C, F lying on the segments II and III of the f–ε curve, the relation (5.17) cannot be satisfied for any value of f, since $w(\varepsilon_2) - w(\varepsilon_1)$ is the area of the horizontally hatched region in Figure 3 and $f(\varepsilon_2 - \varepsilon_1)$ is the necessarily greater area of the rectangle DCFE.

However, if ε_1 and ε_2 correspond to the points B, F lying on the segments I and III of the f–ε curve, as shown in Figure 4, the relation (5.17) can be satisfied by choosing f so that the striated areas above and below the line f = constant are equal. This line f = constant is sometimes called the Maxwell line. We emphasize that the critical value of f and hence those of ε_1 and ε_2 are independent of the detailed shape of segment II of the f–ε curve and depend only on the area under it. It can be shown that the equilibrium state determined in the above manner is stable.

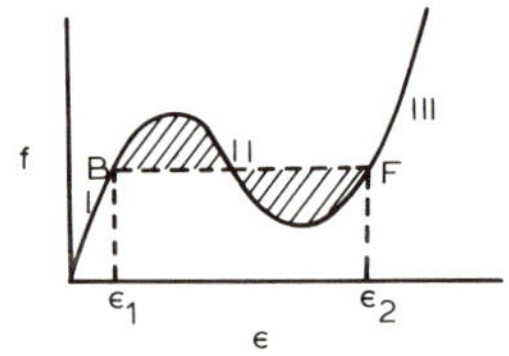

Figure 4.

Since the value of f corresponding to the Maxwell line determines ε_1 and ε_2, if e is specified L_1 and L_2 can be determined from equations (5.13).

This analysis is, of course, dependent on the assumption that an element of the rod can pass freely between the states of strain ε_1 and ε_2. For this to be the case, thermal or other fluctuations must be present of sufficient magnitude to allow the tension in the element to reach the value of f corresponding to the maximum of the f–ε curve, at any rate instantaneously.

It will be recognized that the model outlined above is in accord with that presented in texts on thermodynamics for isothermal reversible first-order phase transitions.

Now suppose that fluctuations of sufficient magnitude, for the processes outlined above to take place, are not present. Then the overall extension of the rod, at which transitions in a part of it to strain values corresponding to points on segment III take place, is greater than that determined by the Maxwell line. In the limiting case when there are no fluctuations, it will reach the value $L\varepsilon_m$. Then, as the overall extension $L\varepsilon_m$ is exceeded the transition takes place dynamically and the calculation of the manner in which it occurs would be rather complicated and, indeed, there is some doubt that it could be carried out purely on the basis of the $f-\varepsilon$ curve of the type considered so far. The reason for this doubt will appear more clearly in the next section, in which is discussed a specific physical situation in which force-induced transitions from one twinning mode to another take place in quartz.

If the force f is applied as a dead load, as soon as it reaches the value f_m, a transition will take place in the whole rod to the point on the segment III of the $f-e$ curve corresponding to this value of f. The kinetic energy of the load will then be

$$f_m L(\varepsilon_2-\varepsilon_1)-[w(\varepsilon_2)-w(\varepsilon_1)].$$

The rod will accordingly oscillate about this state and if the material has any internal friction will eventually reach equilibrium at the point on III corresponding to the load f_m.

We will now show that the situation (iii) in which parts $\mathcal{L}_1$, $\mathcal{L}_2$, $\mathcal{L}_3$ of the rod are in three different states corresponding to points on segments I, II, III respectively of the $f-\varepsilon$ curve does not represent a possible equilibrium condition for the rod.

Let L_1, L_2, L_3 be the total initial lengths of $\mathcal{L}_1$, $\mathcal{L}_2$, $\mathcal{L}_3$ respectively and let ε_1, ε_2, ε_3 be the strains in them in the deformed state in which the total extension of the rod is e. Then,

$$L_1+L_2+L_3=L, \qquad L_1\varepsilon_1+L_2\varepsilon_2+L_3\varepsilon_3=e. \tag{5.18}$$

The total strain-energy W is given by

$$W=L_1w(\varepsilon_1)+L_2w(\varepsilon_2)+L_3w(\varepsilon_3). \tag{5.19}$$

The equilibrium condition $\delta W=0$ for all kinematically possible variations $\delta L_1,\cdots,\delta\varepsilon_3$ yields

$$\frac{dw(\varepsilon_1)}{d\varepsilon_1}=\frac{dw(\varepsilon_2)}{d\varepsilon_2}=\frac{dw(\varepsilon_3)}{d\varepsilon_3}=f, \qquad \text{say,}$$

and

$$w(\varepsilon_2)-w(\varepsilon_1)=f(\varepsilon_2-\varepsilon_1), \qquad w(\varepsilon_3)-w(\varepsilon_2)=f(\varepsilon_3-\varepsilon_2). \tag{5.20}$$

By reasoning similar to that used in discussing case (ii) it is evident that the last two relations in (5.20) cannot be satisfied.

So far we have considered extension of the rod and the concomitant transition of the whole or parts of it from states on segment I to states on segment III of the $f-\varepsilon$ curve. If such transitions have taken place and we now decrease the extension quasistatically, then provided that the fluctuations are large enough for the tension in the rod to decrease below the value of f

corresponding to the minimum on the f–ε curve, the whole process will be reversed and transitions will take place from the state on segment III to the state on segment I determined by the Maxwell line. If the fluctuations are not large enough for this to occur, extension of the rod will have to be further reduced before a transition can take place.

Finally, we note that whether or not the rod can rest in two different states when the tension in it is zero will depend on whether segment III of the f–ε curve does or does not intersect the abscissa of the f–ε curve (see Figure 2). If such an intersection exists, the rod will have an equilibrium state with zero tension provided that the fluctuations are not too large.

6. Force-induced change of twinning mode in quartz

Quartz is crystalline silica, SiO_2. It can exist in two crystalline forms, the so-called α-form, which is normally stable at ordinary temperatures and the β-form which is the stable form above 573 °C.

α-quartz, the form with which we shall be concerned here, belongs to the trigonal-trapezohedral symmetry class (32). It has a single axis of three-fold rotational symmetry, usually called the optic axis, and, in the plane perpendicular to this, three axes of two-fold rotational symmetry, often called electric axes, as shown schematically in Figure 5. The crystal has no plane or center of symmetry.

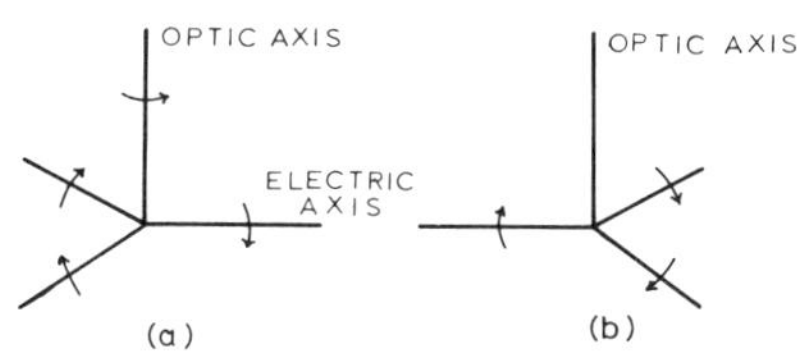

Figure 5.

Structurally, a crystal of α-quartz consists of tetrahedra at the corners of which the oxygen atoms are located. A silicon atom is located at the center of each tetrahedron, and each oxygen atom is chemically bonded to two silicon atoms. If we start at, say, a silicon atom we can trace out a helix, with its axis parallel to the optic axis, on which lie a succession of silicon-oxygen-silicon-oxygen atoms. This helix may form a left-handed or a right-handed screw. As a result of this screw-like structure, if a beam of plane-polarized light is transmitted parallel to the optic axis, its plane of polarization is rotated and we distinguish between left-handed and right-handed quartz accordingly as the plane of polarization is rotated in the sense of a left-handed or right-handed screw. A single crystal of α-quartz may consist in part of left-handed and in part of right-handed quartz and is then said to be optically twinned. We shall not be concerned with this type of twinning here.

There is, however, another type of twinning which occurs in α-quartz. This is called Dauphiné, or electrical twinning, and occurs in the following way. We have a crystal for which the optic axis is, say, the outward drawn normal to the paper in Figure 5. The crystal consists entirely of left-handed quartz or entirely of right-handed quartz. However, in part of the crystal, the electric axes are directed as shown in Figure 5a while in the remainder they are rotated from these directions through 180° about the optic axis, i.e. as shown in Figure 5b.

If a plate of an untwinned crystal is cut with its major surfaces normal to one of the electric axes and compressed by forces acting on these surfaces, an electrical potential difference is developed across the plate, i.e. the crystal is peizo-electric. For an electrically twinned plate the magnitude of the potential difference is reduced as a result of the fact that the potential differences produced in the differently oriented parts of the plate are of opposite sign.

It can be seen by examining a model of the structure of quartz that a crystal oriented as shown in Figure 5b can be converted into one oriented as shown in Figure 5a by relatively small displacements of the individual oxygen atoms relative to the silicon atoms.

It is found experimentally that an electrically twinned crystal of, say, left-handed α-quartz can in certain circumstances be converted to an untwinned crystal of the same hand by the application of appropriate forces. This process is greatly facilitated by raising the temperature of the crystal until it approaches or exceeds the α–β transition temperature and then cooling it before removal of the forces.

An effective method [6] of promoting this transition is to cut from an untwinned crystal of α-quartz a rectangular plate with its length perpendicular to the optic axis. The plate is subjected to torsion by means of couples about the length direction. If the couple is large enough the plate will transform into a crystal of the opposite twin (equivalent to a rotation of the structure through 180° about the optic axis) except in narrow layers at the two free edges. The magnitude of this force depends on the angle θ between the normal to the plate and the optic axis and on the angle ϕ between the projection of this normal on the plane perpendicular to the optic axis and the positive direction of a two-fold symmetry axis.

W. A. Wooster and N. Wooster [6] measured the torsional couple necessary to produce a change from one twinning mode to the other as a function of θ and ϕ. The results they obtained are illustrated in the three-dimensional polar diagram of Figure 6, in which the optic axis is vertical

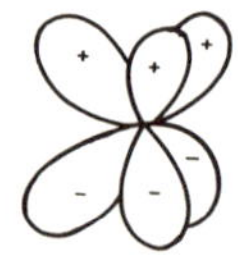

Figure 6.

and the polar distance is proportional to the inverse of the torque necessary to produce the transformation. The upper lobes, labelled with a + sign, relate to transformation in one direction and the lower ones, labelled with a − sign, to transformation in the opposite direction. The maxima on the upper lobes correspond to $(\theta, \phi) = (45°, 30°)$, $(45°, 150°)$, $(45°, 270°)$ and those on the lower lobes to $(\theta, \phi) = (135°, 30°)$, $(135°, 150°)$, $(135°, 270°)$. In addition to these lobes, Wooster and Wooster found six further similar lobes which are not shown in Figure 6. Three of these, with maxima at $(\theta, \phi) = (135°, 90°)$, $(135°, 210°)$, $(135°, 330°)$, relate to transformation in the same direction as the upper lobes shown. The remaining three, with maxima at $(\theta, \phi) = (45°, 90°)$, $(45°, 210°)$, $(45°, 330°)$, relate to transformation in the opposite direction.

In [7] and [8], W. A. Wooster and Thomas proposed a theory to explain these experimental results, based on the assumption that whether or not a transformation will take place, when the plate is subjected to dead-loading by a torque, is determined by the difference in the total energies of the system with the plate in the transformed and untransformed states. We denote these energies by E_T and E_U respectively. Then, if $E_U > E_T$, the transformation will take place; otherwise it will not. It is further assumed that if $E_U > E_T$ the torque required to produce the transformation is a monotonically decreasing function of $E_U - E_T$. This energy difference can be easily calculated from the known elastic constants of α-quartz.

When the plate is subjected to a torque, the stress in the plate is predominantly a shearing stress which increases linearly with distance from the axis of torsion. We denote the appropriate shear compliances for the material of the plate in its transformed and untransformed states by κ_U and κ_T respectively.

Then, it can easily be seen that $E_U - E_T$ is given by an expression of the form

$$E_U - E_T = C(\kappa_T - \kappa_U)M^2, \tag{6.1}$$

where M is the applied torque and C is a positive constant which depends on the dimensions of the plate. The compliances κ_T and κ_U can both be calculated from the known compliance matrix for α-quartz, $\kappa_T - \kappa_U$ being given by an expression of the form

$$\kappa_T - \kappa_U = \kappa \sin 2\theta \sin 3\phi, \tag{6.2}$$

where κ is a constant. It is easily seen that, with (6.2), equation (6.1) yields a three-dimensional polar diagram, in which the radial distance from the pole is proportional to $E_U - E_T$, of the same general form as Figure 6.

The experimental fact, already noted, that the transformation from one twinning mode to another does not take place near the free edges of the plate is easily explained by the theory. To do so we note that when a thin plate is subjected to a torsion, the shearing stress at the free edges must be zero.

The theory outlined above can evidently be generalized, at any rate in principle, to other types of deformation and bodies of other geometries. However, the predictions are not always in accord with the experimental observations. This may be due to the fact that as soon as a change of twinning mode has been initiated at one point of the body, we are faced with a boundary-value problem involving an inhomogeneous body. The stress distribution associated with the prescribed loading may then be very different from that which was assumed to exist either initially or in the final state when complete transformation has taken place. Furthermore, the theory considers only a quasistatic change from one twinning mode to another and does not take account of the dynamics of this transformation.

It can easily be seen that a theory of the dynamics of the transformation from one twinning mode to the other cannot be based on an elastic strain-energy function which depends on the macroscopic deformation gradients alone. Presumably when a load is applied to the crystal, the relative positions of the oxygen and silicon atoms change, for sufficiently small loads, in a reversible fashion, and, at any instant, depend on the current value of the deformation gradient matrix. Accordingly, for sufficiently small loads the stress can be determined from a strain-energy function which depends only on the current value of the deformation gradient matrix. However, when some critical value of the load is reached, this configuration becomes unstable and the atoms snap into a new relative configuration. During this snap-through the relative positions of the oxygen and silicon atoms will not depend in a unique fashion on the macroscopically measured value of the macroscopic deformation gradient matrix and accordingly the stress cannot be derived from a strain-energy function which depends only on the deformation gradient matrix. If the positions of the oxygen atoms relative to the silicon atoms are described by appropriate internal variable fields, then it *may* be possible, in principle, to derive the stress from a strain-energy function which depends on these fields as well as on the deformation gradient matrix. However, for this to be the case it would be necessary that in the snap-through of the atoms from positions appropriate to one twinning mode to positions appropriate to the other twinning mode, no significant amount of energy is irreversibly communicated to internal degrees of freedom of the atoms.

Acknowledgement

This paper was written with the support of the Office of Naval Research under Contract No. N00014-76-C-0235 with Lehigh University.

References

1. B. D. Coleman and W. Noll, Arch. Rat'l Mech. Anal. **3** (1959), 289.
2. C. A. Truesdell and R. A. Toupin, Arch. Rat'l Mech. Anal. **12** (1963), 1.
3. R. S. Rivlin, Phil. Trans. Roy. Soc. A **240** (1948), 459.
4. R. S. Rivlin, Q. Applied Math. **32** (1974), 265.
5. C. A. Truesdell and W. Noll, *The Non-Linear Field Theories of Mechanics*, Handbuch d. Physk. **3**/3 (1965).
6. W. A. Wooster and N. Wooster, Nature **157** (1946), 405.
7. W. A. Wooster, Nature **159** (1947), 94.
8. L. A. Thomas and W. A. Wooster, Proc. Roy. Soc. A **208** (1951), 208.

TWO-DIMENSIONAL APPROXIMATIONS OF THREE-DIMENSIONAL MODELS IN NONLINEAR PLATE THEORY

PHILIPPE G. CIARLET

Laboratoire d'Analyse Numérique, Université Pierre et Marie Curie, Paris

(Received October 6, 1980)

ABSTRACT

The asymptotic expansion method, with the thickness as the parameter, is applied to the nonlinear, three-dimensional, equations for the equilibrium of elastic plates under suitable loads and appropriate boundary conditions. It is shown that the leading term of the expansion is a solution of a system of equations equivalent to a well-known two-dimensional nonlinear plate model, namely the von Kármán equations. The existence of solutions of the two-dimensional problem is established in all cases (by contrast with the three-dimensional model, where no satisfactory existence theory is as yet available). It is also shown that the displacement and the stress corresponding to the leading term of the expansion have the specific form generally assumed *a priori* in the usual derivations of two-dimensional plate models. In particular, the displacement field is of Kirchhoff–Love type. This approach clarifies in particular the nature of the admissible three-dimensional boundary conditions for a given two-dimensional plate model. A discussion is also given regarding the class of admissible three-dimensional models.

1. Introduction

This paper gives a brief description of a method for deriving *known nonlinear two-dimensional plate models* from *general nonlinear three-dimensional elasticity models.* It is based on, and extends as regards the consideration of more general constitutive equations, Ciarlet [1980], where complete proofs can be found.

Our approach is based on the *asymptotic expansion method,* applied to (nonlinear in the present case) problems posed in variational form. *Without any a priori assumption,* either geometrical or mechanical in nature, it is

Proceedings of the IUTAM Symposium on Finite Elasticity, Lehigh University, August 10–15, 1980. Invited paper.

Carlson, D.E. and Shield, R.T. (eds.)
Proceedings of the IUTAM Symposium on Finite Elasticity
ISBN 90 247 2629 8

shown that the first term in the expansion is a solution of a *two-dimensional plate model*, equivalent to the *von Kármán equations.*

A feature of the method is to clearly delineate the type of *boundary conditions* for the three-dimensional model which lead to a specific two-dimensional plate model.

Another aspect of the method is that the displacement and stress components corresponding to the first term in the asymptotic expansion are of the specific forms generally *assumed* in the literature as a result of appropriate *a priori* assumptions. For instance we shall find that the displacement field is necessarily of *Kirchhoff–Love type*, while this is generally an a priori assumption of a geometrical nature.

In other works, the asymptotic expansion method has been shown to apply equally well to:

(i) *linear plate models* [Ciarlet and Destuynder, 1979a], for which it provides in addition a satisfactory *error analysis* [Destuynder, 1979] between the three-dimensional and two-dimensional solutions (the error analysis rests upon methods developed in Lions [1973]);

(ii) *eigenvalue problems* for plates [Ciarlet and Kesavan, 1981];

(iii) *linear shell models* [Destuynder, 1979].

It is also worth mentioning that whereas the asymptotic expansion method is commonly used for linear problems, it is seldom applied to nonlinear problems; in this direction, see however Lions [1973], Rigolot [1977].

Let us review some of the notation used in this paper. The usual partial derivatives will be written $\partial_i v = \partial v/\partial x_i$, $\partial_{ij} v = \partial^2 v/\partial x_i \partial x_j$, etc. If $\mathcal{O}$ is an open subset of $\boldsymbol{R}^n$ we denote by $W^{m,p}(\mathcal{O})$, $m \in \boldsymbol{N}$, $1 \leqq p$, or $H^m(\mathcal{O})$ if $p=2$, the standard Sobolev spaces.

We shall omit the symbol dx in an integral of the form $\int_X f(x)dx$, except in those integrals where the variable of integration is $x_3 \in [-1,1]$, in which case the specific symbol dt will be used.

As a rule, Greek indices, $\alpha, \beta, \mu, \cdots$, take their values in the set $\{1,2\}$; while Latin indices, $i, j, p, \cdots$, take their values in the set $\{1,2,3\}$. The repeated index convention for summation is also systematically used, in conjunction with the above rule.

With each vector-valued function $v = (v_i)$: $\mathcal{O} \subset \boldsymbol{R}^3 \to \boldsymbol{R}^3$, thought of as being a displacement field in $\boldsymbol{R}^3$, we associate the symmetric tensors $\gamma(v) = (\gamma_{ij}(v))$ and $\bar{\gamma}(v) = (\bar{\gamma}_{ij}(v))$: $\mathcal{O} \subset \boldsymbol{R}^3 \to \boldsymbol{R}^9$ respectively defined by

$$\gamma_{ij}(v) = \tfrac{1}{2}(\partial_i v_j + \partial_j v_i),$$

$$\bar{\gamma}_{ij}(v) = \gamma_{ij}(v) + \tfrac{1}{2}\partial_i v_\ell \partial_j v_\ell,$$

which are the *linearized strain tensor*, and the *strain tensor*, respectively.

Finally, if C is a square matrix, we denote by $\operatorname{tr}(C)$ and $\det(C)$ its trace and its determinant, respectively.

2. The three-dimensional model

Let (e_i) be an orthonormal basis in $\boldsymbol{R}^3$, and let ω be a bounded open subset of the plane spanned by (e_α), with a sufficiently smooth boundary γ. Given a constant $\varepsilon > 0$, we let

$$\Omega^\varepsilon = \omega \times]-\varepsilon, \varepsilon[, \qquad \Gamma_0^\varepsilon = \gamma \times [-\varepsilon, \varepsilon],$$

$$\Gamma_+^\varepsilon = \omega \times \{\varepsilon\}, \qquad \Gamma_-^\varepsilon = \omega \times \{-\varepsilon\},$$

so that the boundary of the open subset Ω^ε of $\boldsymbol{R}^3$ is partitioned into the *lateral surface* Γ_0^ε and the *upper and lower faces* Γ_+^ε and Γ_-^ε.

The problem consists in finding the *displacement vector field* $u = (u_i): \bar{\Omega} \to \boldsymbol{R}^3$ and the *second Piola–Kirchhoff tensor field* $\sigma = (\sigma_{ij}): \bar{\Omega} \to \boldsymbol{R}^9$ of a three-dimensional body which occupies the set $\bar{\Omega}$ in the absence of applied forces. Because the *thickness* 2ε of the body is considered to be "small" compared to the dimensions of the set ω, the body is called a *plate*, with *middle surface* ω.

The plate is subjected to three kinds of given forces:

(i) *Body forces* throughout Ω^ε, of density

$$(f_i^\varepsilon) = (0, 0, f_3^\varepsilon);$$

(ii) *Superficial forces* on the upper and lower faces Γ_+^ε and Γ_-^ε, of density

$$(g_i^\varepsilon) = (0, 0, g_3^\varepsilon);$$

(iii) *Superficial forces* along the lateral surface Γ_0^ε, of which only the *resulting density*

$$(h_i^\varepsilon) = (h_1^\varepsilon, h_2^\varepsilon, 0),$$

i.e., after integration across the thickness of the plate (cf. (2.4) below), is known along the boundary γ of the middle surface ω (as a consequence, the functions h_α^ε are given only on γ).

As regards the *boundary conditions involving the displacement field* (u_i), we assume that:

$$\left.\begin{array}{c} u_1 \text{ and } u_2 \text{ are independent of } x_3, \\ u_3 = 0, \end{array}\right\} \quad \text{on } \Gamma_0^\varepsilon.$$

These conditions are readily verified to be complementary to those involving the functions h_α^ε in the variational formulation of the problem (cf. (2.20) below).

Following Truesdell and Noll [1965], or Wang and Truesdell [1973], the associated *equations of finite elastostatics*, which express the elastic equilibrium of the plate, take the following form:

$$-\partial_j(\sigma_{ij}+\sigma_{kj}\partial_k u_i)=f_i^\varepsilon \qquad \text{in } \Omega^\varepsilon, \tag{2.1}$$

$$\sigma_{ij}=\sigma_{ji} \qquad \text{in } \Omega^\varepsilon, \tag{2.2}$$

$$\sigma_{i3}+\sigma_{k3}\partial_k u_i = \pm g_i^\varepsilon \qquad \text{on } \Gamma_\pm^\varepsilon, \tag{2.3}$$

$$\frac{1}{2\varepsilon}\int_{-\varepsilon}^{\varepsilon}(\sigma_{\alpha\beta}+\sigma_{k\beta}\partial_k u_\alpha)\nu_\beta = h_\alpha^\varepsilon \qquad \text{on } \gamma, \tag{2.4}$$

$$u_1,\ u_2 \text{ are independent of } x_3 \text{ on } \Gamma_0^\varepsilon, \tag{2.5}$$

$$u_3=0 \qquad \text{on } \Gamma_0^\varepsilon, \tag{2.6}$$

where $\nu=(\nu_\alpha)$ denotes the unit outer normal vector along γ (and consequently, also along the lateral surface Γ_0^ε).

REMARK 2.1. The reason why we set $f_1^\varepsilon=f_2^\varepsilon=0$, $g_1^\varepsilon=g_2^\varepsilon=0$, and $h_3^\varepsilon=0$, is simply that the consideration of such more general applied forces leads to plate models different from (and more complicated than) the von Kármán equations. ■

REMARK 2.2. If instead of the boundary conditions (2.4)–(2.6), we had chosen the (perhaps more familiar) boundary conditions:

$$(\sigma_{\alpha\beta}+\sigma_{k\beta}\partial_k u_\alpha)\nu_\beta = h_\alpha^\varepsilon \qquad \text{on } \Gamma_0^\varepsilon,$$

$$u_3=0 \qquad \text{on } \Gamma_0^\varepsilon,$$

serious difficulties would arise in the subsequent analysis. In particular, it seems that this type of boundary conditions along the lateral surface does *not* naturally give rise to a two-dimensional plate model. ■

According to the *Rivlin–Ericksen theorem* (cf. Wang and Truesdell [1973]), the most general *constitutive equation* for an *elastic*, *isotropic*, material which satisfies the principle of *frame indifference* is of the form:

$$\sigma=\sqrt{III_c}\{\varphi_0(I_c,II_c,III_c)C^{-1}+\varphi_1(I_c,II_c,III_c)I+\varphi_2(I_c,II_c,III_c)C\}, \tag{2.7}$$

where I denotes the unit matrix,

$$C=I+2\bar{\gamma}, \qquad \text{with} \qquad \bar{\gamma}=\bar{\gamma}(u),$$

denotes the (right) *Cauchy–Green tensor*, I_c, II_c, III_c denote the three *principal invariants of the tensor* C (whose eigenvalues are denoted λ_1, λ_2, λ_3):

$$I_c = \lambda_1 + \lambda_2 + \lambda_3 = C_{ii} = \text{tr}(C),$$

$$II_c = \lambda_1\lambda_2 + \lambda_2\lambda_3 + \lambda_3\lambda_1 = \tfrac{1}{2}\{(\text{tr}\, C)^2 - \text{tr}(C^2)\},$$

$$III_c = \lambda_1\lambda_2\lambda_3 = \det(C) = \tfrac{1}{6}\{(\text{tr}\, C)^3 - 3\text{tr}\, C\, \text{tr}(C^2) + 2\text{tr}(C^3)\},$$

and finally, φ_0, φ_1 and φ_2 are arbitrary functions.

Assuming the functions φ_0, φ_1, φ_2 to be smooth enough, one can write a Taylor expansion of (2.7) around a *natural state* ($\sigma = 0$ for $C = I$) *in terms of the strain tensor* $\bar{\gamma}$. Thus for instance, if we limit ourselves to second order terms, we find a constitutive equation of the form

$$\sigma = \lambda(\text{tr}\,\bar{\gamma})I + 2\mu\bar{\gamma} + a\bar{\gamma}^2 + b(\text{tr}\,\bar{\gamma})\bar{\gamma} + c(\text{tr}\,\bar{\gamma})^2 I + d(\text{tr}\,\bar{\gamma}^2)I + \cdots, \tag{2.8}$$

where λ, μ, a, b, c, d are *constants*. The two constants λ, μ are the *Lamé coefficients of elasticity*; they satisfy the inequalities (cf. Wang and Truesdell [1973])

$$\lambda > 0, \qquad \mu > 0. \tag{2.9}$$

The same type of constitutive equation can also be drawn from the assumption that the material is *hyperelastic*, i.e., that there exists a *strain energy function*

$$\mathcal{W}(F) = W(\sigma_1, \sigma_2, \sigma_3), \tag{2.10}$$

where

$$F = (F_{ij}) = (\partial_j u_i)$$

denotes the *deformation gradient matrix, and*

$$\sigma_1 = \text{tr}\,\bar{\gamma} = \bar{\gamma}_{ii},$$

$$\sigma_2 = \text{tr}(\bar{\gamma}^2) = \bar{\gamma}_{ij}\bar{\gamma}_{ji},$$

$$\sigma_3 = \text{tr}(\bar{\gamma}^3) = \bar{\gamma}_{ij}\bar{\gamma}_{jk}\bar{\gamma}_{ki},$$

in such a way that the *first Piola–Kirchhoff stress tensor*

$$t_{ij} \stackrel{\text{def}}{=} \sigma_{ij} + \sigma_{kj}\partial_k u_i \tag{2.11}$$

satisfies

$$t_{ij} = \frac{\partial \mathcal{W}}{\partial F_{ij}}. \tag{2.12}$$

Then if we express that the *energy*

$$J(v) = \int_{\Omega^\varepsilon} \mathcal{W}(F) - \left(\int_{\Omega^\varepsilon} f_3^\varepsilon v_3 + \int_{\Gamma_+^\varepsilon \cup \Gamma_-^\varepsilon} g_3^\varepsilon v_3 + \int_\gamma \left\{ \int_{-\varepsilon}^{\varepsilon} v_\alpha dx_3 \right\} h_\alpha^\varepsilon \right) \tag{2.13}$$

is *stationary* (i.e., its derivative vanishes) when the functions v span a space of smooth enough functions which satisfy the boundary conditions (2.5)–(2.6), we are naturally led to a constitutive equation. To be more specific, assume that we can expand the strain energy function (2.10) in terms of powers of σ_1, σ_2, σ_3. Then if we limit ourselves to the quadratic and cubic terms in this expansion, i.e., if we write

$$W(\sigma_1, \sigma_2, \sigma_3) = \frac{\lambda}{2}\sigma_1^2 + \mu\sigma_2 + \frac{A}{3}\sigma_1^3 + B\sigma_1\sigma_2 + \frac{C}{3}\sigma_3 + \cdots, \tag{2.14}$$

we find that (cf. John [1971], Novozhilov [1953])

$$\begin{aligned} \sigma &= \lambda\sigma_1 I + 2\mu\bar{\gamma} + (C + 4\mu)\bar{\gamma}^2 + (2B + 2\lambda)\sigma_1\bar{\gamma} \\ &\quad + \left(A - \frac{\lambda}{2}\right)\sigma_1^2 I + (B - \mu)\sigma_2 I + \cdots. \end{aligned} \tag{2.15}$$

In other words, we find a constitutive equation of the same form as in (2.8), but with only 5 arbitrary constants (instead of 6 in (2.8)), because of the assumption of hyperelasticity.

REMARK 2.3. When the higher order terms (represented by three dots) are omitted in (2.15), the resulting constitutive equation is sometimes known as *Murnaghan's law*, after Murnaghan [1937], although it seems to have been first considered by Voigt [1893–1894]. For the actual computations of the third order terms in (2.8), see Novozhilov [1953]. ■

We shall henceforth assume that *the constitutive equation is a polynomial in terms of the components of the strain tensor* $\bar{\gamma}$, i.e., we assume that the expansion (2.8) is finite; hence we do not have to examine questions of convergence in otherwise infinite expansions.

We also make the following assumption, which is *crucial* for our subsequent purposes, and which shall be commented upon later on (in Section 5): *The Lamé coefficients appearing in* (2.8) *are of the form*

$$\lambda^\varepsilon = \varepsilon^{-3}\lambda^1, \qquad \mu^\varepsilon = \varepsilon^{-3}\mu^1, \tag{2.16}$$

where λ^1 *and* μ^1 *are constants independent of* ε, while *the other constants which appear in the constitutive equations* (2.8) *are independent of* ε.

With each tensor $X = (X_{ij})$, we associate the tensor $Y = (Y_{ij}) = AX$ defined by (δ_{ij} is the Kronecker symbol)

$$Y_{ij} = (AX)_{ij} = \left(\frac{1+\nu}{E}\right) X_{ij} - \frac{\nu}{E} X_{pp}\delta_{ij},$$

where the constants E and ν are related to the constants λ^1, μ^1 appearing in (2.16) by the relations

$$\lambda^1 = \frac{E\nu}{(1+\nu)(1-2\nu)}, \qquad \mu^1 = \frac{E}{2(1+\nu)}.$$

Since

$$(A^{-1}Y)_{ij} = \lambda^1 Y_{pp}\delta_{ij} + 2\mu^1 Y_{ij},$$

we can also write the constitutive equation (2.8) as

$$(2.17)\quad \varepsilon^3(A\sigma)_{ij} = \bar{\gamma}_{ij}(u) + \varepsilon^3 \sum_{2 \leq q \leq Q} a_{ijk_1k_2\cdots k_{2q-1}k_{2q}} \bar{\gamma}_{k_1k_2}(u)\cdots \bar{\gamma}_{k_{2q-1}k_{2q}}(u),$$

for appropriate constants $a_{ijk_1k_2\cdots k_{2q-1}k_{2q}}$.

The *three-dimensional problem* is now completely defined, by the data of the *equations of elastic equilibrium* (2.1)–(2.6) and of the *constitutive equation* (2.17).

As regards existence theory for such nonlinear elasticity models, one can extend the analysis given in Ciarlet and Destuynder [1979b] (which relied essentially on the implicit function theorem, L^p-regularity results for linear elliptic systems, and the fact that the Sobolev spaces $W^{1,p}(\Omega)$, $\Omega \subset \boldsymbol{R}^3$, are Banach algebras for $p > 3$), and show in this fashion that for *small enough* applied forces $(f_i) \in (L^p(\Omega))^3$, the *pure* Dirichlet problem:

$$-\partial_j(\sigma_{ij} + \sigma_{kj}\partial_k u_i) = f_i \qquad \text{in } \Omega \subset \boldsymbol{R}^3,$$

$$(A\sigma)_{ij} = \bar{\gamma}_{ij}(u) + \sum_{2 \leq q \leq Q} a_{ijk_1k_2\cdots k_{2q-1}k_{2q}} \bar{\gamma}_{k_1k_2}(u)\cdots \bar{\gamma}_{k_{2q-1}k_{2q}}(u),$$

$$u = 0 \text{ on the boundary of } \Omega$$

(assuming the boundary of Ω is smooth enough), has a solution in the space $(W_0^{1,p}(\Omega) \cap W^{2,p}(\Omega))^3$, for $p > 3$. This is also the approach of Marsden and Hughes [1978, p. 208], Valent [1978].

REMARK 2.4. It is precisely the lack of available regularity results (for the linear elasticity system) in the case of a cylindrical domain such as Ω^ε and of mixed boundary conditions of the form (2.3)–(2.6) which limits the applicability of the method to pure Dirichlet problems and domains with smooth boundaries. ∎

REMARK 2.5. Under the same assumptions, one can also prove the 1–1 character of the mapping

$$\varphi : x \in \Omega \to \varphi(x) = x + u(x),$$

a highly desirable property of the solution. ∎

Regarding existence theory for nonlinear elasticity models, we mention the fundamental results of Ball [1977]. For yet another interesting approach, see Oden [1979].

We notice at this point that a *variational formulation* of equations (2.1)–(2.6), (2.17) consists in expressing that the pair (u, σ), with $u = (u_i)$ and $\sigma = (\sigma_{ij})$, satisfies:

$$u \in V^\varepsilon \overset{\text{def}}{=} \{v = (v_i) \in (W^{1,p}(\Omega^\varepsilon))^3;\ v_1, v_2 \text{ are independent of } x_3 \text{ on } \Gamma_0^\varepsilon,\ v_3 = 0 \text{ on } \Gamma_0^\varepsilon\}, \tag{2.18}$$

$$\sigma \in \Sigma^\varepsilon \overset{\text{def}}{=} \{\tau = (\tau_{ij}) \in (L^2(\Omega))^9;\ \tau_{ij} = \tau_{ji}\}, \tag{2.19}$$

$$\forall v \in V^\varepsilon,$$

$$\int_{\Omega^\varepsilon} \sigma_{ij}\gamma_{ij}(v) + \int_{\Omega^\varepsilon} \sigma_{ij}\partial_i u_l \partial_j v_l = \int_{\Omega^\varepsilon} f_3^\varepsilon v_3 + \int_{\Gamma_+^\varepsilon \cup \Gamma_-^\varepsilon} g_3^\varepsilon v_3 + \int_\gamma \left\{\int_{-\varepsilon}^{\varepsilon} v_\alpha dx_3\right\} h_\alpha^\varepsilon, \tag{2.20}$$

$$\begin{aligned} \forall \tau \in \Sigma^\varepsilon, \quad & \varepsilon^3 \int_{\Omega^\varepsilon} (A\sigma)_{ij}\tau_{ij} - \int_{\Omega^\varepsilon} \tau_{ij}\gamma_{ij}(u) - \frac{1}{2}\int_{\Omega^\varepsilon} \tau_{ij}\partial_i u_l \partial_j u_l \\ & - \varepsilon^3 \sum_{2 \leq q \leq Q} a_{ijk_1k_2\cdots k_{2q-1}k_{2q}} \int_{\Omega^\varepsilon} \bar{\gamma}_{k_1k_2}(u) \cdots \bar{\gamma}_{k_{2q-1},k_{2q}}(u)\tau_{ij} = 0, \end{aligned} \tag{2.21}$$

provided the number p is chosen to be large enough, so that all the integrals make sense.

REMARK 2.6. Specific regularity assumptions on the data f_3^ε, g_3^ε, h_α^ε will be made later on. For the time being, it suffices to assume that they are smooth enough so that all integrals appearing in (2.20)–(2.21) make sense. ■

3. Definition of a "limit" problem for $\varepsilon = 0$

Our first task is to define a problem equvalent to the variational problem (2.20)–(2.21), but now posed over a domain which does *not* depend on ε. Accordingly, we shall successively define appropriate changes of variables and changes of functions. We let

$$\Omega = \omega \times]-1, 1[, \qquad \Gamma_0 = \gamma \times [-1, 1],$$
$$\Gamma_+ = \omega \times \{1\}, \qquad \Gamma_- = \omega \times \{-1\},$$

and with each point $X \in \bar{\Omega}$, we associate the point $X^\varepsilon \in \bar{\Omega}^\varepsilon$ through the correspondence

$$X = (x_1, x_2, x_3) \in \bar{\Omega} \to X^\varepsilon = (x_1, x_2, \varepsilon x_3) \in \bar{\Omega}^\varepsilon.$$

With the space V^ε, Σ^ε of (2.18)–(2.19), we associate the spaces

$$V = \{v = (v_i) \in (W^{1,p}(\Omega))^3;\ v_1, v_2 \text{ are independent of } x_3 \tag{3.1}$$
$$\text{on } \Gamma_0^\varepsilon,\ v_3 = 0 \text{ on } \Gamma_0^\varepsilon\}.$$

$$\Sigma = \{\tau = (\tau_{ij}) \in (L^2(\Omega))^9;\ \tau_{ij} = \tau_{ji}\}. \tag{3.2}$$

With the functions $(v_i) \in V^\varepsilon$, $(\tau_{ij}) \in \Sigma^\varepsilon$, we associate the functions $(v_i^\varepsilon) \in V$, $(\tau_{ij}^\varepsilon) \in \Sigma$ defined by

$$v_\alpha(X^\varepsilon) = \varepsilon^2 v_\alpha^\varepsilon(X), \qquad v_3(X^\varepsilon) = \varepsilon v_3^\varepsilon(X), \tag{3.3}$$

$$\tau_{\alpha\beta}(X^\varepsilon) = \varepsilon^{-1}\tau_{\alpha\beta}^\varepsilon(X), \qquad \tau_{\alpha 3}(X^\varepsilon) = \tau_{\alpha 3}^\varepsilon(X), \qquad \tau_{33}(X^\varepsilon) = \varepsilon\tau_{33}^\varepsilon(X), \tag{3.4}$$

for all corresponding points $X^\varepsilon \in \bar{\Omega}^\varepsilon$ and $X \in \bar{\Omega}$.

As regards the data, we shall assume that there exist functions f_3, g_3, h_α which are *independent of* ε such that

$$f_3^\varepsilon(X^\varepsilon) = f_3(X), \tag{3.5}$$

$$g_3^\varepsilon(X^\varepsilon) = \varepsilon g_3(X), \tag{3.6}$$

$$h_\alpha^\varepsilon(y) = \varepsilon^{-1} h_\alpha(y) \qquad \text{for all points } y \in \gamma. \tag{3.7}$$

The above relations have the basic effects that some integrals appearing in the variational formulation (2.20)–(2.21) of the three-dimensional problem are left unaltered, up to an appropriate multiplicative power of ε. More specifically, one has

$$\int_{\Omega^\varepsilon} \sigma_{ij}\gamma_{ij}(v) = \varepsilon^2 \int_\Omega \sigma_{ij}^\varepsilon \gamma_{ij}(v^\varepsilon),$$

for all corresponding pairs $(v, \sigma) \in V^\varepsilon \times \Sigma^\varepsilon$ and $(v^\varepsilon, \sigma^\varepsilon) \in V \times \Sigma$, and

$$\int_{\Omega^\varepsilon} f_3^\varepsilon v_3 + \int_{\Gamma_+^\varepsilon \cup \Gamma_-^\varepsilon} g_3^\varepsilon v_3 + \int_\gamma \left\{\int_{-\varepsilon}^{\varepsilon} v_\alpha dx_3\right\} h_\alpha^\varepsilon$$
$$= \varepsilon^2 \left\{\int_\Omega f_3 v_3^\varepsilon + \int_{\Gamma_+ \cup \Gamma_-} g_3 v_3^\varepsilon + \int_\gamma \left\{\int_{-1}^{1} v_\alpha^\varepsilon dt\right\} h_\alpha,\right.$$

for all corresponding functions $v \in V^\varepsilon$ and $v^\varepsilon \in V$. Notice that the integrals appearing in the above equations precisely represent the classical *duality* (in elasticity theory) between the stresses and strains on the one hand, and between the forces and displacements on the other.

The justification of the *scaling factor* ε^2 is twofold: first, we want the asymptotic expansion (3.17) below to start with a factor of ε^0 and secondly we want equations (3.18)–(3.19) below to contain all the terms appearing in the equations found by the same process in the *linear* case (cf. Ciarlet and Destuynder [1979a]).

It is then a purely computational matter to establish the following result, whose interest is to formulate the three-dimensional plate problem in a form where *the dependence on the parameter* ε *is very simple*:

THEOREM 3.1. *Let* $(u^\varepsilon, \sigma^\varepsilon) \in V \times \Sigma$ *be constructed from a solution* $(u, \sigma) \in V^\varepsilon \times \Sigma^\varepsilon$ *of* (2.20)–(2.21) *through formulas* (3.3)–(3.4). *Then* $(u^\varepsilon, \sigma^\varepsilon)$ *is a solution of*

$$(3.8)\quad \forall v \in V, \qquad \mathcal{B}(\sigma^\varepsilon, v) + 2\mathcal{C}_0(\sigma^\varepsilon, u^\varepsilon, v) + 2\varepsilon^2\mathcal{C}_2(\sigma^\varepsilon, u^\varepsilon, v) = \mathcal{F}(v),$$

$$(3.9)\quad \begin{aligned} \forall \tau \in \Sigma, \qquad & \mathcal{A}_0(\sigma^\varepsilon, \tau) + \varepsilon^2\mathcal{A}_2(\sigma^\varepsilon, \tau) + \varepsilon^4\mathcal{A}_4(\sigma^\varepsilon, \tau) + \mathcal{B}(\tau, u^\varepsilon) \\ & + \mathcal{C}_0(\tau, u^\varepsilon, u^\varepsilon) + \varepsilon^2\mathcal{C}_2(\tau, u^\varepsilon, u^\varepsilon) + \varepsilon\mathcal{R}(\varepsilon, \tau, u) = 0, \end{aligned}$$

where, for arbitrary elements u, $v \in V$ *and* σ, $\tau \in \Sigma$,

$$(3.10)\qquad \mathcal{B}(\tau, v) = -\int_\Omega \tau_{ij}\gamma_{ij}(v),$$

$$(3.11)\qquad \mathcal{C}_0(\tau, u, v) = -\frac{1}{2}\int_\Omega \tau_{ij}\partial_i u_3 \partial_j v_3,$$

$$(3.12)\qquad \mathcal{C}_2(\tau, u, v) = -\frac{1}{2}\int_\Omega \tau_{ij}\partial_i u_\alpha \partial_j v_\alpha,$$

$$(3.13)\qquad \mathcal{F}(v) = -\left(\int_\Omega f_3 v_3 + \int_{\Gamma_+\cup\Gamma_-} g_3 v_3 + \int_\gamma \left\{\int_{-1}^1 v_\alpha dt\right\} h_\alpha\right),$$

where the functions f_3, g_3, h_α *are those appearing in formulas* (3.5)–(3.7),

$$(3.14)\qquad \mathcal{A}_0(\sigma, \tau) = \int_\Omega \left(\left(\frac{1+\nu}{E}\right)\sigma_{\alpha\beta} - \frac{\nu}{E}\sigma_{\mu\mu}\delta_{\alpha\beta}\right)\tau_{\alpha\beta},$$

$$(3.15)\qquad \mathcal{A}_2(\sigma, \tau) = \int_\Omega \left\{2\left(\frac{1+\nu}{E}\right)\sigma_{\alpha 3}\tau_{\alpha 3} - \frac{\nu}{E}(\sigma_{33}\tau_{\mu\mu} + \sigma_{\mu\mu}\tau_{33})\right\},$$

$$(3.16)\qquad \mathcal{A}_0(\sigma, \tau) = \frac{1}{E}\int_\Omega \sigma_{33}\tau_{33},$$

and $\mathcal{R}(\varepsilon, \tau, u^\varepsilon)$ *is a polynomial with respect to* ε, *whose coefficients, which are integrals over* Ω, *are independent of* ε. ∎

Since the forms $\mathcal{B}$, $\mathcal{C}_0$, $\mathcal{C}_2$, $\mathcal{F}$, $\mathcal{A}_0$, $\mathcal{A}_2$, $\mathcal{A}_4$, are all independent of ε, as well as the coefficients of the *nonnegative* powers of ε in the polynomial $\mathcal{R}(\varepsilon, \tau, u^\varepsilon)$, and since ε is thought of as being a "small" parameter, we are naturally led to define a *formal* series of "approximations" of a solution $(u^\varepsilon, \sigma^\varepsilon)$ of (3.8)–(3.9) by letting *a priori* (the leading term (u, σ) in the following expansion should not be confused with a solution of the original three-dimensional problem):

$$(3.17) \qquad (u^\varepsilon, \sigma^\varepsilon) = (u, \sigma) + \varepsilon(u^1, \sigma^1) + \varepsilon^2(u^2, \sigma^2) + \cdots .$$

Then, following the principle of *the asymptotic expansion method*, we equate to zero the factors of the successive powers ε^p, $p \geqq 0$, in the expressions obtained when the expansion (3.17) is used in (3.8)–(3.9).

In this fashion, we find:

(i) *equations to be satisfied by the first term*;

(ii) *recurrence relations for the following terms* (of course, nothing guarantees at this stage the existence of the terms (u, σ), (u^1, σ^1), etc., let alone the possible convergence of the series (3.17)).

In the sequel, we shall be concerned with the computation of the first term (u, σ) which, according to the above considerations, should satisfy:

$$(3.18) \qquad \forall v \in V, \qquad \mathcal{B}(\sigma, v) + 2\mathcal{C}_0(\sigma, u, v) = \mathcal{F}(v),$$

$$(3.19) \qquad \forall \tau \in \Sigma, \qquad \mathcal{A}_0(\sigma, \tau) + \mathcal{B}(\tau, u) + \mathcal{C}_0(\tau, u, u) = 0.$$

In this respect, our main results consist in:

(i) establishing the *existence* of (at least) one solution to the "*limit*" *problem* (3.18)–(3.19);

(ii) recognizing a known *two-dimensional plate model* in this same limit problem.

4. Equivalence of the "limit" problem with the von Kármán equations

We let

$$g^\varepsilon_{3\pm} = g^\varepsilon_3 \qquad \text{on } \Gamma_\pm,$$

and we denote by ∂_ν the exterior normal derivative operator along the boundary γ of the middle surface ω. We first establish the equivalence of the "limit" problem (3.18)–(3.19) with a *two-dimensional, displacement*, model:

THEOREM 4.1. *Assume that the data have the following regularity*:

$$(4.1) \qquad f_3 \in L^2(\Omega), \qquad g_3 \in L^2(\Gamma_+ \cup \Gamma_-), \qquad h_\alpha \in H^{3/2}(\gamma),$$

and that the functions h_α verify the following compatibility conditions:

$$(4.2) \qquad \int_\gamma h_1 = \int_\gamma h_2 = \int_\gamma (x_1 h_2 - x_2 h_1) = 0.$$

Equations (3.18)–(3.19) *have at least one solution* $(u, \sigma) = ((u_i), (\sigma_{ij}))$ *in the space* $V \times \Sigma$, *which is obtained as follows*:

First, one solves the two-dimensional problem: *Find* $u^0 = (u_i^0)$: $\omega \to \boldsymbol{R}^3$ *such that*

$$\frac{2E}{3(1-\nu^2)}\Delta^2 u_3^0 - 2\sigma_{\alpha\beta}^0(u^0)\partial_{\alpha\beta}u_3^0 = \left(g_{3+} + g_{3-} + \int_{-1}^{1} f_3 dt\right) \quad \textit{in } \omega, \tag{4.3}$$

$$\partial_\alpha \sigma_{\alpha\beta}^0(u^0) = 0 \quad \textit{in } \omega, \tag{4.4}$$

$$u_3^0 = \partial_\nu u_3^0 = 0 \quad \textit{on } \gamma, \tag{4.5}$$

$$\sigma_{\alpha\beta}^0(u^0)\nu_\alpha = h_\beta \quad \textit{on } \gamma, \tag{4.6}$$

where

$$\begin{aligned}\sigma_{\alpha\beta}^0(u^0) &\overset{\text{def}}{=} \frac{E}{(1-\nu^2)}\{(1-\nu)\gamma_{\alpha\beta}(u^0) + \nu\gamma_{\mu\mu}(u^0)\delta_{\alpha\beta}\} \\ &\quad + \frac{E}{2(1-\nu^2)}\{(1-\nu)\partial_\alpha u_3^0 \partial_\beta u_3^0 + \nu\partial_\mu u_3^0 \partial_\mu u_3^0 \delta_{\alpha\beta}\}.\end{aligned} \tag{4.7}$$

This problem has at least one solution $u^0 = ((u_\alpha^0), u_3^0)$ *in the space* $(H^3(\omega))^2 \times (H_0^2(\omega) \cap H^4(\omega))$.

Secondly, one defines, for $(x_1, x_2, x_3) \in \Omega$,

$$u_3(x_1, x_2, x_3) = u_3^0(x_1, x_2), \tag{4.8}$$

$$u_\alpha = u_\alpha^0 - x_3 \partial_\alpha u_3^0, \tag{4.9}$$

$$\sigma_{\alpha\beta} = \sigma_{\alpha\beta}^0(u^0) - \frac{E x_3}{(1-\nu^2)}\{(1-\nu)\partial_{\alpha\beta}u_3^0 + \nu\Delta u_3^0 \delta_{\alpha\beta}\}, \tag{4.10}$$

$$\sigma_{3\beta} = \sigma_{\beta 3} = -\frac{E(1-x_3^2)}{2(1-\nu^2)}\partial_\beta \Delta u_3^0, \tag{4.11}$$

$$\begin{aligned}\sigma_{33} &= \frac{(x_3+1)}{2} g_{3+} + \frac{(x_3-1)}{2} g_{3-} + \left\{\frac{(1+x_3)}{2}\int_{-1}^{1} f_3 dt - \int_{-1}^{x_3} f_3 dt\right\} \\ &\quad + \frac{E x_3(1-x_3^2)}{6(1-\nu^2)}\Delta^2 u_3^0 - \frac{E(1-x_3^2)}{2(1-\nu^2)}\{(1-\nu)\partial_{\alpha\beta}u_3^0 \partial_{\alpha\beta}u_3^0 + \nu(\Delta u_3^0)^2\}.\end{aligned} \tag{4.12}$$

Conversely, any sufficiently regular solution of (3.18), (3.19) *is necessarily of the form* (4.8)–(4.12) *with* $u^0 = (u_i^0)$ *a solution of problem* (4.3)–(4.6). ■

Let us briefly sketch the main steps in the proof of this theorem.

Step 1. In (3.19), we successively choose "trial" functions $\tau \in \Sigma$ of the particular forms

$$\tau = (\tau_{ij}), \quad \text{with } \tau_{\alpha\beta} = \tau_{33} = 0, \tag{4.13}$$

$$\tau = (\tau_{ij}), \quad \text{with } \tau_{\alpha j} = 0. \tag{4.14}$$

Then if we restrict to solutions for which u_3 is *sufficiently regular*, we find that, with the particular choices (4.13) and (4.14), (3.19) is satisfied if and only if:

(4.15) $\begin{cases} \text{the function } u_3 \text{ is independent of the variable } x_3 \text{ and it} \\ \text{can be identified with a function } u_3^0 \in H_0^2(\omega), \end{cases}$

(4.16) $$\exists u_\alpha^0 \in H^1(\omega), \qquad u_\alpha = u_\alpha^0 - x_3 \partial_\alpha u_3^0.$$

Step 2. *Computation of the functions* u_α^0 *and* u_3^0. In (3.18), (3.19), we successively choose "trial" functions of the particular forms

(4.17) $$\begin{cases} \tau_{\alpha\beta} = \tau_{\alpha\beta}^0 \in L^2(\omega), & \tau_{i3} = 0, \\ v_\alpha = v_\alpha^0 \in H^1(\omega), & v_3 = 0, \end{cases}$$

(4.18) $$\begin{cases} \tau_{\alpha\beta} = x_3 \tau_{\alpha\beta}^1, & \tau_{\alpha\beta}^1 \in L^2(\omega), \\ v_\alpha = x_3 \partial_\alpha v, & v_3 = v, \qquad v \in H_0^2(\omega). \end{cases}$$

After some computations (and elimination of the other unknowns) we find that the functions $u_\alpha^0 \in H^1(\omega)$ and $u_3^0 \in H_0^2(\omega)$ should be solution of (4.3)–(4.6).

Step 3. *Computation of the stresses* σ_{ij}. Once the functions $\sigma_{\alpha\beta}^0(u^0)$ and u_3^0 have been computed by solving (4.3)–(4.6), it turns out that (3.19) with $\tau = (\tau_{ij})$, $\tau_{i3} = 0$, and (3.18) are satisfied if and only if the stresses σ_{ij} are given by (4.10)–(4.12).

Step 4. *Existence of a solution to the two-dimensional problem* (4.3)–(4.6), *for data possessing the regularity* (4.1). One can proceed in two ways:

(i) The variational formulation of (4.3)–(4.6) amounts to finding the stationary points of the functional (we let $v^0 = (v_i^0)$)

(4.19) $$\begin{aligned} \mathcal{J}(v^0) = \frac{E}{(1-\nu^2)} \int_\omega & \{\tfrac{1}{3}(\Delta v_3^0)^2 + (1-\nu)\gamma_{\alpha\beta}(v^0)\partial_\alpha v_3^0 \partial_\beta v_3^0 \\ & + \nu \gamma_{\lambda\lambda}(v^0)\partial_\mu v_3^0 \partial_\mu v_3^0 + \tfrac{1}{4}(\partial_\alpha v_3^0 \partial_\alpha v_3^0)^2 \\ & + (1-\nu)\gamma_{\alpha\beta}(v^0)\gamma_{\alpha\beta}(v^0) + \nu\gamma_{\lambda\lambda}(v^0)\gamma_{\mu\mu}(v^0)\} \\ - \int_\omega & \left(g_{3+} + g_{3-} + \int_{-1}^{1} f_3 dt\right) v_3^0 - 2\int_\gamma h_\alpha v_\alpha^0, \end{aligned}$$

when v^0 varies over the space $(H^1(\omega))^2 \times H_0^2(\omega)$. Because of the compatibility conditions (4.2), this functional is also well-defined over the space

(4.20) $$\mathcal{V} = \{(H^1(\omega))^2 / V^0\} \times H_0^2(\omega),$$

where

$$V^0 = \{v = (v_\alpha) \in (H^1(\omega))^2;\ \gamma_{\alpha\beta}(v) = 0\}$$

$$= \{v = (v_\alpha) \in (H^1(\omega))^2; \tag{4.21}$$

$$\exists a_\alpha, b \in \boldsymbol{R},\ v_1 = a_1 - bx_2,\ v_2 = a_2 + bx_1\},$$

and besides, it is now *coercive* over this space (it is not coercive over the space $(H^1(\omega))^2 \times H_0^2(\omega))$, i.e.,

$$\lim_{\|v^0\|_{\mathcal{V}} \to \infty} \mathcal{J}(v^0) = +\infty,$$

provided the norms $\|h_\alpha\|_{L^2(\omega)}$ *are small enough.*

In addition, it can be shown that the functional $\mathcal{J}$ is weakly lower semi-continuous over the space $\mathcal{V}$, and the conclusion follows by standard arguments.

(ii) In order to have an existence theory devoid of any restriction on the magnitude of the functions h_α, one first introduces the so-called *Airy stress function*, as shown in Theorem 4.2 below (a process which again shows the necessity of imposing compatibility conditions on the functions h_1, h_2). Next, one may use the existence theorem of John and Nečas [1975]. One can also eliminate the Airy stress function, following the method of Berger [1977], and show that the resulting problem in the single unknown u_3^0 amounts to finding the stationary point of a specific functional, which has at least one minimum over the space $H_0^2(\omega)$, as in Rabier [1980].

Finally, using standard regularity results for the system of equations of linear, two-dimensional, elasticity, in conjunction with the method described in Lions [1969, p. 56], one can show that the solutions (u, σ) of problem (3.18), (3.19) found in the above process possess the following regularity:

$$u_3^0 \in H_0^2(\omega) \cap H^4(\omega), \qquad u_\alpha^0 \in H_0^1(\omega) \cap H^3(\omega),$$

$$\sigma_{\alpha\beta} \in H^2(\Omega), \qquad \sigma_{3\beta} \in H^1(\Omega), \qquad \sigma_{33} \in L^2(\Omega).$$

REMARK 4.1. Of course, it now remains to go back to the set Ω^ε, i.e., one must define functions *on the set* Ω^ε, which correspond to the functions u_i and σ_{ij} just constructed. For the sake of brevity, we shall skip the corresponding straightforward computations, simply based on formulas (3.3)–(3.7). It suffices to mention that their effect amounts to introducing *appropriate powers* of ε at some places in the above equations. Thus for instance, equation (4.3), expressed with the "new" functions, now reads:

$$\frac{2E\varepsilon^3}{3(1-\nu^2)} \Delta^2 u_3^0 = 2\varepsilon \sigma_{\alpha\beta}^0(u^0) \partial_{\alpha\beta} u_3^0 + \left(g_{3+}^\varepsilon + g_{3-}^\varepsilon + \int_{-\varepsilon}^{\varepsilon} f_3^\varepsilon dx_3 \right). \quad \blacksquare$$

An important conclusion to be drawn from the above theorem is that the expressions found for the functions u_i and σ_{ij} are identical to, or similar to, the *assumed* expressions found in the literature concerning nonlinear plate theory. In particular, we have obtained *Kirchhoff–Love displacement fields*, i.e., of the form (4.8), (4.9), whereas they are usually derived from an a priori assumption of a geometrical nature (cf. e.g. Washizu [1975, Equation (8.60)]).

In the same fashion, the expressions found in (4.7) for the stresses $\sigma^0_{\alpha\beta}$ (i.e., $\sigma_{\alpha\beta}$ for $x_3 = 0$) are standard in nonlinear plate theory, where they are usually derived after *a priori* assumptions have been made regarding which terms should be neglected in the strain tensor corresponding to the two-dimensional problem (cf. e.g. Stoker [1968, pp. 42–47]). Likewise, the expressions found in (4.11) for the stresses $\sigma_{3\beta}$ are similar to those found in Green and Zerna [1968, Equation (7.7.3)], where they are assumed to be quadratic in x_3, etc.

In the second, and final, stage of our analysis, we establish the equivalence of problem (4.3)–(4.6) with the *von Kármán equations* (4.23)–(4.27). This equivalence essentially relies upon the introduction of the so-called *Airy stress function* φ, which satisfies (4.22). We recall that the space V^0 has been defined in (4.21).

In the next theorem, we assume that the set ω is simply connected, and is of Nikodym type, in the sense of Deny and Lions [1953–1954]; for instance, this is the case if the set ω is star-shaped.

Without loss of generality, we also assume that the origin 0 belongs to the boundary γ of the set ω. Given a point y along the boundary γ, we denote by $\gamma(y)$ the arc joining the point 0 to the point y along γ.

THEOREM 4.2. *Assume the data satisfy the regularity assumptions* (4.1) *and the compatibility conditions* (4.2) *and let there be given any solution*

$$u^0 = ((u^0_\alpha), u^0_3) \in (H^3(\omega))^2 \times (H^2_0(\omega) \cap H^4(\omega))$$

of problem (4.3)–(4.6). *Then there exists a function* $\varphi \in H^4(\omega)$, *uniquely determined if we impose* $\varphi(0) = \partial_1\varphi(0) = \partial_2\varphi(0)$, *such that*

$$\partial_{11}\varphi = \sigma^0_{22}(u^0), \qquad \partial_{12}\varphi = -\sigma^0_{12}(u^0), \qquad \partial_{22}\varphi = \sigma^0_{11}(u^0). \tag{4.22}$$

Besides, the pair (φ, u^0_3) *is a solution of the von Kármán equations*:

$$\frac{2E}{3(1-\nu^2)}\Delta^2 u^0_3 = 2[\varphi, u^0_3] + \left(g_{3+} + g_{3-} + \int_{-1}^{1} f_3 dt\right) \quad \textit{in } \omega, \tag{4.23}$$

$$\Delta^2\varphi = -\frac{E}{2}[u^0_3, u^0_3] \quad \textit{in } \omega, \tag{4.24}$$

$$u^0_3 = \partial_\nu u^0_3 = 0 \quad \textit{on } \gamma, \tag{4.25}$$

$$(4.26)\quad \varphi(y) = -y_1 \int_{\gamma(y)} h_2 + y_2 \int_{\gamma(y)} h_1 + \int_{\gamma(y)} (x_1 h_2 - x_2 h_1), \qquad y \in \gamma,$$

$$(4.27)\qquad \partial_\nu \varphi(y) = -\nu_1(y) \int_{\gamma(y)} h_2 + \nu_2(y) \int_{\gamma(y)} h_1, \qquad y \in \gamma,$$

where, for any smooth enough functions v and w,

$$[v, w] = \partial_{11} v \partial_{22} w + \partial_{22} v \partial_{11} w - 2\partial_{12} v \partial_{12} w.$$

Conversely, let there be given any solution

$$(\varphi, u_3^0) \in H^4(\omega) \times (H_0^2(\omega) \cap H^4(\omega))$$

of problem (4.23)–(4.27). *Then, if we define functions $\sigma_{\alpha\beta}^0$ by letting*

$$\sigma_{11}^0 = \partial_{22}\varphi, \qquad \sigma_{12}^0 = \sigma_{21}^0 = -\partial_{12}\varphi, \qquad \sigma_{22}^0 = \partial_{11}\varphi,$$

there exists a unique element (u_α^0) in the space $(H^3(\omega))^2/V^0$ such that

$$\sigma_{\alpha\beta}^0 = \frac{E}{(1-\nu^2)}\{(1-\nu)\gamma_{\alpha\beta}(u^0) + \nu\gamma_{\mu\mu}(u^0)\delta_{\alpha\beta}\}$$

$$+ \frac{E}{2(1-\nu^2)}\{(1-\nu)\partial_\alpha u_3^0 \partial_\beta u_3^0 + \nu \partial_\mu u_3^0 \partial_\mu u_3^0 \delta_{\alpha\beta}\}, \qquad u^0 = ((u_\alpha^0), u_3^0),$$

and besides, the element u^0 is a solution of problem (4.3)–(4.6). ■

REMARK 4.2. A fairly complete mathematical analysis of the von Kármán equations, regarding notably existence theory, multiplicity of solutions, bifurcation theory, etc., is found in Ciarlet and Rabier [1980]. ■

5. Conclusions

(i) The main conclusion is of course that we have been able to *mathematically justify the derivation of a nonlinear plate model from a well-accepted three-dimensional nonlinear elasticity model*, associated with *specific boundary* conditions along the lateral surface of the plate.

(ii) Which *boundary conditions* along the lateral surface are appropriate for the three-dimensional problem is a question of importance since different boundary conditions yield fundamentally different two-dimensional problems (as expected, of course, but this does not seem to be always clear in the literature). In this respect, see notably Ciarlet and Destuynder [1979b], where the case of a "clamped" plate is considered.

(iii) In order that a "limit" problem exist, it has been found that *the various data should simultaneously vary in an appropriate manner as ε approaches* *zero*, as expressed by relations (2.16) and (3.5)–(3.7). These are not the only possible ones, however. For example, the Lamé coefficients λ, μ appearing in (2.8) can stay constant provided relations (3.5)–(3.7) are replaced by the following:

$$f_3^\varepsilon(X^\varepsilon) = \varepsilon^3 f_3(X), \qquad g_3^\varepsilon(X^\varepsilon) = \varepsilon^3 g_3(X),$$

$$h_\alpha^\varepsilon(y) = \varepsilon^2 h_\alpha^\varepsilon(y) \qquad \text{for all } y \in \gamma,$$

and the "higher order constants" appearing in the constitutive equation decay sufficiently rapidly with ε. Then it is readily verified that the *same* "limit problem" (3.18)–(3.19) is retained by an application of the asymptotic expansion method. The above relations are much less realistic however if body forces, such as the weight, are to be taken into account. But they cannot be disposed of: One cannot expect a plate of zero thickness to carry any load!

The interpretation of relations (2.16) is simple: They express that the *rigidity* of the constitutive material of the plate should increase as the thickness of the plate approaches zero, if we are to find a "limit" model compatible with relations (3.5)–(3.7). Incidentally, similar conclusions have been reached in a related linear problem by Caillerie [1980].

The assumption that the coefficients corresponding to the higher-order terms in the constitutive equation (2.17) are constant is in turn made necessary by the requirement that these terms do not appear in, and thus do not affect, the "limit" problem. A different limit problem would otherwise result which could be studied for its own sake. Our aim was however to clearly delineate three-dimensional constitutive equations that correspond to *precisely* the von Kármán equations. Notice also that the above assumption regarding the "higher order constants" is evidently satisfied if the constitutive equation is linear (as a relation between the tensors σ and $\bar{\gamma}(u)$), as was the case in Ciarlet [1980].

(iv) The present analysis suggests that we consider the von Kármán equations, together with the expressions simultaneously found for the unknowns u_i, σ_{ij} as forming a *consistent set of approximations* to the original three-dimensional problem, in the sense that these equations and expressions are all obtained as the solution of a single, three-dimensional problem, namely problem (3.18), (3.19).

Equivalently, if we start out with a solution of either two-dimensional problem, we may think of the expressions giving the unknowns u_i, σ_{ij} as being the natural *extension* of this solution into the space $V \times \Sigma$. Such an extension may prove useful for obtaining existence results for the original three-dimensional problem.

Acknowledgements

The author expresses his thanks to Professors C. Truesdell and S. S. Antman for suggesting the consideration of general constitutive equations. This paper was written while the author was visiting the Mathematics Department of Cornell University, Ithaca, N.Y.; in this respect, the financial support of the Exchange Visitor Programm No. P-1-43 is gratefully acknowledged.

References

J. M. Ball, *Convexity conditions and existence theorems in nonlinear elasticity*, Arch. Rational Mech. Anal. **63** (1977), 337–403.

M. S. Berger, *Nonlinearity and Functional Analysis*, Academic Press, New York, 1977.

D. Caillerie, *The effect of a thin inclusion of high rigidity in an elastic body*, Mathematical Methods in the Applied Sciences **2** (1980), 251–270.

P. G. Ciarlet, *A justification of the von Kármán equations*, Arch. Rational Mech. Anal. **73** (1980), 349–389.

P. G. Ciarlet and P. Destuynder, *A justification of the two-dimensional linear plate model*, J. Mécanique **18** (1979a), 315–344.

P. G. Ciarlet and P. Destuynder, *A justification of a nonlinear model in plate theory*, Comput. Methods Appl. Mech. Engrg. **17/18** (1979b), 227–258.

P. G. Ciarlet and S. Kesavan, *Two-dimensional approximations of three-dimensional eigenvalue problems in plate theory*, Comput. Methods Appl. Mech. Engrg. **26** (1981), 145–172.

P. G. Ciarlet and P. Rabier, *Les Equations de von Kármán*, Lecture Notes in Mathematics, Vol. 826, Springer-Verlag, Heidelberg, 1980.

J. Deny and J.-L. Lions, *Les espaces du type de Beppo-Levi*, Ann. Institut Fourier (Grenoble) **V** (1953–1954), 305–370.

P. Destuynder, *Sur une Justification Mathématique des Théories de Plaques et de Coques en Elasticité Linéaire*, Doctoral Dissertation, Université Pierre et Marie Curie, Paris, 1979.

A. E. Green and W. Zerna, *Theoretical Elasticity*, University Press, Oxford, 1968.

F. John, *Refined interior equations for thin elastic shells*, Comm. Pure Applied Math. **XXIV** (1971), 583–615.

O. John and J. Nečas, *On the solvability of von Kármán equations*, Apl. Mat. **20** (1975), 48–62.

J.-L. Lions, *Quelques Méthodes de Résolution des Problèmes aux Limites Non Linéaires*, Dunod, Paris, 1969.

J.-L. Lions, *Perturbations Singulières dans les Problèmes aux Limites et en Contrôle Optimal*, Lecture Notes in Mathematics, Vol. 323, Springer, Berlin, 1973.

J. E. Marsden and T. J. R. Hughes, *Topics in the mathematical foundations of elasticity*, in *Nonlinear Analysis and Mechanics, Heriot-Watt Symposium, Volume II*, Pitman, London, 1978.

F. D. Murnaghan, *Finite deformations of an elastic solid*, American Journal of Mathematics **59** (1937), 235–260.

V. V. Novozhilov, *Foundations of the Nonlinear Theory of Elasticity*, Graylock Press, Rochester, 1953.

J. T. Oden, *Existence theorems for a class of problems in nonlinear elasticity*, J. Math. Anal. Appl. **69** (1979), 51–83.

P. Rabier, Doctoral Dissertation, Université Pierre et Marie Curie, Paris, 1980.

A. Rigolot, *Déplacements Finis et Petites Déformations des Poutres Droites: Analyse*

Asymptotique de la Solution à Grande Distance des Bases, J. Mécanique Appliquée **1** (1977), 175–206.

J. J. Stoker, *Nonlinear Elasticity*, Gordon and Breach, New York, 1968.

C. Truesdell and W. Noll, *The Non-Linear Field Theories of Mechanics*, Handbuch der Physik, Vol. III/3, Springer, Berlin, 1965.

T. Valent, *Teoremi di esistenza e unicità in elastostatica finita*, Rend. Sem. Mat. Univ. Padova **60** (1978), 165–181.

W. Voigt, *Ueber eine anscheinend notwendige Erweiterung der Theorie der Elasticität*, Nachrichten von der Königlichen Gesellschaft der Wissenschaften zu Göttingen, (1893), pp. 534–552; (1894), pp. 33–42.

C.-C., Wang and C. Truesdell, *Introduction to Rational Elasticity*, Noordhoff, Groningen, 1973.

K. Washizu, *Variational Methods in Elasticity and Plasticity*, Second Edition, Pergamon, Oxford, 1975.

ON NON-UNIVERSAL FINITE ELASTIC DEFORMATIONS

P. K. CURRIE[†] AND M. HAYES
Department of Mathematical Physics, University College, Belfield, Dublin, 4, Ireland

(Received September 11, 1980)

1. Introduction

Ericksen's theorem [1] has had a profound influence on the development of the theory of finite deformations of compressible elastic materials. This theorem states that the only universal controllable deformations (i.e. deformations possible in all compressible elastic materials in the absence of body forces) are homogeneous. Being proved early in the modern period of resurgence of finite elasticity, Ericksen's theorem shaped the subsequent development of the theory. The result has been that until recently little emphasis has been placed on non-homogeneous deformations. The corresponding theorem proved by Ericksen [2] for incompressible materials has had a similar effect on the theory for those materials; most deformations considered belong to the small class of universal deformations.

Detailed examination of universal deformations provides valuable insight into the nature of non-linear elastic response and offers the possibility of comparison with experiment, exemplified in the pioneering work of Rivlin and Saunders [3] for incompressible materials. However, to the student first encountering the subject, the almost exclusive study of universal deformations may give a distorted view of elasticity theory. He is liable to gain the false impression that the *only* deformations possible in an elastic body are the universal deformations, rather than realising that, by definition, non-universal deformations can be discussed only in the context of specific constitutive assumptions.

Examples of non-universal deformations for particular elastic materials are needed, to place Ericksen's theorem in perspective. Various authors have

[†] Now at Koninklijke/Shell Exploratie en Produktie Lab., Rijswijk, The Netherlands.

Proceedings of the IUTAM Symposium on Finite Elasticity, Lehigh University, August 10–15, 1980. Invited paper presented by M. Hayes.

Carlson, D.E. and Shield, R.T. (eds.)
Proceedings of the IUTAM Symposium on Finite Elasticity
ISBN 90 247 2629 8

considered such deformations for both compressible and incompressible materials, e.g. Sensenig [4], Holden [5], Varley and Cumberbatch [6], Ogden [7] and Isherwood and Ogden [8–10]. In some cases [6, 8–10] the authors solve boundary value problems exactly, rather than using the inverse method of specifying the deformation and calculating the required surface tractions. However their analyses are necessarily complicated.

In this note we give examples of constitutive equations for which certain easily-visualised controllable non-universal deformations are possible. We consider both compressible and incompressible materials. We keep the constitutive assumptions sufficiently general to emphasize the fact that the equilibrium equations do not have to be satisfied identically in the strain invariants, but only at the particular deformation considered. However, all of our constitutive assumptions for compressible materials include, as a special case, a member of the well-studied class of Hadamard materials [11–14].

2. Basic equations

A particle $\boldsymbol{X}$ in the undeformed body is displaced to $\boldsymbol{x}$ where

$$\boldsymbol{x} = \boldsymbol{x}(\boldsymbol{X}), \qquad x_i = x_i(X_j). \tag{2.1}$$

The deformation gradient $\boldsymbol{F}$ has components

$$F_{ik} = \partial x_i / \partial X_k, \tag{2.2}$$

and the left Cauchy–Green strain tensor B_{ij} has components[†]

$$B_{ij} = \frac{\partial x_i}{\partial X_k} \frac{\partial x_j}{\partial X_k}. \tag{2.3}$$

Three independent invariants of $\boldsymbol{B}$ are

$$I = B_{ii}, \qquad II = (I^2 - B_{ij}B_{ji})/2, \qquad III = \det \boldsymbol{B}. \tag{2.4}$$

The Cauchy stress t_{ij} for a homogeneous isotropic compressible elastic material has the form [15]

$$t_{ij} = N_0\delta_{ij} + N_1 B_{ij} + N_2 B_{ik}B_{kj}, \tag{2.5}$$

or equivalently

$$t_{ij} = h_0\delta_{ij} + h_1 B_{ij} + h_{-1} B_{ij}^{-1}, \tag{2.6}$$

where N_α, h_α are functions of I, II, III. The stress in the undeformed body is zero if

$$h_0 + h_1 + h_{-1} = 0 \qquad \text{at } I = II = 3,\ III = 1. \tag{2.7}$$

[†] The summation convention is used throughout. Repeated suffixes are summed over 1, 2, 3.

For an incompressible material $III = 1$, and we write

$$t_{ij} = -p\delta_{ij} + h_1 B_{ij} + h_{-1} B_{ij}^{-1}, \tag{2.8}$$

where h_1 and h_{-1} are functions of I and II and where p is a hydrostatic pressure to be determined from the equilibrium equations and the boundary conditions.

The material is said to be hyperelastic if it possesses a strain-energy function. Let this be denoted by $\Sigma(I, II, III)$ measured per unit volume of undeformed body. Then for compressible materials [15],

$$\begin{gathered} h_0 = 2III^{1/2}\frac{\partial \Sigma}{\partial III} + 2III^{-1/2} II \frac{\partial \Sigma}{\partial II}, \\ h_1 = 2III^{-1/2}\frac{\partial \Sigma}{\partial I}, \qquad h_{-1} = -2III^{1/2}\frac{\partial \Sigma}{\partial II}. \end{gathered} \tag{2.9}$$

For incompressible materials $\Sigma = \Sigma(I, II)$, and the first of the relations (2.9) is dropped.

A well-studied special case of the above constitutive assumptions is the hyperelastic Hadamard material for which [11–14]

$$\begin{gathered} 2\Sigma = \nu(I-3) - \mu(II-3) + \int_1^{III} q(\theta)\theta^{-1/2} d\theta, \\ h_0 = q(III) - \frac{\mu II}{III^{1/2}}, \qquad h_1 = \frac{\nu}{III^{1/2}}, \qquad h_{-1} = \mu III^{1/2}, \end{gathered} \tag{2.10}$$

where ν and μ are constants and q an arbitrary function. Ogden [11] has shown that if a Hadamard material is to satisfy the strong-ellipticity condition then

$$\mu \leqq 0, \qquad \nu > 0, \qquad \frac{dq}{dIII} \geqq 0. \tag{2.11}$$

If, in addition, the Ordered-Forces condition is satisfied [12]

$$\mu = 0, \qquad q(III) \leqq 0. \tag{2.12}$$

Finally, in the absence of body forces, the equilibrium equations are

$$\frac{\partial t_{ij}}{\partial x_j} = 0. \tag{2.13}$$

3. Variable shear

Consider first the deformation

$$x = X + f(Y, Z), \qquad y = Y, \qquad z = Z, \tag{3.1}$$

where (x, y, z) and (X, Y, Z) are rectangular Cartesian coordinates of $\boldsymbol{x}$ and $\boldsymbol{X}$. This deformation corresponds for example to a variable shear along the x-axis. The function f is chosen to satisfy

(3.2) $$f_{YY} + f_{ZZ} = 0$$

where $f_Y = \partial f/\partial Y$, etc. We have from (2.2) and (2.3)

$$(\boldsymbol{F}) = \begin{pmatrix} 1 & f_Y & f_Z \\ 0 & 1 & 0 \\ 0 & 0 & 1 \end{pmatrix},$$

(3.3) $$(\boldsymbol{B}) = \begin{pmatrix} 1 + f_Y^2 + f_Z^2 & f_Y & f_Z \\ f_Y & 1 & 0 \\ f_Z & 0 & 1 \end{pmatrix},$$

$$(\boldsymbol{B}^{-1}) = \begin{pmatrix} 1 & -f_Y & -f_Z \\ -f_Y & 1 + f_Y^2 & f_Y f_Z \\ -f_Z & f_Y f_Z & 1 + f_Z^2 \end{pmatrix},$$

and hence from (2.4),

(3.4) $$I = \hat{I} = 3 + f_Y^2 + f_Z^2, \qquad II = \hat{II} = \hat{I}, \qquad III = 1.$$

From (2.6) or (2.8) and (3.3), notice that

(3.5) $$t_{xy} = (\hat{h}_1 - \hat{h}_{-1}) f_Y, \qquad t_{xz} = (\hat{h}_1 - \hat{h}_{-1}) f_Z,$$

where $\hat{h}_{\pm 1} = h_{\pm 1}(\hat{I}, \hat{II}, 1)$. Thus, for materials in which this deformation is possible there is the following universal relation, which is independent of the particular properties of the material,

(3.6) $$\frac{t_{xy}}{t_{xz}} = \frac{f_Y}{f_Z}.$$

Similarly, there is the second universal relation

(3.7) $$\frac{t_{yy} - t_{zz}}{t_{yz}} = \frac{f_Y^2 - f_Z^2}{f_Y f_Z}.$$

However, because of Ericksen's theorem [1, 2], there is no guarantee that the deformation (3.1) is possible in any particular elastic material.

The deformation is possible in the class of compressible materials for which

(3.8) $$\begin{aligned} h_1 &= (I - II) g_1(I, II, III) + n_1(I - II, III) + (III - 1) k_1(I, II, III), \\ h_{-1} &= (I - II) g_2(I, II, III) + n_2(I - II, III) + (III - 1) k_2(I, II, III), \\ h_0 &= -\left(\frac{I + II}{4}\right) n_2 + n_3(I - II, III) + (III - 1) k_3(I, II, III). \end{aligned}$$

Here g_1, g_2, n_1, n_2, n_3, k_1, k_2, k_3 are arbitrary finite-valued functions of the indicated arguments. By (2.7),

$$2n_1(0,1)+2n_3(0,1)=n_2(0,1), \tag{3.9}$$

to ensure zero stress at zero deformation.

With this choice of constitutive equation, it is easily checked that the equilibrium equations (2.13) are satisfied provided f satisfies (3.2).

A special case of the constitutive equation (3.8) is the hyperelastic Hadamard material (2.10) with $\mu=0$. By suitable choice of the functions, conditions (2.11) and (2.12) can be satisfied.

The function f satisfies (3.2). Accordingly it has the form

$$f=\mathrm{Re}\{\alpha(Y+\iota Z)+\beta(Y-\iota Z)\}, \tag{3.10}$$

where α and β are arbitrary functions. It includes, for example,

$f=KY\cdots$ simple shear;

$f=KYZ\cdots$ shear of the hyperbolic cylinders $YZ=\text{constant}$;

$f=K\ln\left(\dfrac{Y^2+Z^2}{a^2}\right)\cdots$ shear of the circular cylinders $Y^2+Z^2=\text{constant}$

in which the cylinder $Y^2+Z^2=a^2$ is held fixed.

Up to this point, it has been assumed that the materials are compressible. If now incompressible materials are considered it is easily checked that the equilibrium equations are satisfied for the deformation (3.1), where f satisfies (3.2), if the constitutive functions are given by

$$\begin{aligned} h_1&=(I-II)\ell_1(I,II)+v_1(I-II),\\ h_{-1}&=(I-II)\ell_2(I,II)+v_2(I-II), \end{aligned} \tag{3.11}$$

and p is chosen to satisfy

$$p=\frac{v_2(0)}{2}(f_Y^2+f_Z^2)+\text{constant}. \tag{3.12}$$

A special case of this is the hyperelastic material

$$\begin{aligned} h_1&=b+\tfrac{1}{2}a(I^2-II^2)+g(I-II),\\ h_{-1}&=a(I-II)II+g(I-II), \end{aligned} \tag{3.13}$$

$$\Sigma=\int_0^{I-II}g(\theta)d\theta-\tfrac{1}{2}aI\,II^2+\tfrac{1}{3}aII^3+\tfrac{1}{6}aI^3+bI,$$

where a, b are constants and g is an arbitrary function. A subcase of this is the Mooney–Rivlin model

$$\Sigma = \alpha(I-3) + \beta(II-3), \tag{3.14}$$

where α, β are constants.

4. Variable uniaxial extension

Consider next the deformation

$$\begin{gathered} x = f(X), \qquad y = Y/g(X), \qquad z = Z/g(X), \\ g(X) = df/dX, \end{gathered} \tag{4.1}$$

where once again Cartesian coordinates are used. This deformation corresponds to a non-uniform extension of a bar, with a corresponding change in the cross-section.

For this deformation, from (2.2) and (2.3)

$$\begin{gathered} (\boldsymbol{F}) = \begin{pmatrix} g & 0 & 0 \\ \dfrac{-g'y}{g} & \dfrac{1}{g} & 0 \\ \dfrac{-g'z}{g} & 0 & \dfrac{1}{g} \end{pmatrix}, \\ (\boldsymbol{B}) = \begin{pmatrix} g^2 & -g'y & -g'z \\ -g'y & \dfrac{1}{g^2}(1+(g'y)^2) & \left(\dfrac{g'}{g}\right)^2 yz \\ -g'z & \left(\dfrac{g'}{g}\right)^2 yz & \dfrac{1}{g^2}(1+(g'z)^2) \end{pmatrix}. \end{gathered} \tag{4.2}$$

Hence, by (2.4)

$$\begin{gathered} I = \hat{I} = g^2 + \frac{1}{g^2}(2 + (g')^2(y^2+z^2)), \\ III = \widehat{III} = \frac{1}{g^2}, \\ II = \widehat{II} = 1 - \widehat{III}^2 + \hat{I}\,\widehat{III}. \end{gathered} \tag{4.3}$$

We take

$$\begin{gathered} h_0 = n(0, III) + J^2 k(J, III) + III \frac{\partial}{\partial III} n(J, III), \\ h_1 = -III^2 \frac{\partial}{\partial III} n(J, III), \qquad h_{-1} = 0, \end{gathered} \tag{4.4}$$

where

$$J = II + III^2 - I\,III - 1. \tag{4.5}$$

n and k are arbitrary, finite-valued functions. From (2.7)

$$n(0,1) = 0. \tag{4.6}$$

It can be checked that the equilibrium equations (2.13) are satisfied provided g satisfies the differential equation

$$\frac{d}{dX}(g'h_1) = \frac{3(g')^2 h_1}{g}\,. \tag{4.7}$$

The Hadamard material (2.10) is a special case of (4.4) if $\mu = 0$ and $q(III) = -\nu(2 + III^{-3/2})/3$. For this Hadamard material, the solution of (4.7) is

$$\begin{aligned} g &= (AX + B)^{-1}, \\ f &= C + A^{-1}\ln(AX + B), \end{aligned} \tag{4.8}$$

for arbitrary constants A, B, C.

5. Torsion of a compressible body

Now consider simple torsion:

$$r = R, \qquad \theta = \Theta + DZ, \qquad z = Z. \tag{5.1}$$

Here (R, Θ, Z) are the material coordinates in a cylindrical polar coordinate system and (r, θ, z) are the spatial coordinates in the same system. The details are set out in Truesdell and Noll [15, Section 57]. We quote:

$$\hat{I} = \widehat{II} = 3 + D^2r^2, \qquad \widehat{III} = 1. \tag{5.2}$$

The physical components of stress are denoted by $t(ij)$ and given by

$$\begin{aligned} t(rr) &= \hat{h}_0 + \hat{h}_1 + \hat{h}_{-1}, \\ t(r\theta) &= t(rz) = 0, \\ t(\theta z) &= Dr(\hat{h}_1 - \hat{h}_{-1}), \\ t(\theta\theta) &= \hat{h}_0 + (1 + D^2r^2)\hat{h}_1 + \hat{h}_{-1}, \\ t(zz) &= \hat{h}_0 + \hat{h}_1 + (1 + D^2r^2)\hat{h}_{-1}. \end{aligned} \tag{5.3}$$

The only equilibrium equation not satisfied identically gives

$$\frac{d}{dr}t(rr) + \frac{t(rr) - t(\theta\theta)}{r} = 0, \tag{5.4}$$

or, using (5.2), (5.3),

$$2\left(\frac{\partial}{\partial I}+\frac{\partial}{\partial II}\right)(\hat{h}_0+\hat{h}_1+\hat{h}_{-1})-\hat{h}_1=0. \tag{5.5}$$

To satisfy this choose the constitutive functions

$$\begin{aligned} h_1 &= (I-II)g(I,II,III)+d_1(I-II,III)+(III-1)e_1(I,II,III),\\ h_0+h_{-1} &= (I+II)d_1/4+d_2(I-II,III)+(III-1)e_2(I,II,III), \end{aligned} \tag{5.6}$$

where g, d_1, d_2, e_1 and e_2 are finite-valued functions of the indicated arguments, with $3d_1(0,1)+2d_2(0,1)=0$, according to (2.7).

A special case of (5.6) is the hyperelastic Hadamard material (2.10) with $\mu=-2\nu$, found by taking

$$\begin{gathered} g=e_1=e_2=0, \qquad d_1=\nu/III^{1/2},\\ d_2=q(III)+\mu III^{1/2}-\nu(I-II)/4III^{1/2}. \end{gathered} \tag{5.7}$$

Conditions (2.11) can be satisfied, but not the more restrictive conditions (2.12).

References

1. J. L. Ericksen, J. Math. Phys. **34** (1955), 126–128.
2. J. L. Ericksen, ZAMP **5** (1954), 466–486.
3. R. S. Rivlin and D. W. Saunders, Phil. Trans. Roy. Soc. Lond. A **243** (1951), 251–288.
4. C. B. Sensenig, Comm. Pure Appl. Math. **18** (1965), 147–161.
5. J. T. Holden, Appl. Sci. Res. **19** (1968), 171–181.
6. E. Varley and E. Cumberbatch, in *Finite Elasticity* (R. S. Rivlin, ed.), Am. Soc. Mech. Eng., New York, 1977, pp. 41–64.
7. R. W. Ogden, Mech. Res. Comm. **4** (1977), 347–352.
8. D. A. Isherwood and R. W. Ogden, Rheol. Acta **16** (1977), 113–122.
9. D. A. Isherwood and R. W. Ogden, Int. J. Solids Structures **13** (1977), 105–123.
10. R. W. Odgen and D. A. Isherwood, Q. J. Mech. Appl. Math. **31** (1978), 219–249.
11. R. W. Ogden, J. Mech. Phys. Solids **18** (1970), 149–163.
12. M. Hayes, Quart. J. Mech. Appl. Math. **21** (1968), 141–146.
13. P. K. Currie and M. Hayes, J. Inst. Maths. Applics. **5** (1969), 140–161.
14. P. K. Currie, ZAMP **22** (1971), 355–359.
15. C. Truesdell and W. Noll, Handbuch der Physik III/1, Springer-Verlag, Berlin, 1965.

NON-LINEAR BOUNDARY VALUE PROBLEM IN THERMOELASTICITY

G. DUVAUT

Université de Paris VI, Mécanique Théorique, Tour 66–4, Place Jussieu, Paris 75230, Cedex 05, France

(Received September 11, 1980)

I. Introduction

In this study we consider an elastic body subjected to a temperature field and in unilateral contact with a rigid body. This type of problem can be considered as a generalization of the Signorini problem to cases involving a temperature field. The purpose of the work is the determination of the right boundary conditions so that the problem be well posed both from a mathematical and a physical point of view. Two results appear: the heat flux through the unilateral boundary must depend on the normal pressure, the second point is that the normal pressure to consider is not the density of normal stress but some local mean value of the density of normal stress. The last two sections are devoted to the study of the limit solution obtained when the heat transfer coefficient at the boundary tends to plus infinity and have been developed after a question of H. Brézis.

II. Physical problem

Let Ω be an open bounded region of $\mathbb{R}^3$, whose boundary $\partial\Omega$ is smooth and made of three disjoined parts Γ_0, Γ_1, Γ_1. On Γ_0 we assume a given temperature that we choose equal to zero, and furthermore the body is clamped. On Γ_1 we prescribe a heat flux and a density of surface forces; on Γ_2 the contact is one-sided which leads to Signorini [3] boundary conditions for the mechanical effects and thermal conditions such that there is no heat flux at points where there is no contact and a given temperature where there is contact.

Proceedings of the IUTAM Symposium on Finite Elasticity, Lehigh University, August 10–15, 1980. Invited paper.

Carlson, D.E. and Shield, R.T. (eds.)
Proceedings of the IUTAM Symposium on Finite Elasticity
ISBN 90 247 2629 8

The equations to be satisfied in Ω are the following [1]:

(1) $$\frac{\partial \sigma_{ij}}{\partial x_j} + f_i = 0 \qquad \text{(equilibrium equations);}$$

(2) $$\sigma_{ij} = a_{ijkh}\varepsilon_{kh}(u) - \beta\Theta\delta_{ij} \qquad \text{(thermoelastic constitutive relations).}$$

The symbol $\{\sigma_{ij}\}$ stands for the stress tensor, and $\varepsilon_{ij}(u)$ is the strain given by

$$\varepsilon_{ij}(u) = \tfrac{1}{2}\left(\frac{\partial u_i}{\partial x_j} + \frac{\partial u_j}{\partial x_i}\right).$$

The temperature field is Θ; a_{ijkh} are the elastic coefficients and β a positive thermal constant. Thermal equilibrium implies

(3) $$-\Delta\Theta = g \qquad \text{in } \Omega.$$

The boundary conditions previously defined are

(4) $$u = 0 \quad \text{and} \quad \Theta = 0 \qquad \text{on } \Gamma_0,$$

(5) $$\sigma_{ij}n_j = F_i, \qquad \frac{\partial\Theta}{\partial n} = G \qquad \text{on } \Gamma_1,$$

(6) $$\sigma_T = 0 \qquad \text{on } \Gamma_2 \text{ (no friction),}$$

(7) $$\left\{\begin{array}{ll} u_N \leqq 0 & \\ u_N < 0 \Rightarrow & \sigma_N = 0 \\ & \dfrac{\partial\Theta}{\partial n} = 0 \\ u_N = 0 \Rightarrow & \sigma_N \leqq 0 \\ & \Theta = \Theta_1 \end{array}\right\} \qquad \text{on } \Gamma_2,$$

where Θ_1 is a given temperature field on Γ_2. In expressing (6) and (7), we have used the following notation: if v is a given vector field on Γ_2, we split it into its normal component v_N and its tangential component v_T by

$$v_N = v \,.\, n, \qquad v_T = v - v_N n,$$

where n is the unit outer normal. According to this notation u_N is the normal component of the displacement vector u, and σ_N and σ_T are the normal and tangential components of the stress vector $\{\sigma_{ij}n_j\}$.

Because of conditions (7) the class of functions (u, Θ) in which we must look for the solution of (1)–(7) is badly defined. This fact makes the written problem difficult to study and in fact we gave it up the first time we considered it in 1972.

Recently a work by M. Comninou and J. Dundurs [2] shows in a particular case that this type of problem may have no solution.

III. One-dimensional case (M. Comninou, J. Dundurs [2])

Let us consider a rectilinear bar AB of length L when it is at rest at 20 degrees Celsius temperature and clamped at its end A. A wall is orthogonal to AB at distance L from A. This wall is assumed to be at 0 degrees. We shall study what will happen according to the temperature Θ_A given at end A.

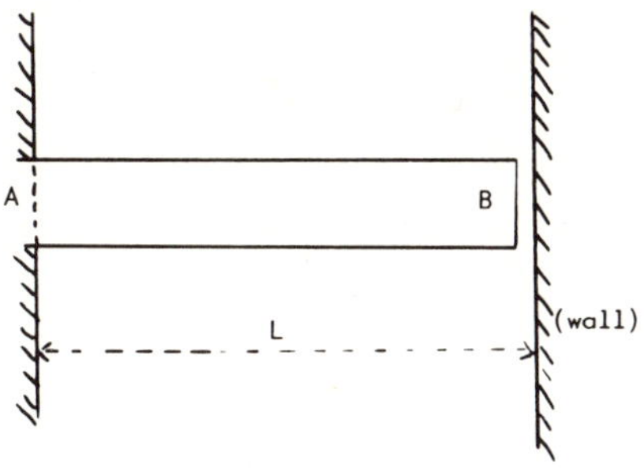

First case. $\Theta_A < 20$. The bar contracts, so there is no contact with the wall at B and the whole bar is at temperature Θ_A.

Second case. $20° < \Theta_A < 40°$. If there is no contact at B with the wall, the whole bar is at temperature Θ_A. This implies that the bar extends and its length is greater than L, and there is contact at B with the wall.

If there is contact at B with the wall, this point will be at temperature zero. The field of temperature in the bar is linear and given by

$$\Theta(x) = \Theta_A \frac{L - x}{L}$$

and, since $\Theta_A < 40°$, most of the bar is in contraction and the total length of the bar is less than L. This implies that there is no contact at B with the wall.

In this case the stated problem has no solution.

Third case. $\Theta_A > 40°$. We check easily that there is a solution with contact at B with the wall.

These results, obtained in [2], show no solution in the second case. But, from an experimental point of view, it appears that there is a steady solution for every given temperature Θ_A. Consequently boundary conditions of type (7) are not those which occur physically. It is shown in [2] that we can modify the boundary conditions in the case of contact and obtain a solution: we accept a temperature difference and impose a heat flux proportional to this temperature difference.

In the three-dimensional case considered in II, we shall modify condition (7) in the previous sense and study the new boundary value problem.

IV. Modified problem

We keep equations and boundary conditions (1)–(6), and we replace (7) by (8) and (9):

$$\text{(8)} \qquad \left\{ \begin{array}{l} u_N \leqq 0 \\ u_N < 0 \Rightarrow \sigma_N = 0 \\ u_N = 0 \Rightarrow \sigma_N \leqq 0 \end{array} \right\} \quad \text{on } \Gamma_2,$$

$$\text{(9)} \qquad -\frac{\partial \Theta}{\partial n} = k(\sigma_N)(\Theta - \Theta_1) \qquad \text{on } \Gamma_2,$$

where the function $\xi \to k(\xi)$ from $\mathbb{R}$ to $\mathbb{R}$ is given and satisfies $k(\xi) = 0$ for $\xi \geqq 0$ and $k(\xi) \geqq 0$ for $\xi \in \mathbb{R}$.

We shall now study the problem (1)–(6), (8), (9) in which the preliminary difficulties encountered with (1)–(7) have disappeared. In particular we get easily a variational property satisfied by any smooth solution (u, Θ).

V. Variational formulation

We introduce

$$\text{(10)} \qquad \left\{ \begin{array}{l} V_1 = \{v \mid v \in (H^1(\Omega))^3, v = 0 \text{ on } \Gamma_0\}, \\ K_1 = \{v \mid v \in V_1, v_N \leqq 0 \text{ on } \Gamma_2\}, \\ V_2 = \{\varphi \mid \varphi \in H^1(\Omega), \varphi = 0 \text{ on } \Gamma_0\}, \end{array} \right.$$

$$\text{(11)} \qquad \left\{ \begin{array}{l} a_1(u, v) = \int_\Omega a_{ijkh}\varepsilon_{kh}(u)\varepsilon_{ij}(v)dx, \\ a_2(\varphi, \psi) = \int_\Omega \nabla\varphi . \nabla\psi dx. \end{array} \right.$$

We have the *variational property*: If (u, Θ) is a smooth solution of the problem (1)–(6), (8), (9), it satisfies

$$\text{(12)} \qquad \left\{ \begin{array}{l} u \in K_1, \\ a_1(u, v-u) - \beta \int_\Omega \Theta \varepsilon_{\ell\ell}(v-u)dx \geqq \int_\Omega f(v-u)dx + \int_{\Gamma_1} F(v-u)d\Gamma, \\ \qquad \forall v \in K_1, \end{array} \right.$$

$$(13)\quad \begin{cases} \theta \in V_2, \\ a_2(\Theta, \varphi) + \int_{\Gamma_2} k(\sigma_N)(\Theta - \Theta_1)\varphi d\Gamma = \int_\Omega g\varphi dx + \int_{\Gamma_1} G\varphi d\Gamma, \quad \forall \varphi \in V_2. \end{cases}$$

We introduce the further notations

$$(14)\quad \begin{cases} L_1(v) = \int_\Omega fvdx + \int_{\Gamma_1} Fvd\Gamma, \\ L_2(\varphi) = \int_\Omega g\varphi dx + \int_{\Gamma_1} G\varphi d\Gamma. \end{cases}$$

Looking at equation (13) we make the following remark. If the solution (u, Θ), instead of being very smooth, belongs only to $K_1 \cap V_2$, all terms of (12) and (13) keep a clear meaning, except

$$Z = \int_{\Gamma_2} k(\sigma_N)(\Theta - \Theta_1)\varphi d\Gamma.$$

In fact $\Theta - \Theta_1$ and φ belong to $H^{1/2}(\Gamma_2)$. The normal component σ_N can be defined as an element of $H^{-1/2}(\Gamma_2)$, but this does not give a pointwise value of σ_N and does not allow us to compute $k(\sigma_N)$. If the problem was purely a mathematical one, it would be possible to define function k as a transformation from $H^{-1/2}(\Gamma_2)$ to $L^2(\Gamma_2)$, and the Z term would then take a precise value because, for $Q \subset R^3$, $H^{1/2}(\Gamma_2) \subset L^4(\Gamma_2)$, which implies $(\Theta - \Theta_1)\varphi \in L^2(\Gamma_2)$.

We shall do something similar, but based on physical arguments. The relation (9) comes from experiment, and means that the heat flux density is proportional to the temperature difference $\Theta - \Theta_1$, the proportionality coefficient being dependent on the normal stress density σ_N. But the normal stress density is a mathematical concept which is measured experimentally as the ratio of a force and a surface element. This means that relation (9) has not been checked with σ_N, but with some σ_N^* which is a regularization of σ_N.

Consequently we shall replace (9) by

$$(15)\quad -\frac{\partial \Theta}{\partial n} = k(\sigma_N^*)(\Theta - \Theta_1),$$

where the transformation

$$\tau \mapsto \tau^*$$

is linear continuous from $H^{-1/2}(\Gamma_2)$ to $L^2(\Gamma_2)$. The term Z is then replaced by

$$\int_{\Gamma_2} k(\sigma_N^*)(\Theta - \Theta_1)\varphi d\Gamma,$$

which has a clear meaning, provided that $k(\sigma_N^*) \in L^2(\Gamma_2)$, which is true if, for example, k is a lipschitz function from $\mathbb{R}$ to $\mathbb{R}$.

We have now the following property: If (u, Θ) is a solution of problem (1)–(6), (8), (15), it satisfies

$$(16) \quad \begin{cases} u \in K_1, \\ a_1(u, v-u) - \beta \displaystyle\int_\Omega \Theta \varepsilon_{\ell\ell}(v-u)dx \geqq L_1(v-u), \qquad \forall v \in K_1, \end{cases}$$

$$(17) \quad \begin{cases} \Theta \in V_2, \\ a_2(\Theta, \varphi) + \displaystyle\int_{\Gamma_2} k(\sigma_N^*)(\Theta - \Theta_1)\varphi d\Gamma = L_2(\varphi), \qquad \forall \varphi \in V_2. \end{cases}$$

The relations (16), (17) give us a weakened formulation of the problem (1)–(6), (8), (15).

VI. Results

The assumptions made on the data are the following:

$$(18) \qquad f_i \in L^2(\Omega), \qquad F_i \in L^2(\Gamma_1), \qquad G \in L^2(\Gamma_1), \qquad g \in L^2(\Omega).$$

$$(19) \quad \begin{cases} a_{ijkh} = a_{jikh} = a_{khij} \in L^\infty(\Omega), \\ \exists \alpha_0 > 0, \qquad a_{ijkh}\varepsilon_{kh}\varepsilon_{ij} \geqq \alpha_0 \varepsilon_{ij}\varepsilon_{ij}, \qquad \forall \varepsilon_{ij} = \varepsilon_{ji}. \end{cases}$$

(20) $\qquad \xi \mapsto k(\xi)$ is a given lipschitz function from $\mathbb{R}$ to $\mathbb{R}^+$.

(21) $\qquad \text{mes}\, \Gamma_0 > 0, \qquad \exists \hat{\Theta}_1 \in V_2$ such that $\hat{\Theta}_1 |_{\Gamma_1} = \Theta_1$.

THEOREM. *With assumptions* (18)–(21), *the problem* (16), (17) *possesses at least one solution. Furthermore, if function k is given by*

$$k(\xi) = \alpha k_1(\xi), \qquad \alpha \in \mathbb{R}^+,$$

there exists $\alpha_1 > 0$ *such that if*

$$0 \leqq \alpha \leqq \alpha_1$$

the solution is unique.

VII. Proof of the theorem

7.1. *Reduction to a fixed point problem*

For any $\psi \in L^2$, if we choose $\Theta = \psi$ in (16), we get a unique solution u of the corresponding equation (16). Successively we have,

$$\varepsilon_{ij}(u) \in L^2(\Omega), \qquad \sigma_{ij} \in L^2(\Omega), \qquad \sigma_N \in H^{-1/2}(\Gamma_2), \qquad \sigma_N^* \in L^2(\Gamma_2).$$

Substituting σ_N^* into (17), we define Θ as the unique solution of the resulting equation (17). We then have defined a function T from $L^2(\Omega)$ to V_2,

$$\psi \in L^2(\Omega) \mapsto \Theta = T(\psi) \in V_2.$$

It is clear that a fixed point of the function T is a solution of (16), (17).

7.2. *Existence of a fixed point*

LEMMA 1. *For any $\psi \in L^2(\Omega)$, $\Theta = T(\psi)$ belongs to a bounded set of V_2, which is a convex compact set of $L^2(\Omega)$.*

PROOF. We choose $\varphi = \Theta - \hat{\Theta}_1$ in (17), and we obtain

$$a_2(\Theta, \Theta - \hat{\Theta}_1) \leqq L_2(\Theta - \hat{\Theta}_1)$$

from which we get

$$\|\Theta\| \leqq C$$

where the constant C is independent of ψ.

LEMMA 2. *The function $\psi \mapsto T(\psi)$ from $L^2(\Omega)$ to V_2 is continuous.*

PROOF. For any ψ and $\psi' \in L^2(\Omega)$ we get u and u', respectively, such that

$$a_1(u'-u, u'-u) \leqq \beta \int_\Omega (\psi' - \psi)\varepsilon_{\ell\ell}(u'-u)dx.$$

which implies

$$\|u' - u\|_{V_1} \leqq C_1 |\psi' - \psi|_{L^2(\Omega)}$$

where C_1 is a constant independent of ψ and ψ'. The corresponding stress tensors $\{\sigma'_{ij}\}$ and $\{\sigma'_{ij}\}$ satisfy

$$|\sigma'_{ij} - \sigma_{ij}|_{L^2(\Omega)} \leqq C_2 |\psi' - \psi|_{L^2(\Omega)}$$

and

$$\frac{\partial}{\partial x_j}(\sigma'_{ij} - \sigma_{ij}) = 0 \qquad \text{in } \Omega.$$

Consequently

$$|\sigma'_N - \sigma_N|_{H^{-1/2}(\Gamma_2)} \leqq C_3 |\psi' - \psi|_{L^2(\Omega)},$$

and then

$$|\sigma'^*_N - \sigma^*_N|_{L^2(\Omega)} \leqq C_4 |\psi' - \psi|_{L^2(\Omega)}.$$

All constants C do not depend on ψ and ψ' in $L^2(\Omega)$. The corresponding function Θ and Θ' will satisfy

$$(22)\quad a_2(\Theta'-\Theta,\Theta'-\Theta)-\int_{\Gamma_2}[k'(\Theta'-\Theta_1)(\Theta-\Theta')+k(\Theta-\Theta_1)(\Theta'-\Theta)]d\Gamma=0$$

with the notation

$$k'=k(\sigma_N'^{*}),\qquad k=k(\sigma_N^{*}).$$

From (22) and assumptions on the $k(\xi)$ function, we obtain that there exists a constant C_5 independent of ψ and ψ' such that

$$(23)\qquad \|\Theta'-\Theta\|_{V_2}\leqq C_5|\psi'-\psi|_{L^2(\Omega)}.$$

From (23) and Lemma 1 it is clear that the function T has at least one fixed point.

7.3. *Uniqueness*

From the reasoning that led to (23), it follows that the constant C_5 is proportional to α,

$$C_5=\alpha C_5',$$

where C_5' does not depend on α. Consequently, if

$$(24)\qquad 0\leqq\alpha<\alpha_1=\frac{1}{C_5'},$$

then the function T is a strict contraction, which proves the existence of a unique fixed point.

VIII. Explicit solution

In the case of the one-dimensional example mentioned in Section III, we can obtain the explicit solution corresponding to the problem where the equations and boundary conditions are the following

$$(25)\qquad \begin{cases} \dfrac{\partial\sigma}{\partial x}=0 & \text{on } \Omega=]0,L[\ , \\ \sigma=a\dfrac{\partial u}{\partial x}-\beta(\Theta-20), & \\ \dfrac{\partial^2\Theta}{\partial x^2}=0 & \text{on } \Omega, \\ \Theta(x)\big|_{x=0}=\Theta_A & (\Theta_A \text{ given}), \end{cases}$$

(26) $$\sigma(L) \leqq 0, \qquad u(L) \leqq 0, \qquad u(L)\sigma(L) = 0,$$

(27) $$-\frac{\partial \Theta}{\partial x}(L) = k(\sigma(L))\Theta(L).$$

The function $k(\xi)$ appearing in the last condition is given from $\mathbb{R}$ to $\mathbb{R}$, decreasing, continuous and equal to zero for $\xi \geqq 0$.

The set of equations (25)–(27) can be solved explicitly.

We get easily from (25) that

(28) $$\sigma(x) = \sigma(L) = a\frac{du}{dx} - \beta(\Theta - 20), \qquad \Theta(x) = \Theta_A + \frac{\Theta(L) - \Theta_A}{L}x;$$

from which we get

(29) $$u(x) = \frac{x}{a}[\sigma(L) + \beta(\Theta_A - 20)] + \frac{\beta x^2}{2aL}(\Theta(L) - \Theta_A), \qquad u(L) = \frac{L}{a}[\sigma(L) + \beta(\Theta_A - 20)] + \frac{\beta L}{2a}[\Theta(L) - \Theta_A].$$

Let us now use condition (26) which implies either $\sigma(L) = 0$ or $u(L) = 0$.

(i) If $\sigma(L) = 0$, equation (27) and $k(0) = 0$ give

$$\frac{\partial \Theta}{\partial x}(L) = 0.$$

Then from (28),

$$\Theta(L) = \Theta_A,$$

and from (29),

$$u(L) = \frac{L\beta}{a}(\Theta_A - 20).$$

Since in this case $u(L)$ must be negative, this solution corresponds to $\Theta_A < 20°$.

(ii) If $u(L) = 0$, the unknowns $\sigma(L)$, $\Theta(L)$ are given by

(30) $$\begin{cases} -\dfrac{\Theta(L) - \Theta_A}{L} = k(\sigma(L))\Theta(L), \\ 0 = \sigma(L) + \beta(\Theta_A - 20) + \dfrac{\beta}{2}(\Theta(L) - \Theta_A). \end{cases}$$

The second of equations (30) leads to

$$\Theta(L) = -\frac{2}{\beta}\sigma(L) - \Theta_A + 40, \tag{31}$$

and using the first of (30) we get

$$k(\sigma(L)) = \frac{1}{L}\,\frac{2\Theta_A - 40 + \dfrac{2}{\beta}\sigma(L)}{40 - \Theta_A - \dfrac{2}{\beta}\sigma(L)}. \tag{32}$$

The unknown $\sigma(L) = \xi$ is then given by studying the points common to the two curves

$$\begin{cases} y = \dfrac{1}{L}\,\dfrac{2\Theta_A - 40 + \dfrac{2}{\beta}\xi}{40 - \Theta_A - \dfrac{2}{\beta}\xi}, \\[2ex] y = k(\xi). \end{cases} \tag{33}$$

According to the values of Θ_A, we have the situations depicted below.

From these we see that

(1) If $\Theta_A < 20°$, the only point common to the curves (33) is

$$\sigma(L) = \xi_1 > 0,$$

which cannot be a solution because of (26). Consequently in the case $\Theta_A < 20°$, the only solution is that obtained in (i), which corresponds to no contact at B and

$$\begin{cases} \Theta(x) = \Theta_A = \Theta(L), \\ \sigma(x) = \sigma(L) = 0, \\ u(x) = \dfrac{\beta x}{a}(\Theta_A - 20) \leqq 0. \end{cases} \tag{34}$$

(2) If $20° < \Theta_A < 40°$, we have one intersection point of curves (33) which corresponds to $\sigma(L)$ satisfying

$$\xi_2 = \beta(20 - \Theta_A) < \sigma(L) < 0.$$

This leads to a solution with contact at B given by

$$\Theta(x) = \Theta_A + \frac{\Theta(L) - \Theta_A}{L}x, \tag{35}$$

where $\Theta(L)$ comes from (30) as

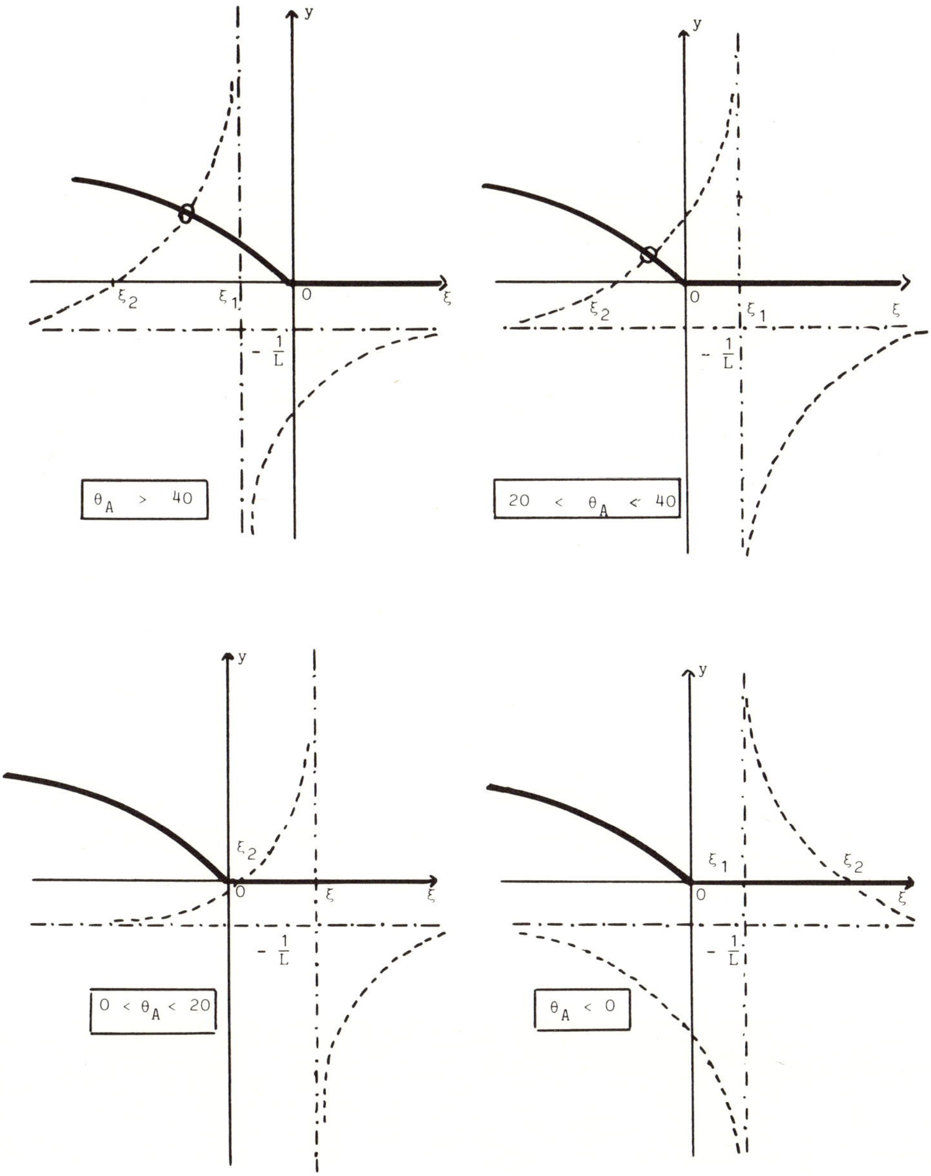

(36) $$\Theta(L)=\frac{\Theta_A}{1+Lk(\sigma(L))}$$

and $u(x)$ is given by (29).

(3) If $\Theta_A \geqq 40°$, we have an intersection point of curves (33) which corresponds to $\sigma(L)$ satisfying

$$\xi_2 = \beta(20 - \Theta_A) < \sigma(L) < \xi_1 = \frac{\beta}{2}(40 - \Theta_A).$$

This leads to a solution with contact at B, and formulas (35), (36) remain valid.

REMARK 1. In the particular case considered, the solution obtained is unique and this comes from the fact that the function $k(\xi)$ is decreasing. In cases where the function $k(\xi)$ could be oscillating for negative values of ξ, we should have several solutions.

REMARK 2. If we assume the function $k(\xi)$ to be of the form

$$k(\xi) = \alpha k_1(\xi),$$

k_1 being independent of α, the solution obtained would of course depend on α. Letting α tend to plus infinity means that the heat transfer coefficient is greater and greater. If we pass to the limit in the obtained solution, we see easily that

(1) If $\Theta_A \leqq 20°$, nothing is changed, the solution being independent of α.

(2) If $20° < \Theta_A < 40°$,

$$\lim_{\alpha \to \infty} \sigma(L) = 0, \qquad u(L) = 0.$$

Using (31) we get

$$\lim_{\alpha \to \infty} \Theta(L) = 40 - \Theta_A, \tag{37}$$

and consequently

$$\lim_{\alpha \to 0} \left[-\frac{\partial \Theta}{\partial x}(L) \right] = \frac{2}{L}(\Theta_A - 20). \tag{30}$$

In this limit case we have contact, without pressure; the temperature difference, given by (37), is not zero and the heat flux is given by (38).

(3) If $\Theta_A \geqq 40°$, we have

$$u(L) = 0, \qquad \lim_{\alpha \to \infty} \sigma(L) = \frac{\beta}{2}(40 - \Theta_A) < 0.$$

Using (31) we get

$$\lim_{\alpha \to \infty} \Theta(L) = 0,$$

and

$$\lim_{\alpha \to \infty} \Theta(x) = \Theta_A \left(1 - \frac{x}{L}\right).$$

The limit heat flux is

$$\lim_{\alpha\to\infty}\left[-\frac{d\Theta}{dx}(L)\right]=\frac{\Theta_A}{L}\,.$$

In this limit case we have contact with a nonzero pressure, no temperature difference and a limit heat flux.

The following pictures give a summary of the limit results obtained for the various values of Θ_A.

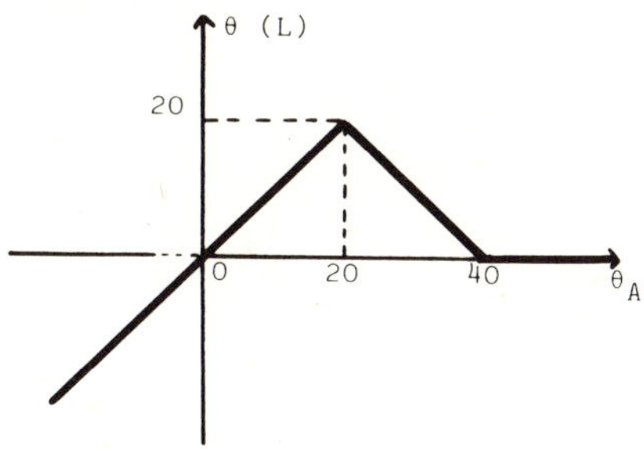

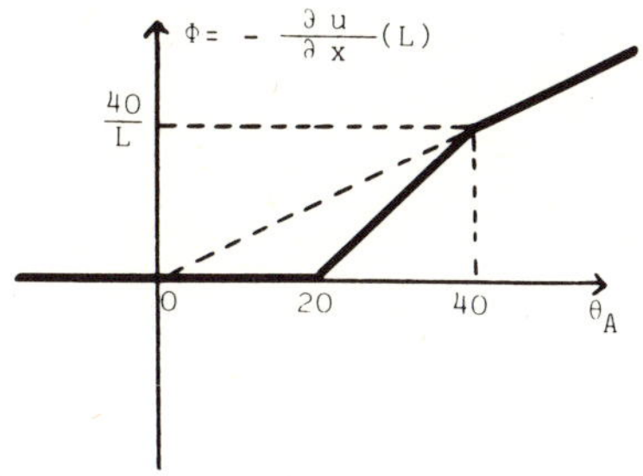

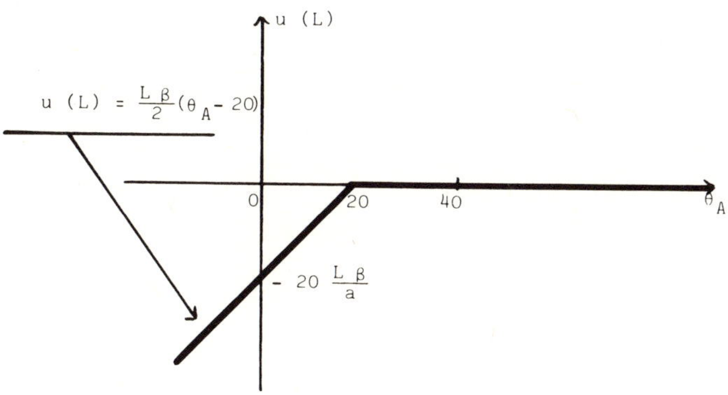

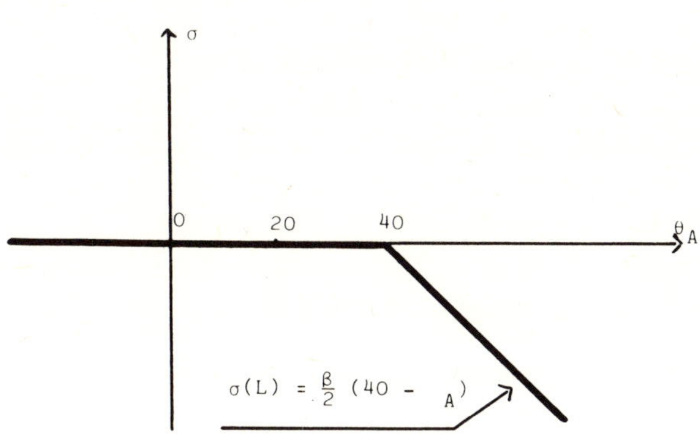

IX. Limit $\alpha \to +\infty$ in the general case

Let us come back to system (16), (17) where $k(\sigma_N^*)$ is replaced by $\alpha k_1(\sigma_N^*)$. The corresponding solution depends on the positive scalar parameter α and will be denoted by $(u_\alpha, \Theta_\alpha)$ and the stress by $\{\sigma_{ij}^\alpha\}$. Let us study the limit of this solution when α tends to plus infinity.

THEOREM 3. *If* $(u_\alpha, \Theta_\alpha)$ *is a solution of* (16), (17) *where* $k(\xi) = \alpha k_1(\xi)$, *we have the following when* α *tends to* $+\infty$ *(at least for a subsequence)*,

(i)

$$\left\{\begin{array}{ll} u_\alpha \to u & \text{in } V_1 \text{ weak}, \\ \sigma_{ij}^\alpha \to \sigma_{ij} & \text{in } L^2(\Omega) \text{ weak}, \\ \Theta_\alpha \to \Theta & \text{in } V_2 \text{ weak}. \end{array}\right. \tag{39}$$

(ii) *The limits* u, σ_{ij}, Θ *satisfy* (1), (2), (3), (4), (5), (6) *and*

$$\left\{\begin{array}{lll} u_N \leqq 0, & \sigma_N \leqq 0, & u_N\sigma_N = 0, \\ u_N < 0 \Rightarrow \sigma_N = 0 & & (\text{nothing on } \Theta), \\ u_N = 0 \Rightarrow \sigma_N \leqq 0 & \left\{\begin{array}{l} \sigma_N = 0 \\ \sigma_N < 0 \Rightarrow \Theta = \Theta_1. \end{array}\right. & (\text{nothing on } \Theta), \end{array}\right. \tag{40}$$

PROOF. Choosing $\varphi = \Theta_\alpha - \hat{\Theta}_1$ in (17), we obtain immediately that

$$\|\Theta_\alpha\|_{V_2} \leqq C. \tag{41}$$

Choosing $v = 0$ in (16), we get that

$$\|u_\alpha\|_{V_1} \leqq C, \tag{42}$$

$$\|\sigma_{ij}^\alpha\|_{L^2(\Omega)} \leqq C, \tag{43}$$

$$\int_{\Gamma_2} k_1(\sigma_N^{(\alpha)*})(\Theta_\alpha - \Theta_1)^2 d\Gamma \leqq \frac{C}{\alpha}, \tag{44}$$

the C's being constants independent of α.

Consequently, at least for a subsequence, we have (39). Furthermore it is clear that (1)–(6) are satisfied by the limit u, σ_{ij}, Θ, and also that

$$u_N \leqq 0, \qquad \sigma_N \leqq 0, \qquad u_N\sigma_N = 0.$$

If we take into account (43), we have the following result:
If, on Γ_2, $u = 0$ and $\sigma_N < 0$, then $\Theta = \Theta_1$, which implies (40).

REMARK. The problem (1)–(6), (40) is not a well-posed problem in the sense that it does not determine Θ uniquely.

References

1. G. Duvaut and J. L. Lions, *Les inéquations en mécanique et en physique*, Dunod, 1972. (Inequalities in Mechanics and Physics, Springer, 1976.)

2. M. Comninou and J. Dundurs, *On the Barber boundary conditions for thermoelastic contact*, J. Appl. Mech. (1977).

3. A. Signorini, *Sopra alcune questioni di elastostatica*, Atti della soc. Ital. per il progreno della Scienze (1933).

CHANGES IN SYMMETRY IN ELASTIC CRYSTALS

J. L. ERICKSEN

Department of Mechanics, Johns Hopkins University, Baltimore, Maryland 21218, USA

(Received September 11, 1980)

1. Introduction

It is not unusual for real material bodies to undergo what we consider to be changes in symmetry. We see water freeze to form solid ice, and know that mechanical forces, as well as temperature, influence such transitions. Similarly, crystals undergo changes in their symmetry, in some phase transitions. There are other phenomena which seem not to involve quite the same kind of change, but which crystallographers rationalize by considerations of symmetry, such as the twinning of crystals. By and large, continuum theory has ignored such phenomena, and conventional theories of material symmetry seem to exclude them. In the abstract, it is not so easy to decide just how we should modify theory to accommodate such phenomena. This induced me to attempt to construct models indicating how it should work, in special cases. For various reasons, the theory of elastic or thermoelastic crystals seemed to me a good place to start. As nonlinear continuum theories go, elasticity and thermoelasticity are relatively well understood. In terms of atomic arrangements, etc., the symmetry of crystals has been studied in some detail. Conceptually, classical molecular theories of elasticity theory are simple enough so that we can ascertain some of the general implications which are more or less independent of specific assumptions about interatomic force laws. As is to be discussed, some progress has been made toward this end, but some subtleties are not yet well understood.

2. Molecular theory

Here, we will consider only the classical molecular theories of Cauchy [1–3]

Proceedings of the IUTAM Symposium on Finite Elasticity, Lehigh University, August 10–15, 1980. Invited paper.

Carlson, D.E. and Shield, R.T. (eds.)
Proceedings of the IUTAM Symposium on Finite Elasticity
ISBN 90 247 2629 8

and Born[†] [4] or, more properly, the obvious extrapolations of these to cover finite deformations. Ericksen [7] discusses such a generalization of Cauchy's theory, in some detail. As he mentions, some workers are now using molecular theory to calculate strain energy functions valid for finite deformations. In such theory, we picture a crystal as consisting of mass points (atoms), subject to central forces[‡] which decay rather rapidly with distance, the crystal filling all of space. The differences between their theories disappear if all atoms are identical and are arranged on a simple lattice, or a 1-lattice, in the terminology introduced by Pitteri [8]. A 1-lattice is generated by applying a translation group to one atom. More precisely, if r_n $(n = 1, 2, 3, \cdots)$ denotes the position vector of the nth atom relative to some one, the collection of these must be representable in the form

$$r_n = n^a e_a = n^1 e_1 + n^2 e_2 + n^3 e_3, \tag{2.1}$$

where the n^a are integers, the e_a being constant, linearly independent (lattice) vectors. Further, there must be an atom at every point which is representable in this form. Clearly, the lattice vectors determine the position in space of all atoms, to within an unimportant translation of the whole crystal. It is easy to see that two sets of lattice vectors describe the same configuration if and only if they are related by a transformation of the form

$$\bar{e}_a = m_a^b e_b, \tag{2.2}$$

where the m_a^b are integers such that

$$\det m = \pm 1, \qquad m \overset{\text{def}}{=} \| m_a^b \|. \tag{2.3}$$

With the obvious rules of composition, or matrix multiplication, these form an infinite discrete group G. It plays a fundamental role in the classical theory of crystallographic groups, and in our considerations.

Within the context of our molecular theories, a general configuration is an n-lattice, consisting of n 1-lattices having the same translation group; the atoms on the different 1-lattices can be identical or different. By convention, we assign the same lattice vectors to all, (2.2) and (2.3) applying to these. A configuration is fixed, to within an insignificant translation of the n-lattice, by giving these and $n-1$ vectors p_m $(m = 1, \cdots, n-1)$; p_m can be interpreted as the position of some atom in the mth 1-lattice relative to one in the nth, for

[†] Critiques of these theories are given by Love [5, Note B] and Stakgold [6].

[‡] Born did consider more general kinds of forces. I have not thought through generalizations of this kind, for finite deformations, so subsequent remarks refer to the theory based on assumption of central forces.

example. Sometimes, these are called polarization vectors. This can mislead, because of the common usage of this name in considerations of electrical phenomena, so I will call them *shifts*.[†] Generally, these can be chosen in infinitely many ways, for a given configuration. This gives rise to a group of transformations relating the different choices, which is discussed by Pitteri [8]. One complication arises here, and it causes some headaches. It is not hard to see that an n-lattice can also be described as an n'-lattice, for any $n' \geqq n$. In practice, workers use descriptions which do not always employ the minimum value of n. Lattice vectors corresponding to different values of n are not related by any $m \in G$. Later, we will illustrate and elaborate this matter.

Concerning such configurations, Cauchy and Born made inequivalent assumptions, Born's being much more palatable. According to either, the energy per unit mass ϕ, in any such configuration, is finite, if the pair potentials decay rapidly enough with distance, and it depends only on the configuration of the crystal. Rather obviously, then, it is a function only of lattice vectors and shifts. The shifts are eliminated, replaced by functions of e_a. Cauchy employed a simple kinematic hypothesis to effect this. Born's work made clear that, in general, the hypothesis is unreasonable. Born's work deals only with infinitesimal deformations, so I give my interpretation of how his idea is to be extrapolated. It is to eliminate the shifts by choosing them to minimize the energy, for fixed values of the lattice vectors. By implication, we have accepted the motion that configurations of interest can all be described as n-lattices, for some fixed value of n. In practice, we guess a value of n, estimated from a conventional description of an unloaded configuration which is observed, so it should be reasonably stable. We are just beginning to scratch the surface, in understanding the subtleties which lie buried here. Later, I will attempt to shed a little light on some factors. Glossing over these subtleties, we should thus reduce ϕ to a function of e_a, and can infer that it has certain properties.[‡] As one might expect from the invariance of pair potentials, it is invariant under the orthogonal group,

$$\phi = \phi(e_a) = \phi(Qe_a), \tag{2.4}$$

for any orthogonal transformation Q. Also, as is perhaps clear from the preceding discussion, it is invariant under G,

$$\phi(e_a) = \phi(m_a^b e_b), \qquad m \in G. \tag{2.5}$$

Actually, Cauchy did not calculate ϕ, but the Cauchy stress σ, which can be

[†] I am indebted to C. Truesdell for suggesting the name.

[‡] For finite deformations of 2-lattices, such theory has been discussed by Ericksen [9] and Parry [10].

done, using his definition of it. Traditionally, this is done after introducing another assumption, but this is not necessary. For either the Cauchy theory of 1-lattices, or the Born theory, my calculations give it, in component form, as

$$\sigma^{ij} = \rho \frac{\partial \phi}{\partial e^j_a} e^i_a, \tag{2.6}$$

where ρ is the mass density, in the present configuration. It follows from (2.4) that σ is a symmetric tensor, with the invariance property indicated by

$$\sigma(Qe_a) = Q\sigma(e_a)Q^{\mathrm{T}}, \tag{2.7}$$

(2.5) implying that

$$\sigma(m^b_a e_b) = \sigma(e_a). \tag{2.8}$$

To get to elasticity theory, Born, following Cauchy, introduced a simple kinematical hypothesis. Pick some configuration, equipped with a set of lattice vectors $\hat{e}_a$, as a reference. If we subject this to a homogeneous macroscopic deformation, with deformation gradient F, the lattice vectors will change, we assume to values given by

$$e_a = F\hat{e}_a. \tag{2.9}$$

This is, for given F, the vectors occurring on the left serve as one of the infinitely many possible choices which might be used to describe the new configuration. Thus (2.9) is, in part, an assumption and, in part, a guideline for selecting particular sets of lattice vectors. Workers do select particular sets of lattice vectors, as well as shifts. Generally, they pay some attention to convention, but different workers have not always made the same choice. Some, like Parry [10, 11] and Pitteri [8], have attempted to construct simple rules for picking special sets agreeing, more or less, with some used in practice.† As is rather obvious, lattice vectors picked out by some more or less conventional rule can change in a discontinuous manner as F varies smoothly, when (2.9) applies to a different selection. This is illustrated by Parry [10]. Thus, it is a somewhat subtle matter to decide whether (2.9) applies and, if so, just how.

If we accept (2.9), we get the elastic strain energy per unit mass as

$$\hat{\phi}(F) = \phi(F\hat{e}_a), \tag{2.10}$$

† As is discussed by Hartshorne and Stuart [12, Chapter 1], different workers have opted for different choices. Referring to choices suggested by Bravais, they remark that "These are not in all cases the ones now preferred by X-ray crystallographers."

the $\hat{e}_a$ being considered as fixed. The Cauchy stress σ is related to this in the way which is commonly assumed in elasticity theory. Using the equation

$$M\hat{e}_a = m_a^b \hat{e}_b, \tag{2.11}$$

we can define a linear transformation M for each $m \in G$. It is easy to show that these form a group G', conjugate to G, a subgroup of the unimodular group. From (2.4) and (2.5) it follows that, for any of the indicated Q's and M's,

$$\hat{\phi}(F) = \hat{\phi}(QF) = \hat{\phi}(FM). \tag{2.12}$$

In general terms, this seems consistent with the general theory of material symmetry of elastic materials, as presented by Truesdell [13, Chapter IV], for example. In his terminology, G' is the peer group, it seems. However, the peer groups commonly used for crystals are the (finite) point groups and G' is an infinite group. Use of the point groups represents extrapolation of successful practice for classical linear theories. Pitteri [8] and by private correspondence, C. M. Kwok, have made known to me different analyses, both implying that the use of G' for finite deformations is compatible with the use of a point group for infinitesimal deformations. This seems to eliminate a possible objection to the use of G' for finite deformations. There is a different issue which I now raise. We have made some rather tacit assumptions and accepted (2.9), in deducing (2.11). In terms of what is known about phase transitions in crystals, it is not too unreasonable to expect that such hypotheses will hold only for a limited range of deformations and that, for a given crystal, there might be more than one such range. Thus, we might have different peer groups applying locally, and molecular theory provides some guidelines for analyzing such possibilities. This idea remains to be explored. Later, we will say a bit more about it.

Cauchy's theory and, in some cases, Born's theory leads to the well-known Cauchy relations, which are better ignored, being contradicted by experiment, for most crystals. To assess the soundness of predictions concerning invariance, we need to better understand what they are. What (2.12) does is to mathematize an old idea, that crystals must have a finite resistance to shear, because certain finite shears take the infinite crystal to an undistinguishable configuration. Rather intuitive discussions of this, such as are given in old works on applied mechanics, e.g. by Nadai [14, p. 34], make no explicit reference to molecular theories of elasticity. One can use more intuitive ideas of this kind to motivate (2.12), without introducing the assumptions of pair potentials etc. used by Cauchy and Born. Here, it is relevant that G contains elements of the simple shearing type, e.g.

$$\text{(2.13)} \qquad \left\| \begin{matrix} 1 & r & 0 \\ 0 & 1 & 0 \\ 0 & 0 & 1 \end{matrix} \right\| ,$$

where r is any integer.

It is one thing to characterize the invariance of the strain energy, a different matter to describe the symmetry of a particular configuration. The literature tends to confuse the two issues, presuming that the invariance of $\hat{\phi}$ is also that which describes the symmetry of some particular equilibrium configurations. The classical theory of point and space groups aims at describing the symmetry of configurations, and it is only the point groups which have been used in elasticity theory. As was mentioned before, we can and do introduce rules for selecting lattice vectors which determine these to within orthogonal transformations. If Q is some orthogonal transformation, e_a and Qe_a will both be possible sets of lattice vectors provided there is some $m \in G$ such that

$$\text{(2.14)} \qquad Qe_a = m_a^b e_b .$$

With e_a fixed, those Q which so qualify form the maximal point group associated with e_a ; any subgroup is also considered as a point group. One gets different point groups for different choices of e_a, in general, and classical theory characterizes all of the possibilities. If we take the lattice vectors $\hat{e}_a$ referred to in (2.9), and compare (2.14) with (2.11), we see that G' contains, as a subgroup, the maximal point group corresponding to $\hat{e}_a$. It is a proper subgroup, since all point groups are finite. Closely related to the point groups are the lattice groups. Theory of the latter is covered by Ericksen [15], who argues that they are superior to point groups, in distinguishing some differences in symmetry. To define these, we use the fact that, for e_a fixed, (2.14) provides a one-to-one map, giving an $m \in G$ for each Q in a possible point group. Such m form a finite subgroup of G, conjugate to the point group, and this is what a lattice group is. Given any particular $m \in G$, we can apply it to e_a to obtain a new set of lattice vectors $\bar{e}_a$, as indicated by (2.2). Using (2.14), it is easy to see that the old point group remains a possible point group for these but that, in general, the corresponding lattice group is different. Basically, this is the reason why lattice groups are better able to discern changes in symmetry.

As a crystal deforms, its lattice vectors will change, and an infinitesimal change can be enough to change a point or lattice group. Let $P(e_a)$ denote the maximal point group associated with the current lattice vectors delivered by (2.9). For $Q \in P(e_a)$, we have, for some $m \in G$,

$$Qe_a = QF\hat{e}_a = m_a^b e_b = F(m_a^b \hat{e}_b) = FM\hat{e}_a, \tag{2.15}$$

for some $M \in G'$, but M need not be in $P(\hat{e}_a)$, for it need not be orthogonal. From (2.7) and (2.8), this imposes a restriction on possible values of σ, viz.

$$\sigma(e_a) = Q\sigma(e_a)Q^{\mathrm{T}}, \qquad Q \in P(e_a). \tag{2.16}$$

With the more traditional theory of material symmetry, we do not get the same restrictions, although (2.16) would apply if M happened to be in $P(\hat{e}_a)$, and the reference configuration is chosen, in a traditional way, as an unstressed configuration. Briefly, we have invariance under two groups, the orthogonal group and G or G'. Roughly, some transformations qualify both ways, and in this category are the point groups, with their "equivalent" lattice groups. This seems to be the best way to interpret the significance of the point groups.

By letting $\hat{\phi}$ depend on temperature and interpreting it as, say, Helmholtz free energy per unit mass, we now have some basis for a thermoelastic analysis of phase transitions. Most such studies involve equilibrium studies of crystals subject to controlled temperature, with the crystals unloaded or loaded by a controlled hydrostatic pressure. In such environments, lattice vectors will change with hydrostatic pressure but experience dictates that it is rather unusual for the symmetry of the equilibrium configurations to change, whether we use point or lattice groups to estimate this. When it does, we have what would be interpreted as a phase transition. If all the reasoning summarized above applies, $\hat{\phi}$ continues to have the same invariance, described by G'. Roughly, this is analogous to what happens in Euler buckling. There, in typical cases, the invariance of the governing equations does not change, but the buckled configurations tend to be less symmetrical than the unbuckled. Some analyses of transitions of this general kind, and a critique of other ideas used by physicists are given by Ericksen [16]. It can be tricky to use these ideas correctly and, in certain kinds of transitions, it is less than clear that they can be used in a sound way. We now turn to some examples, to illustrate these points.

3. Examples

We first consider a crystal consisting of identical atoms, restricted to be in configurations of the body-centered tetragonal variety. The words suggest describing these as 2-lattices, with orthogonal lattice vectors e_a, two being of equal length; we here pay some attention to rather common conventions. Letting

$$C_{ab} \overset{\text{def}}{=} e_a \cdot e_b, \tag{3.1}$$

we then have, say

$$\|C_{ab}\| = \begin{Vmatrix} a & 0 & 0 \\ 0 & a & 0 \\ 0 & 0 & b \end{Vmatrix}, \tag{3.2}$$

with a and b positive. To take care of atoms at the centers of the tetragons, we introduce one shift p, given by

$$2p = e_1 + e_2 + e_3. \tag{3.3}$$

As is fairly clear from (2.14) and (3.1), and discussed in some detail by Ericksen [15], we can calculate the maximal point and lattice group associated with these lattice vectors by determining what elements of G satisfy the matrix equation

$$mCm^{\mathrm{T}} = C, \tag{3.4}$$

and using (2.14). This leads to equations like

$$a[(m_1^1)^2 + (m_1^2)^2] + b(m_1^3)^2 = a, \tag{3.5}$$

and it takes no great ingenuity to determine what sets of integers satisfy these. Omitting details, one finds that if $a \neq b$, (2.14) takes the form

$$\begin{cases} Qe_1 = \pm e_1, \quad Qe_2 = \pm e_2, \quad Qe_3 = \pm e_3, \\ Qe_1 = \pm e_2, \quad Qe_2 = \pm e_1, \quad Qe_3 = \pm e_3, \end{cases} \tag{3.6}$$

where the algebraic signs can be assigned independently. Clearly, the choice of lattice vectors permits the point group to be described in a simple way; this is one of the factors that can influence particular choices of lattice vectors. When $a = b$, the configuration degenerates to the body-centered cubic form. One can calculate that the point group shifts to the form which would be expected for this, including such transformations as

$$Qe_1 = e_1, \qquad Qe_2 = e_3, \qquad Qe_3 = e_2, \tag{3.7}$$

and this contains (3.6) as a proper subgroup. When (3.6) applies, it follows from (2.16) that σ must be of the form

$$\sigma = Al + Be_3 \otimes e_3, \tag{3.8}$$

A and B being scalars. In the degenerate case when $a = b$, we find that

$$\sigma = Al. \tag{3.9}$$

At first glance, this might seem unreasonable. If we start with an unstressed tetragon, and deform it to a cube, we don't expect the cube to be subject to no shear stress. On the other hand, if one ponders the assumed atomic arrangement, it is hard to see how the stress could exhibit less symmetry than is indicated by (3.9). Physically, we should expect to encounter some kind of instability before we even arrive at the cubic configuration.

As long as (3.3) applies, we can describe the same configurations as 1-lattices, using lattice vectors e_a^1 given by

$$e_1^1 = e_1, \qquad e_2^1 = e_2, \qquad e_3^1 = p = (e_1 + e_2 + e_3)/2. \tag{3.9}$$

Since components of e_a^1 relative to e_a are not integers, this is not a transformation covered by (2.2). There is then no guarantee that the ideas of invariance applied to e_a^1 will give equivalent results and, in fact, they do not. Any of the transformations given by (3.6) can be rephrased in terms of e_a^1, albeit more awkwardly. For example,

$$Qe_1 = e_2, \qquad Qe_2 = e_1, \qquad Qe_3 = -e_3,$$

becomes

$$Qe_1^1 = e_2^1, \qquad Qe_2^1 = e_1^1, \qquad Qe_3^1 = e_1^1 + e_2^1 - e_3^1.$$

By such calculations, it is straightforward to show that, if any Q is in the point group corresponding to e_a, it is also on the point group corresponding to e_a^1, including the degenerate case when $a = b$. Examination of point groups associated with e_a^1 indicates that there is an exceptional case when

$$b = 2a. \tag{3.10}$$

When this holds, the point group includes, in particular, the transformation

$$Qe_1^1 = e_3^1 - e_2^1, \qquad Qe_2^1 = e_3^1 - e_3^1, \qquad Qe_3^1 = e_3^1; \tag{3.11}$$

one obtains a group describing cubic symmetry, missed with the 2-lattice description. This special case seems to be well-known to the crystal experts. If one ponders the arrangement of atoms, one sees that such configurations can be described as face-centered cubics, involving a third set of lattice vectors, plus shifts to cover the atoms on the face centers. Again, this description is preferred by at least some, and it does simplify the description of the relevant point group. If we consider that the configurations are varying smoothly with deformation, the sudden shift from tetragonal to cubic lattice vectors will cause problems, if we wish to think of (2.9) as applying. Thus, we do get into difficulties with some of the more conventional descriptions, once we allow crystals to deform. We are better off to use the 1-lattice description, if we wish to properly account for the implications of molecular theory. Of course, it is only a matter of bookkeeping to rephrase the implications in terms of a 2-lattice description, if we wish to do so. Consideration of various other configurations suggests a rule of thumb. When using an n-lattice description, it is best that the value of n be the smallest possible, when we wish to estimate implications of molecular theory.

The rule of thumb can encounter difficulty if, in deforming a crystal, the minimum value of n were to change. In the above example, this could occur if the center atom began to move off center, so that the shift p is no longer given by (3.3), and the 1-lattice description is invalidated. This is one of the kinds of things covered under the general heading of phase transitions involving a change of symmetry. Those who use classical molecular theory to calculate strain energy functions shy away from such problems; it is not so easy to make the stability analysis needed to check such possibilities. Those interested in phase transitions do consider these phenomena. As long as we are within one phase, we can reasonably apply the previous ideas, interpreting (2.12) as applying to deformations which leave us in one phase, relative to a reference configuration selected in that phase. For this to make sense, the domain of $\hat{\phi}$ must be closed under the action of the group G', an assumption which seems to be consistent with molecular theory. In effect, this treats the different phases as different materials. It would make sense to consider them as one, if some configuration can be regarded as a common limit configuration for the phases. We can then take it as a common reference for both phases, with a careful accounting of the invariances appropriate for each phase. This then provides a conceptual basis for constructing a potential which changes invariance as we pass from one phase to another. This is somewhat similar to Kahl's [17] use of the critical point as a common reference for liquid and vapor phases of van der Waal's fluids, except that this transition involves no change of symmetry. Preliminary calculations indicate that analysis of special cases is feasible, and the potential tends to suffer some loss of smoothness at the phase boundary. Such theory is in a primitive state, so we will leave it at this.

The need for a "common configuration" suggests some analogy with the theory of second-order phase transitions involving changes in symmetry, such as is discussed by Landau and Lifshitz [18, Chapter XIV]. Nowadays, physicists consider their theory to be less than completely sound, and still strive to repair it. It is partly for this reason that it seems worthwhile to carefully rethink the whole theory of symmetry and its changes.

Acknowledgment

This work was supported by NSF Grant CME-79 11112.

REFERENCES

1. A.-L. Cauchy, *Sur l'équilibre et le mouvement d'un système de points matériels sollicités par des forces d'attraction ou de répulsion mutuelle*, Ex. de Math. **3** (1828), 227–287.

2. A.-L. Cauchy, *De la pression ou tension dans un système de points matériels*, Ex. de. Math. **3** (1828), 253–277.

3. A.-L. Cauchy, *Sur les équations differentielles d'équilibre ou de mouvement pour un système de points matériels sollicités par des forces d'attraction ou de répulsion mutuelle*, Ex. de. Math. **4** (1829), 342–369.

4. M. Born, *Dynamik der Kristallgitter*, B. G. Teubner, Leipzig–Berlin, 1915.

5. A. E. H. Love, *A Treatise on the Mathematical Theory of Elasticity*, 4th ed., Cambridge University Press, Cambridge, 1927.

6. I. Stakgold, *The Cauchy relations in a molecular theory of elasticity*, Quart. Appl. Math. **8** (1949), 169–186.

7. J. L. Ericksen, *Special Topics in Elastostatics*, in *Advances in Applied Mechanics* (C.-S. Yih, ed.), Vol. 17, Academic Press, New York–San Francisco–London, 1977, pp. 189–244.

8. M. Pitteri, *Reconciliation of local and global symmetries of crystals*, to appear.

9. J. L. Ericksen, *Nonlinear elasticity of diatomic crystals*, Int. J. Solids Structures **6** (1970), 951–957.

10. G. P. Parry, *On diatomic crystals*, Int. J. Solids Structures **14** (1978), 283–287.

11. G. P. Parry, *On the elasticity of monatomic crystals*, Math. Proc. Camb. Phil. Soc. **80** (1976), 189–211.

12. N. H. Hartshorne and A. Stuart, *Practical Optical Crystallography*, American Elsevier Publishing Company, New York, 1964.

13. C. Truesdell, *A First Course in Rational Continuum Mechanics*, Academic Press, New York–San Francisco–London, 1977.

14. A. Nadai, *Plasticity, A Mechanics of the Plastic State of Matter*, McGraw-Hill, New York–London, 1931.

15. J. L. Ericksen, *On the symmetry of deformable crystals*, Arch. Rat'l. Mech. Analysis **72** (1979), 1–13.

16. J. L. Ericksen, *Some phase transitions in crystals*, Arch. Rat'l. Mech. Analysis **73** (1980), 99–124.

17. G. D. Kahl, *Generalization of the Maxwell criterion for van der Waals equation*, Phys. Rev. **155** (1967), 78–80.

18. L. D. Landau and E. M. Lifshitz, *Statistical Physics* (E. Peierls and R. F. Peierls, trans.), Pergamon Press, London–Paris, and Addison–Wesley, Reading, 1958.

A VARIATIONAL APPROACH TO FINITE ELASTICITY

GIUSEPPE GRIOLI
University of Padua, Padua, Italy

(Received September 11, 1980)

ABSTRACT

My main object is not to speak about well known interesting variational theorems in continuum mechanics (see, for instance, [1]) but to show that there exists a certain possibility of a systematic approach to finite elasticity by a variational procedure. The existing variational theorems are generally inadequate for the purpose, although interesting from a heuristic point of view. A systematic variational procedure allows a global approach to finite elasticity, that is an approach in which field and constitutive equations are written in integral form. It seems evident that even when only the field equations have such a form there is the greatest generality and meaning because the general equations of continuum mechanics have their natural origin in integral forms, while some regularity conditions must be satisfied for deducing differential field equations. Further, a global approach generally leads to improved convergence in numerical methods. I shall consider the basic problem of finite elasticity in the presence of unilateral surface constraints, treating either the classical case of non-polar continua or that of Cosserat continua with free rotations. For the latter case the results are not yet exhaustive or complete with the exception of the linear case, but I think it is useful to show some differences in the variational aspect between the two cases. For simplicity I shall consider only elastic continua without inner constraints.

1. Non-polar elastic continua

Let us consider a three-dimensional body and denote by C a reference configuration; C' the configuration at the instant t; σ and σ' their boundaries; n_r and n'_r the unitary normals to σ and σ' (which I suppose to exist everywhere); P and P' corresponding points of C and C' or σ and σ'; y_r and x_r their coordinates with respect to a rectangular reference frame; F'_r and f'_r the density of external forces in C' and on σ' (including in F'_r the inertial

Proceedings of the IUTAM Symposium on Finite Elasticity, Lehigh University, August 10–15, 1980. Invited paper.

Carlson, D.E. and Shield, R.T. (eds.)
Proceedings of the IUTAM Symposium on Finite Elasticity
ISBN 90 247 2629 8

forces in the dynamical case); F_r and f_r the corresponding vectors in C (defined by the equalities: $F_r dC = F'_r dC'$, $f_r d\sigma = f'_r d\sigma'$); X_{rs}, Y_{rs}, K_{rs} respectively the Cauchy stress tensor, the symmetrical Piola–Kirchhoff and the non-symmetrical Piola–Kirchhoff stress tensor; e_{rs} the classical strain tensor.

It is easy to recognize that from the Cauchy equations for continuous media follows the integral property:

$$\int_{C'} X_{rs} T_{/s} dC' = -\int_{C'} T F'_r dC' - \int_{\sigma'} T f'_r d\sigma', \tag{1.1}$$

where T is a differentiable function of x_i and the slash indicates differentiation with respect to x_i.

Assuming $T = x_m$ from (1.1) follows the basic average relation

$$\begin{aligned}\int_{C'} X_{rr} dC' &= \int_C (2e_{rs} + \delta_{rs}) Y_{rs} dC = \int_C K_{rs} x_{r,s} dC \\ &= -\int_{C'} x_r F'_r dC' - \int_{\sigma'} x_r f'_r d\sigma' = -\int_C x_r F_r dC - \int_\sigma x_r f_r d\sigma,\end{aligned} \tag{2.1}$$

where δ_{rs} denotes the Kronecker delta.

If one considers any two stress states $X_{rs}, X'_{rs}, \cdots$, corresponding to the configuration at instant t and to the same intertial and external forces where they are assigned and denoting by δX_{rs} their differences, from (2.1) follows

$$\begin{aligned}\int_{C'} \delta X_{rs} dC' &= \int_C (2e_{rs} + \delta_{rs}) \delta Y_{rs} dC \\ &= \int_C \delta K_{rs} x_{r,s} dC = -\int_{\sigma_2} (y_r + u_r) \delta\phi_r d\sigma,\end{aligned} \tag{3.1}$$

where $u_r = x_r - y_r$ are the components of the displacement PP' and $\delta\phi_r$ the difference between the corresponding constraint forces, ϕ_r, present on the portion σ_2 of σ where constraints are supposed to be present.

It is interesting to show how the average relation (3.1) leads to a variational property for the Cauchy stress $\boldsymbol{X}$.

Let U' be the left stretch tensor, $\boldsymbol{e}'$ the corresponding strain tensor and $\boldsymbol{e}^{(i)}$ the strain tensor relative to the inverse displacement.

The equalities

$$\mathbf{1} + 2\boldsymbol{e}' = \boldsymbol{U}'\boldsymbol{U}', \qquad \mathbf{1} + 2\boldsymbol{e}^{(i)} = (\mathbf{1} + 2\boldsymbol{e}')^{-1} \tag{4.1}$$

are well known.

I denote by $W(\boldsymbol{e})$ the Lagrangean density of potential elastic energy and by $W(\boldsymbol{e}')$ the function of $\boldsymbol{e}'$ obtained by putting $\boldsymbol{e} = \boldsymbol{r}^{(T)}\boldsymbol{e}'\boldsymbol{r}$ in $W(\boldsymbol{e})$, where $\boldsymbol{r}$ is the local rotation. $W(\boldsymbol{e}')$ generally depends on $\boldsymbol{r}$.

Let us consider the equality

$$DL = -W(e')_{/e'} \qquad \left(\text{that is, } DL_{rs} = -\frac{\partial W(e')}{\partial e'_{rs}}\right) \tag{5.1}$$

where D is the Jacobian of the transformation from C to C'.

Relation (5.1) is generally invertible as is the corresponding relation

$$Y = -W(e)_{/e}. \tag{6.1}$$

Therefore one may suppose

$$e' = g(L). \tag{7.1}$$

Putting

$$N(L) = -W(e')[g(L)] - g_{rs}(L)L_{rs}, \tag{8.1}$$

one has

$$e' = -N_{/L} \qquad \left(e'_{rs} = \frac{\partial N}{\partial L_{rs}}\right). \tag{9.1}$$

If is easy to recognize that

$$X = -\frac{1}{D} U'W(e')_{/e'}U' = U'LU'. \tag{10.1}$$

If one denotes by $N(X)$ the function of X (generally depending on the local rotation) obtained by putting, according to (10.1), $L = U'^{-1}XU'^{-1}$, one finds

$$N(X)_{/X} = U'^{-1}N_{/L}U'^{-1}. \tag{11.1}$$

Denoting by I the linear invariant of a tensor and keeping in mind (4.1) (9), (10), (11), one has the set of equalities

$$I(\delta X) = I(U'\delta LU') = I[(1+2e')\delta L] = I[(1-2N_{/L})\delta L], \tag{12.1}$$

which because of (11.1) give rise to the relation

$$I(\delta X) = I\{[(1+2e') - 2N(X)_{/X}]\delta X\}. \tag{13.1}$$

Then, if one puts

$$A' = \int_{C'} [(1+2e^{(i)})^{-1}X - 2N(X)]dC' + \int_{\sigma'_2} x_r\phi'_r d\sigma'_2, \tag{14.1}$$

one finds that *the actual Cauchy stress satisfies the variational condition*

$$\delta A' = 0. \tag{15.1}$$

The variational relation (15.1) means that among all the Cauchy stresses in equilibrium with the external assigned forces and the inertial forces and which give rise to compatible reactions, the actual stress makes the functional A' stationary.

The above property becomes more significant in the case of isotropy. In fact, in this case $W(\boldsymbol{e})$ depends on $\boldsymbol{e}$ through its principal invariants and, consequently, the same applies for $W(\boldsymbol{e}')$. Then the functions $g(\boldsymbol{L})$, $N(\boldsymbol{L})$ and $N(\boldsymbol{X})$ are independent of the local rotation and A' is expressible in terms of $\boldsymbol{X}$ and the strain of the inverse displacement $\boldsymbol{e}^{(i)}$.

The above mentioned property is physically meaningful because it operates on the significant Cauchy stress, but — as I now will show — it is possible to deduce from (3.1) a different Lagrangean variational property which although less significant than the first one is more useful for the analytical point of view and generally for integration problems.

I now consider some consequences of relation (3.1) for hyperelastic bodies. Naturally, other consequences may be obtained by assuming in (1.1) more general functions T, different from $T = x_r$.

Relations (6.1) are generally invertible and permit us to express e in terms of Y. I denote by $W(Y)$ the expression of the second potential energy, obtained using such an expression. Putting

$$A = \int_C (2W(Y) - Y_{rr})dC - \int_{\sigma_2} (y_r + u_r)\phi_r d\sigma_2, \tag{16.1}$$

it is easy to show that relation (3.1) gives rise to the result

$$\delta A = 0, \tag{17.1}$$

whose meaning is evident:

The real stress makes A stationary in the set S of the stresses which satisfy the field equations and on σ_2 give rise to reactions compatible with the constraints. The property is valid also in the dynamical case.

In the static case the above variational property, which generalizes the classical Menabrea's theorem, is invertible, but it is necessary to distinguish the case when a bilateral constraint is present on σ_2 from that of the presence of a unilateral one. In the first case one has: if $\bar{Y}^*$ is a point of stationarity for A in the set S, a displacement $\bar{u}_r$ exists from which the stress is derived. Further $\bar{u}_r$ satisfies the boundary conditions in a weak sense, that is, one has

$$\int_{\sigma_2} (\bar{u}_r - u_r)q_r d\sigma_2 = 0, \tag{18.1}$$

for every vector q_r equilibrated on σ_2. It is evident that if appropriate regularity conditions are fulfilled, $\bar{u}_r = u_r$ follows from (18.1) everywhere on σ_2. For the proof see [7], [8].

I consider in detail the case when on σ_2 a unilateral constraint is present, which is rather complicated. Precisely, I consider only the case when the constraint on σ_2 consists of a plane rigid unilateral supporting constraint, without excluding the presence of friction.

Now it is convenient to represent the stress by the Piola–Kirchhoff stress tensor, K_{rs}. Then from relation (3.1) it follows that

$$\int_C \delta K_{rs} x_{r,s} dC + \int_{\sigma_2} (y_r + u_r)\delta\phi_r d\sigma_2 = 0, \tag{19.1}$$

where the vector u_r is unknown. It is convenient to specify what may happen on σ_2. With this aim let us denote by $\sigma_2^{(1)}$ the portion of σ_2 where in conditions of equilibrium there is contact and by $\sigma_2^{(2)}$ the remaining portion. Therefore, putting

$$\boldsymbol{\phi} = \boldsymbol{\phi}_\pi + \rho \boldsymbol{n}, \qquad (\rho \geqq 0), \tag{20.1}$$

where $\boldsymbol{\phi}_\pi$ is the tangential component of $\boldsymbol{\phi}$, one has

$$\begin{cases} \rho > 0, & \boldsymbol{u} \cdot \boldsymbol{n} = 0 \quad \text{on } \sigma_2^{(1)}, \\ \rho = 0, & \boldsymbol{u} \cdot \boldsymbol{n} \geqq 0 \quad \text{on } \sigma_2^{(2)}. \end{cases} \tag{21.1}$$

Plainly, $\boldsymbol{\phi}_\pi$ is equal to zero if the constraint is frictionless, otherwise it satisfies the Coulomb law

$$|\boldsymbol{\phi}_\pi| \leqq m\rho \qquad (0 \leqq m < 1). \tag{22.1}$$

Every variation $\delta\boldsymbol{\phi}$ of $\boldsymbol{\phi}$ must satisfy the conditions implied by (20.1), (21.1), (22.1). Precisely, observing that $\boldsymbol{\phi}_\pi = 0$ when $\rho = 0$ and supposing that the variation $\delta\rho$ is vanishingly small compared to ρ when $\rho > 0$, it follows that when $\rho > 0$ no condition subsists for ρ. Therefore:

$$\begin{cases} \delta\rho \text{ and } \delta\boldsymbol{\phi}_\pi \text{ are arbitrary} & \text{on } \sigma_2^{(1)}, \\ \delta\rho \geqq 0 \text{ and } |\delta\boldsymbol{\phi}_\pi| \leqq m\delta\rho & \text{on } \sigma_2^{(2)}. \end{cases} \tag{23.1}$$

The constitutive relations

$$K_{rs} = -W_{/x_{r,s}} \qquad \left(K_{rs} = -\frac{\partial W}{\partial x_{r,s}}\right), \tag{24.1}$$

are generally invertible and one may put

$$x_{r,s} = -W(K)_{/K_{rs}} \qquad \left(x_{r,s} = -\frac{\partial W(K)}{\partial K_{rs}}\right), \tag{25.1}$$

where $W(K)$ is the second potential energy expressed through the K_{rs}. Consequently, relation (19.1) becomes

$$\int_C \frac{\partial W(K)}{\partial K_{rs}} \delta K_{rs} dC - \int_{\sigma_2} (y_r + u_r)\delta\phi_r d\sigma_2 = 0. \tag{26.1}$$

Keeping in mind (20.1), (21.1), equality (26.1) becomes

$$\int_C \frac{\partial W(K)}{\partial K_{rs}} \delta K_{rs} dC - \int_{\sigma_2} y_r\delta\phi_r d\sigma_2 - \int_{\sigma_2} \boldsymbol{u}_\pi \cdot \delta\boldsymbol{\phi}_\pi d\sigma_2 - \int_{\sigma_2} \delta\rho \boldsymbol{u} \cdot \boldsymbol{n} d\sigma_2 = 0 \tag{27.1}$$

which, because of (21.1), (23.1), gives rise to the inequality

$$\int_C \frac{\partial W(K)}{\partial K_{rs}} \delta K_{rs} dC - \int_{\sigma_2} y_r\delta\phi_r d\sigma_2 - \int_{\sigma_2} \boldsymbol{u}_\pi \cdot \delta\boldsymbol{\phi}_\pi d\sigma_2 \geqq 0, \tag{28.1}$$

where the sign of equality applies only for the actual stress.

One concludes the following:

The actual stress minimizes the functional

$$B = \int_C W(K)dC - \int_{\sigma_2} (OP + \boldsymbol{u}_\pi)\cdot \boldsymbol{\phi}_\pi d\sigma_2 \tag{29.1}$$

in the set S' of all stresses in equilibrium with the external forces where they are assigned and that give rise to compatible reactions (that is satisfying (22.1) with $\rho \geqq 0$).

Some clarifications are appropriate about the meaning of the above variational property and about the possibility of its inversion. It is well known that in the presence of friction, an infinite number of positions of equilibrium are generally possible under an assigned set of external forces. Therefore it seems natural to investigate the possibility of prefixing the tangential displacement where a unilateral supporting constraint is present. Then I suppose that the vector $\boldsymbol{u}_\pi$ in the expression of functional B defined by (29.1) is assigned and that B has a minimum in the set S' in correspondence to a stress K'_{rs}.

As a consequence of this hypothesis it is possible to show that a displacement $\boldsymbol{u}'$ exists from which the stress is derived. Further, the vector $\boldsymbol{u}'$ satisfies the relations

$$\int_{\sigma_2^{(1)}} \boldsymbol{q}_\pi \cdot (\boldsymbol{u}'_\pi - \boldsymbol{u}_\pi) d\sigma_2^{(1)} = 0, \qquad \int_{\sigma_2^{(1)}} q_n \boldsymbol{n} \cdot \boldsymbol{u}' d\sigma_2^{(1)} = 0 \tag{30.1}$$

on the portion $\sigma_2^{(1)}$ of σ_2 where the reaction corresponding to K'_{rs} is different from zero, while on the other part, $\sigma_2^{(2)}$, of σ_2 we have the following inequality

$$\int_{\sigma_2^{(2)}} [q_n (\boldsymbol{n} \cdot \boldsymbol{u}' + m\,|\boldsymbol{u}'_\pi - \boldsymbol{u}_\pi|\cos h] d\sigma_2^{(2)} \geqq 0. \tag{31.1}$$

In (30.1), (31.1) h is an arbitrary angle, while $\boldsymbol{q} = \boldsymbol{q}_\pi + q_n \boldsymbol{n}$ is a vector equilibrated on σ_2, arbitrary on $\sigma_2^{(1)}$ and satisfying on $\sigma_2^{(2)}$ the conditions

$$(32.1) \qquad q_n \geqq 0, \qquad |\boldsymbol{q}_\pi| \leqq m q_n.$$

It is clear that (30.1) expresses in a weak sense the circumstance that $\boldsymbol{u}'$ satisfies the constraint on $\sigma_2^{(1)}$ and coincides with $\boldsymbol{u}_\pi$. Putting $h = \pi/2$ in the relation (31.1), valid for every value of h, one sees that $\boldsymbol{u}'$ satisfies the constraint also on the portion $\sigma_2^{(2)}$. Further, for $h = \pi$ relation (31.1) gives rise to the inequality

$$(33.1) \qquad \int_{\sigma_2^{(2)}} q_n (\boldsymbol{n} \cdot \boldsymbol{u}' - m\,|\boldsymbol{u}'_\pi - \boldsymbol{u}_\pi|) d\sigma_2^{(2)} \geqq 0,$$

which points out the degree of indeterminateness of the solution.

If regularity conditions allow us to obtain local relations, from (30.1), (33.1) one has $\boldsymbol{n} \cdot \boldsymbol{u}' \geqq 0$ everywhere and, further,

$$(34.1) \qquad \boldsymbol{u}'_\pi - \boldsymbol{u}_\pi = 0 \qquad \text{on } \sigma_2^{(1)},$$

$$(35.1) \qquad |\boldsymbol{u}'_\pi - \boldsymbol{u}_\pi| \leqq \frac{\boldsymbol{n} \cdot \boldsymbol{u}'}{m} \qquad \text{on } \sigma_2^{(2)}.$$

In conclusion we make the following statement:

The solution which minimizes B gives rise to a prefixed targential displacement only on the portion of σ_2 where there is contact, while on the other portion where separation is present the real displacement is limited by inequality (35.1).

If the constraint is frictionless one has $m = 0$, $\boldsymbol{\phi}_\pi = 0$ and the relations (34.1), (35.1) lose their validity. In this case the minimizing solution satisfies the conditions (21.1) almost everywhere and there is complete indetermination of displacement on σ_2.

It is to be observed that the validity of the mentioned results is based on the existence of vectors $\boldsymbol{q}$ equilibrated on σ_2 and satisfying conditions (32.1) on the part of σ_2 where there is separation (if that part exists). This implies that the support constraint is really necessary, that is that the assigned external forces are not in equilibrium, so that contact is present at least on a portion of σ_2.

If the portion σ_2 of the boundary where there is a support constraint is not a plane surface the previous results remain valid under the hypothesis of small deformations (linear elasticity). Fundamental in this field are the results of Fichera [2, 5] who, following A. Signorini, considers as the basic variable the displacement rather than the stress.

In the case of finite deformations the results may remain valid if the constraint is not plane but it requires a complete and difficult examination of the shape of the boundary. To this end there are some very useful studies by Valent [6, 4], who considered the analytical aspect of the problem [6, 9].

2. Elastic Cosserat continua

Now I consider analogous variational problems relating to elastic Cosserat continua with free rotations. As is well known, their geometry is characterized by a displacement field, $\boldsymbol{u}$, and a rotation field, R. Naturally, the matrix R is to be thought of as a function of three parameters. The stress is expressed by the Cauchy stress tensor (generally non-symmetric) and by an Eulerian couple stress tensor, P_{rs}. The strain may be represented by the matrices

$$(1.2) \qquad a_{rs} = x_{i,r}R_{is}, \qquad Z^{(m)}_{rs} = \tfrac{1}{2}R_{ir}R_{is,m} = -Z^{(m)}_{sr}.$$

In the following, it is convenient to consider the Lagrangean auxiliary stress tensors defined by the equalities

$$(2.2) \qquad N_{rs} = A_{ir}X_{mi}R_{ms}, \qquad T^{(m)}_{rs} = e_{rls}R_{il}P_{ip}A_{pm},$$

where A_{ir} is the cofactor of $x_{i,r}$ in the matrix of $x_{i,r}$ and e_{rls} denotes the Ricci tensor in three-dimensional Euclidean space.

It is possible to show that the Lagrangean density of inner force's work, corresponding to a small transformation from C' to $C' + \delta C'$, is expressible by the equality

$$(3.2) \qquad \delta^{(i)}_{1} = \boldsymbol{N}_{rs}\delta a_{rs} + T^{(m)}_{rs}\delta Z^{(m)}_{rs},$$

while the field equations assume the form

$$(4.2) \qquad \begin{cases} N_{mr,m} + 2Z^{(m)}_{rp}N_{mp} = R_{mr}F_m, \\ \tfrac{1}{2}e_{prq}T^{(m)}_{pq,m} + e_{piq}Z^{(m)}_{ri}T^{(m)}_{pq} + e_{prq}a_{mq}N_{mp} = R_{mr}M_m, \end{cases}$$

where M_i denotes the Lagrangean density of the body couples (and of inertial couples in dynamical case).

Equations (4.2) realize a possible form of field equations depending directly on strain and stress, as well as the rotation. Boundary conditions are

$$(5.2) \qquad N_{mr}n_m = R_{mr}f_m, \qquad \tfrac{1}{2}e_{prq}T^{(m)}_{pq}n_m = R_{ir}m_i.$$

Let b_{rs} and $c^{(m)}_{rs} = -c^{(m)}_{sr}$ be prefixed functions of the coordinates y_i and consider the integral relations

$$(6.2) \qquad \begin{cases} \displaystyle\int_C N_{rm}b_{rm}dC = 0, \\ \displaystyle\int_C R_{ri}\,(T^{(m)}_{ti}c^{(m)}_{tr} + b_{tr}N_{ti})dC = 0. \end{cases}$$

If one denotes by S'' the set of all stresses which satisfy the field equations (4.2), (5.2) written for $F_r = M_r = f_r = m_r = 0$, the following theorem subsists.

If b_{rs} and $c_{rs}^{(m)} = -c_{sr}^{(m)}$ express a Cosserat strain according to (1.2), *then they satisfy integral relations* (6.2) *for every stress in the set S''. Vice-versa, if for a fixed rotation, R, b_{rs} and $c_{rs}^{(m)} = -c_{sr}^{(m)}$ satisfy relations* (6.2) *for every stress in the set S'', they represent a Cosserat strain.*

That is a displacement and a rotation exist such that b_{rs} and $c_{rs}^{(m)}$ descend from them according to (1.2). The rotation is just R [12].

Now suppose that the stress is the gradient of a potential energy:

$$N_{rs} = -\frac{\partial W}{\partial a_{rs}}, \qquad T_{rs}^{(m)} = -\frac{\partial W}{\partial Z_{rs}^{(m)}}, \tag{7.2}$$

where W denotes the density of potential energy, a function of a_{rs} and $Z_{rs}^{(m)}$.
Let us consider the function

$$W'(a, Z^{(m)}, N, T^{(m)}) = -W(a, Z^{(m)}) - a_{pq}N_{pq} - Z_{pq}^{(m)}T_{pq}^{(m)}. \tag{8.2}$$

The relations (7.2) express necessary and sufficient conditions for the existence of stationary points $(a, Z^{(m)})$ for the functions W'. Generally such points exist and relations (7.2) are to be considered invertible. I do not consider the difficult question of the unicity of the inversion that exists also in the case of classical nonpolar continua (on this subject see [10]) but I suppose that from (7.2) descend relations of the kind

$$a_{rs} = \alpha_{rs}(N, T^{(m)}), \qquad Z_{rs}^{(m)} = \beta_{rs}^{(m)}(N, Z^{(m)}). \tag{9.2}$$

Putting

$$W^{(N, T^{(m)})} = -W(\alpha, \beta) - \alpha_{pq}N_{pq} - \beta_{pq}^{(m)}T_{pq}^{(m)}, \tag{10.2}$$

one finds

$$a_{rs} = -\frac{\partial W^{(N, T^{(m)})}}{\partial N_{rs}}, \qquad Z_{rs}^{(m)} = -\frac{\partial W^{(N, T^{(m)})}}{\partial T_{rs}^{(m)}}. \tag{11.2}$$

If one puts in the integral relations (6.2), formally,

$$\begin{cases} N_{rs} = \delta N_{rs}, & T_{rs}^{(m)} = \delta T_{rs}^{(m)}, \\ b_{rs} = -\dfrac{\partial W^{(N,T)}}{\partial N_{rs}}, & c_{rs}^{(m)} = -\dfrac{\partial W^{(N,T)}}{\partial T_{rs}^{(m)}}, \end{cases} \tag{12.2}$$

where δN_{rs}, $\delta T_{rs}^{(m)}$ denote variations of the stress satisfying homogeneous forms of the field equations, they give rise to the variational relations

$$(13.2)\qquad \begin{cases} \displaystyle\int_C \frac{\partial W^{(N,T)}}{\partial N_{rs}}\,\delta N_{rs}\,dC = 0, \\ \displaystyle\int_C \left(\frac{\partial W^{(N,T)}}{\partial T^{(m)}_{ts}}\,R_{sq}\delta T^{(m)}_{tq} + \frac{\partial W^{(N,T)}}{\partial N_{ts}}\,R_{sq}\delta N_{tq}\right) dC = 0. \end{cases}$$

Therefore, starting from a real stress on a prefixed configuration the field and constitutive equations of a hyperelastic Cosserat continuum have as a consequence the variational relations expressed by (13.2). This means that the actual stress satisfies (13.2) in the set S''. According to the former theorem one concludes that a stress which satisfies (13.2) in the set S'' derives from the rotation R and from a certain displacement, that is it is the actual stress.

Variational relations (13.2) give rise to a peculiar consequence differing from the case of non-polar continua (classical continua). I mean that if one puts

$$(14.2)\qquad E = \int_C W^{(N,T)}dC, \qquad R'_{rs} = R_{rs} - \delta_{rs},$$

it follows from (13.2) that

$$(15.2)\qquad \delta E = -\int_C \left(\frac{\partial W^{(N,T)}}{\partial T^{(m)}_{ts}}\,\delta T^{(m)}_{tq} + \tfrac{1}{2}e_{rpm}e_{rqs}\frac{\partial W^{(N,T)}}{\partial N_{lm}}\,\delta N_{lp}\right) R'_{sq}dC.$$

The relation (15.2) shows that unlike the non-polar case, the actual stresses do not make the second potential energy stationary.

However, that happens in the linear case because the second member of (15.2) is clearly of higher order than the first one.

In the case of slightly deformable Cosserat continua, the field equations become linear and independent of the strain. Precisely, one has

$$(16.2)\qquad N_{rs} = X_{rs}, \qquad T^{(m)}_{rs} = e_{ris}P_{im}.$$

The actual stress is characterized by a stationary value of the second potential energy. Results analogous to that of the non-polar case subsist also in the presence of constraints. For example, if displacement $\boldsymbol{u}'$ and rotation $\boldsymbol{r}'$ are prefixed on a part, σ_2, of the boundary, keeping in mind that in the linear case these results

$$(17.2)\qquad R_{rs} = e_{rls}Q_l + \delta_{rs}, \qquad a_{rs} = u_{s,r} + e_{rps}Q_p, \qquad Z^{(m)}_{rs} = \tfrac{1}{2}e_{rps}Q_{p,m},$$

where Q_i is the rotation vector, the following theorem subsists:

If the functional

$$(18.2)\qquad E' = \int_C W^{(X,P)}dC - \int_{\sigma_2} [\phi_s(y_s + u'_s) + \tfrac{1}{2}e_{spq}r'_{qp}m'_s]d\sigma_2$$

admits a stationary point (X^*, P^*) in the set of all solutions of the linearized field equations, satisfying the boundary conditions where the external forces and couples are known, that stationary point descends from an actual deformation of the body and satisfies the constraint conditions.

Analogous results to those of the non-polar case subsist also in the presence of a supporting constraint for a linear hyperelastic body, [11].

References

1. E. Reissner, *On a variational theorem for finite elastic deformations*, J. Math. Phys. **32** (1953), 129–135.
2. G. Fichera, *Problemi elastostatici con vincoli unilaterali. II problema di Signorini con ambigue condizioni al contorno*, Memorie Acc. Naz. Lincei Set. VIII, **7** (5) (1964).
3. T. Valent, *Qualche proprietà dei sistemi di vettori applicati. Possibili applicazioni alla teoria matematica dell'elasticità*, Rend. Sem. Math. Univ. Padova **39** (1967), 47–55.
4. T. Valent, *Qualche proprietà e applicazione di sistemi di vettori definiti su una superficie*, ibidem, **41** (1968), 306–318.
5. G. Fichera, *Boundary value problems of elasticity with unilateral constraints*, Handbuch der Physik (B) Vol. VI/2 (1972).
6. T. Valent, *Questioni di esistenza e unicità per il problema di appoggio unilaterale a supporto rigido nel caso di piccole deformazioni*, Rend. Sem. Mat. Univ. Padova **50** (1973), 143–166.
7. G. Grioli, *Proprietà variazionali nella meccanica dei continui*, nota I[a], Rend. Naz. Lincei, Ser. VIII, Vol. LIV, fasc. 5, 1973.
8. G. Grioli, *Proprietà variazionali nella meccanica dei continui*, nota II[a], ibidem, fasc. 6, 1973.
9. T. Valent, *Sull'equilibrio isotermo dei corpi elastici con un vincolo di appoggio unilaterale liscio nel caso di deformazioni finite*, Ann. Mat. pura e applicata (IV) **101** (1974), 1–31.
10. W. T. Koiter, *On the complementary energy theorem in non-linear elasticity theory*, in *Trends in Applications of Pure Mathematics to Mechanics*, Pitman Publishing, 1975, pp. 207–232.
11. G. Grioli, *Una proprietà variazionale nella teoria delle microstrutture*, Rendiconti di Matematica (1) Vol. **8**, Serie VI (1975), 225–235.
12. G. Grioli, *Contributo per una formulazione di tipo integrale della meccanica dei continui di Cosserat*, Ann. Mat. pura e applicata, Ser. IV, Vol. CXI (1976), 175–183.

ON UNIQUENESS IN FINITE ELASTICITY[†]

MORTON E. GURTIN

Department of Mathematics, Carnegie-Mellon University, Pittsburgh, PA 15213, USA

(Received September 26, 1980)

ABSTRACT

This paper lists counterexamples demonstrating lack of uniqueness for the major boundary-value problems of finite elasticity. Uniqueness is then shown to hold in convex, stable sets of deformations.

I. Introduction

Finite elasticity remains one of the more difficult theories of mathematical physics. In this paper we discuss reasons for this difficulty, concentrating mainly on lack of uniqueness. We give heuristic counterexamples for all of the major boundary-value problems; these examples demonstrate that unqualified uniqueness is neither to be expected nor desired.

This discussion leads us to ask: Where in the space of deformations does uniqueness hold? In partial answer to this question we show that uniqueness holds in any convex, stable set of deformations (Gurtin and Spector [1]).

II. The mixed problem

We consider an elastic body $\mathscr{B}$ which we identify with the regular region of $\mathbb{R}^3$ it occupies in a fixed reference configuration. Consider a *deformation* of $\mathscr{B}$; that is, a smooth (i.e, C^1) map $\boldsymbol{f} : \mathscr{B} \to \mathbb{R}^3$ with $\det \boldsymbol{F} > 0$, where

$$\boldsymbol{F} = \nabla \boldsymbol{f}$$

is the deformation gradient. Under $\boldsymbol{f}$ the body experiences a (Piola–Kirchhoff) stress

$$\boldsymbol{S}(\boldsymbol{F}(\boldsymbol{x}), \boldsymbol{x})$$

[†] This paper, with minor modifications, was presented at the 25th Conference of Army Mathematicians, Baltimore, June, 1979.

Proceedings of the IUTAM Symposium on Finite Elasticity, Lehigh University, August 10–15, 1980. Invited paper.

Carlson, D.E. and Shield, R.T. (eds.)
Proceedings of the IUTAM Symposium on Finite Elasticity
ISBN 90 247 2629 8

at each $x \in \mathscr{B}$, where S (with obvious domain) is the smooth *response function* for the body.

We assume that the boundary $\partial\mathscr{B}$ is the union of disjoint sets $\mathscr{D}$ and $\mathscr{S}$, and that the deformation is prescribed on $\mathscr{D}$, the surface traction on $\mathscr{S}$. The *mixed problem* then consists in finding a deformation f that satisfies the *equation of equilibrium*

$$\text{div}\, S(\nabla f) + b = 0, \tag{1}$$

and the *boundary conditions*

$$f = d \quad \text{on } \mathscr{D}, \qquad S(\nabla f)n = s \quad \text{on } \mathscr{S}. \tag{2}$$

Here $S(\nabla f)$ is the field with values $S(\nabla f(x), x)$, b is the prescribed body force, d is the prescribed deformation, and s is the prescribed traction. Note that we have tacitly restricted our attention to *dead loads*, since s and b are functions of x only.

Let f be a class C^2 solution of the mixed problem, and let u be a *variation*; that is, u is a smooth vector field on $\mathscr{B}$ which vanishes on $\mathscr{D}$. Then

$$\begin{aligned}\int_{\mathscr{S}} s \cdot u\, dA = \int_{\partial\mathscr{B}} u \cdot S(\nabla f)n\, dA &= \int_{\mathscr{B}} [S(\nabla f) \cdot \nabla u + u \cdot \text{div}\, S(\nabla f)]\, dV \\ &= \int_{\mathscr{B}} [S(\nabla f) \cdot \nabla u - b \cdot u]\, dV,\end{aligned}$$

and we have the identity

$$\int_{\mathscr{B}} S(\nabla f) \cdot \nabla u\, dV = \int_{\mathscr{S}} s \cdot u\, dA + \int_{\mathscr{B}} b \cdot u\, dV. \tag{3}$$

Conversely, a class C^2 deformation f that satisfies the displacement boundary condition $(2)_1$ and the equation (3) for every variation u will automatically satisfy (1) and $(2)_2$. This motivates the following *weak* statement of the problem: Find a deformation f that satisfies the displacement boundary condition and (3) for every variation u. A deformation with this property will be called a *solution*.

A difficulty intrinsic to finite elasticity concerns the solution space. To be meaningful a solution f must not only satisfy

(a) $\det \nabla f > 0$,

but should also be

(b) one-to-one.

Condition (a) is severe and makes the theory quite difficult, as the collection of fields satisfying (a) is not convex. Further, it is usually not possible to extend the domain of S continuously to tensors F with $\det F = 0$, since $S(F)$ generally becomes infinite as $\det F \to 0$.

Condition (b) is even more severe, since it is *global.* Of course, one can drop this restriction provided one is willing to accept solutions of the form shown in Figure 1.

Figure 1.

An interesting question in global analysis is

$$(a) + \text{what} \Rightarrow (b)?$$

For the displacement problem ($\partial\mathscr{B} = \mathscr{D}$) an answer is furnished by

THEOREM 1 (Meisters and Olech [2]). *Let $\partial\mathscr{B}$ be a sufficiently nice subset of $\mathbb{R}^3$. Let $f : \mathscr{B} \to \mathbb{R}^3$ be a deformation, and suppose that f satisfies the displacement boundary condition $f = \boldsymbol{d}$ on $\partial\mathscr{B}$ with $\boldsymbol{d}$ one-to-one. Then f is one-to-one.*

III. Lack of uniqueness

As the following counterexamples demonstrate, uniqueness in general is not to be expected. We assume in examples (Ab), (Ba), and (Ca) that the reference configuration is *natural*; i.e., that $\boldsymbol{S}(\boldsymbol{I}, \boldsymbol{x}) = \boldsymbol{0}$ for all $\boldsymbol{x} \in \mathscr{B}$.

A. *The traction problem* ($\mathscr{S} = \partial\mathscr{B}$)

(a) A rigid deformation of a solution yields a solution. Rigid deformations of $\mathscr{B}$ which leave the loading invariant also leave a solution invariant.

(b) Consider a thin hemispherical shell with zero surface tractions. Then f = identity is a solution. But there should be a second solution consisting of the everted shell (Armanni [3], Antman [4]). Similar assertions apply to a thin cylindrical tube (Almansi [5]). (An interesting discussion of eversion problems is given by Truesdell [6].)

(c) Consider a rod subject to equal and opposite tractions on its ends. This type of loading should result in the two types of solutions shown in Figure 2 (Ericksen (cf. Wang and Truesdell [7], p. 474)).

(d) Also, in problem (c) we would expect "buckled solutions" when the loads are sufficiently large.

(e) When an incompressible, homogeneous and isotropic cube is loaded in tension by forces which are constant in magnitude and perpendicular to the

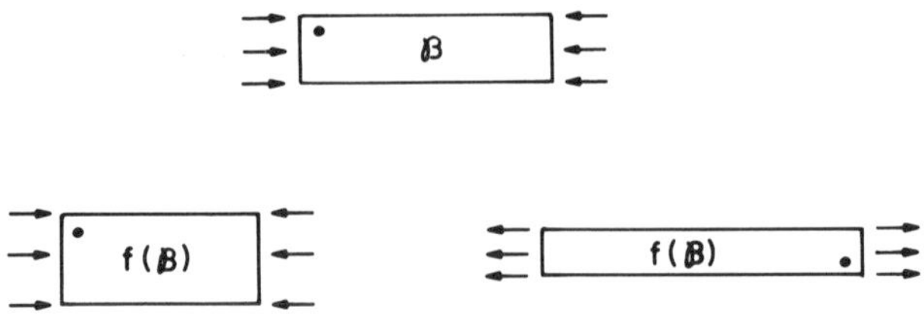

Figure 2.

faces, and when these forces are sufficiently large, there exist seven solutions (Rivlin [8]).

B. *The displacement problem* ($\mathscr{D} = \partial\mathscr{B}$)

(a) Consider a spherical shell with boundary condition $\boldsymbol{f}(\boldsymbol{x}) = \boldsymbol{x}$ on $\partial\mathscr{B}$. One solution, of course, is the identity. But there are other deformations which leave the boundary unmoved, but deform the interior. Indeed, consider the deformation caused by a rotation of the inner boundary by an integral multiple of 2π about an axis through the center of the sphere (John (cf. Truesdell and Noll [9], p. 129)).

(b) Consider an inhomogeneous body consisting of a stiff rod whose cylindrical surface is surrounded by a soft material. For certain sufficiently severe displacements of the boundary we would expect the bar to buckle (Ball [10]).

(c) Consider the anisotropic body shown in Figure 3. Assume that the material is stiff in the vertical direction, soft in the horizontal direction. Let the vertical boundaries be fixed and deform the top and bottom faces toward each other with a vertical displacement which is most severe at the center. Then for sufficiently large displacements we would expect internal buckling as shown in Figure 3 (Noll, private communication).

C. *The genuine mixed problem* ($\mathscr{D} \neq \varnothing$, $\mathscr{S} \neq \varnothing$)

(a) Consider a finite cylindrical rod with sides traction free and ends rigidly fixed. One solution is the identity. Another corresponds to the deformation

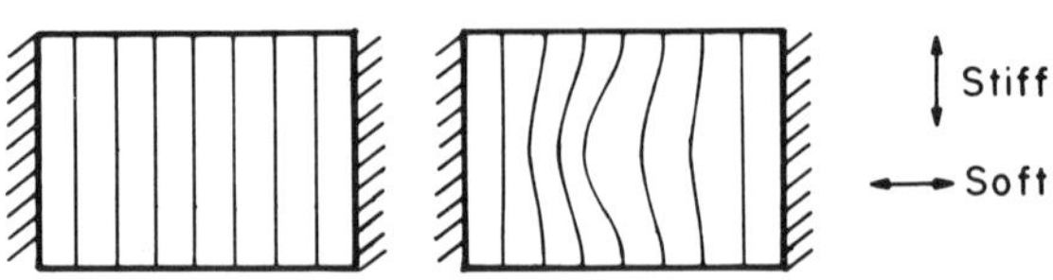

Figure 3.

caused by a rotation (in its plane) of one of the ends by an integral multiple of 2π about an axis through its center (Gurtin [11]).

(b) We would also expect a situation similar to (Ad) for a rod which is loaded at one end, but which has the other end fixed.

IV. Stability and uniqueness

Since unqualified uniqueness is not to be expected, and since many of the foregoing counterexamples involve instabilities, the following questions seem appropriate:

(i) Where in the set of deformations does uniqueness hold?

(ii) Are uniqueness and stability related?

A partial answer to the second question was furnished by Ericksen and Toupin [12] and Hill [13], who showed that Hadamard stability of a stressed state implies uniqueness for infinitesimal deformations superimposed on that state. We now study these questions in further detail. (With the exception of Remarks 2, 5, 6, and 7, the remainder of this section is due to Gurtin and Spector [1].) For convenience, we rule out the traction problem by requiring that $\mathscr{D}$ be *relatively open and non-empty.*

A *process* $\boldsymbol{g}$ is a one-parameter family $\boldsymbol{g}_\sigma$ $(0 \leqq \sigma \leqq \beta)$ of deformations such that

(a) $\dot{\boldsymbol{g}}_\sigma(\boldsymbol{x})$, $\boldsymbol{G}_\sigma(\boldsymbol{x}) = \nabla \boldsymbol{g}_\sigma(\boldsymbol{x})$, and $\dot{\boldsymbol{G}}_\sigma(\boldsymbol{x})$ exist and are jointly continuous in $(\boldsymbol{x}, \sigma)$ on $\mathscr{B} \times [0, \beta)$ (here a superposed dot indicates differentiation with respect to σ, while ∇ is the gradient with respect to $\boldsymbol{x}$);

(b) $\dot{\boldsymbol{g}}_\sigma = \boldsymbol{0}$ on $\mathscr{D}$ for all $\sigma \in [0, \beta)$;

(c) $\dot{\boldsymbol{g}}_\sigma \not\equiv \boldsymbol{0}$.

We say that $\boldsymbol{g}$ *starts from* $\boldsymbol{f}$ if $\boldsymbol{g}_0 = \boldsymbol{f}$.

Central to our notion of stability is the functional

$$P_\sigma(\boldsymbol{g}) = \int_{\mathscr{B}} (\boldsymbol{S}_\sigma - \boldsymbol{S}_0) \cdot \dot{\boldsymbol{G}}_\sigma dV,$$

with

$$\boldsymbol{S}_\sigma = \boldsymbol{S}(\nabla \boldsymbol{g}_\sigma);$$

P_σ represents the *incremental power* needed to sustain the process $\boldsymbol{g}$. A reasonable definition for the stability of a deformation $\boldsymbol{f}$ is that $P_\sigma(\boldsymbol{g})$ be strictly positive near $\sigma = 0$ in any process $\boldsymbol{g}$ starting from $\boldsymbol{f}$. More precisely, $\boldsymbol{f}$ is *stable* if given any process $\boldsymbol{g}_\sigma$ $(0 \leqq \sigma \leqq \beta)$ starting from $\boldsymbol{f}$,

$$P_\sigma(\boldsymbol{g}) > 0$$

for all sufficiently small $\sigma > 0$.

THEOREM 2. *Uniqueness holds in any convex, stable set of deformations.*

PROOF. Consider a straight process

$$\mathbf{g}_\sigma(\mathbf{x}) = \mathbf{f}(\mathbf{x}) + \sigma \mathbf{u}(\mathbf{x})$$

with $\mathbf{f}$ a deformation and $\mathbf{u} \neq \mathbf{0}$ a variation. Assume that $\mathbf{g}$ has values in a stable set Ω. Then, since $\dot{\mathbf{G}}_\sigma = \nabla \mathbf{u}$,

$$P_{\sigma+\delta}(\mathbf{g}) = \int_{\mathscr{B}} (\mathbf{S}_{\sigma+\delta} - \mathbf{S}_0) \cdot \nabla \mathbf{u} \, dV$$

$$= \int_{\mathscr{B}} (\mathbf{S}_{\sigma+\delta} - \mathbf{S}_\sigma) \cdot \nabla \mathbf{u} \, dV + \int_{\mathscr{B}} (\mathbf{S}_\sigma - \mathbf{S}_0) \cdot \nabla \mathbf{u} \, dV.$$

But (for $\delta > 0$ sufficiently small) the first integral is > 0, since $\mathbf{g}_\sigma \in \Omega$ is stable, and the second integral is $P_\sigma(\mathbf{g})$; thus

$$P_{\sigma+\delta}(\mathbf{g}) > P_\sigma(\mathbf{g})$$

and

(4) $\qquad P_\sigma(\mathbf{g})$ is *strictly increasing* in σ.

Now let Ω be convex and stable, and let $\mathbf{f}, \mathbf{h} \in \Omega$ with $\mathbf{f} \neq \mathbf{h}$ be two solutions. Then

$$\mathbf{u} = \mathbf{h} - \mathbf{f}$$

is a variation and

$$\int_{\mathscr{B}} \mathbf{S}(\nabla \mathbf{f}) \cdot \nabla \mathbf{u} \, dV = \int_{\mathscr{S}} \mathbf{s} \cdot \mathbf{u} \, dA + \int_{\mathscr{B}} \mathbf{b} \cdot \mathbf{u} \, dV,$$

$$\int_{\mathscr{B}} \mathbf{S}(\nabla \mathbf{h}) \cdot \nabla \mathbf{u} \, dV = \int_{\mathscr{S}} \mathbf{s} \cdot \mathbf{u} \, dA + \int_{\mathscr{B}} \mathbf{b} \cdot \mathbf{u} \, dV,$$

so that

$$\int_{\mathscr{B}} [\mathbf{S}(\nabla \mathbf{h}) - \mathbf{S}(\nabla \mathbf{f})] \cdot \nabla \mathbf{u} \, dV = 0. \tag{5}$$

Consider the straight path

$$\mathbf{g}_\sigma(\mathbf{x}) = \mathbf{f}(\mathbf{x}) + \sigma[\mathbf{h}(\mathbf{x}) - \mathbf{f}(\mathbf{x})] \qquad (0 \leqq \sigma \leqq 1)$$

from $\mathbf{f}$ to $\mathbf{h}$. Then $\mathbf{g}$ lies in Ω, because Ω is convex, and a simple calculation shows that

$$P_1(\mathbf{g}) = \int_{\mathscr{B}} [\mathbf{S}(\nabla \mathbf{h}) - \mathbf{S}(\nabla \mathbf{f})] \cdot \nabla \mathbf{u} \, dV.$$

The result (4) implies $P_1(\boldsymbol{g}) > 0$, which contradicts (5). Thus $\boldsymbol{f} \equiv \boldsymbol{h}$ and the proof is complete.

COROLLARY. *Let* $\boldsymbol{f}$ *and* $\boldsymbol{h}$ *be solutions of the mixed problem. Then the straight path from* $\boldsymbol{f}$ *to* $\boldsymbol{h}$ *(provided it lies in the space of deformations) cannot be stable.*

Consider a straight rod placed between two parallel plates which are moved toward each other until the rod buckles. (a) and (b) in Figure 4 denote two possible buckled states. If the buckling is not too severe, the straight line connecting these states will lie in the space of deformations. The corollary asserts that at least one deformation on this path is not stable; a strong candidate for such a deformation is the intermediate state (c).

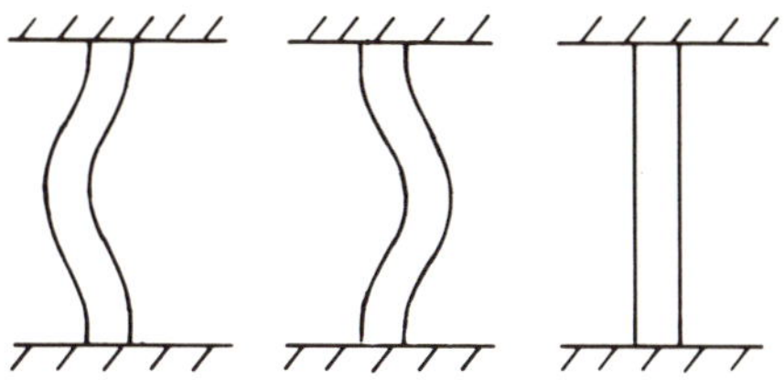

(a) (b) (c)

Figure 4.

REMARK 1. Using Theorem 2 as a basis, Gurtin and Spector [1] have established uniqueness:

(a) in a neighborhood of a uniformly stable deformation;

(b) in a neighborhood of a natural configuration whose elasticity tensor is positive definite;

(c) (for the displacement problem) in a neighborhood of a homogeneous, strongly-elliptic configuration.

The results (b) and (c) are similar in nature to results established previously by Stoppelli [14] and van Buren [15], while (c) is due to John [16]. (See also Valent [17].)

REMARK 2. As Capriz and Podio-Guidugli [18] have noted, and as is clear from the proof, for the validity of Theorem 2 it suffices to define stability only for straight processes.

REMARK 3. The foregoing results have been extended to more general types of loading by Gurtin and Spector [1] and Spector [19], to a Sobolev-space setting by Valent [17], and to nonlinear viscoelastic materials by Gurtin, Reynolds and Spector [20].

REMARK 4. One can show, using (4), that if Ω is a convex, stable set of deformations, then the underlying operator for the mixed problem is strictly monotone on Ω.

REMARK 5 (cf. Gurtin [21]). Let the body be hyperelastic with total potential energy

$$\Phi(f) = \int_{\mathscr{B}} \psi(\nabla f) dV - \int_{\mathscr{S}} s \cdot f dA - \int_{\mathscr{B}} \boldsymbol{b} \cdot f dV,$$

where ψ is the stored energy:

$$\boldsymbol{S}(\boldsymbol{F}, \boldsymbol{x}) = \frac{\partial}{\partial \boldsymbol{F}} \psi(\boldsymbol{F}, \boldsymbol{x}).$$

Let f be a *solution*, and let $\boldsymbol{g}$ be a process starting from f. Then

$$P_\sigma(\boldsymbol{g}) = \frac{d}{d\sigma} \Phi(\boldsymbol{g}_\sigma).$$

Thus, if f is stable,

$$\Phi(f) < \Phi(\boldsymbol{g}_\sigma)$$

for all sufficiently small $\sigma > 0$.

REMARK 6. Assume that the body is hyperelastic. Let $\boldsymbol{g}$ be a *straight* process and define

$$Q_\sigma(\boldsymbol{g}) = \frac{d}{d\sigma} \Phi(\boldsymbol{g}_\sigma).$$

Then a simple computation shows that

$$Q_{\sigma+\delta}(\boldsymbol{g}) - Q_\sigma(\boldsymbol{g}) = P_{\sigma+\delta}(\boldsymbol{g}) - P_\sigma(\boldsymbol{g});$$

hence (4) implies that for $\boldsymbol{g}$ in a stable set $Q_\sigma(\boldsymbol{g})$ is a strictly increasing function of σ. Thus Φ is a (strictly) convex function on any convex, stable set of deformations.

REMARK 7. The extension of Theorem 2 to incompressible bodies remains an open question. The problem is that the deformation gradient lies on the *manifold* $\{\det \boldsymbol{F} = 1\}$ and one cannot connect deformations by straight lines, even locally.

REFERENCES

1. M. E. Gurtin and S. J. Spector, *On stability and uniqueness in finite elasticity*, Arch. Rational Mech. Anal. **70** (1979), 153–165.

2. G. H. Meisters and C. Olech, *Locally one-to-one mappings and a classical theorem on Schlicht functions*, Duke Math. J. **30** (1963), 63–80.

3. G. Armanni, *Sulle deformazioni finite dei solidi elastici isotropi*, Nuovo Cimento (6) **10** (1915), 427–447.

4. S. S. Antman, *The eversion of thick spherical shells*, Arch. Rational Mech. Anal. **70** (1979), 113–123.

5. E. Almansi, *La teoria delle distorsioni e le deformazioni finite dei solidi elastici*, Rend. Accad. Lincei (5) **25** (1916), 191–192.

6. C. Truesdell, *Some challenges to analysis by rational thermomechanics*, in *Contemporary Developments in Continuum Mechanics and Partial Differential Equations*, North-Holland, Amsterdam, 1978, pp. 495–603.

7. C.-C. Wang and C. Truesdell, *Introduction to Rational Elasticity*, Noordhoff, Leyden, 1973.

8. R. S. Rivlin, *Stability of pure homogeneous deformations of an elastic cube under dead loading*, Q. Appl. Math. **32** (1974), 265–271.

9. C. Truesdell and W. Noll, *The non-linear field theories of mechanics*, Handbuch der Physik, III/3, Springer-Verlag, Berlin, 1965.

10. J. M. Ball, *Constitutive inequalities and existence theorems in nonlinear elastostatics*, in *Nonlinear Analysis and Mechanics*, Heriot-Watt Symposium 1, Pitman, London, 1977, pp. 187–241.

11. M. E. Gurtin, *On the nonlinear theory of elasticity*, in *Contemporary Developments in Continuum Mechanics and Partial Differential Equations*, North-Holland, Amsterdam, 1978, pp. 237–253.

12. J. L. Ericksen and R. A. Toupin, *Implications of Hadamard's condition for elastic stability with respect to uniqueness theorems*, Canad. J. Math. **8** (1956), 432–436.

13. R. Hill, *On uniqueness and stability in the theory of finite elastic strain*, J. Mech. Phys. Solids **5** (1957), 229–241.

14. F. Stoppelli, *Un teorema di esistenza e di unicita relativo alle equazioni dell'elastostatica isoterma per deformazioni finite*, Ricerche mat. **3** (1954), 247–267.

15. W. van Buren, *On the existence and uniqueness of solutions to boundary value problems in finite elasticity*, Thesis, Department of Mathematics, Carnegie-Mellon University. Research Report 68–ID7-MEKMA-RI, Westinghouse Research Laboratories, Pittsburgh, Pa., 1968.

16. F. John, *Uniqueness of nonlinear elastic equilibrium for prescribed boundary displacements and sufficiently small strains*, Comm. Pure Appl. Math. **25** (1972), 617–634.

[17] T. Valent, *Local theorems of existence and uniqueness in finite elastostatics*, J. Elasticity, to appear.

18. G. Capriz and P. Podio-Guidugli, *Questions of uniqueness in finite elasticity*, to appear.

19. S. J. Spector, *On uniqueness in finite elasticity with general loading*, J. Elasticity **10** (1980), 145–161.

20. M. E. Gurtin, D. W. Reynolds and S. J. Spector, *On uniqueness and bifurcation in nonlinear viscoelasticity*, Arch. Rational Mech. Anal. **72** (1980), 303–313.

21. M. E. Gurtin, *Topics in Finite Elasticity*, Society for Industrial and Applied Mathematics, Philadelphia, 1980.

ON PHYSICAL AND MATERIAL CONSERVATION LAWS

A. G. HERRMANN

Division of Applied Mechanics, Stanford University, Stanford, California 94305, USA

(Received September 17, 1980)

ABSTRACT

Using the variational principle with varying boundaries, it is shown that both physical and material conservation laws can be obtained from the same expression. The basic quantities used in those conservation laws are: physical momentum (stress) and material momentum. Two different representations are employed to study the equal importance of both of them in the theory of elasticity.

Introduction

The variational calculus in mechanics proved to be successful in many problems starting with classical dynamics through field theories, to composite materials. In recent years interest in continuum mechanics gravitated towards defects and fracture. In connection with this trend, new conservation laws have been studied on one hand, and so-called configurational forces and energy release rates, as a result of nonconservation laws, on the other.

The pioneering work in this field has been done by Eshelby in 1951 [1] and followed by his own developments [2–4], as well as by others [5–13]. The basic idea seems to be to use the Lagrangian description of continua and the Lagrangian formulation of motion. The basic field used as action-dependent by Eshelby and most other approaches was the displacement field. In such formulation the stress and strain tensors retain their basic importance and the field momentum tensor (or Eshelby's energy-momentum tensor) plays an auxiliary role; in general it can be expressed through stress and strain (or displacement gradient).

Here we would like to use the material momentum tensor introduced in [14] and place it on equal footing with the physical momentum tensor (which

Proceedings of the IUTAM Symposium on Finite Elasticity, Lehigh University, August 10–15, 1980. Invited paper.

Carlson, D.E. and Shield, R.T. (eds.)
Proceedings of the IUTAM Symposium on Finite Elasticity
ISBN 90 247 2629 8

in many cases is simply the stress tensor). By dualism (of some sort) of material and physical coordinates we would like to show how to obtain conservation laws in a general framework, performing explicitly transformations of both independent and dependent variables.

Equations of motion and conservation laws

In classical particle mechanics the Lagrange equations of motion can be obtained from variational principles — namely Hamilton's principle of stationary action A:

$$A = \int_{t_1}^{t_2} L[t; \boldsymbol{x}(t), \dot{\boldsymbol{x}}(t)]dt,$$

where t denotes time, $\boldsymbol{x}(t)$ the actual position of a particle and $\dot{\boldsymbol{x}}(t)$ the time derivative or the velocity of a particle. L is the Lagrangian of the system. By varying the dependent variable $\boldsymbol{x}$, we obtain the Euler–Lagrange equations, or the equations of motion. Conservation laws of classical dynamics are connected with both transformation of dependent and independent variables; namely translation in time is associated with conservation of energy, while translation of $\boldsymbol{x}$ with conservation of momentum, and rotation of $\boldsymbol{x}$ with conservation of angular momentum. Thus, from the point of view of conservation laws, there exists an equivalence among transformations of dependent and independent variables. This is in perfect agreement with a general theorem of Noether: any infinitesimal transformation of either the action variables or the independent variables involving a constant parameter α, which leaves the Lagrangian unchanged, leads automatically to a certain conservation law [15].

The roots of this theorem can be found in the dynamics of rigid bodies. If a certain generalized coordinate does not appear in the Lagrangian function, the momentum connected with this variable, called "cyclic" or "ignorable", is constant during the motion. This approach is partly developed in [17], where conservation of energy of the electromagnetic field is derived by infinitesimal transformation of fields, and a parameter α appearing in this transformation is treated as an additional variable.

Continuum mechanics followed the structure of classical particle dynamics: it used the same conservation laws. The introduction of new independent variables, namely the material coordinates $\boldsymbol{X}$ which are simply continualized indices of particles, was reflected only by the presence of additional terms, usually divergencies with respect to $\boldsymbol{X}$, in the well-known equations.

A typical variational formulation has the form of an action integral:

(1) $$A = \int dt \int d^3XL,$$

where L is now a Lagrangian density. What, however, has not been exploited until Eshelby introduced forces on singularities, were the conservation laws connected with possible transformations of material space.

Adopting the field-theoretical point of view, conservation laws are usually related to possible transformations of 4-dimensional $(\boldsymbol{x}, t)$ space, but x-space coordinates belong now, by contrast to classical mechanics, to independent variables, so the action integral has the form

(2) $$A = \int dt \int d^3x\mathscr{L},$$

where $\mathscr{L}$ is a Lagrangian density in $(\boldsymbol{x}, t)$ space, depending, in general on $\boldsymbol{x}, t$; and field variables and their derivatives with respect to $\boldsymbol{x}$ and t.

As is well known, in general nonlinear elasticity we can use either $\boldsymbol{X}$ (Lagrangian description) or $\boldsymbol{x}$ (Eulerian description) as independent variables. Let us consider an infinite material continuum, whose motion can be derived from the principle of stationary action of

(3) $$A = \int dt \int L(X_k, t; x_i, v_i, x_{i,j})d^3X,$$

where

$$v_i = \frac{dx_i}{dt} \equiv \frac{\partial x_i}{\partial t}\bigg|_{X_k = \text{const.}}, \qquad x_{i,j} = \frac{\partial x_i}{\partial X_j}.$$

The equation of motion has the form

(4) $$\frac{d}{dt} p_i + \frac{\partial}{\partial X_k} p_{ik} = \frac{\partial L}{\partial x_i},$$

where

(5) $$p_i = \frac{\partial L}{\partial v_i}, \qquad p_{ik} = \frac{\partial L}{\partial x_{i,k}}.$$

If we reverse completely the roles of $\boldsymbol{x}$ and $\boldsymbol{X}$, we can write an action integral (similar to the field-theoretical one (2)), in the form:

(6) $$A = \int dt \int \mathscr{L}(x_i\,; t; X_k, V_k, X_{i,j})d^3x, \qquad V_i = \frac{\partial X_i}{\partial t}, \qquad X_{i;k} = \frac{\partial X_i}{\partial x_k}.$$

The Euler–Lagrange equations following from stationary action are:

(7) $$\frac{\partial}{\partial t} B_i + \frac{\partial}{\partial x_k} B_{ik} = \frac{\partial \mathscr{L}}{\partial X_i},$$

where

$$B_i = \frac{\partial \mathscr{L}}{\partial V_i}, \qquad B_{ik} = \frac{\partial \mathscr{L}}{\partial X_{i;k}} \tag{8}$$

are the material momentum vector and tensor, respectively. It can be easily seen that equations (7) and (4) are not equivalent. Transforming (4) into the $(\boldsymbol{x}, t)$ space, we obtain

$$\frac{\partial}{\partial t} P_i + \frac{\partial}{\partial x_k} P_{ik} = -\left(\frac{\partial \mathscr{L}}{\partial x_i}\right)_{\text{exp}} \tag{9}$$

where

$$P_i = \frac{\partial \mathscr{L}}{\partial V_j} X_{j;i}, \qquad P_{ik} = \frac{\partial \mathscr{L}}{\partial X_{j;k}} X_{j;i} - \mathscr{L}\delta_{ik}, \tag{10}$$

while transforming (7) into the $(\boldsymbol{X}, t)$ space we obtain

$$\frac{d}{dt} b_i + \frac{\partial}{\partial X_k} b_{ik} = -\left(\frac{\partial L}{\partial X_i}\right)_{\text{exp}}, \tag{11}$$

where

$$b_i = \frac{\partial L}{\partial v_j} x_{j,i}, \qquad b_{ik} = \frac{\partial L}{\partial x_{j,k}} x_{j,i} - L\delta_{ik}, \tag{12}$$

and $(\partial L/\partial X_i)_{\text{exp}}$, $(\partial \mathscr{L}/\partial x_k)_{\text{exp}}$ denote the explicit derivatives with respect to X_i or x_k, respectively.

Physically, the nonequivalence of (4) and (7) is obvious: in the first case we consider possible motions of a fixed volume of the material, while in the other we are interested in the flow of material through a fixed physical volume in $\boldsymbol{x}$ space.

Let us observe, however, that (9) could have been derived directly from the form (3) by differentiation of L with respect to X_k and use of (4). Similarly, (11) is obtainable from (6) by differentiation of $\mathscr{L}$ with respect to x_i and use of (7). So (9) is a conservation law connected with the action (3), while (11) plays the same role for (6). The form of the right-hand side of (9) suggests that it represents balance of momentum, while (11) by symmetry can be interpreted as balance of material momentum. Usually, momentum in a variational approach is connected with translations. Before we present a simple direct way to study other conservation laws, we will prove the latter statement in the framework of a general field-theoretical formulation.

Variational principle for continua with varying boundaries

To derive the equations of motion it is sufficient to consider variations of fields with fixed boundaries. If, e.g., ϕ is a field variable, we consider a variation

$$\delta\phi = \varepsilon\eta,$$

where η is a function vanishing on the boundary. In general, we can consider transformations, which produce variations depending on the field itself and because of that we cannot assume any more a fixed boundary. To treat the problem generally, such that it would be applicable for a nonlinear dynamic case, let A be given by equation (3). We assume certain transformations of the independent variables X_k, t. Then the total variation of the field x_i consists of two parts: a local one, called sometimes the shape variation, which in general depends on the structure of the field itself, and a second "convective one", related directly to the transformation of the independent variables. The situation is analogous to the one faced in the motion of fluids, when we want to calculate the time rate of change of a quantity f attached to a specific particle: the total change consists of the local rate $\partial f/\partial t$ and the convective part $v(\partial f/\partial x)$. As we can easily check

$$\frac{d}{dt}\frac{\partial}{\partial x_k} \neq \frac{\partial}{\partial x_k}\frac{d}{dt},$$

and the same holds true for total field variation and differentiation with respect to independent variables. If we denote by ϕ any dependent variable or its derivative (e.g., $\phi = x_i$ or $\Phi = x_{i,k}$), then

$$\delta\Phi = \Phi'(X', t') - \Phi(X, t) \tag{8}$$

and

$$\delta\frac{\partial\phi}{\partial X_k} \neq \frac{\partial\delta\phi}{\partial X_k}.$$

Thus the step usually made to derive the equations of motion is not admissible here. In order to proceed further we need to split $\delta\Phi$ into two parts mentioned above:

$$\delta\Phi = \delta_*\Phi + \frac{d\Phi}{dt}\,\delta t + \frac{\partial\Phi}{\partial X_k}\,\delta X_k,$$

where $\delta_*\Phi$ denotes the shape variation at a fixed point:

$$\delta_*\Phi(X_k, t) = \phi'(X_k, t) - \phi(X_k, t).$$

The arguments X_k, t are not important, because the transformation under consideration is infinitesimal; we could instead use X'_k, t'. What is important, however, is the fact that the arguments in ϕ' and ϕ have to be identical. (This corresponds to the local rate $\partial f/\partial t$ for the material derivative.)

The advantage of introducing $\delta_* \Phi$ resides in the fact that

$$\frac{d(\delta_* \Phi)}{dt} = \delta_* \frac{d\Phi}{dt} \quad \text{and} \quad \frac{\partial(\delta_* \Phi)}{\partial X_k} = \delta_* \frac{\partial \Phi}{\partial X_k} .$$

Due to the coordinate transformation, the 4-dimensional volume $dt\ d^3X$ changes too. If we recall that the considered transformation is to be infinitesimal, the Jacobian will reduce to

$$j \approx 1 + \frac{d(\delta t)}{dt} + \frac{\partial(\delta X_k)}{\partial X_k} .$$

We can now write a general formula for the variation of the action (3):

$$\delta A = \int dt \int d^3X \left\{ \frac{d}{dt}\left[L\delta t + \frac{\partial L}{\partial v_i} \delta_* x_i \right] + \frac{\partial}{\partial X_k}\left[L\delta X_k + \frac{\partial L}{\partial x_{i,k}} \delta_* x_i \right] - \delta_* x_i \left[\frac{\partial}{\partial X_k}\left(\frac{\partial L}{\partial x_{i,k}} \right) + \frac{d}{dt}\left(\frac{\partial L}{\partial v_i} \right) - \frac{\partial L}{\partial x_i} \right] \right\} . \tag{13}$$

If we admit now arbitrary variations of fields x_i vanishing on the boundary, keeping independent variables fixed we can rederive equation (4) from the principle of stationary action. (In this case, of course, $\delta x_i = \delta_* x_i$.) However, formula (13) is of much broader use, because of explicit appearance of all variations, both of fields and independent variables. It forms a basis for studying all possible conservation laws, following from admissible (in the sense of leaving the Lagrangian invariant) transformations performed on independent and dependent action variables.

Let us examine some of them. First let us consider translations in physical and material space; they can be described by the following sets of expressions for the variations:

(1) translations in physical space:

$$\delta x_i = \varepsilon_i, \qquad \delta X_k = 0, \qquad \delta t = 0, \qquad \delta_* x_i = \delta x_i = \varepsilon_i ;$$

(2) translations in material space:

$$\delta X_i = \gamma_i, \qquad \delta t = 0, \qquad \delta x_i = 0, \qquad \delta_* x_i = -\gamma_k x_{i,k} ;$$

where ε_i and γ_i are infinitesimal (constant) parameters connected with the transformations.

If L is to be invariant under these transformations, the coefficients of ε_i (or γ_i) have to vanish, which yields

(14) $$\frac{d}{dt} p_i + \frac{\partial}{\partial X_k} p_{ik} = 0,$$

where p_i, p_{ik} are given by (5); and

(14a) $$\frac{d}{dt} b_i + \frac{\partial}{\partial X_k} b_{ik} = 0,$$

where b_i and b_{ik} are given by (12).

Combining (14) with equations of motion (4) we recover the well-known fact that the Lagrangian is invariant under translations if it does not depend on the corresponding coordinates. Equations (14) and (14a) thus express conservation of physical and material momentum, respectively.

Next, let us consider rotations:

(1) in physical space:

$$\delta x_i = \varepsilon_{ijk}\lambda_j x_k, \qquad \delta X_k = \delta t = 0, \qquad \delta_* x_i = \delta x_i\,;$$

(2) in material space:

$$\delta X_i = \varepsilon_{ijk}\omega_j X_k, \qquad \delta t = 0, \qquad \delta x_i = 0, \qquad \delta_* x_i = -x_{i,m}\varepsilon_{mjk}\omega_j X_k.$$

In the first case we obtain conservation of physical moment of momentum

(15) $$\frac{d}{dt}[\varepsilon_{ijk}p_i x_k] + \frac{\partial}{\partial X_k}[p_{ik}\varepsilon_{ijm}x_m] = 0,$$

while in the second we have

(16) $$\frac{d}{dt}[\varepsilon_{ijk}b_i X_k] + \frac{\partial}{\partial X_k}[\varepsilon_{ijk}b_{ik}x_m] = 0.$$

Let us observe that combining (16) with conservation of material momentum (14a) there results the symmetry of the material momentum tensor, i.e.,

$$b_{ik} = b_{ki},$$

but use of (14) with (15) will not lead to a symmetry of p_{ik}, but rather to a dynamic generalization of similar relations given in [3]. Only for linear elasticity do we derive from (14) and (15) the symmetry $p_{ik} = p_{ki}$ [16].

Concluding this part, let us remark that the complete procedure can be carried out in $(\boldsymbol{x}, t)$ space, starting with the action in the form (6). We can write the corresponding expression for δA directly from (13) by replacing L by $\mathscr{L}$, x_i by X_i, and X_i by x_i, etc. The equations representing conservation of physical and material moments of momenta, respectively, have now the forms:

(17) $$\frac{\partial}{\partial t}[\varepsilon_{ijk}P_i x_k] + \frac{\partial}{\partial x_k}[\varepsilon_{ijm}P_{ik}x_m] = 0,$$

and

$$\frac{\partial}{\partial t}[\varepsilon_{ijk}B_iX_k]+\frac{\partial}{\partial x_k}[\varepsilon_{ijm}B_{ik}X_m]=0. \tag{18}$$

If the physical momentum is conserved (the corresponding equation would be (9) with vanishing right-hand side), then from (17) its symmetry would follow, i.e.,

$$P_{ik} = P_{ki}.$$

The symmetry of B_{ik}, however, would not follow from conservation of material momentum and (18).

The conclusions which can be drawn are the following: b_{ik} plays the same role with respect to material space as P_{ik} with respect to the physical one even though the physical momentum in material space is p_{ik} and not b_{ik}. We can see, for example, that a simple transformation of (17) leads to (15), involving p_{ik}, and not (16), involving b_{ik}.

This shows that the material momentum tensor b_{ik} is as important in studying conservation laws as the physical momentum P_{ik}. The basic conservation of material momentum is automatically satisfied if we study properties of a perfect continuum; it is becoming of essential importance for imperfect continua containing defects or nonhomogeneities.

Concluding remarks

The variational principle in the last section represents a general expression for studying equations of motion and conservation laws of elastic bodies. The general form of the Lagrangian allows to apply the proposed procedure to both linear and nonlinear elasticity. The explicit form of variation (13) is the basis for studying physical and material conservation laws. It becomes clear that the "dualism" of physical and material coordinates is essential for this development.

There exists an alternate method to derive, in unified fashion, balance laws in continuum mechanics, i.e. equations of motion and conservation laws [16]. This method is based on simpler procedures involving transformations applied to the fields (or dependent variables) only. It provides, perhaps, better insight into the physics of continua with defects where conservation laws become rather balance laws.

ACKNOWLEDGMENT

This research was supported in part by Air Force Office of Scientific Research Grant AF80–0149 to Stanford University.

References

1. J. D. Eshelby, *The force on an elastic singularity*, Phil. Trans. Roy. Soc. London **A244** (1951), 87.
2. J. D. Eshelby, *The continuum theory of lattice defects*, Solid State Physics, Vol. 3 (1956).
3. J. D. Eshelby, *The elastic energy-momentum tensor*, J. Elasticity **5** (1975), 321–335.
4. J. D. Eshelby, *Energy relations and the energy-momentum tensor in continuum mechanics*, in *Inelastic Behavior of Solids* (M. F. Kanninen et al., eds.), McGraw-Hill, New York, 1970.
5. Wilhelm Günther, *Uber einige Randintegrale der Elastomechanik*, in *Adhandlungen der Branschweigischen Wissenschaftlichen Gesellschaft*, Vol. XIV, Verlag Friedr. Vieweg and Sohn, Branschweig, 1962.
6. J. R. Rice, *A path-independent integral and the approximate analysis of strain concentrations by notches and cracks*, J. Appl. Mech. **35** (1968), 379–386.
7. J. K. Knowles and E. Sternberg, *On a class of conservation laws in linearized and finite elastostatics*, Arch. Rat. Mech. Anal. **44** (1972), 187–211.
8. B. Budiansky and J. R. Rice, *Conservation laws and energy-release rates*, J. Appl. Mech. **40** (1973), 201–203.
9. R. T. Shield, *Conservation laws in finite elasticity*, in *Finite Elasticity* (R. S. Rivlin, ed.), American Society of Mechanical Engineers, Applied Mechanics Division, Vol. 27, 1977, pp. 1–10.
10. M. E. Gurtin, *On a path-independent integral for elastodynamics*, Int. J. Fracture **12** (1976), 643–644.
11. D. C. Fletcher, *Conservation laws in linear elastodynamics*, Arch. Rat. Mech. Anal. **60** (1976), 329–353.
12. D. Rogula, *Forces in material space*, Arch. of Mech. **29** (5) (1977), 705–715.
13. G. Herrmann, *Some applications of invariant variational principles in mechanics of solids*, Proceedings of the IUTAM Conference, Northwestern University, 1978, to appear.
14. A. G. Hermann, *On Conservation laws of continuum mechanics*, Int. J. Solids and Structures, to appear.
15. E. Noether, *Invariant variational problems*, Transp. Th. Stat. Phys. (M. Tavel, trans.), **1** (1971), 186–207; also see: Nachr. Ges. Göttingen (Math-Phys. Klasse) **3** (1918), 235.
16. A. G. Hermann, *Material momentum tensor and path-independent integrals of fracture mechanics*, to appear.
17. C. Lanczos, *The Variational Principles of Mechanics*, University of Toronto Press, 1970.

FINITE DEFORMATION OF THICK-WALLED INNER TUBES AND TYRES UNDER INFLATION AND ROTATION

JAMES M. HILL

Department of Mathematics, The University of Wollongong, Wollongong, New South Wales, Australia

(Received September 11, 1980)

ABSTRACT

The finite deformation of thick-walled inner tubes and tyres under inflation by internal pressure and rotation about its axis of symmetry is clearly an important technological problem. For the general homogeneous isotropic incompressible hyperelastic material two problems are solved for a torus using a perturbation scheme. In both cases it is assumed that the radius of the larger circle which generates the torus is small in comparison to the overall radius of the torus. The problems considered are those of inflation and rotation of toroidal tubes which are firstly a perfect torus in the undeformed state and secondly a perfect torus in the deformed configuration. The solutions of both problems are approximated in the first instance by the well known solution for the uniform inflation and rotation of a circular cylindrical tube under uniform internal pressure. Formulae are given for the first-order correction for the general perfectly elastic material. The results obtained are illustrated with numerical results for three strain-energy functions, namely the Mooney, Gent and Thomas and Ogden materials. For both problems, the zero order approximation for extremely large deformations for the Mooney material appears to exhibit a non-physical discontinuity corresponding to the inner tube being turned inside out. This non-physical behaviour is not observed for the neo-Hookean, Gent and Thomas and Ogden strain-energy functions.

1. Introduction

The inflation by uniform internal pressure together with rotation of toroidal rubber inner tubes and tyres is evidently an important technological problem. Kydoniefs and Spencer [1] give an approximate solution to the inflation problem alone of a thick-walled perfectly elastic material which is a torus in

Proceedings of the IUTAM Symposium on Finite Elasticity, Lehigh University, August 10–15, 1980. Invited paper.

Carlson, D.E. and Shield, R.T. (eds.)
Proceedings of the IUTAM Symposium on Finite Elasticity
ISBN 90 247 2629 8

its deformed state. These authors formulate the problem in general terms and then obtain explicit solutions for the special case of the neo-Hookean material. Their perturbation solution is approximate in the sense that the radius of the larger circles which generate the torus is assumed to be small in comparison to the overall radius of the torus so that squares of the ratio ε of these radii may be neglected in comparison to unity. In a subsequent paper Kydoniefs [2] gives a similar approximate solution to the problem of inflation by uniform internal pressure as well as rotation about the axis of symmetry of the torus. In [2] the deformed configuration is still assumed to be a perfect torus and again the problem is fully solved only for the neo-Hookean material. However the special problem of the deformation of a revolving torus under no internal pressure is solved for the general strain-energy function. In a recent paper [3] the analysis of [1] is extended to the general homogeneous isotropic incompressible hyperelastic material and a new problem is considered, namely the uniform inflation by internal pressure of a thick-walled inner tube or tyre which is a perfect torus in its undeformed state. The purpose of this paper is to similarly extend the results of [2] for the general perfectly elastic material.

We consider two problems for the general homogeneous isotropic incompressible hyperelastic material. Firstly we examine the inflation by uniform internal pressure of a revolving body which is a perfect torus in its undeformed state. Secondly we consider the problem studied by Kydoniefs [2]. In both cases in the limit of ε zero the problem reduces to the uniform inflation and rotation of a circular cylindrical tube under uniform internal pressure, the solution of which has been given by Rivlin [4]. Moreover for both problems solutions of the order ε equations can be obtained for the general hyperelastic material so that specialization to particular strain-energy functions is unnecessary. These solutions as described in [3] are deduced from closed form solutions given previously by Hill [5] for small deformations superimposed upon the inflation of a circular cyclindrical tube.

In this paper we make use of a number of results given in [3]. For example in the following section we merely state the basic exact governing equations without derivation. For the first problem we use coordinates for the undeformed body as independent variables while for the second problem coordinates for the deformed configuration are employed as the independent variables. Consequently we give two forms of the equilibrium equations in Section 2. In Section 3 we describe the perturbation scheme for inflation and rotation of a body which is a torus in its undeformed state. Similarly in Section 4 we give the solution for inflation and rotation of an inner tube or tyre which is a torus in its deformed state. In both of Sections 3 and 4 we employ formulae derived in [3]. In Sections 5, 6 and 7 we give detailed numerical results for both problems for the Mooney, Gent and Thomas and Ogden strain-energy functions. Finally in this section we note that for the uniform inflation by internal pressure of elastic toroidal membranes we refer the reader to Kydoniefs and Spencer [6] and Kydoniefs [7].

2. Basic equations in toroidal coordinates

In this section we simply state the exact governing equations for an inflated torus rotating about its axis of symmetry with constant angular velocity ω. Wherever possible for the principal quantities we employ a notation consistent with that of Kydoniefs [2]. In the first part of this section we detail results required for the problem considered in Section 3 while the formulae of the latter part of the section apply to Section 4. For material and spatial rectangular cartesian coordinates (X, Y, Z) and (x, y, z) respectively we introduce material and spatial toroidal coordinates (R, Θ, Φ) and (r, θ, ϕ) defined by

$$
\begin{aligned}
R &= \{[(X^2+Y^2)^{1/2}-b\}^2+Z^2\}^{1/2},\\
\Theta &= \tan^{-1}(Y/X), \qquad\qquad (2.1)\\
\Phi &= \tan^{-1}\{Z/[(X^2+Y^2)^{1/2}-b]\},
\end{aligned}
$$

and

$$
\begin{aligned}
r &= \{[(x^2+y^2)^{1/2}-c]^2+z^2\}^{1/2},\\
\theta &= \tan^{-1}(y/x), \qquad\qquad (2.2)\\
\phi &= \tan^{-1}\{z/[(x^2+y^2)^{1/2}-c]\},
\end{aligned}
$$

where b and c are constants which denote the distance from the origin to the centre of the concentric circles which generate the torus (see Figures 1 and 2). The metric tensors for (2.1) and (2.2) are given respectively by

$$
(2.3)\quad G_{KL} = \begin{bmatrix} 1 & 0 & 0 \\ 0 & (b+R\cos\Phi)^2 & 0 \\ 0 & 0 & R^2 \end{bmatrix}, \quad g_{ij} = \begin{bmatrix} 1 & 0 & 0 \\ 0 & (c+r\cos\phi)^2 & 0 \\ 0 & 0 & r^2 \end{bmatrix},
$$

with the usual notation (see for example Hill [8]).

We consider a torus which in its undeformed state is obtained by rotating the region between two concentric circles of radii R_1 and R_2 about the Z-axis (see Figure 1). If the torus is now simultaneously inflated by a uniform internal pressure P_1 and rotated about the Z-axis with constant angular velocity ω we assume an axially symmetric deformation of the form

$$
(2.4)\qquad r = r(R,\Phi), \qquad \theta = \Theta, \qquad \phi = \phi(R,\Phi),
$$

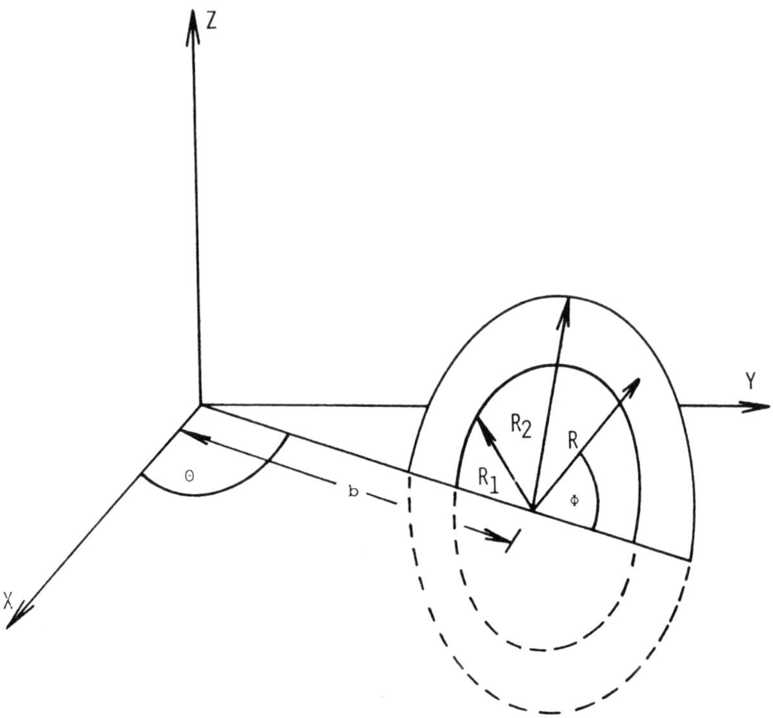

Figure 1. Coordinates for the undeformed configuration.

where (r, θ, ϕ) are defined by (2.2) for some constant c while (R, Θ, Φ) are defined by (2.1) for a given constant b. For an incompressible material (2.4) satisfies

$$j = \frac{\partial(r, \phi)}{\partial(R, \Phi)} = r_R\phi_\Phi - r_\Phi\phi_R = \frac{R(b + R\cos\Phi)}{r(c + r\cos\phi)}, \tag{2.5}$$

where subscripts denote partial derivatives. From the usual Finger stress–strain relations for a homogeneous isotropic incompressible hyperelastic material we can show using the Cayley–Hamilton theorem that the contravariant components of the Cauchy stress tensor are given by

$$\begin{aligned} t^{11} &= -p_1 + \phi_1\left(r_R^2 + \frac{r_\Phi^2}{R^2}\right), \qquad t^{33} = -\frac{p_1}{r^2} + \phi_1\left(\phi_R^2 + \frac{\phi_\Phi^2}{R^2}\right), \\ t^{13} &= \phi_1\left(r_R\phi_R + \frac{r_\Phi\phi_\Phi}{R^2}\right), \qquad t^{22} = -\frac{p_1}{(c + r\cos\phi)^2} + \frac{\phi_2}{(b + R\cos\Phi)^2}, \end{aligned} \tag{2.6}$$

and all other components are zero. In (2.6) p_1 is the pressure function while ϕ_1 and ϕ_2 are response functions which are defined by

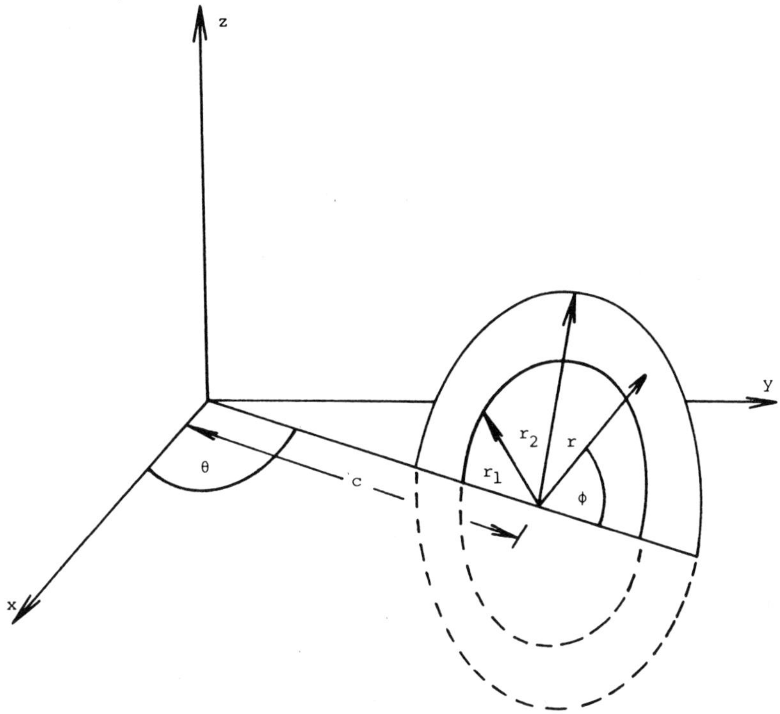

Figure 2. Coordinates for the deformed configuration.

$$\phi_1 = 2\left(\frac{\partial\Sigma}{\partial I_1} + \alpha^2 \frac{\partial\Sigma}{\partial I_2}\right) = 2\,\frac{\partial\Sigma}{\partial I}\,,$$

(2.7)

$$\phi_2 = 2\left\{\frac{\partial\Sigma}{\partial I_1} + \left(I - \frac{1}{\alpha^4}\right)\frac{\partial\Sigma}{\partial I_2}\right\} = 2\,\frac{\partial\Sigma}{\partial\alpha^2}\,,$$

where $\Sigma(I_1, I_2)$ is the strain-energy function of the material and I_1, I_2 are the usual first two invariants of the inverse Cauchy deformation tensor. These are given by

(2.8) $$I_1 = I + \alpha^2, \qquad I_2 = \alpha^2 I + \alpha^{-2},$$

where I and α are defined by

(2.9) $$I = r_R^2 + \frac{r_\Phi^2}{R^2} + r^2\left(\phi_R^2 + \frac{\phi_\Phi^2}{R^2}\right), \qquad \alpha = \left(\frac{c + r\cos\phi}{b + R\cos\Phi}\right).$$

From the momentum equations which are given explicitly in [2] and the relations (2.6) we can show that the deformation (2.4) can be formally reduced to the equilibrium problem,

$$q_{1r} = \phi_1 \left\{ \nabla^2 r - r\left(\phi_R^2 + \frac{\phi_\Phi^2}{R^2}\right)\right\} + \phi_{1R} r_R + \frac{\phi_{1\Phi} r_\Phi}{R^2} - \phi_2 \frac{(c + r\cos\phi)}{(b + R\cos\Phi)^2}\cos\phi,$$

(2.10)

$$\frac{q_{1\phi}}{r^2} = \phi_1 \left\{ \nabla^2 \phi + \frac{2}{r}\left(r_R\phi_R + \frac{r_\Phi\phi_\Phi}{R^2}\right)\right\} + \phi_{1R}\phi_R + \frac{\phi_{1\Phi}\phi_\Phi}{R^2} + \phi_2 \frac{(c + r\cos\phi)}{r(b + R\cos\Phi)^2}\sin\phi,$$

where the Laplacian ∇^2 is given by

$$(2.11)\quad \nabla^2 = \frac{\partial^2}{\partial R^2} + \frac{1}{R}\frac{\partial}{\partial R} + \frac{1}{R^2}\frac{\partial^2}{\partial \Phi^2} + \frac{\cos\Phi}{(b + R\cos\Phi)}\frac{\partial}{\partial R} - \frac{\sin\Phi}{R(b + R\cos\Phi)}\frac{\partial}{\partial\Phi},$$

and q_1 is defined by

$$(2.12)\qquad q_1 = p_1 - \rho\omega^2 r\cos\phi(c + 2^{-1} r\cos\phi).$$

From (2.6) and the first Piola–Kirchhoff stress tensor we can in the usual way show that the boundary conditions of a normal pressure P_1 acting outwardly over a surface given originally by $R =$ constant become

$$(2.13)\qquad (p_1 - P_1)\phi_\Phi - j\phi_1 r_R = 0, \qquad (p_1 - P_1) r_\Phi + j\phi_1 r^2 \phi_R = 0,$$

where j is given by (2.5). Further if da denotes an element of area of the deformed body, $\boldsymbol{n}$ the unit outward drawn normal to da and $\boldsymbol{t}$ the Cauchy stress vector, then by considering the equilibrium of a sector cut off by the planes $\theta = \pm\theta_0$ and resolving forces in the x direction we can deduce that

$$2\sin\theta_0 \iint_a (c + r\cos\phi) t^2 da = \iiint_v (c + r\cos\phi)\rho\omega^2\cos\theta dv$$
$$+ \iint_{a^*} P_1[(n^1\cos\phi - rn^3\sin\phi)\cos\theta - (c + r\cos\phi)n^2\sin\theta]da,$$

where a denotes the cross-section $\theta = \theta_0$, v the volume between $\theta = \pm\theta_0$ and a^* denotes the surface between $\theta = \pm\theta_0$ given originally by $R = R_1$. This equation simplifies to give

$$\int_0^{2\pi}\int_{R_1}^{R_2} (c + r\cos\phi) t^{22} R(b + R\cos\Phi) dR d\Phi$$

$$(2.14)\qquad = \rho\omega^2 \int_0^{2\pi}\int_{R_1}^{R_2} (c + r\cos\phi) R(b + R\cos\Phi) dR d\Phi$$

$$+ P_1 \int_0^{2\pi} (c + r\cos\phi)(r\sin\phi)_\Phi d\Phi,$$

where it is understood that $R = R_1$ in the final integral. In Section 3 we make use of (2.14) to check the calculations.

The second problem we consider is a torus which in its deformed state is obtained by rotating the region between two concentric circles of radii r_1 and

r_2 about the z-axis (see Figure 2). If the torus is simultaneously inflated by a uniform internal pressure P_2 and rotated about the z-axis with constant angular velocity ω then we consider the axially symmetric deformation (2.4) but with (r, ϕ) as the independent variables, that is

$$R = R(r, \phi), \qquad \Theta = \theta, \qquad \Phi = \Phi(r, \phi), \tag{2.15}$$

where here (r, θ, ϕ) are defined by (2.2) for a prescribed constant c while (R, Θ, Φ) are defined by (2.1) for some constant b as yet unspecified. Instead of (2.5) we have for an incompressible material

$$\frac{1}{j} = \frac{\partial(R, \Phi)}{\partial(r, \phi)} = R_{,r}\Phi_{,\phi} - R_{,\phi}\Phi_{,r} = \frac{r(c + r\cos\phi)}{R(b + R\cos\Phi)} . \tag{2.16}$$

Again from the Finger stress–strain relations for a perfectly incompressible elastic material we can deduce by means of the Cayley–Hamilton theorem that the covariant components of the Cauchy stress tensor become

$$\begin{aligned} t_{11} &= -p_2 - \psi_1(R_{,r}^2 + R^2\Phi_{,r}^2), \qquad & t_{33} &= -r^2 p_2 - \psi_1(R_{,\phi}^2 + R^2\Phi_{,\phi}^2), \\ t_{13} &= -\psi_1(R_{,r}R_{,\phi} + R^2\Phi_{,r}\Phi_{,\phi}), \qquad & t_{22} &= -(c + r\cos\phi)^2 p_2 - \psi_2(b + R\cos\Phi)^2, \end{aligned} \tag{2.17}$$

and all other components are zero. In (2.17) p_2 is a pressure function related to that appearing in (2.6) by

$$p_1 - p_2 = 2I\frac{\partial\Sigma}{\partial I} . \tag{2.18}$$

Also in (2.17) the response functions ψ_1 and ψ_2 are given by

$$\psi_1 = 2\left(\frac{\partial\Sigma}{\partial I_1}\beta^2 + \frac{\partial\Sigma}{\partial I_2}\right) = 2\frac{\partial\Sigma}{\partial J}, \quad \psi_2 = 2\left\{\left(J - \frac{1}{\beta^4}\right)\frac{\partial\Sigma}{\partial I_1} + \frac{\partial\Sigma}{\partial I_2}\right\} = 2\frac{\partial\Sigma}{\partial\beta^2}, \tag{2.19}$$

where here I_1 and I_2 are given by

$$I_1 = \beta^2 J + \beta^{-2}, \qquad I_2 = J + \beta^2, \tag{2.20}$$

and J and β are defined by

$$J = R_{,r}^2 + \frac{R_{,\phi}^2}{r^2} + R^2\left(\Phi_{,r}^2 + \frac{\Phi_{,\phi}^2}{r^2}\right), \qquad \beta = \left(\frac{b + R\cos\Phi}{c + r\cos\phi}\right). \tag{2.21}$$

We remark that clearly we have $J = \alpha^2 I$ and $\alpha\beta = 1$. In this case from (2.17) the momentum equations given in [2], yield after some simplifications the equivalent equilibrium problem,

$$\begin{aligned} q_{2R} &= \psi_1\left\{\nabla_1^2 R - R\left(\Phi_{,r}^2 + \frac{\Phi_{,\phi}^2}{r^2}\right)\right\} + \psi_{1,r}R_{,r} + \frac{\psi_{1\phi}R_{\phi}}{r^2} - \psi_2\frac{(b + R\cos\Phi)}{(c + r\cos\phi)^2}\cos\Phi, \\ \frac{q_{2\Phi}}{R^2} &= \psi_1\left\{\nabla_1^2\Phi + \frac{2}{R}\left(R_{,r}\Phi_{,r} + \frac{R_{\phi}\Phi_{\phi}}{r^2}\right)\right\} + \psi_{1,r}\Phi_{,r} + \frac{\psi_{1\phi}\Phi_{\phi}}{r^2} + \frac{\psi_2(b + R\cos\Phi)}{R(c + r\cos\phi)^2}\sin\Phi, \end{aligned} \tag{2.22}$$

where the function q_2 is related to p_2 by

$$q_2 = -p_2 - \Sigma + \rho\omega^2 r\cos\phi(c + 2^{-1} r\cos\phi), \tag{2.23}$$

and where the Laplacian ∇_1^2 is here given by

$$\nabla_1^2 = \frac{\partial^2}{\partial r^2} + \frac{1}{r}\frac{\partial}{\partial r} + \frac{1}{r^2}\frac{\partial^2}{\partial\phi^2} + \frac{\cos\phi}{(c + r\cos\phi)}\frac{\partial}{\partial r} - \frac{\sin\phi}{r(c + r\cos\phi)}\frac{\partial}{\partial\phi}. \tag{2.24}$$

Finally in this section by considering the equilibrium of a sector of the torus cut off by the planes $\theta = \pm\theta_0$ and resolving forces in the x-direction we can deduce, as described in [2], that

$$\begin{aligned} 2\sin\theta_0 &\int_0^{2\pi}\int_{r_1}^{r_2} (c + r\cos\phi)^2 t^{22} r\,dr\,d\phi \\ &= \int_0^{2\pi}\int_{-\theta_0}^{\theta_0}\int_{r_1}^{r_2} \rho\omega^2 (c + r\cos\phi)^2 r\cos\theta\,dr\,d\theta\,d\phi \\ &\quad + \int_{-\theta_0}^{\theta_0}\int_0^{2\pi} P_2\cos\phi\cos\theta\, r_1(c + r_1\cos\phi)\,d\phi\,d\theta, \end{aligned}$$

which becomes

$$\int_0^{2\pi}\int_{r_1}^{r_2} (c + r\cos\phi)^2 t^{22} r\,dr\,d\phi = \pi\rho\omega^2 (r_2^2 - r_1^2)[c^2 + 4^{-1}(r_1^2 + r_2^2)] + \pi P_2 r_1^2. \tag{2.25}$$

As noted in [2] we can make use of (2.25) to check the calculations of Section 4.

3. Deformation from a torus in its undeformed state

We suppose that a torus which in its undeformed state is obtained by rotating the region between two concentric circles of radii R_1 and R_2 about the Z-axis is such that $\varepsilon = R_2/b$ is small compared with unity. We introduce

$$\xi = \frac{R}{R_2}, \qquad \eta = \frac{r}{R_2}, \qquad \lambda = \frac{R_1}{R_2}, \qquad \gamma = \frac{c}{b}, \tag{3.1}$$

so that $\lambda \leqq \xi \leqq 1$ and the deformation (2.4) becomes

$$\eta = \eta(\xi, \Phi), \qquad \theta = \Phi, \qquad \phi = \phi(\xi, \Phi), \tag{3.2}$$

which we now suppose is an analytic function of the parameter ε. The basic equations of Section 2 become

$$\frac{\partial(\eta, \phi)}{\partial(\xi, \Phi)} = \frac{\xi(1 + \varepsilon\xi\cos\Phi)}{\eta(\gamma + \varepsilon\eta\cos\phi)}, \tag{3.3}$$

$$(3.4) \qquad I = \eta_\xi^2 + \frac{\eta_\Phi^2}{\xi^2} + \eta^2\left(\phi_\xi^2 + \frac{\phi_\Phi^2}{\xi^2}\right), \qquad \alpha = \left(\frac{\gamma + \varepsilon\eta\cos\phi)}{1 + \varepsilon\xi\cos\Phi}\right),$$

$$q_{1\eta} = \phi_1\left\{\nabla^2\eta - \eta\left(\phi_\xi^2 + \frac{\phi_\Phi^2}{\xi^2}\right)\right\} + \phi_{1\xi}\eta_\xi + \frac{\phi_{1\Phi}\eta_\Phi}{\xi^2} - \varepsilon\phi_2\frac{(\gamma + \varepsilon\eta\cos\phi)}{(1 + \varepsilon\xi\cos\Phi)^2}\cos\phi,$$

$$(3.5) \qquad \begin{aligned} \frac{q_{1\phi}}{\eta^2} &= \phi_1\left\{\nabla^2\phi + \frac{2}{\eta}\left(\eta_\xi\phi_\xi + \frac{\eta_\Phi\phi_\Phi}{\xi^2}\right)\right\} + \phi_{1\xi}\phi_\xi + \frac{\phi_{1\Phi}\phi_\Phi}{\xi^2} \\ &\quad + \varepsilon\phi_2\frac{(\gamma + \varepsilon\eta\cos\phi)}{\eta(1 + \varepsilon\xi\cos\Phi)^2}\sin\phi, \end{aligned}$$

where here ∇^2 is given by

$$(3.6) \qquad \nabla^2 = \frac{\partial^2}{\partial\xi^2} + \frac{1}{\xi}\frac{\partial}{\partial\xi} + \frac{1}{\xi^2}\frac{\partial^2}{\partial\Phi^2} + \frac{\varepsilon\cos\Phi}{(1 + \varepsilon\xi\cos\Phi)}\frac{\partial}{\partial\xi} - \frac{\varepsilon\sin\Phi}{\xi(1 + \varepsilon\xi\cos\Phi)}\frac{\partial}{\partial\Phi}.$$

Further the boundary conditions (2.13) and (2.14) become

$$(3.7) \qquad (p_1 - P_1)\phi_\Phi - j\phi_1\eta_\xi = 0, \qquad (p_1 - P_1)\eta_\Phi + j\phi_1\eta^2\phi_\xi = 0,$$

$$(3.8) \qquad \begin{aligned} &\varepsilon\int_0^{2\pi}\int_\lambda^1 (\gamma + \varepsilon\eta\cos\phi)\left[-\frac{p_1}{(\gamma + \varepsilon\eta\cos\phi)^2} + \frac{\phi_2}{(1 + \varepsilon\xi\cos\Phi)^2}\right] \\ &\quad \times \xi(1 + \varepsilon\xi\cos\Phi)d\xi d\Phi \\ &= \rho b^2\omega^2\varepsilon\int_0^{2\pi}\int_\lambda^1 (\gamma + \varepsilon\eta\cos\phi)\xi(1 + \varepsilon\xi\cos\Phi)d\xi d\Phi \\ &\quad + P_1\int_0^{2\pi}(\gamma + \varepsilon\eta\cos\phi)(\eta\sin\phi)_\Phi d\Phi, \end{aligned}$$

where now j in (3.7) is given by the Jacobian in (3.3) and in (3.8) we have made use of $(2.6)_4$. It is apparent from the above equations (with the exception of (3.8)) that in the limit of ε zero the problem reduces to the uniform inflation by internal pressure and rotation of a circular cylindrical tube under plane strain conditions (see Rivlin [4]). We therefore seek an approximate solution of the above equations of the form

$$(3.9) \qquad \begin{aligned} \eta &= \gamma^{-1/2}(\xi^2 + K)^{1/2} + \varepsilon u(\xi)\cos\Phi + O(\varepsilon^2), \\ \phi &= \Phi + \varepsilon\gamma^{1/2}v(\xi)\sin\Phi + O(\varepsilon^2), \\ q_1 &= \gamma^{-1}p_0(\xi) + \varepsilon\gamma^{-1/2}q(\xi)\cos\Phi + O(\varepsilon^2), \\ p_1 &= \gamma^{-1}p_0(\xi) + \varepsilon\gamma^{-1/2}p(\xi)\cos\Phi + O(\varepsilon^2), \end{aligned}$$

where K is a constant and the functions $p(\xi)$ and $q(\xi)$ are related by

$$(3.10) \qquad q(\xi) = p(\xi) - \gamma\rho b^2\omega^2(\xi^2 + K)^{1/2}.$$

For the zero order terms we can deduce from (3.4), (3.5) and (3.6)

$$I_0 = \frac{1}{\gamma}\left[\frac{\xi^2}{(\xi^2+K)} + \frac{(\xi^2+K)}{\xi^2}\right], \qquad \alpha_0 = \gamma, \tag{3.11}$$

$$p_0(\xi) = \frac{\xi^2\phi_{10}(\xi)}{(\xi^2+K)} - \int_\lambda^\xi \frac{K(2t^2+K)\phi_{10}(t)}{t(t^2+K)^2}\,dt + \pi_0, \tag{3.12}$$

where π_0 is a constant and the subscript zero denotes evaluation at the zero order deformation. From (3.7) with internal pressure P_1 at $R = R_1$ and external pressure zero at $R = R_2$ we deduce that $\pi_0 = \gamma P_1$ and that

$$P_1 = \int_\lambda^1 \frac{K(2\xi^2+K)\phi_{10}(\xi)d\xi}{\gamma\xi(\xi^2+K)^2}\,. \tag{3.13}$$

If we introduce the functions $w(\xi)$ and $W(\xi)$ defined by

$$\begin{aligned} w(\xi) &= q(\xi) + \frac{\gamma\xi f(\xi)}{(\xi^2+K)^{1/2}}\left\{\frac{\partial\phi_{10}}{\partial\alpha_0} - \frac{2(\xi^2+K)}{\gamma^2\xi^2}\frac{\partial\phi_{10}}{\partial I_0}\right\}, \\ W(\xi) &= p(\xi) + \frac{\gamma\xi f(\xi)}{(\xi^2+K)^{1/2}}\left\{\frac{\partial\phi_{10}}{\partial\alpha_0} - \frac{2(\xi^2+K)}{\gamma^2\xi^2}\frac{\partial\phi_{10}}{\partial I_0}\right\}, \end{aligned} \tag{3.14}$$

where $f(\xi)$ is given by

$$f(\xi) = \xi(\xi^2+K)^{-1/2}\gamma^{-1/2}[\xi - \gamma^{-3/2}(\xi^2+K)^{1/2}], \tag{3.15}$$

then exactly as described in [3] we can deduce the following solutions for the first order terms. We obtain

$$\begin{aligned} u(\xi) &= \bar{u}(\xi) + \sum_{i=1}^4 \gamma_i u_i(\xi), \\ v(\xi) &= \bar{v}(\xi) + \sum_{i=1}^4 \gamma_i v_i(\xi), \\ w(\xi) &= \bar{w}(\xi) + \sum_{i=1}^4 \gamma_i w_i(\xi), \end{aligned} \tag{3.16}$$

where γ_i $(i = 1,2,3,4)$ denote arbitrary constants and $u_i(\xi)$, $v_i(\xi)$ and $w_i(\xi)$ $(i = 1,2,3,4)$ are the linearly independent solutions of a fourth order system of ordinary differential equations. These solutions which were originally derived in [5] are given by

$$\begin{aligned} u_1(\xi) &= \frac{\xi}{(\xi^2+K)^{1/2}}\int_1^\xi \frac{t^2(t^2+K)^{1/2}}{\phi_{10}(t)}\,dt - \int_1^\xi \frac{t^3}{\phi_{10}(t)}\,dt, \\ u_2(\xi) &= \frac{\xi}{(\xi^2+K)^{1/2}}\int_1^\xi \frac{t(t^2+K)}{\phi_{10}(t)}\,dt - \int_1^\xi \frac{t^2(t^2+K)^{1/2}}{\phi_{10}(t)}\,dt, \\ u_3(\xi) &= \frac{\xi}{(\xi^2+K)^{1/2}}, \qquad u_4(\xi) = 1, \end{aligned} \tag{3.17a}$$

$$v_1(\xi)=\frac{1}{(\xi^2+K)^{1/2}}\int_1^\xi \frac{t^3}{\phi_{10}(t)}\,dt-\frac{1}{\xi}\int_1^\xi \frac{t^2(t^2+K)^{1/2}}{\phi_{10}(t)}\,dt,$$

(3.17)$_b$
$$v_2(\xi)=\frac{1}{(\xi^2+K)^{1/2}}\int_1^\xi \frac{t^2(t^2+K)^{1/2}}{\phi_{10}(t)}\,dt-\frac{1}{\xi}\int_1^\xi \frac{t(t^2+K)}{\phi_{10}(t)}\,dt,$$

$$v_3(\xi)=-\frac{1}{\xi},\qquad v_4(\xi)=-\frac{1}{(\xi^2+K)^{1/2}},$$

while for $i=1,2,3,4$,

(3.18)
$$w_i(\xi)=\frac{\xi^2(\xi^2+K)^{1/2}}{K}\,\phi_{10}'(\xi)u_i'(\xi)+\frac{\xi}{(\xi^2+K)^{1/2}}\,\phi_{10}(\xi)[u_i(\xi)-(\xi^2+K)^{1/2}v_i(\xi)]'+\frac{\delta_{i1}K^2}{(\xi^2+K)^{1/2}},$$

where primes denote differentiation with respect to ξ and δ_{ij} is the usual Kronecker delta. Further the particular integrals $\bar{u}(\xi)$, $\bar{v}(\xi)$ and $\bar{w}(\xi)$ in (3.16) can be shown to be given by

(3.19)
$$\bar{u}(\xi)=-\int_1^\xi \frac{(t^2+K)^{1/2}f(t)}{K}\left[\frac{\xi t}{(\xi^2+K)^{1/2}}-(t^2+K)^{1/2}\right]dt+\int_1^\xi \frac{tg(t)}{\phi_{10}(t)}\left[\frac{\xi(t^2+K)^{1/2}}{(\xi^2+K)^{1/2}}-t\right]dt,$$

$$\bar{v}(\xi)=\int_1^\xi \frac{(t^2+K)^{1/2}f(t)}{K}\left[\frac{t}{\xi}-\left(\frac{t^2+K}{\xi^2+K}\right)^{1/2}\right]dt-\int_1^\xi \frac{tg(t)}{\phi_{10}(t)}\left[\frac{(t^2+K)^{1/2}}{\xi}-\frac{t}{(\xi^2+K)^{1/2}}\right]dt,$$

$$\bar{w}(\xi)=Kh(\xi)+\left\{\frac{\xi^2\phi_{10}'(\xi)}{(\xi^2+K)}-\frac{K^2\phi_{10}(\xi)}{\xi(\xi^2+K)^2}\right\}\times\int_1^\xi \frac{t(t^2+K)^{1/2}}{K\phi_{10}(t)}\,[Kg(t)-f(t)\phi_{10}(t)]dt,$$

where the functions $g(\xi)$ and $h(\xi)$ are defined by

(3.20)
$$g(\xi)=\xi A(\xi)+(\xi^2+K)^{1/2}B(\xi),\quad h(\xi)=(\xi^2+K)^{1/2}A(\xi)+\xi B(\xi).$$

The functions $A(\xi)$ and $B(\xi)$ are shown in [3] to be given by

(3.21)
$$A(\xi)=\left[\frac{\xi^2(\xi^2+K)}{K}\,\phi_{10}'(\xi)+2\xi\phi_{10}(\xi)\right]\frac{f(\xi)}{K^2}+\frac{1}{K^2}\int_1^\xi\left\{\frac{\phi_{10}(t)}{\gamma^2}\left[\frac{t^2}{(t^2+K)}+\frac{(t^2+K)}{t^2}\right]-2\gamma\phi_{20}(t)\right\}t\,dt,$$

$$
\begin{aligned}
B(\xi) = & -\left[\frac{\xi^3(\xi^2+K)^{1/2}}{K}\,\phi_{10}'(\xi) + \frac{\xi^2\phi_{10}(\xi)}{(\xi^2+K)^{1/2}} + \gamma^3(\xi^2+K)^{1/2}\phi_{20}(\xi)\right]\frac{f(\xi)}{K^2} \\
& -\frac{\gamma^{3/2}}{K^2}\int_1^{\xi}\left\{\frac{\phi_{10}(t)}{\gamma^2}\left[\frac{t^2}{(t^2+K)} + \frac{(t^2+K)}{t^2}\right] - 2\gamma\phi_{20}(t)\right\} t\,dt. \qquad (3.21)
\end{aligned}
$$

Now from (3.7) and (3.9) we can deduce that the first order boundary conditions become

$$
\left.
\begin{aligned}
W(\xi) &= \frac{\xi^2(\xi^2+K)^{1/2}}{K}\,\phi_{10}'(\xi)u'(\xi) + \frac{2\xi}{(\xi^2+K)^{1/2}}\,\phi_{10}(\xi)u'(\xi) \\
v'(\xi) &= \frac{\xi u(\xi)}{(\xi^2+K)^{3/2}}
\end{aligned}
\right\} \text{ at } \xi = \lambda,\, 1, \qquad (3.22)
$$

where $W(\xi)$ is defined by $(3.14)_2$. The four conditions (3.22) constitute four equations for the determination of the four constants γ_i $(i = 1, 2, 3, 4)$. However since γ_4 is not involved in (3.22) the four conditions overdetermine the constants γ_1, γ_2 and γ_3 and a non-trivial solution exists only if the four equations (3.22) are compatible. The compatibility condition is most easily determined using the first order incompressibility condition, (3.10), (3.14) and (3.22). We find that

$$
\begin{aligned}
w(\xi) & - \frac{\xi^2(\xi^2+K)^{1/2}}{K}\,\phi_{10}'(\xi)u'(\xi) - \frac{\xi}{(\xi^2+K)^{1/2}}\,\phi_{10}(\xi)[u(\xi) - (\xi^2+K)^{1/2}v(\xi)]' \\
& = \frac{\xi\phi_{10}(\xi)f(\xi)}{(\xi^2+K)^{1/2}} - \gamma\rho b^2\omega^2(\xi^2+K)^{1/2} \qquad \text{at } \xi = \lambda,\, 1, \qquad (3.23)
\end{aligned}
$$

which from (3.16)–(3.20) simplifies to yield

$$
\begin{aligned}
A(\xi) + \gamma_1 = & \left[\frac{\xi^2(\xi^2+K)}{K}\,\phi_{10}'(\xi) + 2\xi\phi_{10}(\xi)\right]\frac{f(\xi)}{K^2} \\
& - \frac{\gamma\rho b^2\omega^2(\xi^2+K)}{K^2} \qquad \text{at } \xi = \lambda,\, 1. \qquad (3.24)
\end{aligned}
$$

Thus from (3.21) and (3.24) we deduce that

$$
\gamma_1 = -\gamma K^{-2}\rho b^2\omega^2(1+K), \qquad (3.25)
$$

and that the compatibility condition becomes

$$
\int_\lambda^1\left\{\frac{\phi_{10}(\xi)}{\gamma^2}\left[\frac{\xi^2}{(\xi^2+K)} + \frac{(\xi^2+K)}{\xi^2}\right] - 2\gamma\phi_{20}(\xi)\right\}\xi\,d\xi + \gamma\rho b^2\omega^2(1-\lambda^2) = 0. \qquad (3.26)
$$

We can confirm (3.26), since from (3.8) we can deduce that

$$
2\int_\lambda^1\left[-\frac{p_0(\xi)}{\gamma^2} + \gamma\phi_{20}(\xi)\right]\xi\,d\xi = \gamma\rho b^2\omega^2(1-\lambda^2) + \frac{P_1}{\gamma}(\lambda^2+K), \qquad (3.27)
$$

which using (3.12), (3.13) and changing orders of integration can be shown to be equivalent to (3.26).

If we define the integrals J_1, J_2 and J_3 by

$$J_1 = \int_\lambda^1 \frac{\xi(\xi^2+K)^{1/2}}{K\phi_{10}(\xi)}[Kg(\xi)-f(\xi)\phi_{10}(\xi)]d\xi,$$

(3.28)

$$J_2 = \int_\lambda^1 \frac{\xi(\xi^2+K)}{\phi_{10}(\xi)}\,d\xi, \qquad J_3 = \int_\lambda^1 \frac{\xi^2(\xi^2+K)^{1/2}}{\phi_{10}(\xi)}\,d\xi,$$

and the functions $a(\xi)$, $b(\xi)$ and $c(\xi)$ by

$$a(\xi) = \frac{\xi^2(\xi^2+K)^{3/2}g(\xi)}{\phi_{10}(\xi)(2\xi^2+K)}, \qquad b(\xi) = \frac{\xi^2(\xi^2+K)^2}{\phi_{10}(\xi)(2\xi^2+K)},$$

(3.29)

$$c(\xi) = \frac{\xi^3(\xi^2+K)^{3/2}}{\phi_{10}(\xi)(2\xi^2+K)},$$

then from the boundary conditions (3.22) we can show that the constants γ_2 and γ_3 are given by

$$\gamma_2 = \frac{[a(\lambda)-a(1)+J_1]+\gamma_1[c(\lambda)-c(1)+J_3]}{[b(1)-b(\lambda)-J_2]},$$

(3.30)

$$\gamma_3 = \frac{[b(1)a(\lambda)-a(1)b(\lambda)+b(1)J_1-a(1)J_2]}{[b(1)-b(\lambda)-J_2]}$$

$$+\frac{\gamma_1[b(1)c(\lambda)-c(1)b(\lambda)+b(1)J_3-c(1)J_2]}{[b(1)-b(\lambda)-J_2]}.$$

For a given pressure P_1 and angular velocity ω, (3.13) together with either of (3.26) or (3.27) determine the constants γ and K. It therefore remains only to specify the constant γ_4. Following Kydoniefs [2] we set $u(1)=0$ so that we have simply $\gamma_4 = -\gamma_3/(1+K)^{1/2}$ and the deformed configuration is obtained by rotating the region between eccentric circles of radii r_1 and r_2 about the z-axis where

(3.31) $$r_1 = R_2\gamma^{-1/2}(\lambda^2+K)^{1/2}, \qquad r_2 = R_2\gamma^{-1/2}(1+K)^{1/2},$$

and then the eccentricity of these circles is $\varepsilon R_2 u(\lambda)$.

Finally in this section we note that if we define the integral E, related to the elastic energy of the zero order deformation, by

(3.32) $$E = \int_\lambda^1 \Sigma_0(\xi)\xi d\xi,$$

and consider E as a function of K and γ then we find that (3.13) and (3.26) can be written more simply as

$$\frac{\partial E}{\partial K} = \frac{P_1}{2}, \qquad \frac{\partial E}{\partial \gamma} = \frac{\gamma}{2}\rho b^2\omega^2(1-\lambda^2). \tag{3.33}$$

Moreover since (3.13) can be expressed as

$$P_1 = -\frac{1}{K}\int_\lambda^1 \xi^2\Sigma_0'(\xi)d\xi, \tag{3.34}$$

we have the additional relation

$$2E - KP_1 = [\xi^2\Sigma_0(\xi)]_\lambda^1. \tag{3.35}$$

4. Deformation to a torus in its deformed state

In this section we consider the problem of an inflated rotating tyre first studied by Kydoniefs [2]. By introducing the same symbols as those employed in the previous section we can with slight changes make use of the equations given there. We suppose that a torus in its deformed state is obtained by rotating the region between two concentric circles of radii r_1 and r_2 about the z-axis and is such that $\varepsilon = r_2/c$ is small compared with unity. We introduce

$$\xi = \frac{r}{r_2}, \qquad \eta = \frac{R}{r_2}, \qquad \lambda = \frac{r_1}{r_2}, \qquad \gamma = \frac{b}{c}, \tag{4.1}$$

so that again we have $\lambda \leqq \xi \leqq 1$. The deformation (2.15) becomes

$$\eta = \eta(\xi,\phi), \qquad \Theta = \theta, \qquad \Phi = \Phi(\xi,\phi), \tag{4.2}$$

and we again assume an analytic dependence of (4.2) on the parameter ε. From Section 2 we see that the basic equations for (4.2) are those given by (3.3)–(3.6) with ϕ and Φ interchanged, q_2 in place of q_1 and response functions ψ_1 and ψ_2 which we consider to be functions of J and β defined by (2.21). Moreover, in the following, for the equations of the previous section we read J and β in place of I and α respectively. The pressure boundary conditions and the condition (2.25) become

$$\begin{aligned} t^{11} &= -P_2, \qquad t^{13} = 0 \qquad \text{at } \xi = \lambda, \\ t^{11} &= 0, \qquad t^{13} = 0 \qquad \text{at } \xi = 1, \end{aligned} \tag{4.3}$$

$$\begin{aligned} &\int_0^{2\pi}\int_\lambda^1 c^2(1+\varepsilon\xi\cos\phi)^2 t^{22}\xi d\xi d\phi \\ &\quad = \pi\rho c^2\omega^2(1-\lambda^2)[1+4^{-1}\varepsilon^2(1+\lambda^2)] + \pi P_2\lambda^2, \end{aligned} \tag{4.4}$$

where the components of the Cauchy stress tensor are obtained from (2.17).

Following [2] we seek an approximate solution of the form

$$
\begin{aligned}
&\eta = \gamma^{-1/2}(\xi^2 + K)^{1/2} + \varepsilon u(\xi)\cos\phi + O(\varepsilon^2),\\
&\Phi = \phi + \varepsilon\gamma^{1/2}v(\xi)\sin\phi + O(\varepsilon^2),\\
&q_2 = \gamma^{-1}p_0(\xi) + \varepsilon\gamma^{-1/2}q(\xi)\cos\phi + O(\varepsilon^2),\\
&p_3 = \gamma^{-1}p_0(\xi) + \varepsilon\gamma^{-1/2}p(\xi)\cos\phi + O(\varepsilon^2),
\end{aligned}
\tag{4.5}
$$

where K is a constant and we are using the abbreviation p_3 for $-(p_2+\Sigma)$. Further from (2.23) and (4.5) we see that the functions $p(\xi)$ and $q(\xi)$ are related by

$$
q(\xi) = p(\xi) + \gamma^{1/2}\rho c^2\omega^2\xi. \tag{4.6}
$$

For the zero order deformation $p_0(\xi)$ is given by (3.12) with ψ_{10} in place of ϕ_{10}. From (2.17), (2.23), (3.12), (4.3) and (4.5) we deduce that $\pi_0 = -\gamma[P_2 + \Sigma_0(\lambda)]$ and that

$$
P_2 = -\int_\lambda^1 \frac{K(2\xi^2+K)\psi_{10}(\xi)}{\gamma\xi^3(\xi^2+K)}\,d\xi. \tag{4.7}
$$

The first order deformation is again governed by a fourth order system of ordinary differential equations which is precisely that arising in the previous section with ψ_{10} and ψ_{20} in place of ϕ_{10} and ϕ_{20} respectively and J_0 and β_0 in place of I_0 and α_0 respectively. If we introduce $w(\xi)$ and $W(\xi)$ as defined by (3.14) then the solutions for $u(\xi)$, $v(\xi)$ and $w(\xi)$ are exactly as detailed in the previous section with appropriate changes.

Now from (2.17), (2.23), (4.3) and (4.5) we find that the first order boundary conditions become

$$
\begin{aligned}
W(\xi) = {}&\frac{\xi^2(\xi^2+K)^{1/2}}{K}\psi_{10}'(\xi)u'(\xi) + \left[\frac{\xi}{(\xi^2+K)^{1/2}} + \frac{(\xi^2+K)^{3/2}}{\xi^3}\right]\psi_{10}(\xi)u'(\xi)\\
&-\left[\frac{(\xi^2+K)^{3/2}}{\xi^3}\psi_{10}(\xi) - \frac{\gamma^3(\xi^2+K)^{1/2}}{\xi}\psi_{20}(\xi)\right]f(\xi), \qquad \text{at } \xi = \lambda, 1,\\
v'(\xi) = {}&\frac{\xi}{(\xi^2+K)^{3/2}}u(\xi) \qquad \text{at } \xi = \lambda, 1,
\end{aligned}
\tag{4.8}
$$

where $f(\xi)$ is defined by (3.15). If we multiply $(4.8)_2$ by $(\xi^2+K)\psi_{10}(\xi)/\xi$ and then add the two boundary conditions (4.8) we obtain from the solutions of the previous section

$$
\begin{aligned}
B(\xi) + \gamma_2 = {}&-\left[\frac{\xi^3(\xi^2+K)^{1/2}}{K}\psi_{10}'(\xi) + \frac{\xi^2\psi_{10}(\xi)}{(\xi^2+K)^{1/2}} + \gamma^3(\xi^2+K)^{1/2}\psi_{20}(\xi)\right]\frac{f(\xi)}{K^2}\\
&-\frac{\gamma^{1/2}\rho c^2\omega^2\xi^2}{K^2}, \qquad \text{at } \xi = \lambda, 1,
\end{aligned}
\tag{4.9}
$$

where we have made use of (3.14) with appropriate changes and (4.6). Thus in this case from $(3.21)_2$ and (4.9) we deduce that

$$\gamma_2 = -\gamma^{1/2}K^{-2}\rho c^2\omega^2, \tag{4.10}$$

and that the compatibility condition becomes

$$\int_\lambda^1 \left\{\frac{\psi_{10}(\xi)}{\gamma^2}\left[\frac{\xi^2}{(\xi^2+K)}+\frac{(\xi^2+K)}{\xi^2}\right]-2\gamma\psi_{20}(\xi)\right\}\xi d\xi = \gamma^{-1}\rho c^2\omega^2(1-\lambda^2). \tag{4.11}$$

In order to show that (4.11) is consistent with the zero order equation obtained from (4.4), namely

$$2\int_\lambda^1 [p_0(\xi)+\gamma\Sigma_0(\xi)-\gamma^3\psi_{20}(\xi)]\xi d\xi = \gamma\rho c^2\omega^2(1-\lambda^2)+P_2\gamma\lambda^2, \tag{4.12}$$

we need the result

$$p_0(\xi)+\gamma\Sigma_0(\xi) = \frac{\xi^2\psi_{10}(\xi)}{(\xi^2+K)}+\int_\xi^1 \frac{K(2t^2+K)\psi_{10}(t)}{t^3(t^2+K)}\,dt. \tag{4.13}$$

Using (4.7) and (4.13) in (4.12) and interchanging orders of integration we can confirm (4.11).

From the solutions of the previous section and the boundary conditions $(4.8)_2$ we can readily deduce that the constants γ_1 and γ_3 are given by

$$\gamma_1 = \frac{[a(\lambda)-a(1)+J_1]+\gamma_2[b(\lambda)-b(1)+J_2]}{[c(1)-c(\lambda)-J_3]},$$

$$\gamma_3 = \frac{[a(\lambda)c(1)-c(\lambda)a(1)+c(1)J_1-a(1)J_3]}{[c(1)-c(\lambda)-J_3]} + \frac{\gamma_2[b(\lambda)c(1)-c(\lambda)b(1)+c(1)J_2-b(1)J_3]}{[c(1)-c(\lambda)-J_3]}. \tag{4.14}$$

The functions and integrals appearing in (4.14) are precisely those defined by (3.28) and (3.29) with ψ_{10} and ψ_{20} in place of ϕ_{10} and ϕ_{20} respectively. As in the previous section the constant γ_4 is determined by the condition $u(1)=0$ so that again we have $\gamma_4 = -\gamma_3/(1+K)^{1/2}$ and the undeformed configuration is obtained by rotating the region between eccentric circles of radii R_1 and R_2 about the Z-axis where

$$R_1 = r_2\gamma^{-1/2}(\lambda^2+K)^{1/2}, \qquad R_2 = r_2\gamma^{-1/2}(1+K)^{1/2}, \tag{4.15}$$

and then the eccentricity of these circles is $\varepsilon r_2 u(\lambda)$.

Finally in this section we note the relations corresponding to (3.32), (3.33) and (3.35). With E defined by (3.32) we first observe that (4.7) can be written alternatively as

$$(4.16)\qquad P_2 = [\Sigma_0(1) - \Sigma_0(\lambda)] - \int_\lambda^1 \frac{K(2\xi^2+K)}{\gamma\xi(\xi^2+K)^2}\,\psi_{10}(\xi)d\xi,$$

and thus (4.7) and (4.11) become respectively

$$(4.17)\qquad \frac{\partial E}{\partial K} = -\frac{P_2}{2} + \frac{[\Sigma_0(1)-\Sigma_0(\lambda)]}{2}, \qquad \frac{\partial E}{\partial \gamma} = -\frac{\rho c^2\omega^2(1-\lambda^2)}{2\gamma},$$

while the equation corresponding to (3.35) becomes

$$(4.18)\qquad 2E + KP_2 = [(\xi^2+K)\Sigma_0(\xi)]_\lambda^1.$$

In the final three sections we illustrate the results obtained for three standard strain-energy functions.

5. Mooney material

The strain-energy function for a Mooney material is given by

$$(5.1)\qquad \Sigma = C_1(I_1-3) + C_2(I_2-3),$$

where C_1 and C_2 are material constants. As usual we let $\Gamma = C_2/C_1$. From (5.1) we can show that (3.13) and (3.26) become

$$(5.2)\qquad \begin{aligned} \frac{P_1}{C_1} &= \frac{(1+\gamma^2\Gamma)}{\gamma}\left\{\frac{K(1-\lambda^2)}{(\lambda^2+K)(1+K)} - 2\log\left[\lambda\left|\frac{1+K}{\lambda^2+K}\right|^{1/2}\right]\right\},\\ \frac{\sigma\gamma^3}{(1-\gamma^2\Gamma)} &= \frac{(\gamma^3-1)(\gamma+\Gamma)}{\gamma(1-\gamma^2\Gamma)} + \frac{K}{(1-\lambda^2)}\log\left[\lambda\left|\frac{1+K}{\lambda^2+K}\right|^{1/2}\right], \end{aligned}$$

where σ is the non-dimensional constant given by

$$(5.3)\qquad \sigma = \frac{\rho b^2\omega^2}{2C_1}.$$

In order to illustrate (5.2) we have adopted three standard values for σ, namely $\sigma_0 = 0$, $\sigma_1 = 0.681$ and $\sigma_2 = 0.875$. These latter two values of σ have been chosen so that the results for rotating and non-rotating tubes are sufficiently distinguishable when graphed. For a tyre of overall outer radius 30 cm and shear modulus ($\mu = 2C_1$) of approximately 14 kgf/cm^2, the values σ_1 and σ_2 correspond to speeds of approximately 360 and 460 kilometers per hour respectively. In order to understand the solutions of (5.2), Figure 3 shows the variation of $\log|(1+K)/(1+K/\lambda^2)|^{1/2}$. Figure 4 shows the variation of P_1/C_1 and K with γ for a Mooney material with $\Gamma = 0.3$ and $\lambda = 0.9$. We see that K is positive for $\gamma < \Gamma^{-1/2}$ and negative for $\gamma > \Gamma^{-1/2}$. The discontinuity in K can be readily confirmed by examining $(5.2)_2$ and Figure 3. It corresponds to the inner tube being turned inside out which is clearly unrealistic. We note however that the pressure remains continuous.

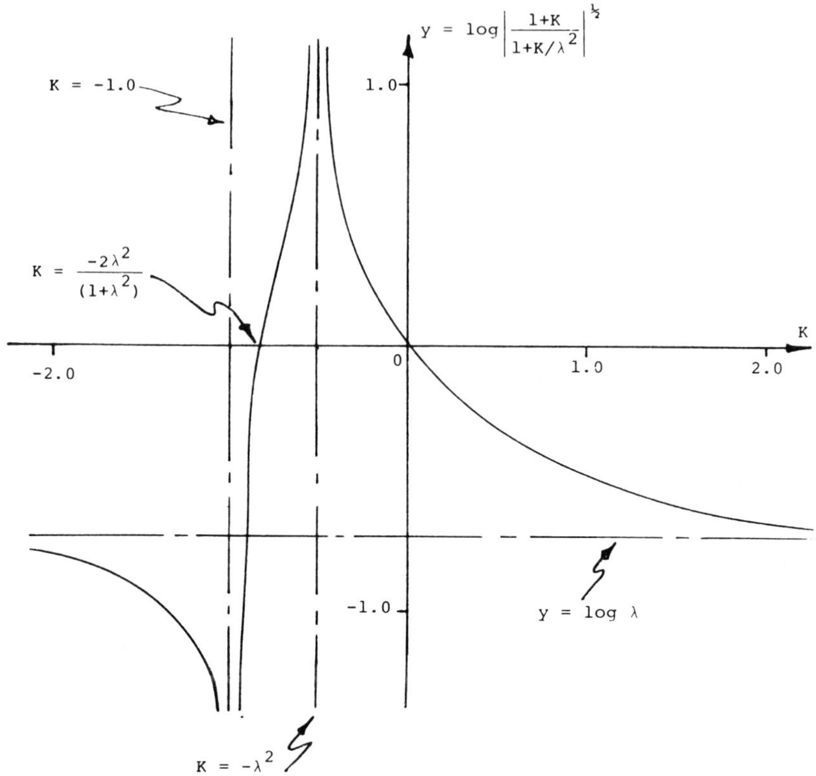

Figure 3. Graph of $y = \log|(1+K)/(1+K/\lambda^2)|^{\frac{1}{2}}$.

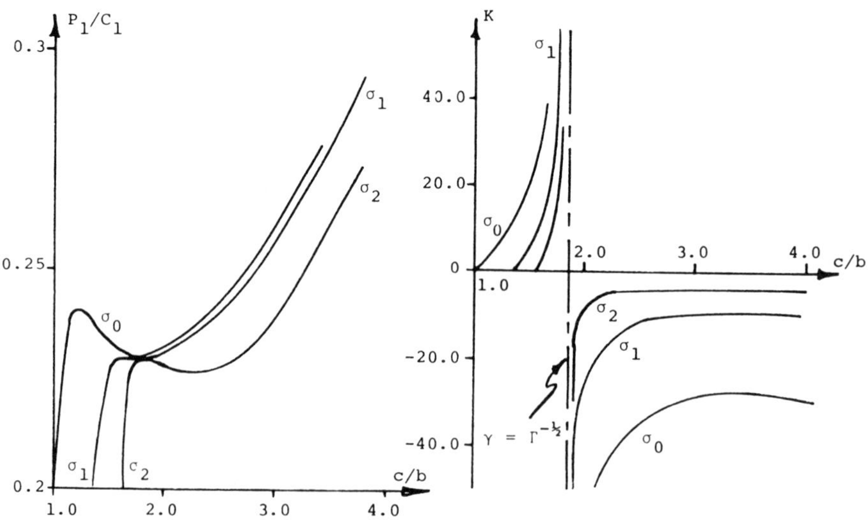

Figure 4. Graphs of P_1/C_1 and K for various values of c/b for a Mooney material with $\Gamma = 0.3$.

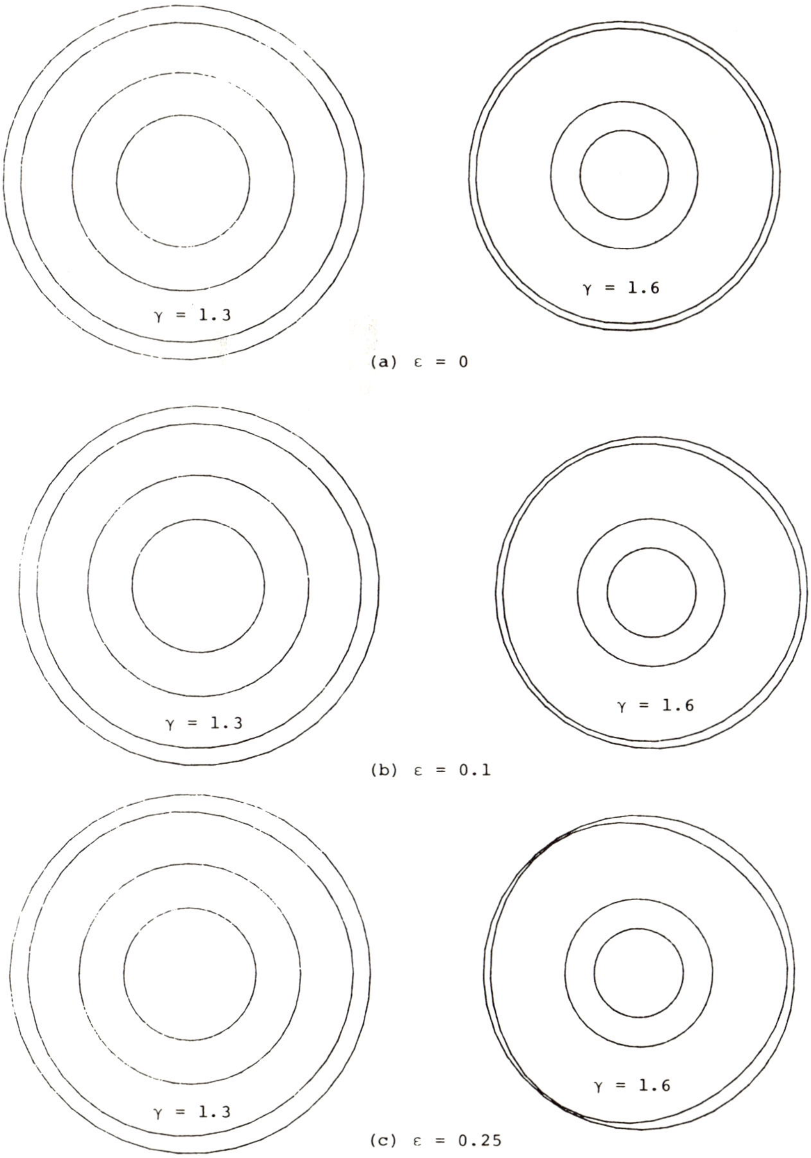

Figure 5. Cross-sections for zero rotation, $\Gamma = 0.1$, $\lambda = 0.6$, $\gamma = 1.3$ and 1.6 and various values of ε (scaled so that the two inner circles represents original cross-section).

Figure 5 shows the cross-sections of the tube for zero angular velocity, $\Gamma = 0.1$, $\lambda = 0.6$, $\varepsilon = 0$, 0.1 and 0.25 and $\gamma = 1.3$ and 1.6. These curves are plotted from

$$(5.4)\qquad \begin{aligned}\frac{x}{R_2} &= \frac{|\xi^2+K|^{1/2}}{\gamma^{1/2}}\cos\Phi + \varepsilon\{u(\xi)\cos^2\Phi - |\xi^2+K|^{1/2}v(\xi)\sin^2\Phi\},\\ \frac{y}{R_2} &= \frac{|\xi^2+K|^{1/2}}{\gamma^{1/2}}\sin\Phi + \varepsilon\{u(\xi) + |\xi^2+K|^{1/2}v(\xi)\}\sin\Phi\cos\Phi,\end{aligned}$$

where $u(\xi)$ and $v(\xi)$ are obtained from Section 3 and we are consistently interpreting $(\xi^2+K)^{1/2}$ as $|\xi^2+K|^{1/2}$ in order to accommodate the case where K is large and negative. In fact the inner curve moves through the outer curve prior to $\gamma = \Gamma^{-1/2}$ (approximately at $\gamma = 2.0$ for $\varepsilon = 0.1$ and $\gamma = 1.7$ for $\varepsilon = 0.25$). We note that the curves in Figure 5 are scaled such that the inner two circles represent the original cross-section of the torus through a Θ = constant plane. We observe that the inner tube thickens on its outside ($\Phi = 0$) as γ increases. We note also that for small γ the cross-sections for ε non-zero are not substantially different from those for the zero order approximation ($\varepsilon = 0$).

For the problem of the deformation to a torus we can show for a Mooney material that (4.7) and (4.11) become

$$(5.5)\qquad \begin{aligned}\frac{P_2}{C_1} &= \frac{(\gamma^2+\Gamma)}{\gamma}\left\{\frac{K(\lambda^2-1)}{\lambda^2} + 2\log\left[\lambda\left|\frac{1+K}{\lambda^2+K}\right|^{1/2}\right]\right\},\\ \frac{\sigma^*\gamma}{(\Gamma-\gamma^2)} &= \frac{(1-\gamma^3)(1+\gamma\Gamma)}{\gamma(\Gamma-\gamma^2)} - \frac{K}{(1-\lambda^2)}\log\left[\lambda\left|\frac{1+K}{\lambda^2+K}\right|^{1/2}\right],\end{aligned}$$

where σ^* is the non-dimensional constant given by (5.3) with c in place of b. The variation of P_2/C_1 and $-K$ with c/b is shown in Figure 6 for a Mooney material with $\Gamma = 0.1$ and $\lambda = 0.9$. The values of σ^* taken in Figure 6 are the three standard values adopted for σ. We note that for this problem, for sufficiently large values of c/b (c/b greater than $\Gamma^{-1/2}$) numerical results appear to indicate both P_2/C_1 and K are discontinuous functions of c/b.

Finally in this section we note that the zero order problems "to" and "from" a torus are clearly related. If we let λ^*, γ^* and K^* be the values of λ, γ and K for the problem of deformation to a torus and λ, γ and K are the values for the same material for the problem of deformation from a torus then we can confirm that we have

$$(5.6)\qquad \lambda^{*2} = \left(\frac{\lambda^2+K}{1+K}\right),\qquad \gamma^* = \frac{1}{\gamma},\qquad K^* = \frac{-K}{(1+K)}.$$

If we substitute λ^*, γ^* and K^* as given by (5.6) into (5.5) then since $\sigma^* = \gamma^2\sigma$ we obtain precisely (5.2) with P_2 in place of P_1. Evidently the zero order problems for all materials are related in this manner. Moreover we observe from (5.6) that firstly K may be infinite while K^* is well defined and secondly

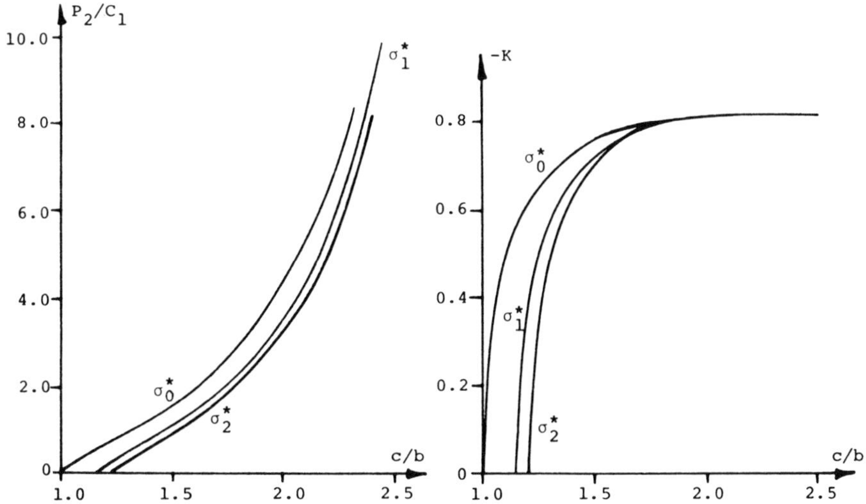

Figure 6. Graphs of P_2/C_1 and $-K$ for various values of c/b for a Mooney material with $\Gamma = 0.1$.

K must be such that $(\lambda^2 + K)/(1 + K)$ is always positive. This latter observation clearly excludes the region $-1 < K < -\lambda^2$.

6. Gent and Thomas material

The strain-energy function proposed by Gent and Thomas [9] is given by

$$\Sigma = k_1(I_1 - 3) + k_2 \log(I_2/3), \tag{6.1}$$

where k_1 and k_2 are material constants which Gent and Thomas [9] related to the Mooney constants C_1 and C_2 by

$$k_1 = C_1 + 0.247C_2, \qquad k_2 = 2.18C_2. \tag{6.2}$$

If we introduce $k = k_2/k_1$ then from (6.2) we have

$$k = \frac{2.18\Gamma}{1 + 0.247\Gamma}, \tag{6.3}$$

where $\Gamma = C_2/C_1$. Thus the values of k corresponding to Γ equal to 0, 0.1, 0.2 and 0.3 are 0, 0.2128, 0.4155 and 0.6089 respectively. From (6.1), (3.13) and (3.26) we obtain

$$\begin{aligned} \frac{P_1}{k_1} = \frac{1}{\gamma} & \left\{ \frac{K(1-\lambda^2)}{(1+K)(\lambda^2+K)} - 2\log\left[\lambda \left| \frac{1+K}{\lambda^2+K} \right|^{1/2}\right] \right. \\ & \left. - k\left\{ 2\log\left| \frac{1+K}{\lambda^2+K} \right|^{1/2} - \tfrac{1}{2}\log\left[\frac{K^2\gamma^3 + (2\gamma^3+1)(1+K)}{K^2\gamma^3 + (2\gamma^3+1)\lambda^2(\lambda^2+K)} \right] \right\} \right. \end{aligned} \tag{6.4}$$

$$+k\left(\frac{2\gamma^3-1}{2\gamma^3+1}\right)^{1/2}\left[\tan^{-1}\frac{(1+2K^{-1})(2\gamma^3+1)^{1/2}}{(2\gamma^3-1)^{1/2}}\right.$$
$$\left.-\tan^{-1}\frac{(1+2\lambda^2K^{-1})(2\gamma^3+1)^{1/2}}{(2\gamma^3-1)^{1/2}}\right],$$

(6.4)

$$\sigma\gamma^3=(\gamma^3-1)\left[1+\frac{k\gamma}{(2\gamma^3+1)}\right]+\frac{K}{(1-\lambda^2)}\log\left[\lambda\left|\frac{1+K}{\lambda^2+K}\right|^{1/2}\right]$$
$$+\frac{3kK\gamma^4}{(1-\lambda^2)(2\gamma^3+1)^{3/2}(2\gamma^3-1)^{1/2}}\left[\tan^{-1}\frac{(1+2K^{-1})(2\gamma^3+1)^{1/2}}{(2\gamma^3-1)^{1/2}}\right.$$
$$\left.-\tan^{-1}\frac{(1+2\lambda^2K^{-1})(2\gamma^3+1)^{1/2}}{(2\gamma^3-1)^{1/2}}\right],$$

where here σ is given by (5.3) with k_1 in place of C_1. The variation of $P_1/2k_1$ with c/b is shown in Figure 7 for the three previously adopted values for σ and for $\lambda=0.9$. We notice that for zero rotation ($\sigma=\sigma_0$) the variation in pressure with c/b for increasing k is qualitatively different to that given in [3] for the corresponding Mooney materials. We note also that for the curves given in Figure 7 the corresponding graphs of K are all smooth, positive and monotonically increasing functions of γ.

For completeness we note the equations corresponding to (4.7) and (4.11) for the Gent and Thomas material for the problem of the deformation to a torus. These are

(6.5)

$$\frac{P_2}{k_1}=\gamma\left\{\frac{K(\lambda^2-1)}{\lambda^2}+2\log\left[\lambda\left|\frac{1+K}{\lambda^2+K}\right|^{1/2}\right]\right\}$$
$$-k\left\{2\log\lambda+\tfrac{1}{2}\log\left[\frac{K^2+(2+\gamma^3)(1+K)}{K^2+(2+\gamma^3)\lambda^2(\lambda^2+K)}\right]\right\}$$

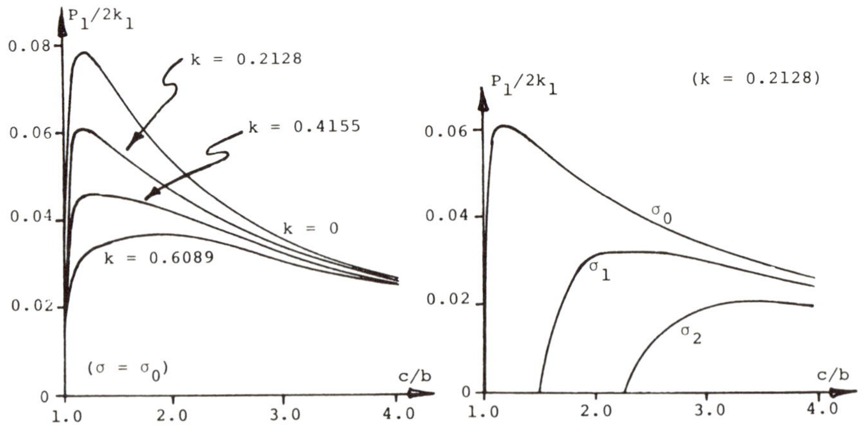

Figure 7. Graphs of $P_1/2k_1$ for various values of c/b for a Gent and Thomas material with the shown values of k (K is a positive monotonically increasing function of γ).

$$-k\left(\frac{2-\gamma^3}{2+\gamma^3}\right)^{1/2}\left[\tan^{-1}\frac{(1+2K^{-1})(2+\gamma^3)^{1/2}}{(2-\gamma^3)^{1/2}}\right.$$

$$\left.-\tan^{-1}\frac{(1+2\lambda^2K^{-1})(2+\gamma^3)^{1/2}}{(2-\gamma^3)^{1/2}}\right],$$

(6.5)
$$\frac{\sigma^*}{\gamma}=\frac{(1-\gamma^3)}{\gamma^3}\left[1+\frac{k\gamma^2}{(2+\gamma^3)}\right]+\frac{K}{(1-\lambda^2)}\log\left[\lambda\left|\frac{1+K}{\lambda^2+K}\right|^{1/2}\right]$$

$$+\frac{3kK\gamma^2}{(1-\lambda^2)(2+\gamma^3)^{3/2}(2-\gamma^3)^{1/2}}$$

$$\times\left[\tan^{-1}\frac{(1+2K^{-1})(2+\gamma^3)^{1/2}}{(2-\gamma^3)^{1/2}}-\tan^{-1}\frac{(1+2\lambda^2K^{-1})(2+\gamma^3)^{1/2}}{(2-\gamma^3)^{1/2}}\right],$$

where σ^* is $\rho c^2\omega^2/2k_1$. As a check on the calculations leading to (6.4) and (6.5) we can make use of the transformation (5.6) to deduce one set of equations from the other. We note that in this case the variation of P_2/k_1 and $-K$ with c/b are similar to that indicated in Figure 6. Namely K appears to be restricted to the range $-\lambda^2 < K < 0$ and moreover the variation in pressure with c/b for increasing k is qualitatively the same as that given in [3] for the corresponding Mooney materials. The agreement for this problem, in contrast to the different qualitative behaviour for the previous problem, is not altogether unexpected since the relations (6.2) are chosen so that the Mooney and logarithmic forms agree in simple extension but are widely different in compression (see Gent and Thomas [9]). Finally in this section we note that for both problems for the Gent and Thomas material K does not appear to exhibit the discontinuity characteristic of the Mooney material.

7. Ogden material

For the material proposed by Ogden [10], the strain-energy function and shear modulus are given by

(7.1)
$$\Sigma=\sum_{n=1}^{3}\frac{\mu_n}{\alpha_n}(\lambda_1^{\alpha_n}+\lambda_2^{\alpha_n}+\lambda_3^{\alpha_n}-3),\qquad \mu=\frac{1}{2}\sum_{n=1}^{3}\mu_n\alpha_n,$$

where λ_i $(i=1,2,3)$ denote the principal stretches of the deformation and μ_n and α_n $(n=1,2,3)$ are constants. We shall consider three sets of constants, namely those given by Ogden [10] and two further sets proposed by Chadwick, Creasy and Hart [11]. These constants are

(i) $\left.\begin{matrix}\alpha_1=1.3 & \alpha_2=-2.0 & \alpha_3=5.0\\ \mu_1=6.3 & \mu_2=-0.1 & \mu_3=0.012\end{matrix}\right\}$ $\mu=4.225\ (\text{kgf/cm}^2)$,

$$\text{(ii)}\quad \left.\begin{array}{lll} \alpha_1 = 2.0 & \alpha_2 = -1.25 & \alpha_3 = 7.82 \\ \mu_1 = 3.0 & \mu_2 = -0.81 & \mu_3 = 3.7 \times 10^{-5} \end{array}\right\} \quad \mu = 3.506\ (\text{kgf/cm}^2),$$

$$\text{(iii)}\quad \left.\begin{array}{lll} \alpha_1 = 2.0 & \alpha_2 = -2.0 & \alpha_3 = 8.7 \\ \mu_1 = 3.24 & \mu_2 = -0.1 & \mu_3 = 6.2 \times 10^{-6} \end{array}\right\} \quad \mu = 3.340\ (\text{kgf/cm}^2),$$

and we note that (ii) and (iii) are modifications of the neo-Hookean and Mooney theories respectively.

For the zero order problem of the deformation from a torus we can show that the zero order principal stretches are given by

$$\lambda_1 = \frac{\xi}{\gamma^{1/2}(\xi^2 + K)^{1/2}}, \qquad \lambda_2 = \frac{(\xi^2 + K)^{1/2}}{\gamma^{1/2}\xi}, \qquad \lambda_3 = \gamma. \tag{7.2}$$

From (7.1), (7.2) we find that (3.13) and (3.26) become

$$\begin{aligned} \frac{P_1}{\mu} &= \frac{1}{2\mu}\sum_{n=1}^{3}\frac{\mu_n}{\gamma^{\alpha_n/2}}\left\{\phi^*\left(\frac{K}{1+K};\frac{\alpha_n}{2}\right) - \phi^*\left(\frac{K}{\lambda^2+K};\frac{\alpha_n}{2}\right)\right\}, \\ \sigma\gamma^2 &= \frac{K}{2\mu(1-\lambda^2)}\sum_{n=1}^{3}\frac{\mu_n}{\gamma^{\alpha_n/2}}\left\{\Phi^*\left(\frac{K}{1+K};\frac{\alpha_n}{2}\right) - \Phi^*\left(\frac{K}{\lambda^2+K};\frac{\alpha_n}{2}\right)\right\} \\ &\quad + \frac{1}{\mu}\sum_{n=1}^{3}\frac{\mu_n}{\gamma^{\alpha_n/2}}(\gamma^{3\alpha_n/2} - 1), \end{aligned} \tag{7.3}$$

where σ is $\rho b^2\omega^2/\mu$ and the functions $\phi^*(x;\alpha)$ and $\Phi^*(x;\alpha)$ for α non-zero are defined as follows,

$$\begin{aligned} \phi^*(x;\alpha) &= \int_{1-x}^{1}\left(\frac{u^{\alpha} - u^{-\alpha}}{1-u}\right)du, \\ \Phi^*(x;\alpha) &= \int_{1-x}^{1}\left(\frac{u^{\alpha/2} - u^{-\alpha/2}}{1-u}\right)^2 du \quad (x < 1). \end{aligned} \tag{7.4}$$

The variation of P_1/μ and K with c/b is shown in Figure 8 for the case of zero rotation and for the above three sets of constants with $\lambda = 0.9$.

Again for completeness we note the details for the zero order problem of inflation to a torus. In this case we can show that the zero order principal stretches are given by

$$\lambda_1 = \frac{\gamma^{1/2}\xi}{(\xi^2 + K)^{1/2}}, \qquad \lambda_2 = \frac{\gamma^{1/2}(\xi^2 + K)^{1/2}}{\xi}, \qquad \lambda_3 = \frac{1}{\gamma}. \tag{7.5}$$

Further the equations corresponding to (4.7) and (4.11) become

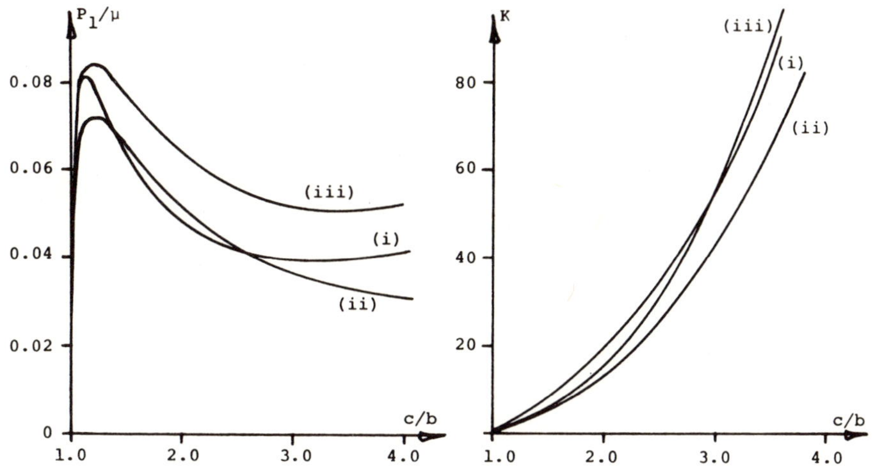

Figure 8. Graphs of P_1/μ and K for various values of c/b for Ogden materials and zero rotation.

$$\begin{aligned}
&\frac{P_2}{\mu} = \frac{1}{2\mu} \sum_{n=1}^{3} \mu_n \gamma^{\alpha_n/2} \left\{ \phi^* \left(-K; \frac{\alpha_n}{2} \right) - \phi^* \left(-\frac{K}{\lambda^2}; \frac{\alpha_n}{2} \right) \right\}, \\
(7.6) \quad &\sigma^* = -\frac{K}{2\mu(1-\lambda^2)} \sum_{n=1}^{3} \mu_n \gamma^{\alpha_n/2} \left\{ \Phi^* \left(-K; \frac{\alpha_n}{2} \right) - \Phi^* \left(-\frac{K}{\lambda^2}; \frac{\alpha_n}{2} \right) \right\} \\
&\quad + \frac{1}{\mu} \sum_{n=1}^{3} \frac{\mu_n}{\gamma^{\alpha_n}} (1 - \gamma^{3\alpha_n/2}),
\end{aligned}$$

where σ^* is $\rho c^2 \omega^2/\mu$ and $\phi^*(x;\alpha)$ and $\Phi^*(x;\alpha)$ are as previously defined. Again it is a simple matter to show from either (7.3) or (7.6) that the other equation is obtained from the transformation (5.6). We note that for both problems for the Ogden materials K appears to be a smooth continuous function of c/b.

Acknowledgements

The author wishes to express his gratitude to Ian Piper for undertaking the numerical work. He is also grateful to Geoffrey Aldis for helpful discussions with this and related work.

References

1. A. D. Kydoniefs and A. J. M. Spencer, *The finite inflation of an elastic torus*, Int. J. Engng. Sci. **3** (1965), 173–195.

2. A. D. Kydoniefs, *Finite deformation of an elastic torus under rotation and inflation*, Int. J. Engng. Sci. **4** (1966), 125–154.

3. J. M. Hill, *The finite inflation of a thick-walled elastic torus*, Q. Jl. Mech. Appl. Math. **33** (1980), 471–490.

4. R. S. Rivlin, *Large elastic deformations of isotropic materials, Part VI, Further results in the theory of torsion, shear and flexure*, Phil. Trans. Roy. Soc. Lond. **A242** (1949), 173–195.

5. J. M. Hill, *Closed form solutions for small deformations superimposed upon the simultaneous inflation and extension of a cylindrical tube*, J. Elasticity **6** (1976), 113–123.

6. A. D. Kydoniefs and A. J. M. Spencer, *The finite inflation of an elastic toroidal membrane of circular cross section*, Int. J. Engng. Sci. **5** (1967), 367–391.

7. A. D. Kydoniefs, *The finite inflation of an elastic toroidal membrane*, Int. J. Engng. Sci. **5** (1967), 477–494.

8. J. M. Hill, *Partial solutions of finite elasticity — three dimensional deformations*, Z. Angew. Math. Phys. **24** (1973), 609–618.

9. A. N. Gent and A. G. Thomas, *Forms for the stored (strain) energy function for vulcanized rubber*, J. Polym. Sci. **28** (1958), 625–628.

10. R. W. Ogden, *Large deformation isotropic elasticity — On the correlation of theory and experiment for incompressible rubberlike solids*, Proc. R. Soc. Lond. **A326** (1972), 565–584.

11. P. Chadwick, C. F. M. Creasy and V. G. Hart, *The deformation of rubber cylinders and tubes by rotation*, J. Austral. Math. Soc. (Series B) **20** (1977), 62–96.

FINITE STRAIN J_2 DEFORMATION THEORY

J. W. HUTCHINSON

Division of Applied Sciences, Harvard University, Cambridge, Massachusetts 02138, USA

K. W. NEALE

Department of Civil Engineering, University of Sherbrooke, Sherbrooke, Quebec, Canada

(Received November 7, 1980)

ABSTRACT

A finite strain version of the J_2 deformation theory of plasticity is given. The material model is an isotropic, nonlinearly elastic solid. The range of states is investigated for which the equations governing incremental responses are elliptic.

1. Introduction

A particular finite-elasticity constitutive law which can be considered as a prototype model for certain limited classes of time-independent deformations of plastic solids is the subject of this paper. This law was first formulated by Hutchinson and Neale [1] in connection with an investigation of localized necking failures in thin sheet metals. It has subsequently been employed in a number of other studies of finite-strain and bifurcation phenomena in plastic solids (e.g. [2–4]).

In small-strain plasticity the most commonly employed constitutive laws are the "J_2 flow theory" and "J_2 deformation theory" relations. In both of these constitutive theories the plastic strain increments satisfy incompressibility, and they are connected to a general multiaxial stress state through J_2 — the second invariant of the deviatoric stress tensor. J_2 flow theory involves relations between stress increments and strain increments, which lead to path-dependence of total stress and total strain for arbitrary stress histories. In contrast, J_2 deformation theory is a small-strain nonlinear elasticity constitutive law. Although deformation theory is clearly inadequate for characterizing the most general path-dependent features of plastic behavior,

Proceedings of the IUTAM Symposium on Finite Elasticity, Lehigh University, August 10–15, 1980. Invited paper presented by J. W. Hutchinson.

Carlson, D.E. and Shield, R.T. (eds.)
Proceedings of the IUTAM Symposium on Finite Elasticity
ISBN 90 247 2629 8

there are nonetheless some restricted classes of plastic behavior for which its use can be rigorously justified. For example, small-strain J_2 deformation theory is simply the integrated result of the corresponding J_2 flow theory if the loading history is "proportional", i.e., if all stress components are increased monotonically in fixed proportion to one another. An example involving non-proportional loading increments arises in classical bifurcation analyses, where the use of deformation theory can be justified by showing it to be equivalent to a flow theory which permits the development of yield surface vertices [5]. The wide-spread use of the small strain deformation theory suggests a corresponding role for a finite strain version.

2. Finite strain J_2 deformation theory

The finite strain J_2 deformation theory developed in [1] is a nonlinear elastic law, where the solid is assumed to be isotropic and incompressible. Its development makes extensive use of Hill's theory [6] for finitely deformed isotropic elastic solids.

For an isotropic nonlinear elastic solid, the principal directions of Cauchy stress $\boldsymbol{\sigma}$ must coincide with the axes of the Eulerian strain ellipsoid. Also, to fully specify the state of strain in a material element we need only know the three principal stretches λ_i relative to some reference configuration and the principal directions of strain. Thus, the constitutive law is completely determined once the relations between the principal components of Cauchy stress σ_i and principal stretches λ_i are known. The incremental form of the constitutive law can be obtained using Hill's "principal-axes techniques" [6].

The strain measure adopted is the logarithmic strain tensor $\boldsymbol{\varepsilon}$ which, by definition, is coaxial with the Lagrangian strain ellipsoid and has principal values

$$\varepsilon_i = \ln \lambda_i. \tag{2.1}$$

The logarithmic strain rates $\dot{\varepsilon}_1 = \dot{\lambda}_1/\lambda_1$, etc. are then the Eulerian strain-rate components $\dot{\varepsilon}_{ij}$ on the axes of the Eulerian strain ellipsoid. For incompressible deformations, the constraint $\lambda_1\lambda_2\lambda_3 = 1$ with the choice (2.1) implies the simple condition $\varepsilon_1 + \varepsilon_2 + \varepsilon_3 = 0$ as well as $\dot{\varepsilon}_{kk} = 0$. Inherent advantages of the logarithmic strain measure in setting up constitutive inequalities for both elastic and elastic-plastic solids have been discussed by Hill [6, 7]. This strain measure ("natural strain") has conventionally been used over the years by metallurgists to report (true) stress-strain data for metals.

By analogy with small-strain J_2 deformation theory we introduce the following stress invariant

$$\sigma_e = (3J_2)^{1/2} = (3s_i s_i/2)^{1/2}, \tag{2.2}$$

where $s_i = \sigma_i - (\sigma_1 + \sigma_2 + \sigma_3)/3$ are the principal components of the Cauchy stress deviator. We commonly refer to σ_e as the "effective stress". An "effective strain" ε_e is defined as follows

$$\varepsilon_e = (2\varepsilon_i \varepsilon_i/3)^{1/2}. \tag{2.3}$$

For simple tension in the 1-direction σ_e and ε_e correspond to the axial stress σ_1 and strain ε_1, respectively. The strain energy and complementary strain energy functions, $W(\varepsilon_e)$ and $W^c(\sigma_e)$, are assumed to be functions of only ε_e and σ_e, respectively. The constitutive law has the following form

$$\begin{aligned} \varepsilon_i &= \frac{\partial W^c}{\partial \sigma_i} = \frac{3}{2}\frac{\varepsilon_e}{\sigma_e} s_i = \frac{3}{2}\frac{1}{E_s} s_i, \\ \sigma_i &= \frac{\partial W}{\partial \varepsilon_i} - p = \frac{2}{3} E_s \varepsilon_i - p, \end{aligned} \tag{2.4}$$

where $E_s = \sigma_e/\varepsilon_e$ denotes the secant modulus, obtainable from the uniaxial tension curve, and p is an arbitrary hydrostatic pressure. Note that $\dot{W} = \sigma_e \dot{\varepsilon}_e = s_i \dot{\varepsilon}_i = \sigma_i \dot{\varepsilon}_i$.

For certain applications (e.g., bifurcation analyses) it is convenient to express the constitutive law in incremental or rate form. Using the definitions for effective stress σ_e, effective strain ε_e and secant modulus E_s, we obtain the following from (2.4)

$$\begin{aligned} \dot{s}_i &= \dot{\sigma}_i + \dot{p} \\ &= \frac{2}{3} E_s \dot{\varepsilon}_i - s_i (E_s - E_t) \frac{s_k \dot{\varepsilon}_k}{\sigma_e^2}, \end{aligned} \tag{2.5}$$

where $E_t = d\sigma_e/d\varepsilon_e$ is the tangent modulus. For the shear components of stress-rate and Eulerian strain-rate $\dot{\varepsilon}_{ij}$ on the principal (Eulerian ellipsoid) axes, Hill's method of principal axes [6] gives

$$\overset{*}{\sigma}_{12} = (\sigma_1 - \sigma_2)\coth(\varepsilon_1 - \varepsilon_2)\dot{\varepsilon}_{12}, \qquad \text{etc.} \tag{2.6}$$

where the asterisk denotes the Jaumann or co-rotational stress-rate and coth is the hyperbolic cotangent. Thus, with reference to Cartesian base vectors coaxial with the principal stress axes, we can express the above as

$$\overset{*}{\sigma}_{ij} = L_{ijkl}\dot{\varepsilon}_{kl} + \dot{p}\delta_{ij}, \tag{2.7}$$

where

$$L_{ijkl} = \frac{2}{3} E_s \left[\frac{1}{2}(\delta_{ik}\delta_{jl} + \delta_{jk}\delta_{il}) - \frac{1}{3}\delta_{ij}\delta_{kl}\right] - (E_s - E_t)\frac{s_{ij}s_{kl}}{\sigma_e^2} + Q_{ijkl}. \tag{2.8}$$

The tensor $\boldsymbol{Q}$ is symmetric under $i \leftrightarrow j$, $k \leftrightarrow l$, and $ij \leftrightarrow kl$; and its only non-zero components in principal axes are the "shearing" terms

$$Q_{1212} = \frac{1}{3} E_s[(\varepsilon_1 - \varepsilon_2)\coth(\varepsilon_1 - \varepsilon_2) - 1], \quad \text{etc.} \tag{2.9}$$

These quantities are inherently non-negative. Note that the instantaneous moduli $\boldsymbol{L}$ in (2.8) and components of $\boldsymbol{Q}$ share the same indicial symmetries.

Recently, the above law has been extended somewhat to account for a slight degree of compressibility [8, 9]. Again, by analogy with the corresponding small-strain J_2 deformation theory, the total strain is written as the sum of an "elastic" part

$$\varepsilon_i^e = \frac{1+\nu}{E}\tau_i - \frac{\nu}{E}\tau_{kk}, \tag{2.10}$$

plus a "plastic" part

$$\varepsilon_i^p = \frac{3}{2}\left(\frac{1}{E_s} - \frac{1}{E}\right) t_i, \tag{2.11}$$

where the τ_i are the principal values of Kirchhoff stress, $\tau_{kk} \equiv \tau_1 + \tau_2 + \tau_3$ and $t_i = \tau_i - \tau_{kk}/3$. Thus,

$$\begin{aligned} \varepsilon_i &= \frac{\partial W^c}{\partial \tau_i} = \frac{1+\nu_s}{E_s}\tau_i - \frac{\nu_s}{E_s}\tau_{kk}, \\ \tau_i &= \frac{\partial W}{\partial \varepsilon_i} = \frac{E_s}{1+\nu_s}\left[\varepsilon_i + \frac{\nu_s}{1-2\nu_s}\varepsilon_{kk}\right], \end{aligned} \tag{2.12}$$

where

$$\frac{\nu_s}{E_s} = \frac{1}{2}\left[1 - (1-2\nu)\frac{E_s}{E}\right], \tag{2.13}$$

and $\varepsilon_{kk} = \varepsilon_1 + \varepsilon_2 + \varepsilon_3$. In the above, Poisson's ratio ν and Young's modulus E are assumed to be fixed constants. The secant modulus is now $E_s = \tau_e/\varepsilon_e$ where $\tau_e = (3t_it_i/2)^{1/2}$ denotes the effective Kirchhoff stress. The total effective strain is $\varepsilon_e = \tau_e/E + \varepsilon_e^p$ where

$$\varepsilon_e^p = (2\varepsilon_i^p \varepsilon_i^p/3)^{1/2}. \tag{2.14}$$

For simple tension in the 1-direction, $\tau_e = \tau_1$ and $\varepsilon_e = \varepsilon_1$. Note that the finite "plastic" strains (2.11) satisfy incompressibility, as do the total finite strains when $\nu = 1/2$.

The rate form of the above constitutive law can be obtained as described previously. With reference to the Eulerian ellipsoid axes we have

$$\overset{*}{\tau}_{ij} = L_{ijkl}\dot{\varepsilon}_{kl}, \tag{2.15}$$

with

$$L_{ijkl} = \frac{E_s}{1+\nu_s}\left[\frac{1}{2}(\delta_{ik}\delta_{jl} + \delta_{jk}\delta_{il}) + \frac{\nu_s}{1-2\nu_s}\,\delta_{ij}\delta_{kl}\right] - \frac{3}{2(1+\nu_s)}\left[h_s(E_s - E_t)\frac{s_{ij}s_{kl}}{\tau_e^2} - Q_{ijkl}\right],$$

(2.16)

$$h_s = \frac{E_s}{E_s - (1-2\nu_s)E_t/3}\,.$$

The tangent modulus is now defined as $E_t = d\tau_e/d\varepsilon_e$.

Both the incompressible and compressible versions of J_2 finite-strain deformation theory assume identical effective stress–effective strain relations in tension and compression. Furthermore, the only stress invariant affecting plastic response is J_2. These assumptions are generally considered to be good first-order approximations in metal plasticity. For the special case where the principal axes remain fixed relative to the material *and* proportional loading (i.e., when the principal stress components increase monotonically and in fixed proportion to one another), the finite-strain J_2 deformation theory is exactly the integrated result of finite-strain J_2 flow theory. As in the small strain theory, applicability of the theory to polycrystalline metals must be suspect when significant deviations from proportional plastic straining arise. A path-dependent version of finite strain J_2 deformation theory has been proposed by Stören and Rice [10] with the primary purpose of modeling nearly proportional responses of a material which develops a yield surface vertex. That theory has the form (2.16) but with $\boldsymbol{Q}$ deleted. For histories in which the principal stress axes remain fixed relative to the material the two versions of the theory coincide.

An alternative approach for incorporating the logarithmic strain measure $\boldsymbol{\varepsilon}$ in finite elasticity constitutive laws has recently been developed by Fitzgerald [11], who has presented a general tensorial formulation for the logarithmic strain components in arbitrary axes. A strain energy function depending on the invariants of $\boldsymbol{\varepsilon}$ can then be assumed to give a hyperelastic law in arbitrary axes, thus eliminating the need for principal axes methods. Fitzgerald's formulation is completely general and includes our J_2 law as a special case.

3. Loss of ellipticity for incompressible J_2 deformation theory

Conditions for ellipticity of the equations governing incremental deformations superimposed on finite homogeneous deformations have been given by a number of authors. Recent studies include those by Hill [12, 13], Hill and Hutchinson [14], Knowles and Sternberg [15], Rice [16], and Sawyers and

Rivlin [17–19]. Here we quickly rederive a set of necessary conditions for strong ellipticity of incompressible, isotropic hyperelastic solids which was originally obtained by Sawyers and Rivlin [17, 18]. These conditions are then specialized to the J_2 deformation theory material. With the aid of numerical calculations for a specific family of materials, it is noted that the necessary conditions may also be sufficient to guarantee strong ellipticity for J_2 deformation theory, although this has not been shown.

In the sequel, Cartesian axes x_i are chosen to coincide with the axes of the principal stresses and strains of the underlying homogeneous state. Strong ellipticity of the equations governing quasi-static, superimposed incremental deformations requires

$$c_{ijkl}\nu_i\eta_j\nu_k\eta_l > 0 \tag{3.1}$$

for *all* mutually orthogonal unit vectors $\boldsymbol{\nu}$ and $\boldsymbol{\eta}$. The so-called acoustic tensor of moduli $\boldsymbol{c}$ is related to the moduli tensor $\boldsymbol{L}$ in (2.7) by

$$c_{ijkl} = L_{ijkl} + \frac{1}{2}\sigma_{ik}\delta_{lj} - \frac{1}{2}\sigma_{il}\delta_{kj} - \frac{1}{2}\sigma_{jk}\delta_{il} - \frac{1}{2}\sigma_{jl}\delta_{ik}, \tag{3.2}$$

and both tensors share the indicial symmetries $c_{ijkl} = c_{klij}$ for hyperelastic solids. Condition (3.1) excludes quasi-static shearing discontinuities characteristic of a planar shear band with normal $\boldsymbol{\nu}$ and shearing direction $\boldsymbol{\eta}$ in the plane of the band. It also ensures that all plane waves with propagation direction $\boldsymbol{\nu}$ and particle velocity parallel to $\boldsymbol{\eta}$ have real wave speeds.

Following Sawyers and Rivlin [17, 18], we can obtain necessary conditions for strong ellipticity by restricting $\boldsymbol{\nu}$ to lie in one of the planes of the principal stress axes. Let $\boldsymbol{\nu}$ lie in the x_1x_2-plane at angle ψ from the x_1-axis so that

$$\boldsymbol{\nu} = (\cos\psi, \sin\psi, 0) \quad \text{and} \quad \boldsymbol{\eta} = (-a\sin\psi, a\cos\psi, b) \tag{3.3}$$

where $a^2 + b^2 = 1$. The shearing direction $\boldsymbol{\eta}$ does not lie in the x_1x_2-plane unless $b = 0$. With (3.3), condition (3.1) becomes

$$\begin{aligned} &a^2\{c_{1212}\cos^4\psi + [c_{1111} + c_{2222} - 2c_{1122} - 2c_{1221}]\cos^2\psi\sin^2\psi + c_{2121}\sin^4\psi\} \\ &\quad + b^2\{c_{1313}\cos^2\psi + c_{2323}\sin^2\psi\} > 0 \end{aligned} \tag{3.4}$$

for all ψ and all a such that $|a| \leqq 1$ with $b^2 = 1 - a^2$. In arriving at (3.4) we have used the fact that components such as c_{1113} vanish in the principal axes.

With $b = 0$, (3.4) is satisfied for all ψ if and only if

$$c_{1212} > 0 \quad \text{and} \quad c_{2121} > 0, \tag{3.5}$$

and

$$c_{1111} + c_{2222} - 2c_{1122} > 2c_{1221} - 2\sqrt{c_{1212}c_{2121}}. \tag{3.6}$$

Conditions (3.5) and (3.6) are necessary and sufficient for strong ellipticity of incremental plane strain deformation in the x_1x_2-plane — i.e., for restricted displacement increments of the form $v_1(x_1, x_2)$, $v_2(x_1, x_2)$ and $v_3 = 0$. The choice $\boldsymbol{a} = 0$ in (3.4) requires

$$c_{1313} > 0 \quad \text{and} \quad c_{2323} > 0. \tag{3.7}$$

Conditions (3.5)–(3.7) are equivalent to (3.4). One notes immediately that satisfaction of (3.5) and (3.6) and their equivalents for each of the other two principal planes renders the third set of conditions (3.7) extraneous. In other words, strong ellipticity for incremental plane strain deformations parallel to each of the principal planes ensures that (3.1) is satisfied for any $\boldsymbol{\nu}$ which lies in a principal plane whether or not $\boldsymbol{\eta}$ lies in one of the principal planes. This was established by Sawyers and Rivlin [17, 18]. As they noted [18], these conditions are sufficient for strong ellipticity when the underlying state has two equal principal strains since then any $\boldsymbol{\nu}$ necessarily lies in a principal plane.

At this point it is convenient to introduce two shearing moduli μ and μ^*, used by Hill and Hutchinson [14], governing incremental plane strain deformations in the x_1x_2-plane. For such deformations (2.7) can be written as

$$\overset{*}{\sigma}_{11} - \overset{*}{\sigma}_{22} = 2\mu^*(\dot{\varepsilon}_{11} - \dot{\varepsilon}_{22}), \qquad \overset{*}{\sigma}_{12} = 2\mu\dot{\varepsilon}_{12}, \qquad (\dot{\varepsilon}_{11} + \dot{\varepsilon}_{22} = 0), \tag{3.8}$$

where

$$4\mu^* = L_{1111} + L_{2222} - 2L_{1122} \quad \text{and} \quad \mu = L_{1212}. \tag{3.9}$$

With the aid of (3.2), conditions (3.5) and (3.6) can be expressed as

$$\mu > |\Delta\sigma|/2, \tag{3.10}$$

and

$$2\mu^* > \mu - (\mu^2 - \Delta\sigma^2/4)^{1/2}, \tag{3.11}$$

where $\Delta\sigma = \sigma_1 - \sigma_2$. This is the form of the ellipticity conditions given by Hill and Hutchinson [14] for incremental plane strain deformations of a broad class of incrementally linear materials satisfying (3.8).

The six conditions of the form (3.5) or, equivalently, the three conditions (3.10) are always satisfied by J_2 deformation theory as long as the secant modulus E_s is positive. For example, one can show that

$$c_{2121} = \mu - \frac{1}{2}\Delta\sigma = \frac{1}{3}E_s\frac{\Delta\varepsilon e^{-\Delta\varepsilon}}{\sinh(\Delta\varepsilon)}, \tag{3.12}$$

and

$$c_{1212} = \mu + \frac{1}{2}\Delta\sigma = \frac{1}{3}E_s\frac{\Delta\varepsilon e^{\Delta\varepsilon}}{\sinh(\Delta\varepsilon)}, \tag{3.13}$$

where $\Delta\sigma = \sigma_1 - \sigma_2$ and $\Delta\varepsilon = \varepsilon_1 - \varepsilon_2$.

We have not yet succeeded in showing that the three remaining conditions of the form (3.11) are sufficient for strong ellipticity for J_2 deformation theory. Sawyers and Rivlin [18] have shown that their conditions, which are equivalent to those of the form (3.10) and (3.11), are both necessary and sufficient for strong ellipticity for two special classes of solids, each of which has a strain energy function which depends on only one strain invariant. The energy density of J_2 deformation theory depends on only one strain invariant, ε_e, but it is not included in either of the special classes of solids for which Sawyers and Rivlin have established sufficiency. It is known [19] that incompressible, isotropic hyperelastic materials do exist for which conditions (3.10) and (3.11) are not sufficient for strong ellipticity.

A limited numerical study of the sufficiency of the three conditions of the form (3.11) has been carried out for J_2 deformation theory with a power-law relation between true stress and natural strain of the form

$$\sigma_e = K\varepsilon_e^N, \tag{3.14}$$

where the hardening index N is restricted to the range $0 < N \leqq 1$. The secant and tangent moduli diminish with increasing strain according to

$$E_s = K\varepsilon_e^{N-1} \quad \text{and} \quad E_t = NE_s. \tag{3.15}$$

Using (2.8) and (3.9), one can show that the three conditions of the form (3.11) become

$$\Delta\varepsilon^2 < 4\left[1 - \frac{1}{3}(N-1)\left(\frac{\Delta\varepsilon}{\varepsilon_e}\right)^2\right]\left[(\Delta\varepsilon \coth\Delta\varepsilon - 1) + \frac{1}{3}(N-1)\left(\frac{\Delta\varepsilon}{\varepsilon_e}\right)^2\right], \tag{3.16}$$

where $\Delta\varepsilon$ is identified with the principal strain differences $\varepsilon_1 - \varepsilon_2$, $\varepsilon_1 - \varepsilon_3$, and $\varepsilon_2 - \varepsilon_3$. The conditions (3.16) are first violated by the maximum principal strain difference corresponding to a shear band with normal and shearing direction lying in the plane of the maximum principal strain difference.

The boundary of the region of principal strains states within which the three conditions (3.16) are satisfied is shown in Figure 1 for $N = 0.1$, 0.5 and 1. Since $\varepsilon_3 = -\varepsilon_1 - \varepsilon_2$, the region is fully depicted by its trace in the $\varepsilon_1\varepsilon_2$-plane. The region is symmetric with respect to the 45° lines in Figure 1 and only one quarter of the region is shown. The line on which $\varepsilon_1 = \varepsilon_3 = -\frac{1}{2}\varepsilon_2$ marks the switch in the maximum principal strain difference from $\varepsilon_2 - \varepsilon_1$ to $\varepsilon_2 - \varepsilon_3$. For $N = 1$, (3.16) reduces to

$$\Delta\varepsilon^2 < 4(\Delta\varepsilon \coth\Delta\varepsilon - 1), \tag{3.17}$$

which is equivalent to

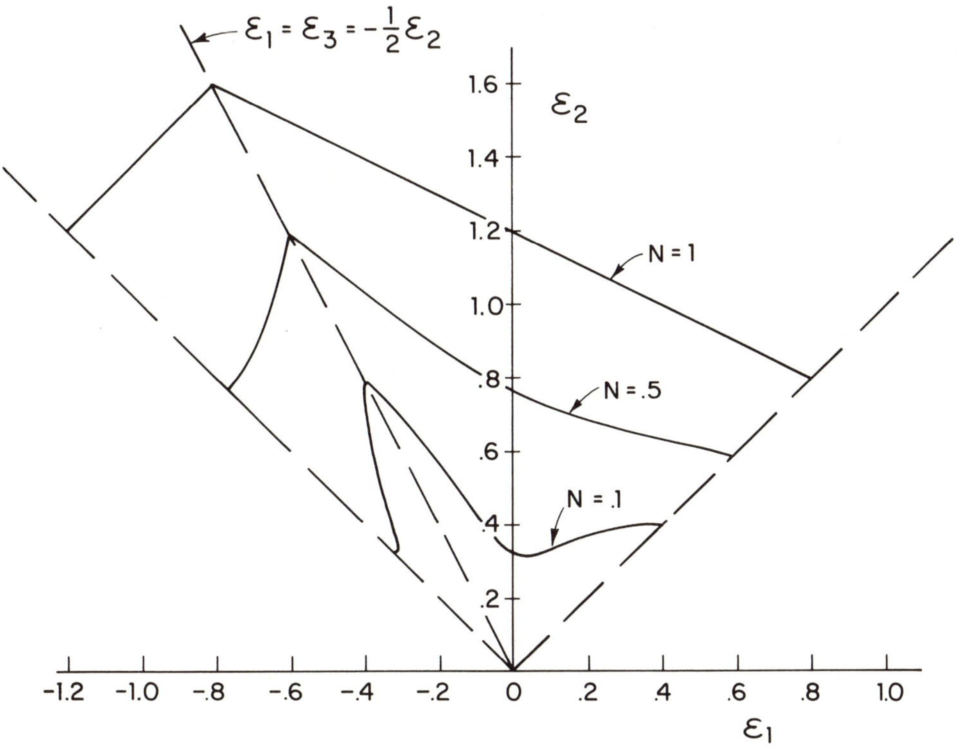

Figure 1. Boundary of the elliptic region for power-law material for three values of the exponent N. The elliptic region is symmetric with respect to the lines $\varepsilon_1 = \varepsilon_2$ and $\varepsilon_1 = -\varepsilon_2$.

$$|\Delta\varepsilon| < 2.399, \tag{3.18}$$

and which leads to the boundary with straight line segments in Figure 1.

Numerical calculations have been carried out to ascertain whether (3.1) is satisfied for all mutually orthogonal $\boldsymbol{\nu}$ and $\boldsymbol{\eta}$ when the strain states lie within the boundaries of Figure 1. Orientations of $\boldsymbol{\nu}$ and $\boldsymbol{\eta}$ with respect to the principal axes were specified by three Euler angles and these angular coordinates were taken to range over their full range consistent with the symmetry of the underlying state. For each strain state considered, (3.1) was checked at more than 10^4 orientations. For each of $N = 0.1$ 0.5 and 1, the strain states considered were those at 5° intervals measured from the 45° line in Figure 1 at strains which were 0.999 times the corresponding values at the boundary as ascertained by (3.16). In no case was (3.1) violated. Although strain states further within the boundaries of Figure 1 were not considered, no violation of (3.1) is expected since the incremental moduli $\boldsymbol{L}$ increase with decreasing strain when $N < 1$.

Our numerical checks of (3.1) for the example of the power-law solid suggest that the three conditions (3.16) may be both necessary and sufficient for strong ellipticity for this material. It is an open question as to whether this actually is the case and whether, more generally, the three conditions of the form (3.11) are sufficient for gauging strong ellipticity of any J_2 deformation theory solid.

Acknowledgement

The work of J.W.H. was supported in part by the National Science Foundation under Grant ENG78-10756, and by the Division of Applied Sciences, Harvard University. The work was conducted while K.W.N. was on sabbatical leave at Harvard University, and was supported in part by the Faculty of Applied Sciences at the University of Sherbrooke, and by the Division of Applied Sciences, Harvard University.

References

1. J. W. Hutchinson and K. W. Neale, *Sheet necking — II. Time-independent behavior*, in *Mechanics of Sheet Metal Forming* (D. P. Koistinen and N-M. Wang, eds.), Plenum, 1978, p. 127.

2. J. W. Hutchinson and V. Tvergaard, *Surface instabilities on statically strained plastic solids*, International Journal of Mechanical Sciences **22** (1980), 339.

3. J. L. Bassani, D. Durban and J. W. Hutchinson, *Bifurcations at a spherical hole in an infinite elastoplastic medium*, Mathematical Proceedings of the Cambridge Philosophical Society **87** (1980), 339.

4. N. Triantafyllidis, *Bifurcation phenomena in pure bending*, J. Mech. Phys. Solids. **28** (1980), 221.

5. B. Budiansky, *A reassessment of deformation theories of plasticity*, Journal of Applied Mechanics **26** (1959), 259.

6. R. Hill, *Constitutive inequalities for isotropic elastic solids under finite strain*, Proceedings of the Royal Society of London **A314** (1970), 457.

7. R. Hill, *On constitutive inequalities for simple materials*, Journal of the Mechanics and Physics of Solids **16** (1968), 229, 315.

8. K. W. Neale, *Phenomenological constitutive laws in finite plasticity*, Solid Mechanics Archives **6** (1981), 79.

9. V. Tvergaard, A. Needleman and K. K. Lo, *Flow localization in the plane strain tensile test*, Journal of the Mechanics and Physics of Solids **29** (1981), 115.

10. S. Stören and J. R. Rice, *Localized necking in thin sheets*, Journal of the Mechanics and Physics of Solids **23** (1975), 421.

11. J. E. Fitzgerald, *A tensorial Hencky measure of strain and strain rate for finite deformations*, in *Developments in Theoretical and Applied Mechanics* (J. E. Stoneking, ed.), Vol 10, 1980, p. 635.

12. R. Hill, *Acceleration waves in solids*, Journal of the Mechanics and Physics of Solids **10** (1962), 1.

13. R. Hill, *On the theory of plane strain in finitely deformed compressible materials*, Mathematical Proceedings of the Cambridge Philosophical Society **86** (1979), 161.

14. R. Hill and J. W. Hutchinson, *Bifurcation phenomena in the plane tension test*, Journal of the Mechanics and Physics of Solids **23** (1975), 239.

15. J. K. Knowles and E. Sternberg, *On the ellipticity of the equations of nonlinear elastostatics for a special material*, Journal of Elasticity **5** (1975), 341.

16. J. R. Rice, *The localization of plastic deformations*, in *Theoretical and Applied Mechanics* (W. T. Koiter, ed.), North-Holland, 1976, p. 207.

17. K. N. Sawyers and R. S. Rivlin, *Instability of an elastic material*, International Journal of Solids and Structures **9** (1973), 607.

18. K. N. Sawyers and R. S. Rivlin, *On the speed of propagation of waves in a deformed elastic material*, Journal of Applied Mathematics and Physics (ZAMP) **28** (1977), 1045.

19. K. N. Sawyers and R. S. Rivlin, *A note on the Hadamard criterion for an incompressible isotropic elastic material*, Mechanics Research Communications **5** (1978), 211.

INSTABILITY OF FINITE AMPLITUDE ELASTIC WAVES

FRITZ JOHN

Courant Institute of Mathematical Sciences, 251 Mercer Street, New York, NY 10012, USA

(Received September 11, 1980)

The equations of motion for hyper-elastic material form a hyperbolic system of second order quasi-linear partial differential equations

$$\frac{\partial^2 u_i}{\partial t^2} = \sum_{k,r,s} c_{ikrs} \frac{\partial^2 u_r}{\partial x_k \partial x_s} \tag{1}$$

for the components u_i of the displacement vector $u = u(x,t)$. The coefficients c_{ikrs} are functions of the matrix u' whose elements are the displacement gradients $u_{i,k} = \partial u_i / \partial x_k$. In order to avoid assumptions about the behavior of the material for large stresses or strains we require the $c_{ikrs}(u')$ to be defined only for u' in the neighborhood of the origin. More generally we consider hyperbolic systems (1) of N equations for a vector $u(x,t)$ with N components u_i defined for $x \in \mathbb{R}^n$, $t \geqq 0$, the coefficients c_{ikrs} depending on u' and u_t. The initial value problem consists in finding a solution of (1) for prescribed values of $u(x,0)$, $u_t(x,0)$.

We call the system (1) *stable with respect to the vectors* $f(x)$, $g(x)$ if for every $\delta > 0$ there exists an $\varepsilon_0 > 0$ such that for $|\varepsilon| \leqq \varepsilon_0$ the initial value problem with data

$$u(x,0) = \varepsilon f(x); \qquad u_t(x,0) = \varepsilon g(x) \tag{2}$$

has a solution $u(x,t)$ of class C^2 for $x \in \mathbb{R}^n$, $t \geqq 0$ for which

$$|u'(x,t)| = \operatorname*{Max}_{i,k} |u_{i,k}(x,t)| \leqq \delta \qquad \text{for } x \in \mathbb{R}^n, \quad t \geqq 0. \tag{3}$$

We call (1) *stable*, if it is stable with respect to any f, $g \in C_0^\infty(\mathbb{R}^n)$ (that is any infinitely often differentiable f, g with compact support in $\mathbb{R}^n$). The system (1)

Proceedings of the IUTAM Symposium on Finite Elasticity, Lehigh University, August 10–15, 1980, Invited paper.

Carlson, D.E. and Shield, R.T. (eds.)
Proceedings of the IUTAM Symposium on Finite Elasticity
ISBN 90 247 2629 8

is unstable, if not stable. In that case there exist $\delta > 0$, f, $g \in C_0^\infty(\mathbb{R}^n)$ and arbitrarily small $|\varepsilon|$ such that the solution u of (1), (2) either ceases to exist ("blows up") for some $x \in \mathbb{R}^n$, $t \geqq 0$, or involves values $|u'(x,t)| > \delta$.

The system (1) is known to be stable in the linear hyperbolic case, where the c_{ikrs} are constants. The system (1) is unstable for $n = 1$, $N \geqq 1$, provided the system is genuinely nonlinear in a certain sense; (see [1]). More precisely, in that case (1) is unstable with respect to any nontrivial f, $g \in C_0^\infty(\mathbb{R})$; for $\varepsilon \neq 0$ and $|\varepsilon|$ sufficiently small the solution u of (1), (2) blows up in finite time, though u' stays less than δ. Recently, Klainerman [2] proved stability for a single hyperbolic equation in 6 or more space variables, that is for $N = 1$, $n \geqq 6$.

We are interested here in the case of the equations of motion for an isotropic, homogeneous, hyper-elastic material in $\mathbb{R}^3$. The motions most easily discussed are the *plane waves* (see [3]), for which f, g depend only on a linear combination of the x_k. The deciding factor here is the genuine non-linearity of the system. We have stability for f, g corresponding to shear waves (see [4]), and generally instability for irrotational waves. However, initial data for plane waves do not have compact support in $\mathbb{R}^3$, and their behavior does not tell us anything about waves that are due to small localized disturbances.

It will be shown here that certain materials (the so-called *harmonic* ones) are unstable in the sense that the corresponding equations of motion are unstable. The proof is based on the fact that certain wave motions in those materials can be described by a scalar non-linear wave equation in 3-space of the form

$$v_{tt} - \Delta v = F(v_{tt}). \tag{4}$$

Here $F(s)$ is a function satisfying

$$F(0) = F'(0) = 0; \qquad F''(s) \geqq c > 0 \qquad \text{for all } s. \tag{5}$$

It was proved in [5] that every non-trivial solution $v \in C^3$ of (4) for which $v(x,0)$, $v_t(x,0)$, $v_{tt}(x,0)$ have compact support in $\mathbb{R}^3$, blows up in finite time.[†]

In order to connect equation (4) with elastic waves we have to recall some notions about isotropic homogeneous hyper-elastic solids. Motion is described by the displacement *vector* $u(x,t)$ which takes the point with rest position x into the point $x + u(x,t)$ at the time t. The corresponding Jacobian matrix is $I + u' = (\delta_{ik} + u_{i,k})$, where I = unit matrix.

[†] This shows incidentally that Klainerman's theorem cannot be extended to three space dimensions, for the quasi-linear equation $u_{tt} - \Delta u = F'(u_t)u_{tt}$ with $u(x,0) = \varepsilon f(x)$, $u_t(x,0) = 0$, where $f \in C_0(\mathbb{R}^3)$ does not vanish identically; then $v(x,t) = \int_0^t u(x,s)ds$ satisfies (4), while $v(x,0) = 0$, $v_t(x,0) = \varepsilon f(x)$, $v_{tt}(x,0) = 0$ have compact support; thus v blows up and u cannot exist for all $x \in \mathbb{R}^3$, $t \geqq 0$.

The *rotation matrix* ω and *strain matrix*[†] σ can be defined by the polar decomposition of $I+u'$:

(6) $$I+u'=(I+\omega)(I+\sigma);$$

(7) $$(I+\omega^*)(I+\omega)=I; \qquad \sigma^*=\sigma; \qquad I+\sigma>0.$$

The *strain invariants* A, B, C are the elementary symmetric functions of the eigenvalues of σ, so that

(8) $$\sigma^3-A\sigma^2+B\sigma-CI=0.$$

The elastic material is characterised by a function $W(A,B,C)$, the strain energy per unit mass, which can also be expressed as a function[‡] of the displacement gradient u'. The coefficients c_{ikrs} in (1) are given by

(9) $$c_{ikrs}=\frac{\partial^2 W}{\partial u_{i,k}\partial u_{r,s}}\,.$$

A motion with displacement $u(x,t)$ is called *pseudo-irrotational*, if u' is symmetric:

(10) $$u_{i,j}-u_{j,i}=0.$$

Pseudo-irrotationality is not preserved in general hyper-elastic media.[§] A medium preserving pseudo-irrotationality is called *harmonic*. Harmonic media can also be characterised as those for which the energy function $W(A,B,C)$ has the special form[§§]

[†] Here * stands for transposition. Any monotone function of the matrix σ, such as the more customary $\sigma+(\sigma^2/2)$, could be used equally well as a measure for the strain. The choice made here is convenient for introducing harmonic materials.

[‡] The strain matrix σ, and hence also its invariants, are real analytic algebraic functions of u' for $\det(I+u')>0$. This is seen from the representation of σ as a Cauchy integral

$$\sigma=\frac{1}{2\pi i}\int[\lambda I-(I+u'^*)(I+u')]^{-1}(\sqrt{\lambda}-1)d\lambda,$$

where the path of integration in the λ-plane encloses all singularities of the integrand, except the branch point $\lambda=0$, and $\sqrt{\lambda}$ denotes the principal value.

[§] Necessary for preservation would be that symmetry of u' implies that of $\partial^2u'/\partial t^2$ when expressed with the help of (1).

[§§] See [3]. In the paper the letters r, s, t are used to denote the invariants of the matrix $I+\sigma$, so that

$$A=r-3; \qquad B=s-2r+3; \qquad C=t-s+r-1.$$

The function denoted in [6] by $W=F(r)+as+bt$ is to be identified with the expression

$$\rho(G(r-3)+\alpha(s-2r+3)+\beta(t-s+r-1))$$

of the present paper, where $\rho=$ density in the unstrained state. The term "harmonic" media recalls the fact that plane strain equilibrium problems can be solved in terms of harmonic functions; see [6].

(11) $$W = G(A) + \alpha B + \beta C,$$

with an arbitrary function G and constants α, β. For consistency with the classical linear theory we postulate that

(12) $$G(0) = G'(0) = 0; \qquad G''(0) = \frac{\lambda + 2\mu}{\rho}; \qquad \alpha = \frac{-2\mu}{\rho},$$

where λ, μ are the Lamé constants; hyperbolicity in the linear theory requires that

(13) $$G''(0) > 0, \qquad \alpha < 0.$$

We shall assume that units in space and time are chosen in such a way that

(14) $$G''(0) = 1.$$

The equations of motion for a harmonic material are

(15) $$\frac{\partial^2 u_i}{\partial t^2} = \sum_k \frac{\partial}{\partial x_k} [(G'(A) + (A+1))(\delta_{ik} + \omega_{ik}) - (\delta_{ik} + u_{i,k})].$$

For a pseudo-irrotational motion in $\mathbb{R}^3$ there exists a displacement potential $v(x, t)$ such that

(16a) $$u_i = \frac{\partial v}{\partial x_i} = v_{,i},$$

(16b) $$\omega_{ik} = 0; \qquad u_{i,k} = v_{,ik}, \qquad A = \Delta v.$$

The (15) becomes

(17) $$\frac{\partial}{\partial x_i} (v_{tt} - G'(\Delta v)) = 0.$$

For displacements $u(x, t)$ of compact support in x we can normalise v so that also v has compact support in x. Then (17) implies the scalar equation

(18) $$v_{tt} = G'(\Delta v),$$

which describes pseudo-irrotational waves in harmonic media.

Now equation (18) can be put into the form (4), at least locally. Assume that

(19) $$G'''(0) \neq 0.$$

There exists then a positive γ such that

(20) $$G''(s) > 0; \qquad G'''(s) \neq 0 \qquad \text{for } |s| \leqq \gamma.$$

For $|s| < \gamma$ the equation

(21) $$S = G'(s)$$

can be solved for s. We write its solution in the form

(22a) $$s = S - F(S) \qquad \text{if } G'''(0) > 0,$$

(22b) $$s = S + F(-S) \qquad \text{if } G'''(0) < 0.$$

Then in the case $G'''(0) > 0$ the function v satisfies

(23) $$v_{tt} - \Delta v = F(v_{tt})$$

as long as $|\Delta v| < \gamma$, while in the case $G'''(0) < 0$ the same holds for the function $w = -v$. Here by (12) and (14)

(24) $$F(0) = F'(0) = 0; \qquad F''(S) = \frac{G'''}{G''^3} \geqq c > 0$$

for some constant c, as long as $|s| < \gamma$. It follows from the theorem on p that if $u(x, 0)$, $u_t(x, 0)$ have compact support, then either $u = 0$, or u ceases to exist beyond a certain finite time T, or $|\operatorname{div} u| = |\Delta v|$ exceeds the value γ and $|u'| > \gamma/3$ at some time. This proves the instability of a harmonic material satisfying (19).[†] We only have to take initial displacements of the form (2) where the vectors f, g are gradients of functions in $C_0^\infty(\mathbb{R}^3)$.

Instability is not confined to harmonic media. There exist special pseudo-irrotational waves for all homogeneous isotropic hyper-elastic materials, namely the spherical waves. These can be described by a function $\phi(r, t)$, odd in r, that only depends on the distance $r = |x|$ from the origin and on the time t. Here[‡]

(25) $$u_i = \frac{1}{r} x_i \phi(r, t).$$

If the displacement vector is of compact support in x, we can again introduce a displacement potential

(26) $$v = v(r, t) = \int_{-\infty}^{r} \phi(s, t)\, ds,$$

which will be an even function of r and of compact support in r. Then $u_i = v_{,i}$. Due to the isotropic character of the medium waves originating from initial conditions with spherical symmetry stay spherical.[§] We shall find a wider class of media that exhibits instability for spherical motions.

[†] Condition (19) excludes the "standard" harmonic materials for which $G(s) = s^2/2$ and (18) reduces to the linear wave equation $v_{tt} = \Delta v$. Condition (19) expresses that pseudo-irrotational wave speed changes with density in Lagrange coordinates.

[‡] Spherical waves are irrotational in the sense that the velocity vector $\partial u/\partial t$ derives from a potential in Euler coordinates.

[§] At least as long as we have uniqueness for the initial value problem, which is certainly the case for sufficiently small displacement gradients.

For the spherical motion (25) the matrix $u' = \sigma$ has the eigenvalues p, p, q where

$$p = \frac{1}{r}\phi; \qquad q = \phi_r, \tag{27}$$

$$A = 2p + q; \qquad B = 2pq + p^2; \qquad C = p^2 q. \tag{28}$$

Since the characteristic equation (8) has the double root p, its discriminant

$$D(A,B,C) = (27C + 2A^3 - 9AB)^2 - 4(A^2 - 3B)^3 \tag{29}$$

vanishes. Relations (28) give a parameter representation for the surface Σ in ABC-space with equation $D(A,B,C) = 0$. Spherical waves only involve the strain energy function $W(A,B,C)$ on the surface, that is only involve the expression

$$V(p,q) = W(2p + q, 2pq + p^2, p^2 q). \tag{30}$$

The equations of motion for a spherical wave reduce to a single equation[†] for the scalar $\phi(r,t)$:

$$\phi_{tt} = -\frac{1}{r} V_p + \frac{1}{r^2}\frac{\partial}{\partial r}(r^2 V_q). \tag{31}$$

Elastic materials that have the same V, that is whose energy functions agree on the surface Σ, have the same spherical waves. In particular materials with an energy function W of the form

$$W(A,B,C) = G(A) + \alpha B + \beta C + E(A,B,C)D(A,B,C), \tag{32}$$

with an arbitrary sufficiently regular $E(A,B,C)$ have the same spherical waves as the harmonic material (11). Since spherical waves are pseudo-irrotational, these waves will exhibit the same instability behavior as in harmonic materials, provided condition (19) is satisfied.

The results of this paper are confined to materials with strain energy functions of type (32), for which reduction to an equation of the form (4) is possible.[‡] It seems likely that the same phenomenon of instability can be

[†] Which can easily be derived from Hamilton's principle

$$\delta \iint \left(\phi_t^2 - V\left(\frac{1}{r}\phi, \phi_r\right)\right) \rho r^2 dr dt = 0.$$

[‡] To get an idea of the degree of generality of such W, we can expand W into a formal Taylor series keeping in mind, that A, B, C are respectively linear, quadratic, and cubic in the elements of σ. For W of the form (32) we then have up to 6-th order in σ the expansion

$$W = \frac{1}{2}G''(0)A^2 + \alpha B + \frac{1}{6}G'''(0)A^3 + \beta C + \frac{1}{24}G^{(iv)}(0)A^4$$
$$+ \frac{1}{120}G^{(v)}(0)A^5 + \frac{1}{720}G^{(vi)}(0)A^6 + E(0,0,0)D(A,B,C) + \cdots.$$

established for much more general materials, using other techniques of proof. The instability discussed here is mainly of theoretical interest. It presupposes a medium extending without limits in all directions. It applies only to strict solutions; it might not occur if shocks are present. It would be felt only after a very long time, at least of order $1/(\varepsilon \log(1/\varepsilon))^4$, as shown in [7].

Acknowledgment

This article represents work performed at the Courant Institute of Mathematical Sciences, supported by the Office of Naval Research under Contract No. N00014-76-C-0301.

References

1. F. John, *Formation of singularities in one-dimensional nonlinear wave propagation*, Comm. Pure Appl. Math. **27** (1974), 377–405.
2. S. Klainerman, *Global existence for nonlinear wave equations*, Comm. Pure Appl. Math.? (1980), 000.
3. F. John, *Plane elastic waves of finite amplitude, Hadamard materials and harmonic materials*, Comm. Pure Appl. Math. **19** (1966), 309–341.
4. M. M. Carroll, *Some results on finite amplitude elastic waves*, Acta Mech. **3** (1967), 167–341.
5. F. John, *Blow-up for quasi-linear wave equations in three space dimensions*, Comm. Pure Appl. Math., to appear.
6. F. John, *Plane strain problems for a perfectly elastic material of harmonic type*, Comm. Pure Appl. Math. **13** (1969), 239–296.
7. F. John, *Finite amplitude waves in a homogeneous isotropic elastic solid*, Comm. Pure Appl. Math. **30** (1977), 421–446.

LOCALIZED SHEAR NEAR THE TIP OF A CRACK IN FINITE ELASTOSTATICS

JAMES K. KNOWLES

Division of Engineering and Applied Science, California Institute of Technology, Pasadena, California 91125, USA

(Received September 11, 1980)

ABSTRACT

This paper describes some recent results concerning crack problems in finite anti-plane shear for a class of incompressible elastic materials for which the associated equilibrium equations lose ellipticity at sufficiently severe deformations. The principal feature of the elastic fields arising in these problems is the presence near a crack-tip of curves bearing discontinuities in displacement gradient and stress.

Introduction

Among the several recent investigations of crack problems in finite elastostatics, some have been concerned with elastic materials which lose equilibrium ellipticity at sufficiently severe deformations [1–3]. For materials of this kind, the elastic field near a crack-tip is quite different from that associated with materials which retain equilibrium ellipticity at all deformations. Perhaps the most conspicuous feature of crack problems for materials exhibiting a loss of ellipticity is the occurrence near the crack-tips of curves across which the stress and the displacement gradient are discontinuous. Thus, in a certain sense, the familiar crack-tip point-singularity which occurs for materials which always retain ellipticity "spreads out" along these curves. The present paper summarizes the results of these investigations.

All three studies [1–3] are concerned initially with finite anti-plane shear (or "Mode III") deformations of an infinite solid containing a finite crack, the solid being deformed at infinity to a state of simple shear. In all cases treated, the material is assumed to belong to a special class of incompressible, homogeneous, isotropic, elastic materials which permit nontrivial states of

Proceedings of the IUTAM Symposium on Finite Elasticity, Lehigh University, August 10–15, 1980. Invited paper.

Carlson, D.E. and Shield, R.T. (eds.)
Proceedings of the IUTAM Symposium on Finite Elasticity
ISBN 90 247 2629 8

finite anti-plane shear. These materials are characterized by a strain energy density $W = W(I_1)$ which depends only on the *first* of the two fundamental scalar invariants I_1, I_2 of the Cauchy–Green deformation tensors; we shall refer to them as "I_1-materials" here.

When the amount of shear applied at infinity is small, the field near either crack-tip can be determined on the basis of an asymptotic scheme in which the crack of finite length is replaced by a semi-infinite one, and the *far* field is required to match the elastostatic field *near* the crack-tip predicted by the solution of the *original* problem according to linearized theory. It is this *small-scale nonlinear crack-problem* which is ultimately studied in [1–3] for various I_1-materials which behave appropriately near the undeformed state but lose equilibrium ellipticity at severe deformations.

Crack problems

The geometry associated with the underlying Mode III problem of the crack of finite length is illustrated in Figure 1, which shows an undeformed cross-section of the cracked solid. The out-of-plane displacement $u = u(x_1, x_2)$ in the anti-plane shear field satisfies the differential equation

$$\nabla \cdot (W'(I_1)\nabla u) = 0 \tag{1}$$

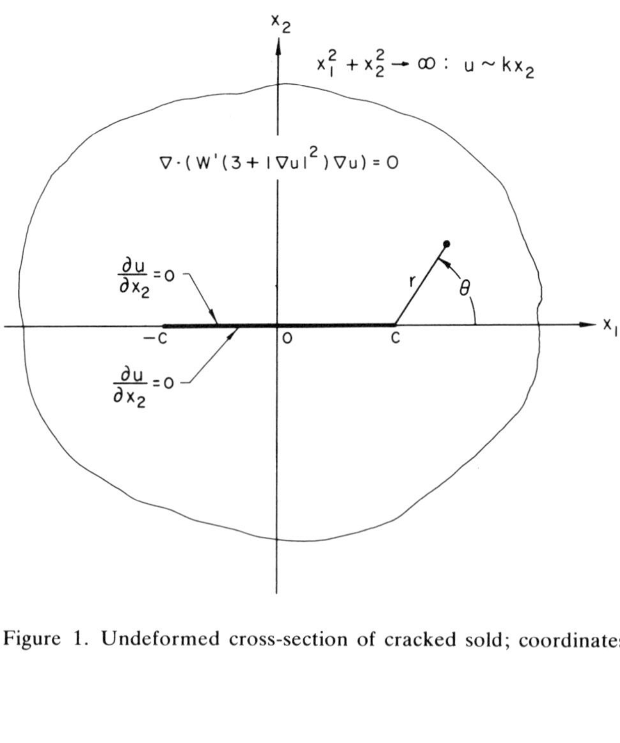

Figure 1. Undeformed cross-section of cracked sold; coordinates.

away from the crack-tips, where

$$I_1 = 3 + |\nabla u|^2, \tag{2}$$

∇ is the gradient operator with respect to the material coordinates x_1, x_2, and W' is the derivative of W with respect to I_1. On the faces of the crack, the requirement of vanishing traction leads to

$$\frac{\partial u}{\partial x_2}(x_1, 0\pm) = 0. \tag{3}$$

At infinity, u is required to correspond to a state of simple shear:

$$u \sim kx_2 \qquad \text{as} \qquad x_1^2 + x_2^2 \to \infty; \tag{4}$$

the constant $k \geqq 0$ represents the given amount of applied shear. Finally, u must be bounded at the crack-tips. (For details of the arguments leading to the formulation given above, the reader may refer to [4]; for additional background, [5] is appropriate.)

In the absence of the crack, a solution of the differential equation (1) on the entire x_1, x_2-plane which satisfies (4) is of course given by $u \equiv kx_2$, corresponding to simple shear throughout the body. The Cauchy shear stress τ_{32} arising from this simple shear is given by

$$\tau_{32} = 2W'(3 + k^2)k \equiv \tau(k), \tag{5}$$

(see [4]). It is clear from (5) that the strain energy density W for an I_1-material is essentially determined by the shear stress response function $\tau(k)$.

One can show [4] that the differential equation (1) is elliptic at a solution u and at a point (x_1, x_2) if $\tau'(k) > 0$ at $k = |\nabla u(x_1, x_2)|$, provided $W'(I_1) > 0$ for all $I_1 \geqq 3$, as is always assumed here. Thus for an I_1-material whose shear stress response function $\tau(k)$ is monotone strictly increasing for all k, (1) is always elliptic.

The local structure near a crack-tip of the solution to the boundary-value problem (1)–(4), and of the associated stresses, was analyzed in [4] for a subclass of the I_1-materials (the "power law materials") for which ellipticity always prevails. For these, there are singularities at the crack-tips in ∇u and in at least some of the stresses, the order of the singularities in r, as well as the angular distribution, being material-dependent. (This is in marked constrast to the results based on linearized theory, in which all singularities are of order $r^{-1/2}$.) Away from the crack-tips, the field is presumably infinitely smooth.

When (1) is linearized (or, in fact, for the Neo-Hookean material, for which $W' =$ constant), the problem (1)–(4) becomes an elementary one for Laplace's equation; near the right crack-tip, one finds (see [6]) that

(6) $$u \sim k(2cr)^{1/2}\sin(\theta/2), \qquad r \to 0,$$

and that the stresses are given by

(7) $$\tau_{31} \sim -\mu k c(2cr)^{-1/2}\sin(\theta/2), \qquad \tau_{32} \sim \mu k c(2cr)^{-1/2}\cos(\theta/2).$$

In (6) and (7), $2c$ is the crack-length and $\mu = 2W'(3)$ is the infinitesimal shear modulus.

In general, the singular behavior predicted by the linearized theory according to (6), (7) does not properly describe the behavior of the solution of the nonlinear problem near the crack-tips. It is plausible, however, to assume that, for sufficiently small values of the applied shear k, the estimates (6), (7) are realistic near, but not *too* near, a crack-tip, while in the *immediate* neighborhood of the tip, nonlinear effects must be dominant. A magnification of this state of affairs (the case of "small scale yielding" in plasticity) suggests the consideration of the following problem: find a solution u of (1) in the region exterior to the *semi-infinite* crack (Figure 2) which satisfies (3) for $-\infty < x_1 < 0$, which is bounded near the crack-tip, and which fulfills the *matching condition*

(8) $$u \sim k(2cr)^{1/2}\sin(\theta/2) \qquad \text{as } r \to \infty.$$

It is this *small-scale nonlinear crack problem* which is analyzed in [1–3]; see also [4], [7].

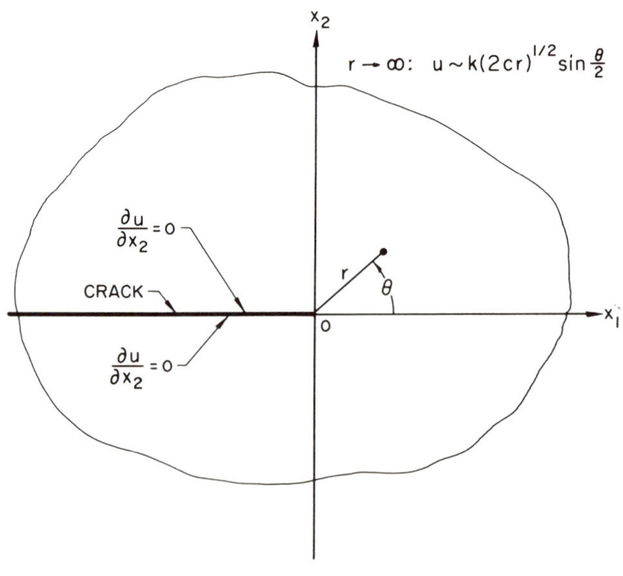

Figure 2. Semi-infinite crack; small-scale nonlinear crack problem.

Since breakdowns in smoothness are to be expected when (1) fails to be elliptic everywhere, and since this eventuality is a principal issue for the situations to be encountered here, it is necessary to relax the conventional smoothness requirements to be imposed on (1). We require in fact that u be merely a *weak* solution of the small-scale nonlinear crack problem. In particular, although u must be continuous everywhere except possibly at the crack-tip, its first and second derivatives need only be piecewise continuous. Across a curve, if any, which bears a discontinuity in ∇u, the traction must remain continuous in order to preserve equilibrium.

Special materials

In [1], $W(I_1)$ was chosen in such a way that the associated shear stress response function $\tau(k)$ for simple shear (see (5)) has the form shown in Figure 3. This particular choice of $\tau(k)$ makes it possible to construct an explicit, exact global solution of the small-scale nonlinear crack problem.

In [2], a subclass of I_1-materials is considered for which $\tau(k)$ behaves qualitatively as shown in Figure 4; this subclass is delineated by a number of restrictions on $\tau(k)$ which are set out in detail in Section 1 of [2]. In particular the ultimate stress τ_∞ in simple shear may or may not vanish. No explicit form for $\tau(k)$ is assumed in [2], and so the analysis is necessarily qualitative.

Finally, in [3], Abeyaratne considers I_1-materials for which the shear stress response curve is as shown in Figure 5. The assumed form for $\tau(k)$ is now

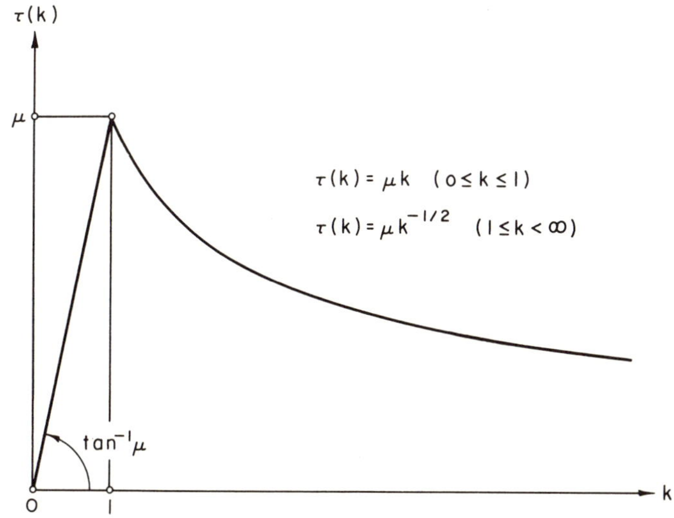

Figure 3. Response curve in simple shear for the material considered in [1].

again fully explicit, and again an exact and explicit solution to the nonlinear small-scale crack problem is constructed.

Since each of the response curves in Figures 3–5 represents a function $\tau(k)$ which *decreases* over a certain range of k, the differential equation (1) need not remain elliptic everywhere in the field, and the small-scale nonlinear crack problem is thus a boundary-value problem of mixed type. This is of course

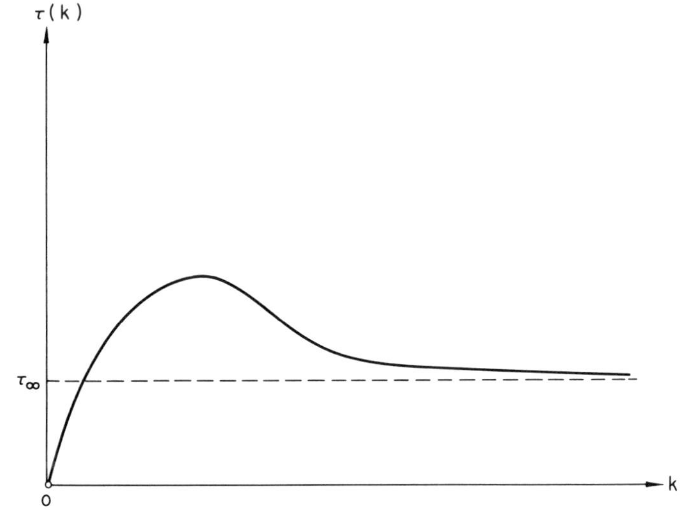

Figure 4. Response curve in simple shear for the materials in the class considered in [2].

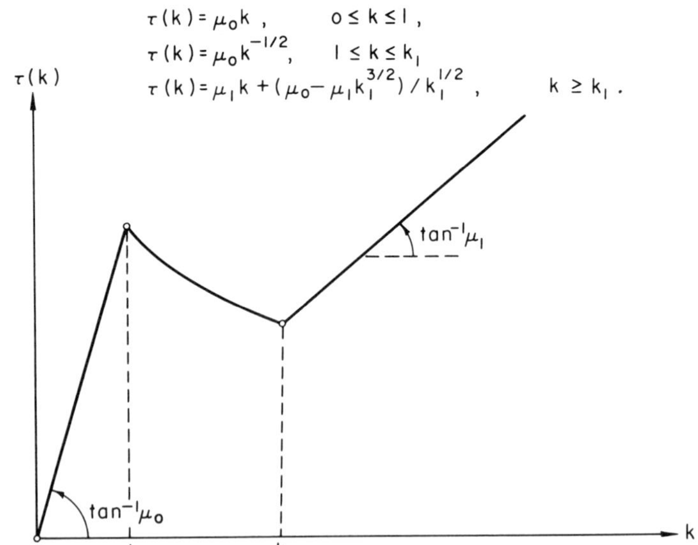

Figure 5. Response curve in simple shear for the material considered in [3].

also true of the original problem for the finite crack, which in fact is closely analogous to the famous problem of mixed type associated with steady flow of a compressible fluid past a flat plate at 90° angle of attack. A more detailed discussion of this analogy may be found in [4].

Results

As one might expect from the extensive literature devoted to the theory of compressible flows, the differential equation (1) can be treated profitably with the help of the hodograph transformation. Applications of this procedure in anti-plane shear for solids seem to have been made first by Hult and McClintock [8] and, more extensively, by Rice [7]. All of the results in [1–3] are obtained on the basis of this procedure; for details, the reader may consult these references as well as [4]. We shall cite only the results here, omitting all of the supporting analysis.

It is a simple matter to show that, if the physical coordinates x_1, x_2 of Figure 2 and the displacement u are rendered nondimensional by dividing each of them by ck^2, the amount of shear k applied at infinity does not appear in the resulting dimensionless version of the small-scale nonlinear crack-problem for *any* I_1-material. We shall use an overbar (e.g., $\bar{x}_1$, $\bar{r}$) to indicate dimensionless quantities obtained from their physical counterparts by this scaling. We note in passing that the disappearance of k from the dimensionless boundary-value problem shows that the length scale on which nonlinear effects near the crack-tip are significant when k is small is of second order in k.

For the material of Figure 3, the exact solution of the (dimensionless) small-scale nonlinear crack-problem is given by (see Figure 6)

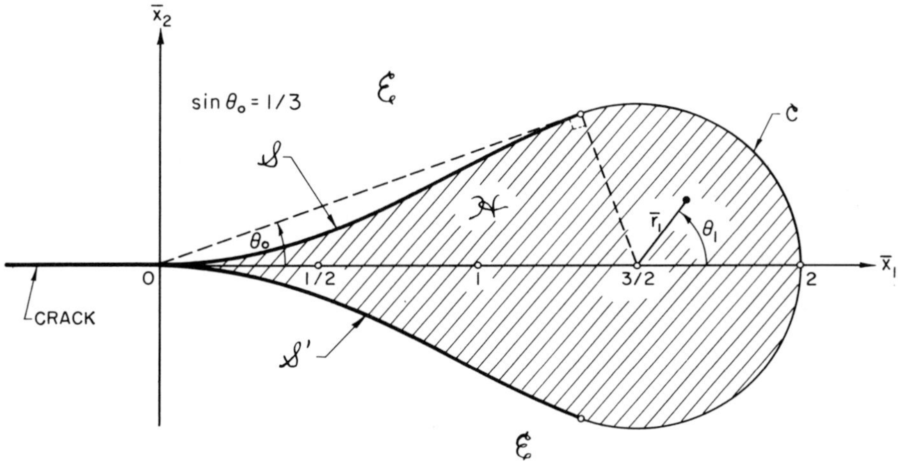

Figure 6. Elliptic and hyperbolic domains for solution of small-scale nonlinear crack problem; material of Figure 3.

$$(9)\qquad \bar{u} = \begin{cases} (2\bar{r}_1)^{1/2}\sin(\theta_1/2) & \text{on } \mathscr{E}, \\ \dfrac{f(\theta)}{\bar{r}} & \text{on } \mathscr{H}, \end{cases}$$

where

$$(10)\qquad f(\theta) = \frac{1}{\sqrt{2}}\,\frac{[3\cos\theta + (9\cos^2\theta - 8)^{1/2}]^{3/2}}{[\cos\theta + (9\cos^2\theta - 8)^{1/2}]^{1/2}}\sin\theta.$$

Here $\bar{r}$, θ are polar coordinates centered at the origin, while $\bar{r}_1$, θ_1 are polar coordinates centered at the point $\bar{x}_1 = 3/2$, $\bar{x}_2 = 0$. $\mathscr{H}$ is the domain whose boundary is comprised of the two curves $\mathscr{S}$ and $\mathscr{S}'$ and the circular arc $\mathscr{C}$ of radius 1/2 centered at (3/2, 0). For the solution given by (9), the differential equation (1) is hyperbolic on $\mathscr{H}$, with the crack itself deleted. The symmetrically located curves $\mathscr{S}$ and $\mathscr{S}'$ (which we call elastostatic shocks) are curves across which displacement and traction are continuous, but displacement gradient and stress suffer jump discontinuities. Although these curves are determined numerically (rather than explicitly) in [1], their unique existence and properties are demonstrated analytically. In particular, near the origin on $\mathscr{S}$, for example,

$$(11)\qquad \bar{r} \sim \frac{4}{\sqrt{3}}\sin\theta, \qquad \theta \to 0,$$

so that $\mathscr{S}$ (and, by symmetry, $\mathscr{S}'$) is asymptotically tangent to the $\bar{x}_1$-axis as the origin is approached. (This fact, together with the behavior of $f(\theta)$ as $\theta \to 0$, keeps $\bar{u}$ bounded near the crack-tip, despite the ominously unbounded appearance of $\bar{u}$ near $\bar{r} = 0$ as given by (9).)

Across the circular arc $\mathscr{C}$, $\bar{u}$ and $\bar{\nabla}\bar{u}$ are continuous; there *are* discontinuities in the stress gradients across $\mathscr{C}$, but these are traceable to the kink in the response curve of the material (Figure 3).

It is interesting to note that, in the elliptic domain, the solution (8) coincides exactly and globally with the near field of linearized theory, *except that the crack-tip is shifted from the origin to* (3/2, 0).

Explicit formulas for the stresses follow from (9) and the basic field equations; we omit these here. The angular dependence of $\bar{u}$ and the polar components τ_{rz}, $\tau_{\theta z}$ of Cauchy shear stress is displayed in Figures 7, 8 and 9. The displacement $\bar{u}$, although bounded at the crack-tip, is not continuous there. The shear stresses are bounded at the crack-tip; the axial Cauchy stress τ_{33} is unbounded as the crack-tip is approached from within the hyperbolic region, but bounded on the elliptic domain.

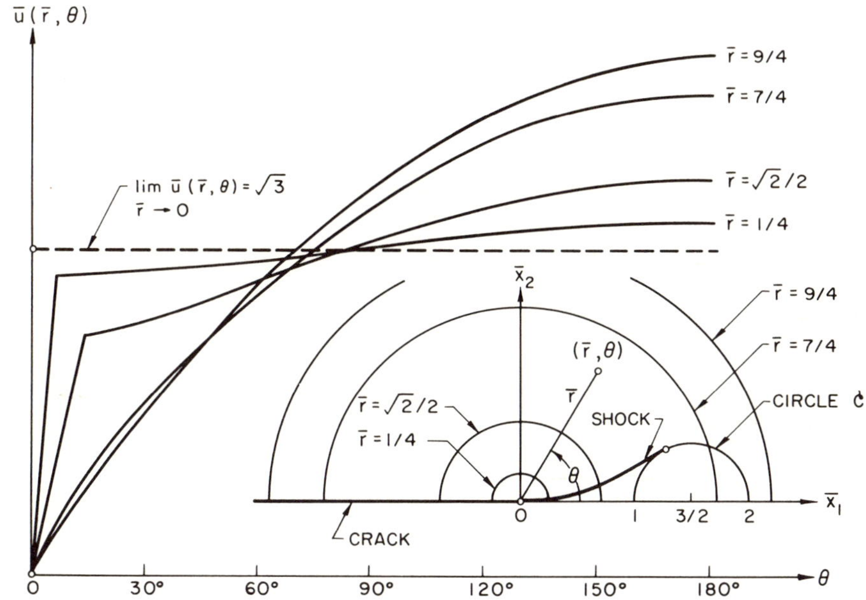

Figure 7. Angular dependence of displacement $\bar{u}$; material of Figure 3.

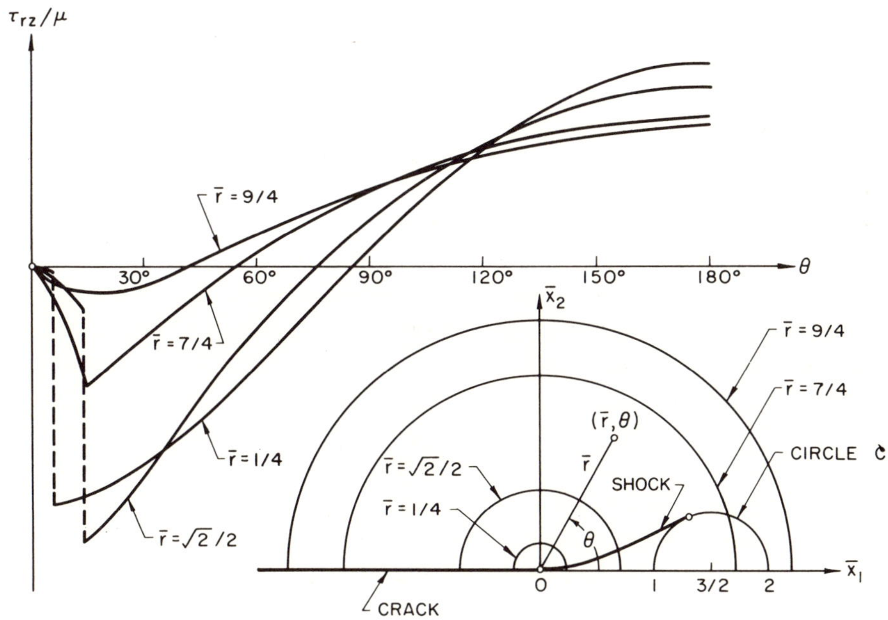

Figure 8. Angular dependence of shear stress τ_{rz} ; material of Figure 3.

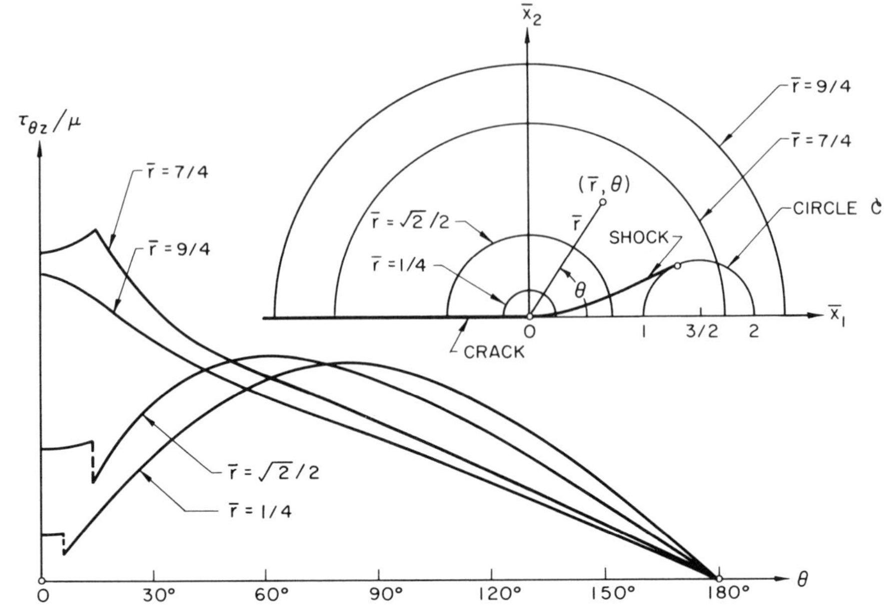

Figure 9. Angular dependence of shear stress $\tau_{\theta z}$; material of Figure 3.

Suppose the loading parameter k were imagined to increase quasi-statically with time, and consider a portion P of the deformed body whose undeformed preimage lies astraddle the shock $\mathcal{S}$. One finds that, because of the discontinuity in $\bar{\nabla}\bar{u}$ across $\mathcal{S}$, the power supplied to P by the tractions acting on its boundary does not coincide with the rate of increase of energy stored in P. It was verified in [1] that this power in fact exceeds the rate of increase of stored energy, so that the shock consumes, rather than creates, energy, at least as long as the body is being loaded, rather than unloaded.

The question of whether a *smoother* solution exists to the small-scale nonlinear crack problem, as well as questions pertaining to the stability of the solution constructed, were not dealt with in [1].

Broadly speaking, the analysis in [2] confirms that, for response curves such as the one shown in Figure 4 which rise monotonically to a smooth maximum and then decline monotonically, the qualitative character of the weak solution found is similar to that of the field described above. There are differences of detail: for example, the shocks $\mathcal{S}$ and $\mathcal{S}'$, upon approaching the origin, now make an angle with the $\bar{x}_1$-axis which depends on the ultimate stress τ_∞, vanishing when $\tau_\infty = 0$.

Again, the shear stresses are bounded near the crack-tips, while the axial stress is unbounded as the crack-tip is approached from within the hyperbolic region, but bounded otherwise.

In order to carry out the analysis in [2], it was necessary to assume, among other restrictions on the shear stress response function $\tau(k)$, that the curve respresenting τ^2 as a function of k^2 has only one point of inflection for values of k^2 which exceed the value at which τ^2 has its maximum. Although the consequences of relinquishing this rather refined and certainly mysterious material restriction were not studied in [2], it seems that they might well include the possibility that each of the shocks $\mathscr{S}$ and $\mathscr{S}'$ forks into *two* curves part way along its length.

In [3], Abeyaratne deals with the more interesting case of a material whose shear stress response curve is that shown in Figure 5. When the parameters governing the detailed behavior of the curve in Figure 5 are conveniently related, he finds an explicit solution of the corresponding small-scale non-linear crack problem. Now there are *two* elliptic domains $\mathscr{E}_1$ and $\mathscr{E}_2$, one of which completely surrounds the crack-tip, and one hyperbolic domain $\mathscr{H}$ (see Figure 10). The most striking difference between Abeyaratne's results and those reported in [1, 2] is that, as seen in Figure 9, the shocks $\mathscr{S}$ and $\mathscr{S}'$ no longer issue from the crack-tip, but rather originate on the crack faces. Over a portion of its length, $\mathscr{S}$ (or $\mathscr{S}'$) separates two "elliptic" solutions, while

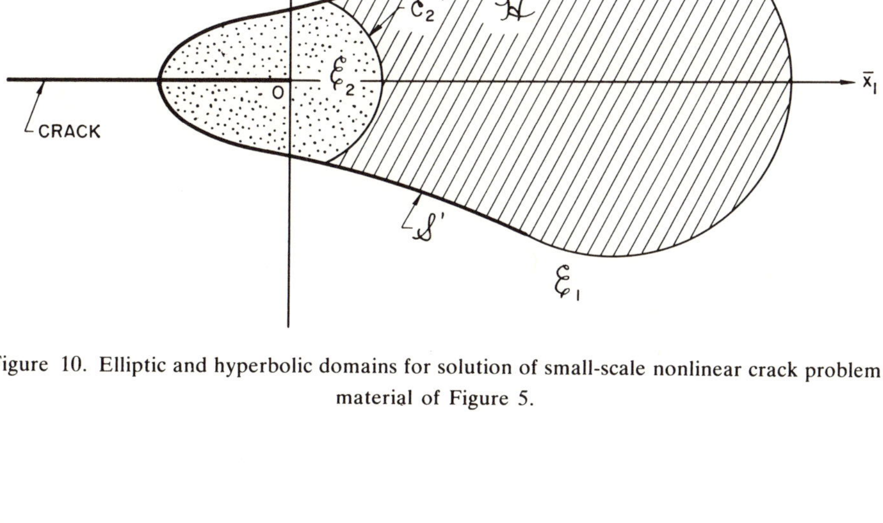

Figure 10. Elliptic and hyperbolic domains for solution of small-scale nonlinear crack problem; material of Figure 5.

elsewhere it separates an elliptic solution from a hyperbolic one. Again, the circular arcs $\mathscr{C}_1$ and $\mathscr{C}_2$ do *not* carry discontinuities in stress or displacement gradient.

For the material described by Figure 5, all stresses are unbounded at the crack-tip. In fact, as one would expect, the local structure of the field in the *immediate* vicinity of the crack-tip is precisely of the form given in (6), (7), *except* for the amplitude constants.

Detailed numerical results for the angular distributions of the displacement and stresses are given in [3].

As in [1, 2], stability questions and the issue of the existence of a smoother solution are not investigated.

Acknowledgement

Most of the results described in this paper were obtained in the course of an investigation supported by the Office of Naval Research.

References

1. J. K. Knowles and E. Sternberg, *Discontinuous deformation gradients near the tip of a crack in finite anti-plane shear: an example*, Journal of Elasticity **10** (1980), 81.
2. J. K. Knowles and E. Sternberg, *Anti-plane shear fields with discontinuous deformation gradients near the tip of a crack in finite elastostatics*. Journal of Elasticity, **11** (1981), 129.
3. R. Abeyaratne, *Discontinuous deformation gradients away from the tip of a crack in anti-plane shear*, Journal of Elasticity, **11** (1981), 373.
4. J. K. Knowles, *The finite anti-plane shear field near the tip of a crack for a class of incompressible elastic solids*, International Journal of Fracture **13** (1977), 611.
5. J. K. Knowles, *On finite anti-plane shear for incompressible elastic materials*, The Journal of the Australian Mathematical Society, Series B, **19** (1977), 400.
6. I. Sneddon and M. Lowengrub, *Crack Problems in the Classical Theory of Elasticity*, Wiley, New York, 1969.
7. J. R. Rice, *Stresses due to a sharp notch in a work-hardening elastic-plastic material loaded by longitudinal shear*, Journal of Applied Mechanics **34** (1967), 287.
8. J. A. H. Hult and F. A. McClintock, *Elastic-plastic stress and strain distributions around sharp notches under repeated shear*, Proceedings, Ninth International Congress of Applied Mechanics, Brussels, **8** (1956), 51.

STRESS–STRAIN–TEMPERATURE CURVES IN PSEUDOELASTIC BODIES

INGO MÜLLER

Hermann-Föttinger-Institut — FB 9, Technische Universität Berlin, 1 Berlin 12, FRG

(Received September 11, 1980)

ABSTRACT

Pseudoelastic bodies are characterized by a strong dependence of the stress–strain relations on temperature. At low temperatures they behave much like plastic bodies with a yield limit and residual deformation upon unloading, but at high temperatures the plastic deformation is recovered and the pseudoelastic body returns to its initial configuration. This paper presents a model which is capable of simulating the complex behaviour of a pseudoelastic body. The model consists of a superposition of many identical elements whose statistical mechanical treatment allows us to calculate the free energy of the body and this free energy has the form typical for a body that may undergo a first-order phase transition. The results bear a strong resemblance to the theory of ferroelectric bodies.

1. Observations

The form of the stress–strain curves of a pseudoelastic body is strongly affected by temperature in a manner illustrated by Figures 1a through 1f. These are the main qualitative features in dead-loading experiments as abstracted from a book [1] on the subject:

(a) At low temperatures the bodies behave much like plastic bodies, in particular:

(i) there is a virginal elastic curve;

(ii) there is a yield limit;

(iii) there is a residual strain upon unloading;

(iv) the yield limit decreases with increasing temperature.

All this is reminiscent of a plastic body, but

(v) there is also a second elastic branch at high strains which is not observed in a plastic body of course.

Proceedings of the IUTAM Symposium in Finite Elasticity, Lehigh University, August 10–15, 1980. Invited paper.

Carlson, D.E. and Shield, R.T. (eds.)
Proceedings of the IUTAM Symposium on Finite Elasticity
ISBN 90 247 2629 8

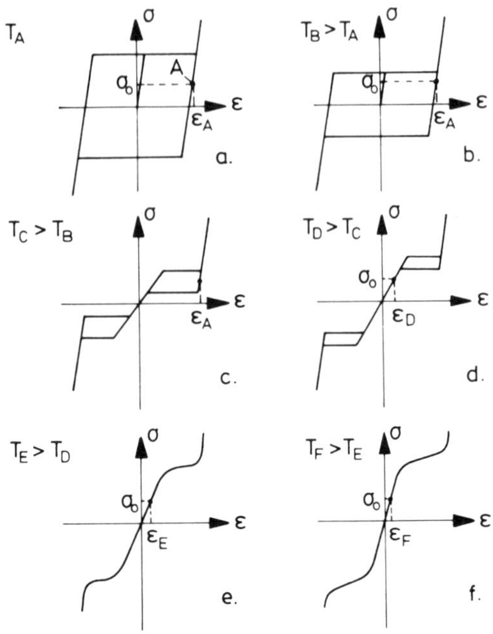

Figure 1.

(b) At intermediate temperatures:

(i) the slope of the elastic branch through the origin grows with increasing temperature;

(ii) the yield limit grows with increasing temperature;

(iii) after yielding there can be further loading and partial unloading along a second elastic branch;

(iv) upon unloading after yielding there is a recovery limit at which the body creeps back to the original elastic branch;

(v) the recovery limit increases with increasing temperature;

(vi) yield limit and recovery limit grow closer together with increasing temperature.

It is this behaviour at intermediate temperatures which has given rise to the name "pseudoelastic", because the body is elastic in that it returns to the original configuration, but it is only *pseudo* elastic, since there occurs a hysteresis in the stress–strain curves.

(c) At high temperatures

(i) from a critical temperature on upwards the body exhibits true (non-linear) elastic behaviour;

(ii) at the critical temperature there is one horizontal point of inflection;

(iii) at low strains the slopes of the stress–strain curves increase with increasing temperature.

As a consequence of the stress–strain behaviour described above, one can observe the following spectacular effect: If a pseudoelastic body at a low temperature is subjected to a load bigger than the yield load and then unloaded there is a residual deformation as in a plastic body. When the body is subsequently heated to an intermediate or high temperature, it creeps back to its original configuration, because at that temperature there is only one strain, viz. $\varepsilon = 0$, corresponding to zero load. This behaviour has given rise to the name *memory material* for a pseudoelastic body.

Another even more spectacular aspect of a pseudoelastic body can also be read off from the stress–strain curves of Figure 1: At a given stress σ_0 and at the low temperature T_A we observe the body in state A with strain ε_A (say). An increase of temperature to T_B and T_C will obviously leave the strain unchanged, but when the temperature is increased to T_D, T_E and T_F, the strain will decrease to ε_D, ε_E and ε_F respectively. When the temperature is now lowered, we observe that the original strain is restored, but there is a hysteresis in the strain–temperature curve which reflects the hysteresis in the stress–strain curve; in particular, at T_C the strain will be much lower when the temperature is decreased than when it is increased. Figure 2 shows a qualitative picture of the strain–temperature curve. Inspection of that curve

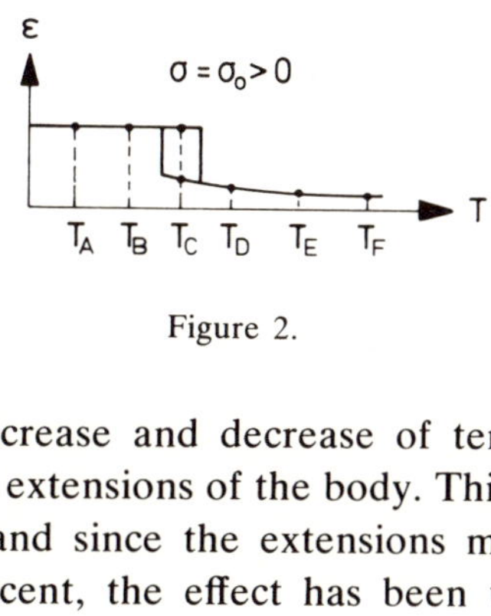

Figure 2.

shows that a repeated increase and decrease of temperature will lead to repeated contractions and extensions of the body. This effect is known as the *two-way memory* effect and since the extensions may be of the order of magnitude of several percent, the effect has been used to construct heat engines.

2. Model for a pseudoelastic body

It is the purpose of this paper to construct a model that is capable of simulating the above-described stress–strain–temperature behaviour of a pseudoelastic body. The model is an Ersatz model whose details were suggested by observations of the metallurgists who have found out that the pseudoelastic behaviour is caused by strain-induced or temperature-induced phase transitions between austenite and martensite and between different twins of martensite.

The basic element of the model is a lattice particle, that is a little piece of the metallic lattice of the body which has three equilibrium configurations shown in Figure 3 and denoted by $M_\pm$ for the martensitic twins and by A for austenite. We may consider M_+ and M_- as sheared versions of A with shear lengths $\pm J$ respectively. For all other shear lengths Δ the particle is not in equilibrium and the postulated potential energy Φ corresponding to a particular Δ is represented in the upper part of Figure 3. From this plot of $\Phi(\Delta)$ we conclude that the martensitic configurations are stable while the austenitic one is metastable. There are energetic barriers between the three minima of $\Phi(\Delta)$.

To construct the model of the pseudoelastic body we arrange the lattice particles in layers and stack these as shown in Figure 4. Figures 4a and 4b represent different realizations of the reference configuration: a uniform one where all lattice particles are of type A and a non-uniform one where the particles in different layers alternate between M_- and M_+.

Figure 5 shows how the model with alternating lattice layers of types M_+ and M_- reacts to a load at low temperatures. When a vertical load P is

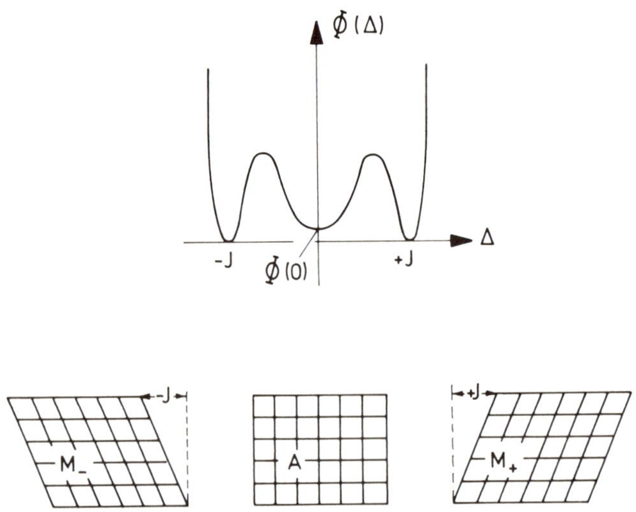

Figure 3.

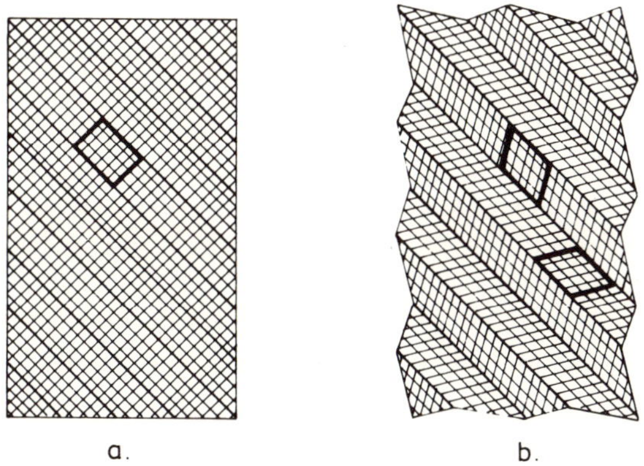

Figure 4.

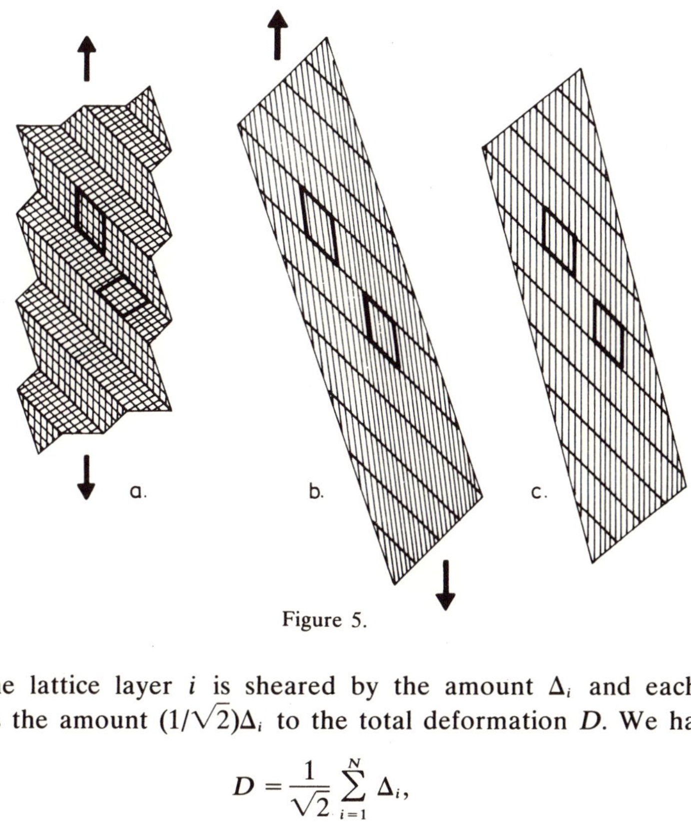

Figure 5.

applied, the lattice layer i is sheared by the amount Δ_i and each layer contributes the amount $(1/\sqrt{2})\Delta_i$ to the total deformation D. We have

$$D = \frac{1}{\sqrt{2}} \sum_{i=1}^{N} \Delta_i, \tag{2.1}$$

where the index i ranges over the N lattice layers. As long as the load is too small to lift the lattice particles of type M_+ over the energy barriers, this deformation is elastic. But when the load is strong enough to overcome those barriers there is a sudden big increase in deformation as all lattice particles of type M_+ flip to the M_- configuration and beyond. Subsequent removal of the load will leave us with a body uniformly built up of M_- particles, so that there is a residual deformation.

The above discussion illustrates quite suggestively how the model simulates a pseudoelastic body under a load at low temperatures. But this discussion serves only to acquaint the reader with the model, because we are specifically interested in studying the effects of higher temperatures. At higher temperatures the lattice particles do not lie still in their equilibrium positions but they perform a random motion about these positions. Because of that motion it is possible for a particle to change its type from M_+ to M_- (say) all by itself, i.e. without the help of a load, and that possibility becomes more probable as temperature goes up. However, it is difficult to take that random motion into account and predict the behaviour of the model under load in a simple suggestive manner. For that purpose we must rely upon the concepts of statistical mechanics and thermodynamics which are lying ready for use in cases like this one.

3. Statistical mechanics of the body

Let N_Δ be the number of lattice layers with the shear displacement Δ. Thus

$$p_\Delta = \frac{N_\Delta}{N} \tag{3.1}$$

is the probability that a layer has that displacement and we have

$$\sum_{\Delta=-\infty}^{\infty} p_\Delta = 1 \quad \text{and} \quad \sum_{\Delta=-\infty}^{\infty} \Delta p_\Delta = \sqrt{2}\,\frac{D}{N}, \tag{3.2}$$

the latter from (2.1).

The potential energy E_{Pot} and the entropy H of the model are given by

$$E_{\text{Pot}} = \sum_{\Delta=-\infty}^{\infty} \Phi(\Delta) N_\Delta \quad \text{and} \quad H = k \ln \frac{N!}{\Pi_\Delta N_\Delta!}, \tag{3.3}$$

or, in terms of p_Δ, when we use the Stirling formula on $N!$ and $N_\Delta!$

$$E_{\text{Pot}} = N \sum_\Delta \Phi(\Delta) p_\Delta \quad \text{and} \quad H = -Nk \sum_\Delta (\ln p_\Delta) p_\Delta. \tag{3.4}$$

Once E_{Pot} and H are known, we can form the free energy $\Psi_{\text{Pot}} = E_{\text{Pot}} - TH$ and thus obtain the load P by differentiation

$$P = \frac{\partial \Psi_{\text{Pot}}(D, T)}{\partial D}. \tag{3.5}$$

However, in order to determine E_{Pot} and H we must know the probability distribution p_Δ first.

p_Δ can be calculated approximately as follows: We replace the true potential $\Phi(\Delta)$ of Figure 3 by the simple box potential

$$\Phi_0(\Delta) = \begin{cases} 0 \\ \infty \end{cases} \quad \text{for} \quad \begin{matrix} |\Delta| < J \\ |\Delta| > J. \end{matrix} \tag{3.6}$$

This obviously entails $p_\Delta = 0$ for $|\Delta| > J$, and for $|\Delta| < J$ we determine p_Δ by maximising the entropy $(3.4)_2$ under the constraints (3.2). Thus we obtain

$$p_\Delta = e^{1-\alpha} \cdot e^{-\beta\Delta}, \tag{3.7}$$

where α and β are the Lagrange multipliers on the constraints $(3.2)_{1,2}$ respectively. These multipliers are determined by insertion of p_Δ into the constraints and a brief calculation shows that

$$p_\Delta = \frac{1}{2ZJ} \frac{\beta J}{\sinh \beta J} e^{-\beta\Delta} \quad \text{and} \quad \mathscr{L}(\beta J) = -\sqrt{2}\frac{D}{NJ}. \tag{3.8}$$

The function $\mathscr{L}(x)$ in (3.8) stands for $(\operatorname{ctgh} x - 1/x)$ and this function is called the Langevin function. The factor Z determines the number of displacements in the range between Δ and $\Delta + d\Delta$. While $(3.8)_2$ cannot be solved analytically for β, it determines a one-to-one correspondence between β and D, so that p_Δ is seen to be a function of Δ and D.

Elimination of p_Δ between (3.8) and (3.4) leads to the following expressions for E_{Pot} and H:

$$\begin{aligned} E_{\text{Pot}} &= N\frac{1}{2}\frac{\beta J}{\sinh \beta J}\frac{1}{J}\int_{-J}^{J} \Phi(\Delta)e^{-\beta\Delta}d\Delta, \\ H &= Nk\left(\ln(2ZJ) + \beta J\sqrt{2}\frac{D}{NJ} + \ln\frac{\sinh \beta J}{\beta J}\right). \end{aligned} \tag{3.9}$$

Hence it follows for the free energy that

$$\begin{aligned} \Psi_{\text{Pot}} = N\Bigg[&\frac{1}{2}\frac{\beta J}{\sinh \beta J}\frac{1}{J}\int_{-J}^{J} \Phi(\Delta)e^{-\beta\Delta}d\Delta \\ &- kT\left(\ln(2ZJ) + \beta J\sqrt{2}\frac{D}{NJ} + \ln\frac{\sinh \beta J}{\beta J}\right)\Bigg], \end{aligned} \tag{3.10}$$

and this is a function of deformation D and temperature T; note that β, by $(3.8)_2$, is a function of D.

In this approximate calculation of Ψ_{Pot} we have determined p_Δ for the potential (3.6) and yet we used this p_Δ to calculate E_{Pot} for the true potential

$\Phi(\Delta)$ rather than for $\Phi_0(\Delta)$. This procedure can be justified as the first step to an iterative scheme as Wilmanski and Müller have shown in the paper [2], of which this note gives a brief summary.

4. Free energy vs. deformation and load vs. deformation

To evaluate (3.10) we must specify $\Phi(\Delta)$ in the range $|\Delta| < J$. Almost every function of the qualitative shape shown in Figure 3 will produce the results discussed below and therefore I do not define $\Phi(\Delta)$ analytically and refer the reader to [2] for such details. Once a choice has been made for $\Phi(\Delta)$ we may calculate Ψ_{Pot} as a function of D and T and Figure 6 presents a graphical representation of that function.

The curves of Figure 6 refer to different temperatures and their dashed parts represent states of the model that are mechanically unstable. On the right and left solid branch the model is in the phases M_+ and M_-, respectively; on the central solid branch, if there is one, the body is in the austenitic phase.

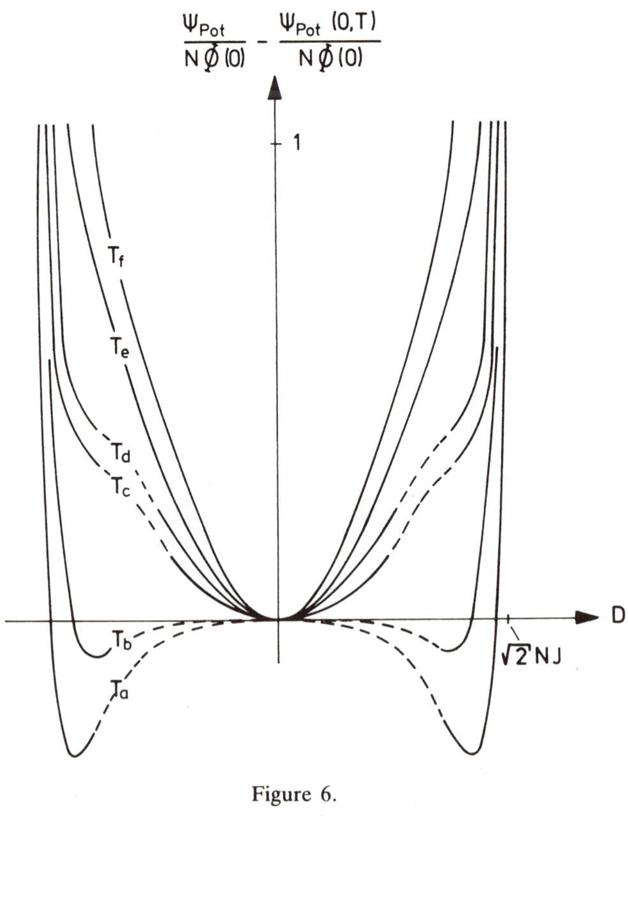

Figure 6.

Obviously therefore the austenitic phase cannot exist at low temperatures, while at intermediate temperatures it exists for small deformations.

More suggestive than the free energy vs. deformation curves are the load-deformation curves which are obtained by differentiation according to (3.5). The solid and dashed lines in Figure 7 show the result.

For discussion we start with Figure 7c at $P = 0$, $D = 0$, where the model is austenitic. In a dead-loading experiment the deformation grows gradually until the maximum is reached. Upon further loading the body creeps along the horizontal dotted line until it reaches the right solid branch on which it may move up and down. Unloading brings the model to the minimum and from there it creeps back horizontally along the dotted line. In a similar manner a dead-loading experiment traces out horizontal dotted lines in Figures 7a, b and d. Comparison of the solid and dotted lines of Figure 7 with the observed curves of Figure 1 shows complete qualitative agreement. Thus we conclude that the model simulates the observed behaviour well.

After this discussion it would seem that we cannot find the body in a state within the hysteresis loops. And this is quite so in a dead-loading experiment after we have surpassed the yield limit once. However, initially before the

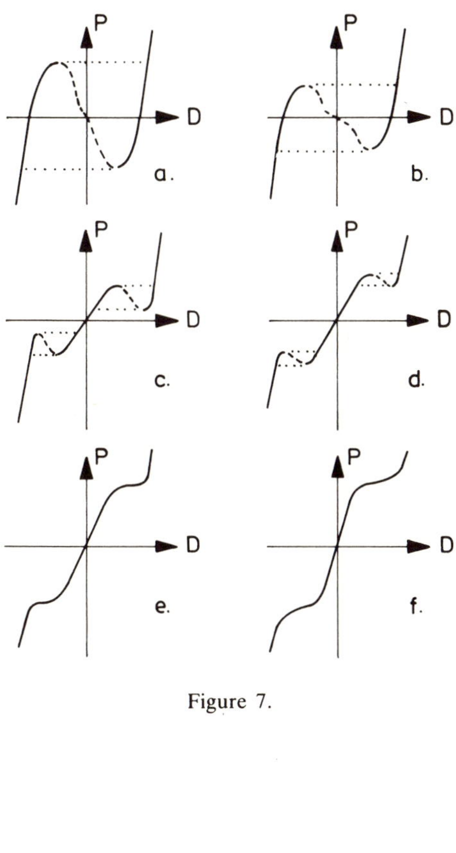

Figure 7.

loading starts, we do generally find the body at the origin and at low temperatures this means that we find it within the hysteresis loop. In this state is consists of an equal share of M_+ and M_- layers as shown in Figure 4b, and this configuration has been created as follows:

When the material was produced, it was cooled down from a high temperature, where it consisted of austenite with all lattice particles roaming freely about the potential well so that their average position lay in the neighbourhood of the central minimum of the potential $\Phi(\Delta)$. Upon cooling, the structure of the bottom of the potential well asserted itself and the particles preferred the positions of the two lateral minima, because they are deeper. And since there is no preference between the two, the body settled down as half M_+ and half M_-. When the body is loaded in this state it will react as described in Figure 5, initially tracing out the virginal curve indicated in Figures 1a and 1b.

It is easy to see that by cooling the body under a load, rather than unloaded, we may produce an arbitrary initial fraction of M_- and M_+ particles. And subsequent loading will produce elastic lines inside the hysteresis region which are roughly parallel to the virginal curve.

Almost all states within the hysteresis loops are not in phase equilibrium, even though they are known to be quite stable in the ordinary sense of the word. Indeed, as long as the free energy vs. deformation curve has parts of negative curvature, the body may assume lower free energies than on that curve by splitting into two phases and by changing their fraction upon deformation. In this manner the states lie on the subtangent t, see Figure 8a, and the two phases are the states 1 and 2. In the load vs. deformation diagram the corresponding states lie on the horizontal line that cuts through the hysteresis loop in such a manner that the shaded areas in Figure 8b are equal in size.

It is well known that ordinary phase transitions, like the liquid-vapor

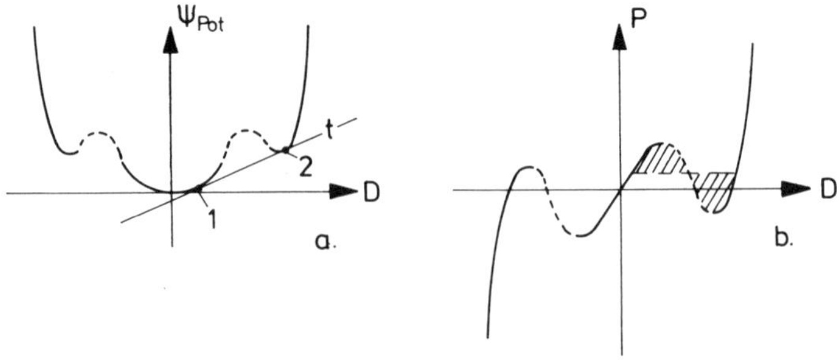

Figure 8.

transition, do not produce a hysteresis, while we have seen that the transitions between M_+ and M_- and between austenite and martensite do produce a hysteresis. The reason is simple: Since a lattice particle must change its shape when it wants to change its phase, it will have to overcome internal stresses, and these stresses prevent it from establishing phase equilibrium. Obviously no such internal stresses are present in the liquid or the vapor.

5. A close analogy — ferroelectricity

The discussion so far has shown that the model simulates the observed properties of a pseudoelastic body well. In addition the model predicts certain features which, to my knowledge, have not been observed so far. These predictions concern the temperatures between those where the stress–strain curves are of the types shown in Figures 1b and 1c. The theory predicts that in this temperature range the free energy vs. deformation curves are of the shapes shown in Figures 9a and 9c so that the corresponding load vs. deformation curves are the solid and dashed parts of Figures 9b and 9c.

The load vs. deformation curves of Figure 9 must be supplemented by the dotted horizontal lines to produce the curves that are traced out in a dead-loading experiment. We conclude that at both temperatures austenite prevails in the origin. But while at the lower temperature the yield limit for austenite is very small, at the higher temperature it is bigger than the yield limit for the martensitic twins.

The behaviour of the model as laid down in Figures 6, 7 and 9 is indicative of a system that undergoes a phase transition of the first order in Landau's nomenclature (e.g. see Burfoot [3]). Indeed, ferroelastic bodies can have phase transitions like that which is put in evidence by plots of the polarization π versus the electric field E schematically shown in Figure 10.

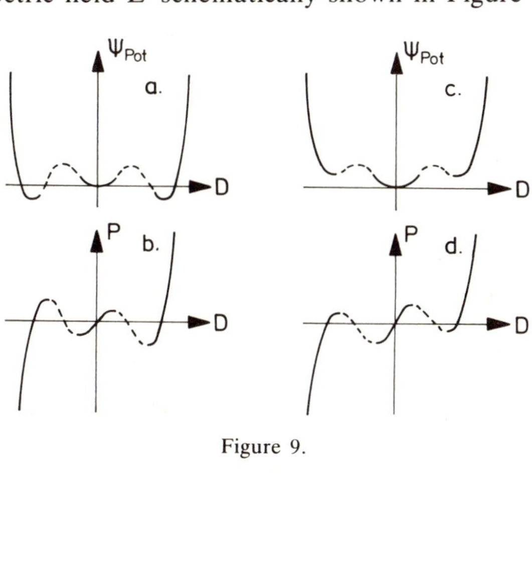

Figure 9.

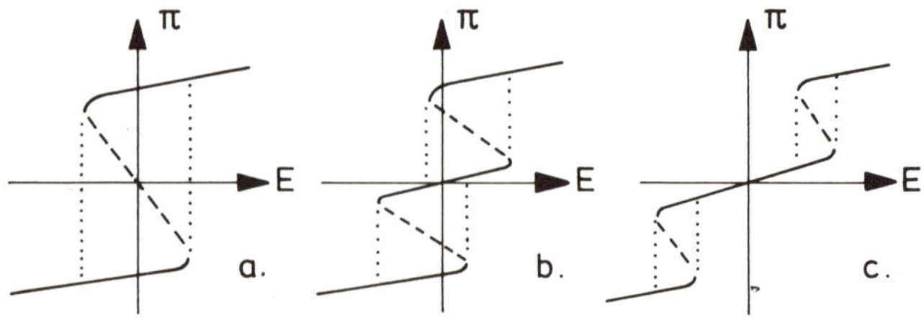

Figure 10.

The similarity between these diagrams and those of Figures 1a, 9d, and 1c is evident when polarization is replaced by deformation and electric field by load.

It seems that Falk [4] was the first to recognize that the stress–strain curves of pseudoelastic bodies indicate a first order phase transition in Landau's sense.

Acknowledgement

The author gratefully acknowledges the support of his attendance in the IUTAM Symposium on Finite Elasticity by the Deutsche Forschungsgemeinschaft.

References

1. J. Perkins (ed.), *Shape Memory Effects in Alloys*, Plenum Press, New York and London, 1975.

2. I. Müller and K. Wilmanski, *A model for phase transition in pseudoelastic bodies*, Il Nuovo Cimento **57B** (1980), 283.

3. J. C. Burfoot, *Ferroelastics, An Introduction to the Physical Principles*, van Nostrand Co., London, 1967.

4. F. Falk, *Wie hängt die Freie Energie einer Memory Legierung von der Verzerrung und der Temperatur ab*? ZAMM Sonderband on GAMM meeting 1979 in Wiesbaden, to appear.

PENALTY METHODS FOR CONSTRAINED PROBLEMS IN NONLINEAR ELASTICITY

J. T. ODEN

Department of Aerospace Engineering and Engineering Mechanics, The University of Texas at Austin, Austin, Texas 78712, USA

(Received September 11, 1980)

ABSTRACT

This paper presents a general theory of exterior penalty methods for constrained optimization problems with applications to the analysis and approximation of certain problems in linear and nonlinear elasticity. Particular attention is given to the role of a "generalized LBB-condition" which governs the behavior of penalty approximations of Lagrange multipliers.

1. Introduction

The idea of exterior penalty methods as a device for reducing constrained optimization problems to a sequence of unconstrained problems is generally attributed to Richard Courant who used the idea in 1941 to resolve a boundary-value problem in which boundary-conditions were treated as constraints. Since then, penalty methods have had an extremely important role in the study of constrained optimization problems in $\mathbb{R}^n$. In recent years, they have also provided a basis for the study of the existence and regularity of solutions to a large class of nonlinear boundary-value problems; more recently, they have been used as a basis for the development of powerful approximation theories and numerical methods.

In the present paper, a general theory of exterior penalty methods for constrained optimization problems in reflexive Banach spaces is given and applications of this theory to several classes of linear and nonlinear elasticity problems with constraints are described. Special attention is given to the issue

Proceedings of the IUTAM Symposium on Finite Elasticity, Lehigh University, August 10–15, 1980. Invited paper.

Carlson, D.E. and Shield, R.T. (eds.)
Proceedings of the IUTAM Symposium on Finite Elasticity
ISBN 90 247 2629 8

of using penalty methods to also provide an approximation to Lagrange multipliers associated with the constraints. This aspect of the theory is especially important in applications to problems in elasticity as the multipliers frequently represent quantities of physical interest (e.g. hydrostatic pressures, contact pressures, etc.)

A key condition, sufficient to guarantee the existence of Lagrange multipliers in many applications, is proposed which is referred to as the "generalized LBB-condition" in reference to related work on linear boundary-value problems with linear constraints by Ladyszhenskaya [12], Babuska [1–3] and Brezzi [6]. This condition, which is presented here in a general and abstract form appropriate for penalty functionals on reflexive Banach spaces, guarantees the boundedness of series $\{p_\varepsilon\}$ approximating the Lagrangian multipliers p for penalty parameters $\varepsilon > 0$. Applications to the study of boundary-value problems in elasticity and to the approximation of such problems by finite element methods are discussed. It is shown, for example, that the success of reduced-integration-penalty methods for finite element approximations of problems with constraints strongly depends upon whether or not a discrete LBB-condition holds for the finite-dimensional problem.

Following this introductory section, some general results on exterior penalty methods for an abstract and general class of constrained optimization problems are given. This is followed by sections on the relationships between penalty and Lagrange multiplier methods for the case of differentiable functionals. A general LBB-condition is given in Section 4 together with an existence theorem which establishes the importance of such conditions in calculating multipliers. Section 5 of the paper describes applications of the theory to four types of constrained problems in elasticity: (1) finite deformation of incompressible elastic bodies, (2) equilibrium problems in linear elasticity involving incompressible elastic bodies, (3) large transverse deflections and buckling of Von Karman plates with unilateral constraints, and (4) contact problems involving the unilateral contact of a linear elastic body with a rigid frictionless foundation. Approximations of exterior penalty formulations by finite elements are outlined briefly in Section 6, together with the notion of selective reduced integration. Several applications of the theory are given in the final section of the paper.

2. Penalty methods for nonconvex optimization

For orientation, we will examine certain ideas concerning penalty methods for treating optimization problems involving possibly nonconvex functionals. The general problem we wish to consider is this:

$$(2.1)\quad \begin{cases} \text{Given } V \text{ a reflexive Banach space, } K \text{ a nonempty subset of } V, \\ \text{and } \Pi \text{ an extended real-valued functional defined on } V, \text{ find} \\ u \in K \text{ such that} \\ \qquad \inf_{v \in K} \Pi(v) = \Pi(u). \end{cases}$$

In other words, we wish to find a minimizer of Π subject to the constraint $u \in K$.

It is well known that there exists at least one solution u to the optimization problem (2.1) whenever the following conditions hold:

$$(2.2)\quad \begin{cases} \text{(i) } K \text{ is weakly sequentially closed (i.e. if } u_m \in K \text{ and if} \\ u_m \rightharpoonup u \text{ weakly in } V, \text{ then } u \in K); \\ \text{(ii) } \Pi \text{ is proper } (\not\equiv +\infty) \text{ and coercive on } K \text{ (i.e. } \Pi(v) \to +\infty \\ \text{as } \|v\|_V \to \infty,\ v \in K); \\ \text{(iii) } \Pi \text{ is weakly sequentially lower semicontinuous (i.e. if} \\ u_m \rightharpoonup u \text{ weakly, then } \liminf_{m\to\infty} \Pi(u_m) \geqq \Pi(u)). \end{cases}$$

The *exterior penalty method* for the constrained optimization problem (2.1) involves introducing a penalty functional $P : V \to \mathbb{R}$ with the following properties:

$$(2.3)\quad \begin{cases} \text{(i) } P \text{ is positive semidefinite on } K \text{ in the sense that } P(v) \geqq 0, \\ P(v) = 0 \text{ if } v \in K \text{ and } v \notin K \Rightarrow P(v) > 0; \\ \text{(ii) } P \text{ is weakly sequentially lower semicontinuous,} \end{cases}$$

We then construct the penalized functional

$$(2.4)\qquad \Pi_\varepsilon : V \to \mathbb{R}; \qquad \Pi_\varepsilon(v) = \Pi(v) + \frac{1}{\varepsilon} P(v),$$

where ε is an arbitrary positive number.

Whenever Π satisfies conditions (2.2), it is clear that Π_ε is also proper, coercive, and weakly lower semicontinuous on all of V. Thus, there exists a solution $u_\varepsilon \in V$ to the minimization problem

$$(2.5)\qquad \inf_{v \in V} \Pi_\varepsilon(v) = \Pi_\varepsilon(u_\varepsilon)$$

for every $\varepsilon > 0$.

The importance of the exterior penalty functional Π_ε can be appreciated by taking note of the following properties:

(1) $\Pi_\varepsilon(v) \geqq \Pi_\varepsilon(u_\varepsilon) = \Pi(u_\varepsilon) + (1/\varepsilon)P(u_\varepsilon) \geqq \Pi(u_\varepsilon)$.

Hence,

$$\Pi(u_\varepsilon) \leqq \Pi(v) + \frac{1}{\varepsilon} P(v), \qquad \forall v \in V.$$

Picking $v \in K \Rightarrow P(v)=0$; therefore,

$$\Pi(u_\varepsilon) \leqq \Pi(v), \qquad v \in K, \tag{2.6}$$

which means that $\Pi(u_\varepsilon)$ is bounded. But since Π is coercive, we must have $\|u_\varepsilon\|_V \leqq C$, where C is a constant independent of ε.

(2) In view of (2.6), there exists a subsequence of solutions to (2.5), also denoted u_ε, obtained as ε tends to zero, and an element $u \in V$ such that $u_\varepsilon \rightharpoonup u$ weakly in V (by virtue of the reflexivity of V).

(3) For the subsequence u_ε in (2), note that

$$\Pi(u_\varepsilon) \leqq \Pi_\varepsilon(u_\varepsilon) \leqq \inf_{v \in K} \Pi_\varepsilon(v) = \inf_{v \in K} \Pi(v),$$

since $P(v)=0$ for $v \in K$. Moreover,

$$\liminf_{\varepsilon \to 0} \Pi(u_\varepsilon) \geqq \Pi(u),$$

because Π is weakly sequentially lower semicontinuous. Hence,

$$\Pi(u) \leqq \inf_{v \in K} \Pi(v).$$

(4) Finally, observe that $\Pi_\varepsilon(u_\varepsilon) = \Pi(u_\varepsilon) + \varepsilon^{-1} P(u_\varepsilon) \leqq \Pi(v_0)$ for some $v_0 \in K$. Hence,

$$P(u_\varepsilon) \leqq \varepsilon(\Pi(v_0) - \Pi(u_\varepsilon)).$$

But, in view of property (ii) of (2.3) and the fact that $\Pi(v_0)$ and $\Pi(u_\varepsilon)$ are bounded independently of ε,

$$0 \leqq P(u) \leqq \liminf_{\varepsilon \to 0} P(u_\varepsilon) \leqq \lim_{\varepsilon \to 0} (\Pi(v_0) - \Pi(u_\varepsilon))\varepsilon = 0,$$

i.e.,

$$P(u) = 0,$$

which (because of (2.3(i))) $\Rightarrow u \in K$.

In summary, we have proved the following theorem:

THEOREM 1. *Let conditions* (2.2) *and* (2.3) *hold. Then, for every* $\varepsilon > 0$ *there exists at least one solution* u_ε *to the penalized optimization problem* (2.5). *Moreover, there exists a subsequence of solutions to* (2.5) *obtained as* ε *tends to zero which converges weakly in* V *to a solution* u *of the constrained optimization problem* (2.1). □

The advantages of an exterior penalty formulation of the optimization problem (2.1) are clear: the minimizers of Π_ε can be sought in the entire space

V; the constraint set K enters the problem only in the construction of the penalty functional P. Minimizers u_ε of Π_ε can be selected so as to approximate an actual minimizer of Π arbitrarily closely in the weak topology of V by taking ε sufficiently small.

3. Relation to Lagrange multipliers

In many instances, there exists an important relationship between exterior penalty methods and Lagrange multiplier methods. To fix ideas, we now consider cases in which the following conditions are in force:

(3.1)

(i) Π is Gâteaux-differentiable on V; i.e. there exists $D\Pi(u) \in V'$, where V' is the topological dual of V, such that for $v \in V$,

$$\lim_{t\to 0^+} \frac{1}{t}[\Pi(u+tv)-\Pi(u)] = \langle D\Pi(u), v\rangle,$$

where $\langle\cdot,\cdot\rangle$ denotes duality pairing on $V' \times V$.

(ii) The constraint set $K \subset V$ is of the form

$$K = \{v \in V \mid B(v) = 0\},$$

where B is a nonlinear operator satisfying the conditions:

(ii.1) B is weakly sequentially continuous from V into a reflexive Banach space Q,

$$B : V \to Q;$$

$$u_m \rightharpoonup v \quad \text{in} \quad V \Rightarrow B(v_m) \rightharpoonup B(v) \quad \text{in} \quad Q;$$

(ii.2) B is C^1-differentiable, in the sense that an operator $\nabla B(u) \in \mathscr{L}(V, Q)$ exists such that

$$\lim_{t\to 0^+} \frac{1}{t}[q, B(u+tv)-B(u)] = [q, \nabla B(u)\cdot v],$$

$\forall v \in V$, where $[\cdot,\cdot]$ denotes duality pairing on $Q' \times Q$ and $q \in Q'$.

Under these circumstances, the method of Lagrange multipliers is often used as an alternate formulation of the optimization problem (2.1). We then introduce a Lagrangian $L : V \times Q' \to \mathbb{R}$ defined by

(3.2) $$L(v,q) = \Pi(v) - [q, B(v)]; \qquad v \in V, \qquad q \in Q'.$$

Whenever conditions (3.1) hold, a stationary value $(u,p) \in V \times Q'$ of L is characterized by

$$(3.3)\qquad \begin{cases} \langle D\Pi(u), v\rangle - [p, \nabla B(u)\cdot v] = 0, & \forall v \in V, \\ [q, B(u)] = 0, & \forall q \in Q'. \end{cases}$$

To construct an exterior penalty formulation of this problem, let us introduce a functional G which has the following properties:

$$(3.4)\qquad \begin{cases} \text{(i) } G: Q \to \mathbb{R},\ G \text{ is convex and } G\text{-differentiable on a convex set containing the range of the operator } B; \\ \text{(ii) } G(\phi) \geqq 0,\ G(\phi) = 0 \text{ iff } \phi = 0,\ (\phi \in Q). \end{cases}$$

Under conditions (3.4), we easily verify that the functional

$$(3.5)\qquad P: V \to \mathbb{R}; \qquad P = G \circ B$$

is a penalty functional. Indeed, it is positive semidefinite on K and it is weakly sequentially lower semicontinuous: if $v_m \rightharpoonup v$, then $B(v_m) \rightharpoonup B(v)$, and, by (3.4(i)), G is weakly lower semicontinuous. Hence

$$\liminf_{m\to\infty} P(v_m) = \liminf_{m\to\infty} G(B(v_m)) \geqq G(B(v)) = P(v).$$

Therefore, conditions (2.3) hold. Moreover, by applying the chain rule of differentiation, we see that $DP(u) \in V'$ exists and

$$(3.6)\qquad \langle DP(u), v\rangle = [G'(B(u)), \nabla B(u)\cdot v]$$

for all $u, v \in V$.

Let us now construct the penalized functional Π_ε of (2.4) with Π and P now satisfying (3.1) and (3.5). We easily verify that if u_ε is a minimizer of Π_ε, then u_ε is a solution of the nonlinear equation,

$$(3.7)\qquad \langle D\Pi(u_\varepsilon), v\rangle + \varepsilon^{-1}[G'(B(u_\varepsilon)), \nabla B(u_\varepsilon)\cdot v] = 0, \qquad \forall v \in V.$$

Comparing (3.7) with $(3.3)_1$, we see that it might be reasonable to take, as an approximation of the Lagrange multiplier p, the function

$$(3.8)\qquad p_\varepsilon = -\frac{1}{\varepsilon} G'(B(u_\varepsilon)), \qquad \varepsilon > 0.$$

Then (3.7) becomes

$$(3.9)\qquad \langle D\Pi(u_\varepsilon), v\rangle - [p_\varepsilon, \nabla B(u_\varepsilon)\cdot v] = 0, \qquad \forall v \in V.$$

We examine conditions under which the functional p_ε in (3.8) approximates p in the next section.

REMARK 3.1. The close relationship (indeed, the equivalence) of the penalty problem (3.7) with that obtained by "perturbed Lagrange methods" is noteworthy. Suppose that Q is a strictly convex Banach space with a strictly convex dual Q'. On such spaces it is possible to define a guage function Φ which is a strictly increasing function from $\mathbb{R}^+$ into $\mathbb{R}^+$ such that $\Phi(0) = 0$ and $\Phi(r) \to +\infty$ as $r \to +\infty$ (see Kikuchi [10]). The operator $j : Q \to Q'$ is a duality map if

$$\left\{ \begin{array}{l} \langle j(q), q \rangle = |j(q)|_* |q|, \\ |j(q)|_* = \Phi(|q|), \end{array} \right. \tag{3.10}$$

where $|\cdot|$ is the norm on Q and $|\cdot|_*$ the norm on Q'. The duality map is strictly monotone, continuous, and coercive and has a continuous inverse j^{-1}, the guage function of which is Φ^{-1}. Moreover, it can be shown that the functional

$$q \to \psi(|q|), \qquad \psi(r) = \int_0^r \Phi(s)ds \tag{3.11}$$

is Frechét differentiable and that, $\forall q, \bar{q} \in Q$,

$$\langle D\psi(|q|), \bar{q} \rangle = \langle j(q), \bar{q} \rangle. \tag{3.12}$$

In order to produce a Lagrangian coercive with respect to the Lagrange multipliers (in other words, to manufacture a "more concave" functional with respect to q) the perturbed Lagrangian L_ε is often introduced, where, for $\varepsilon > 0$, $v \in V$, and $q \in Q'$,

$$L_\varepsilon(v, q) = L(v, q) - \varepsilon\psi(|q|_*), \tag{3.13}$$

in which L is as given in (3.2), ψ is now defined using the guage function on Q', and $|\cdot|_*$ is the norm on Q'. A stationary value $(u_\varepsilon, p_\varepsilon)$ of L_ε is then characterized by the equations,

$$\left\{ \begin{array}{ll} \langle D\Pi(u_\varepsilon), v \rangle - [p_\varepsilon, \nabla B(u_\varepsilon)\cdot v] = 0, & \forall v \in V, \\ [q, \varepsilon j^{-1} p_\varepsilon + B(u_\varepsilon)] = 0, & \forall q \in Q', \end{array} \right. \tag{3.14}$$

wherein j is the duality map from Q onto Q'. Thus,

$$p_\varepsilon = -\frac{1}{\varepsilon} j(B(u_\varepsilon)),$$

and (3.9) becomes

$$\langle D\Pi(u_\varepsilon), v \rangle + \varepsilon^{-1}[j(B(u_\varepsilon)), \nabla B(u_\varepsilon)\cdot v] = 0, \qquad \forall v \in V.$$

We recognize (3.11) as an exterior penalty formulation of essentially the same structure as (3.7). □

4. The generalized LBB-condition

The question which arises naturally at this point is, under what conditions are the functionals p_ε in (3.8) approximations of the Lagrange multipliers p of (3.3)? We shall describe here a general and apparently fundamental condition which is sufficient to guarantee the existence of $\varepsilon_i > 0$, $i \in \mathbb{N}$, $\varepsilon_i \to 0$, such that $p_{\varepsilon_i} \rightharpoonup p$ weakly in Q'. The condition generalizes and abstracts that of Ladyszhenskaya [12] proposed in the study of the existence of hydrostatic pressures in Stokesian flows, the "inf sup" conditions of Babuska [1, 2] for linear elliptic boundary-value problems, the stability condition for saddle-point problems established by Brezzi [6], and the condition on admissible smooth hydrostatic pressure in finite elasticity established by Le Tallec and Oden [13]. We shall refer to it as the *generalized* LBB-*condition.* Let $\nabla B(u)^*$ denote the transpose of $\nabla B(u)$ at u:

$$(4.1)\qquad \begin{cases} \nabla B(u)^* \in \mathscr{L}(Q', V'): \qquad \forall q \in Q' \text{ and } \forall v \in V, \\ \langle \nabla B(u)^* \cdot q, v\rangle = [q, \nabla B(u) \cdot v], \end{cases}$$

and consider the norm on the quotient space,

$$(4.2)\qquad \inf_{q_0 \in \ker \nabla B(u)^*} \|q + q_0\|_{Q'} = \|q\|_{Q'/\ker \nabla B(u)^*}.$$

A generalized LBB-condition for problem (3.3) or (3.10) (or for the spaces V and Q and the constraint operator $B: V \to Q$) holds at $u \in V$ if

$$(4.3)\qquad \begin{cases} \text{There exists a constant } \alpha > 0 \text{ such that} \\ \alpha \|q\|_{Q'/\ker \nabla B(u)^*} \leqq \sup_{v \in V} \dfrac{|[q, \nabla B(u) \cdot v]|}{\|v\|_V}, \qquad \forall q \in Q'. \end{cases}$$

We then have:

THEOREM 2. *Let the following conditions hold:*

(i) *conditions* (2.2), (3.1), *and* (3.4);

(ii) $D\Pi: V \to V'$ *is a bounded operator;*

(iii) u_ε, $\varepsilon > 0$, *is a minimizer of* $\Pi_\varepsilon = \Pi + \varepsilon^{-1}P$, *where* P *is given by* (3.5) (*i.e.,* u_ε *is a solution of* (3.7), *and* $u_\varepsilon \rightharpoonup u$ *weakly in* V, *where* u *is a minimizer of* Π *in* K, *so that* u *is a solution of* (2.1) *and let* $u_\varepsilon \to u \in \operatorname{int} K$).

(iv) $\ker \nabla B^*(u) \supseteq \ker \nabla B^*(u_\varepsilon)$, $\forall \varepsilon > 0$.

In addition, let condition (4.3) *hold at each* u_ε *and let* p_ε *be defined by* (3.8). *Then there exists a subsequence* $\{p_{\varepsilon'}\}$ *of* $\{p_\varepsilon\}$ *which converges weakly in* Q' *to a function* p, *where* (u, p) *is a solution of* (3.3).

PROOF. Let u_ε denote a subsequence of minimizers of Π_ε which converges weakly in V to a minimizer $u \in K$ of Π. Since u_ε satisfies (3.9),

$$|[p_\varepsilon, \nabla B(u_\varepsilon)\cdot v]| = |\langle D\Pi(u_\varepsilon), v\rangle| \leqq C_0 \| v \|_V$$

for all $v \in V$, where C_0 is independent of ε. Thus,

$$\begin{aligned} \alpha' \| p_\varepsilon \|_{Q'/\ker \nabla B(u)^*} &\leqq \alpha \| p_\varepsilon \|_{Q'/\ker \nabla B(u_\varepsilon)^*} \\ &\leqq \sup_{v \in V} \frac{|\langle D\Pi(u_\varepsilon), v\rangle|}{\| v \|_V} \leqq C_0. \end{aligned}$$

Hence, $\| p_\varepsilon \|_{Q'/\ker \nabla B(u)^*}$ is uniformly bounded in ε. The assertion then follows from the reflexivity of Q'. □

5. Applications

We now describe several applications of the theory outlined in the previous sections to boundary-value problems in elasticity.

5.1. *Finite deformations of incompressible elastic bodies*

As a first example, we consider the equilibrium of an incompressible, isotropic, hyperelastic body $\Omega \subset \mathbb{R}^n$ subjected to dead loading. The total potential energy functional is of the form

$$\Pi(\boldsymbol{v}) = \int_\Omega \sigma(\nabla \boldsymbol{v})dx - f(\boldsymbol{v}), \tag{5.1}$$

where $\boldsymbol{v}$ is an element of a suitable class of admissible motions, σ is the stored energy function for the material, and

$$f(\boldsymbol{v}) = \int_\Omega \boldsymbol{f}\cdot \boldsymbol{v}dx + \int_{\Gamma_\sigma} \boldsymbol{t}\cdot \boldsymbol{v}ds, \tag{5.2}$$

wherein $\boldsymbol{f}$ is the body force per unit volume and $\boldsymbol{t}$ is the surface traction prescribed on $\Gamma_\sigma \subset \partial\Omega$. We seek the motions $\boldsymbol{u}$ in a suitable class V which minimize Π subject to the incompressibility constraint

$$\det \nabla \boldsymbol{u} = 1 \qquad \text{in } \Omega, \tag{5.3}$$

where $\nabla \boldsymbol{u}$ is the material gradient of $\boldsymbol{u}$.

For illustration purposes, we consider a form of the stored energy function suggested by Isihara, Hashitsume and Tatibana [9] for certain rubber-like materials:

$$2\sigma = E_1(I-3) + E_2(II-3) + E_3(I-3)^2. \tag{5.4}$$

Here E_1, E_2, and E_3 are material constants and I and II are principal invariants of the deformation tensor, $\boldsymbol{C} = \nabla\boldsymbol{u}^{\mathrm{T}}\cdot\nabla\boldsymbol{u}$. We confine our attention to plane deformations, and in this case we have

$$\tag{5.5} \begin{cases} I = \operatorname{trace} \nabla\boldsymbol{u}^{\mathrm{T}}\nabla\boldsymbol{u}, \\ II = \operatorname{trace}(\operatorname{adj}\nabla\boldsymbol{u}^{\mathrm{T}}\operatorname{adj}\nabla\boldsymbol{u}) = I + J^2 - 1, \\ J = \det\nabla\boldsymbol{u}, \end{cases}$$

where $\operatorname{adj}\nabla\boldsymbol{u}$ denotes the transpose of the cofactors of $\nabla\boldsymbol{u}$. Then if σ is given by (5.2), the energy function Π assumes the form

$$\tag{5.6} \begin{aligned} &\Pi(\boldsymbol{v}) \\ &= \frac{1}{2}\int_\Omega [(E_1+E_2)(I(\boldsymbol{v})-3)+E_3(I(\boldsymbol{v})-3)^2+E_2(J^2(\boldsymbol{v})-1)]dx - f(\boldsymbol{v}). \end{aligned}$$

To apply the theory outlined in Sections 2–4 to this case of problems, we proceed as follows:

1. Let Ω be a bounded, C^2-domain in $\mathbb{R}^2$ and take as the space of admissible motions

$$\tag{5.7} \begin{cases} V = (W^{1,4}(\Omega))^2/\boldsymbol{P}_0, \\ \|\boldsymbol{v}\|_V^4 = \|\boldsymbol{v}\|_{1,4}^4 = \displaystyle\int_\Omega \sum_{i,\alpha=1}^{2} |v_{i,\alpha}|^4 dx, \\ \boldsymbol{P}_0 = \text{space of rigid translations of } \Omega. \end{cases}$$

We then consider the traction problem in finite plane strain with $\Gamma_\sigma = \partial\Omega$ and $\boldsymbol{f}$ and $\boldsymbol{t}$ such that

$$\tag{5.8} \int_\Omega \boldsymbol{f}\cdot\boldsymbol{r}dx + \int_{\partial\Omega}\boldsymbol{t}\cdot\boldsymbol{r}ds = 0, \qquad \forall\boldsymbol{r}\in\boldsymbol{P}_0,$$

under the assumption that $\boldsymbol{f}\in L^2(\Omega)$, $\boldsymbol{t}\in L^2(\partial\Omega)$. The space V equipped with the above norm, is a reflexive Banach space.

2. Π is differentiable on V; indeed,

$$\tag{5.9} \begin{aligned} \langle D\Pi(\boldsymbol{u}),\boldsymbol{v}\rangle = \int_\Omega &[(E_1-6E_3+E_2)\mathbf{1}+2E_2I(\boldsymbol{u})\nabla\boldsymbol{u} \\ &+E_2J(\boldsymbol{u})\operatorname{adj}\nabla\boldsymbol{u}^{\mathrm{T}}]:\nabla\boldsymbol{v}dx - f(\boldsymbol{v}), \end{aligned}$$

with $f\in V'$ and $\langle\cdot,\cdot\rangle$ denoting duality pairing on $[(W^{1,4}(\Omega))^2]'\times((W^{1,4}(\Omega))^2$. We easily verify that $D\Pi$ is bounded; indeed,

$$\tag{5.10} \|D\Pi(\boldsymbol{u})\|_{1,4}^* \leqq |E_2+2E_3|\,\|\boldsymbol{u}\|_{1,4}^3 + |E_1+E_2-6E_3|\operatorname{meas}(\Omega)^{1/2}\|\boldsymbol{u}\|_{1,4}.$$

3. The operator $u \to J(u) = \det \nabla u$ is weakly sequentially continuous from $(W^{1,4}(\Omega))^2$ into $L^2(\Omega)$ (see, e.g., Reshetnyak [25], Ball [4], or Rostamian [26]). Moreover, the total potential energy Π can be expressed as the sum,

$$\Pi = \Pi_1 + \Pi_2 : \Pi_1(v) = \frac{1}{2}\int_\Omega [(E_1 + E_2)(I(v) - 3) + E_3(I(v) - 3)^2]dx - f(v),$$

$$\Pi_2(v) = \frac{1}{2}\int_\Omega E_2(J(v) - 1)^2 dx.$$

We observe that whenever

$$E_1 + E_2 \geqq 4E_3, \qquad E_2 \geqq 0, \qquad E_3 \geqq 0, \tag{5.11}$$

Π_1 is differentiable and $D\Pi_1$ is monotone; hence Π_1 is convex and, therefore, weakly lower semicontinuous. Likewise, Π_2 is a convex differentiable function of J and J is weakly sequentially continuous. Hence, Π_2 is weakly lower semicontinuous. It follows that Π is weakly lower semicontinuous on the space V of (5.7).

4. A straightforward calculation reveals that if $E_2 \geqq 0$ and $E_3 > 0$, then, $\forall v \in V$,

$$\begin{aligned}\Pi(v) \geqq \tfrac{1}{2}E_3\| v \|_{1,4}^4 + \tfrac{1}{2}(E_1 + E_2 - 6E_3)\| v \|_{1,2}^2 \\ - C_1(f_1, t)\| y \|_{1,4} - C_2(\Omega, f, t),\end{aligned} \tag{5.12}$$

where C_1 and C_2 are positive constants. From this we conclude that Π is proper and coercive on V.

5. The constraint set K, given by

$$K = \{v \in V \,|\, \det \nabla v = 1 \text{ a.e. in } \Omega\} \tag{5.13}$$

is weakly sequentially closed in V. In this case,

$$\begin{aligned} &B(u) = \det \nabla u - 1 \in L^2(\Omega); \\ &\nabla B(u) = \operatorname{adj} \nabla u^{\mathrm{T}}, \qquad [q, \nabla B(u) \cdot v] = \int_\Omega q \operatorname{adj} \nabla u^{\mathrm{T}} : \nabla v dx. \end{aligned} \tag{5.14}$$

6. It follows from steps 1–5 above and the theory given in Section 2 that, if conditions (5.11) hold, then for every $\varepsilon > 0$, there exists a minimizer u_ε of the penalized functional

$$\Pi_\varepsilon(v) = \Pi(v) + \frac{1}{2\varepsilon}\int_\Omega (J(v) - 1)^2 dx. \tag{5.15}$$

Moreover, $\{\varepsilon_i\}$, $i \in \mathbb{N}$, exist such that $u_{\varepsilon_i} \to u$ weakly in V, where u is a minimizer of Π. Moreover, as an approximation of the hydrostatic pressure, we take

$$p_\varepsilon = -\frac{1}{\varepsilon}(\det \nabla \boldsymbol{u}_\varepsilon - 1). \tag{5.16}$$

7. There exists a subsequence of $\{p_\varepsilon\}$ converging weakly to a hydrostatic pressure p satisfying $\langle D\Pi(\boldsymbol{u}), \boldsymbol{v}\rangle - \int_\Omega p \operatorname{adj} \nabla \boldsymbol{u}^{\mathrm{T}} : \nabla \boldsymbol{v} dx = f(\boldsymbol{v})$, $\forall \boldsymbol{v} \in V$, if the following LBB-condition is satisfied:

$$\begin{cases} \exists \alpha > 0 \text{ such that } \forall q \in L^2(\Omega), \\ \alpha \|q\|_{0,2} \leqq \sup_{\boldsymbol{v} \in V} \dfrac{\int_\Omega q \operatorname{adj} \nabla \boldsymbol{u}^{\mathrm{T}} : \nabla \boldsymbol{v} dx}{\|\boldsymbol{v}\|_{1,4}}. \end{cases} \tag{5.17}$$

The generalized LBB-condition (5.17) is recognized as the condition described by Le Tallec and Oden [13] in their studies of the existence of hydrostatic pressures in finite deformations of incompressible elastic bodies.

5.2. *Large deflections and buckling of thin elastic plates with unilateral supports*

Using the Von Karman theory of thin plates, we consider the problem of buckling and large transverse deflections of a clamped, elastic plate the lateral displacement of which is constrained by the presence of a rigid, frictionless plane parallel to and a distance b from the undeformed middle plane of the plate. In this case

$$\Pi(\boldsymbol{u}, w) = \tfrac{1}{2}\{A(w, w) + B(\boldsymbol{u}, \boldsymbol{u}) + a(\boldsymbol{u}; w, w)\} - F(w), \tag{5.18}$$

where $\boldsymbol{u} = (u_1, u_2)$ are the "membrane" components of displacement in the plane of the (undeformed) plate, w is the transverse deflection,

$$\begin{cases} A(w, z) = D\int_\Omega (\Delta w \Delta z - (1-\nu)[w, z])dx, \\ [w, z] = w_{,11}z_{,22} + w_{,22}z_{,11} - 2w_{,12}z_{,21}, \\ B(\boldsymbol{u}, \boldsymbol{v}) = t\int_\Omega E_{\alpha\beta\lambda\mu} u_{\lambda,\mu} v_{\alpha,\beta} dx, \qquad 1 \leqq \alpha, \beta, \lambda, \mu \leqq 2, \\ a(\boldsymbol{u}; w, z) = t\int_\Omega \sigma_{\alpha\beta}(\boldsymbol{u}, w) w_{,\alpha} z_{,\beta} dx, \\ \sigma_{\alpha\beta}(\boldsymbol{u}, w) = E_{\alpha\beta\lambda\mu}\left(u_{\lambda,\mu} + \dfrac{1}{2} w_{,\lambda} w_{,\mu}\right), \\ F(w) = \int_\Omega f w dx. \end{cases} \tag{5.19}$$

Here $D = Et^3/2(1-\nu^2)$ is the usual flexural rigidity, the elasticities $E_{\alpha\beta\lambda\mu}$ are bounded L^∞-functions such that $E_{\alpha\beta\lambda\mu}\varepsilon_{\lambda\mu}\varepsilon_{\alpha\beta} \geqq \alpha_0\varepsilon_{\alpha\beta}\varepsilon_{\alpha\beta}$ a.e. in Ω $\forall\varepsilon$ such that $\varepsilon_{\alpha\beta} = \varepsilon_{\beta\alpha}$, $(\alpha_0 > 0)$, and t is the plate thickness.

It can be shown (see, e.g., Ohtake, Oden and Kikuchi [23]) that Π is coercive and weakly lower semicontinuous on the space

$$V = (W_0^{1,2}(\Omega))^2 \times W_0^{2,2}(\Omega), \tag{5.20}$$

$$K = \{z \in W_0^{2,2}(\Omega) \mid z \geqq -b \text{ in } \Omega\}. \tag{5.21}$$

In this case, we use as a penalty functional,

$$P(w) = \frac{1}{2}\int_\Omega (w+b)_-^2\,dx, \tag{5.22}$$

wherein $(\ )_-$ denotes the function $(\phi)_- = \min(\phi, 0)$.

It is possible to show that all of the conditions for the existence of minimizers $(\boldsymbol{u}_\varepsilon, w_\varepsilon) \in V$ of $\Pi_\varepsilon = \Pi + \varepsilon^{-1}P$ listed earlier hold for the functions in (5.18) and (5.22) and the space V, Π and that $(\boldsymbol{u}_\varepsilon, w_\varepsilon) \rightharpoonup (\boldsymbol{u}, w)$ strongly in V where $(\boldsymbol{u}, w)$ is a minimizer of Π on the set K of (5.21).

5.3. *Incompressible linearly elastic materials*

The case in which

$$\left\{\begin{aligned} &\Pi_\varepsilon(\boldsymbol{v}) = \mu\int_\Omega \varepsilon_{ij}^D(\boldsymbol{v})\varepsilon_{ij}^D(\boldsymbol{v})\,dx + \frac{\varepsilon^{-1}}{2}\int_\Omega (\operatorname{div}\boldsymbol{v})^2\,dx - f(\boldsymbol{v}),\\ &V = \{\boldsymbol{v} = (v_1, v_2, v_3) \in (H^1(\Omega))^3 \mid \boldsymbol{v} = \boldsymbol{0} \text{ a.e. on } \Gamma_D \subset \partial\Omega\},\\ &\|v\|_1^2 = \int_\Omega v_{i,j}v_{i,j}\,dx, \end{aligned}\right. \tag{5.23}$$

wherein μ is the shear modulus, ε_{ij}^D the deviatoric strains due to the displacement $\boldsymbol{v}$, and $f \in V'$ is of the form (5.2), corresponds to the total potential energy of a compressible isotropic body with bulk modulus ε^{-1}, $\varepsilon > 0$. As $\varepsilon \to 0$, we obtain, as the limiting case, an incompressible elastic body. The mean stress,

$$p_\varepsilon = -\varepsilon^{-1}\operatorname{div}\boldsymbol{u}_\varepsilon,$$

where $\boldsymbol{u}_\varepsilon$ is a minimizer of Π_ε, is an obvious approximation of the hydrostatic pressure $p \in L^2(\Omega)$. In this case, it can be shown that (Oden, Kikuchi and Song [21])

$$\|\boldsymbol{u} - \boldsymbol{u}_\varepsilon\|_1 + \|p - p_\varepsilon\|_0 \leqq C\varepsilon,$$

and that the following LBB-condition holds:

$$(5.24)\qquad \begin{cases} \exists \alpha > 0 \text{ such that} \\ \|q\|_{0/\ker B^*} \leqq \sup\limits_{v \in V} \dfrac{(q, \operatorname{div} v)}{\|v\|_1} \qquad \forall q \in L^2(\Omega), \end{cases}$$

where $\ker B^* = \{0\}$ if meas $(\partial\Omega - \Gamma_D) > 0$ and $\ker B^* = \{\text{constants}\}$ if $\partial\Omega = \Gamma_D$.

All of the conditions for the existence of solutions (minimizers) and pressures (Lagrange multipliers) listed earlier are, thus, satisfied in this example.

5.4. *Unilateral contact problems in linear elasticity*

Let us again consider a linearly elastic body characterized by (5.23) (with ε^{-1} replaced by κ, the bulk modulus of the material and with $\Pi = \Pi_\kappa$). Now we suppose that the body may come in contact with a rigid frictionless foundation along a material contact surface Γ_C. The constraint is then

$$(5.25)\qquad K = \{v \in V \mid \gamma_n(v) \leqq s \text{ in } Q\},$$

where γ_n is the normal trace operator mapping V onto Q, s is the initial "gap" between the body and the foundation and is given as part of the data in Q, and $Q = H^{1/2}(\Gamma_C)$ (see Kikuchi and Oden [11] for further details). In this case, we can take as the penalty functional

$$(5.26)\qquad P(v) = \frac{1}{2} |(\gamma_n(v) - s)_+|_Q^2,$$

where $(\phi)_+ = \max(\phi, 0)$. All of the conditions sufficient to guarantee the existence of a convergent sequence of minimizers of Π_ε hold in this case. In addition, the LBB-condition,

$$(5.27)\qquad \begin{cases} \exists \beta > 0 \text{ such that } \forall q \in Q' \\ \beta \|q\|_{Q'} \leqq \sup\limits_{v \in V} \dfrac{\langle q, \gamma_n(v)\rangle}{\|v\|_1} \end{cases}$$

holds. Thus, subsequences of the penalty approximation, $p_\varepsilon = -\varepsilon^{-1} j(\gamma_n(u_\varepsilon) - s)_+$ (with j the Riesz map from Q onto Q') converge to the actual contact pressure p developed on Γ_C.

6. Approximations

The study of discrete approximations of penalty formulations of constrained variational problems is a somewhat more difficult undertaking than the analyses given in previous sections because of the necessity of establishing that the key conditions for the success of these methods are preserved when one passes to finite-dimensional subspaces of V and Q. As will be established below, a crucial problem in devising acceptable approximations is whether or not an LBB-condition holds on subspaces of V and Q.

To fix ideas, let V be a subset of a Sobolev space $(W^{m,p}(\Omega))^n$ and suppose that we construct, using regular families of conforming finite elements, family $\{V_h\}_{h>0}$ of finite-dimensional subspaces of V which are endowed with interpolation properties of the type

$$\|\boldsymbol{v} - P_h\boldsymbol{v}\|_{m,p,\Omega} \leqq Ch^{k+1-m}\|\boldsymbol{v}\|_{k+1,p,\Omega}, \tag{6.1}$$

where P_h is a projection from $(W^{k+1,p}(\Omega))^n$ onto $(W^{m,p}(\Omega))^n$ satisfying $P_h\phi = \phi \ \forall\phi \in (\mathscr{P}_k(\Omega))^n$, $\mathscr{P}_k(\Omega)$ being the space of polynomials of degree $\leqq k$ defined on Ω, and h is the maximum diameter of an element in the finite element mesh approximating Ω (see Ciarlet [7] or Oden and Reddy [20]). Then a finite-element approximation of the general penalty problem (3.7) is obtained by posing this problem on V_h rather than V. There is one major problem: this direct approximation of the penalized variational problem is, for small ε, "overconstrained", e.g. the approximate solution $\boldsymbol{u}_\varepsilon^h$ may approach zero for fixed h as $\varepsilon \to 0$. Thus, the most obvious method of approximation will, in general, fail!

There are two ways (at least) to overcome this difficulty: (1) choose the penalty parameter ε as a function of the mesh parameter h, or (2) use an *inexact* numerical integration scheme to evaluate the penalty functional P (here we have in mind any of the penalty functionals in the applications listed in the previous section). We generally reject option (1) on the grounds that for reasonable meshes ε must be taken so large that the constraint is not adequately satisfied. Thus, we choose the second alternative which brings us to the so-called method of *selective* reduced integration (see, e.g., [5, 8, 14, 15, 21, 27, 28]).

As an example of such methods, consider the problem of finite deformation of an incompressible elastic body discussed in Section 5.1. We then introduce an approximation $I(\ \ )$ of the penalty term such that

$$\begin{cases} \langle DP(\boldsymbol{u}^h), \boldsymbol{v}^h\rangle = I[J(\boldsymbol{u}^h) - 1; \boldsymbol{u}^h, \boldsymbol{v}^h] \\ \qquad\qquad \approx \int_\Omega (J(\boldsymbol{u}^h) - 1)\operatorname{adj}\nabla\boldsymbol{u}^{h\mathrm{T}} : \nabla\boldsymbol{v}^h dx; \\ \lim_{h\to 0} I(J(\boldsymbol{u}^h) - 1; \boldsymbol{u}^h, \boldsymbol{v}^h) = \int_\Omega (J(\boldsymbol{u}) - 1)\operatorname{adj}\nabla\boldsymbol{u}^{\mathrm{T}} : \nabla\boldsymbol{v} dx \end{cases} \tag{6.2}$$

for $\boldsymbol{u}^h$, $\boldsymbol{v}^h \in V_h$. The reduced-integration-penalty (RIP) approximation of the equilibrium problem for (5.1) and (5.3) then consists of seeking $\boldsymbol{u}_\varepsilon^h \in V_h$ such that

$$(6.3) \qquad \langle D\Pi(\boldsymbol{u}_\varepsilon^h), \boldsymbol{v}^h \rangle + \varepsilon^{-1} I(J(\boldsymbol{u}_\varepsilon^h) - 1; \boldsymbol{u}_\varepsilon^h, \boldsymbol{v}^h) = 0, \qquad \forall \boldsymbol{v}^h \in V_h.$$

While it is not immediately obvious upon inspecting the form of problem (6.3), there is intrinsic to each such RIP formulation an approximation Q_h of the space Q characterizing the constraint. In particular, for the discrete problem (6.3), we can choose $I(\quad)$ and Q_h such that

$$(6.4) \quad \begin{cases} \text{(i) For every } \forall q^h \in Q'_h,\ \boldsymbol{u}^h,\ \boldsymbol{v}^h \in V_h, \\ \qquad [q^h, \operatorname{adj} \nabla \boldsymbol{u}^{hT} : \nabla \boldsymbol{v}^h] = I(q^h; \boldsymbol{u}^h; \boldsymbol{v}^h); \\ \text{(ii) a discrete LBB-condition holds; i.e., there exists} \\ \qquad \alpha_h > 0 \text{ such that } \forall q^h \in Q'_h, \\ \qquad \alpha_h \|q^h\|_{Q'} \leqq \sup\limits_{\boldsymbol{v}^h \in V} \dfrac{I(q^h; \boldsymbol{u}^h, \boldsymbol{v}^h)}{\|\boldsymbol{v}^h\|_V}. \end{cases}$$

Under these conditions, one can generally solve (6.3) for $\boldsymbol{u}_\varepsilon^h$, compute an approximate p_ε^h (e.g. $I[p_\varepsilon^h; \boldsymbol{u}_\varepsilon^h, \boldsymbol{v}^h] = I[-\varepsilon^{-1}(J(\boldsymbol{u}_\varepsilon^h) - 1); \boldsymbol{u}_\varepsilon^h, \boldsymbol{v}^h]$) and obtain an approximation $(\boldsymbol{u}_\varepsilon^h, p_\varepsilon^h)$ which converges to an exact solution of the problem as $\varepsilon, h \to 0$.

For a more concrete example, consider the case of an incompressible linearly elastic material for which Π and V are as described in Section 5.3. In this case, we take $I(\cdot)$ to be the Gaussian quadrature rule,

$$(6.5) \qquad I(f) = \sum_{e=1}^{E} I_e(f); \qquad I_e(f) = \sum_{j=1}^{G} W_j^e f(\boldsymbol{\xi}_j^e),$$

where $f \in C^0(\bar{\Omega})$, E is the number of elements in the finite element mesh, W_j^e are the quadrature weights for finite element $\Omega_e \subset \Omega$, and $\boldsymbol{\xi}_j^e$ are the quadrature points in element Ω_e. In this case, we choose Q_h to be a space satisfying the discrete LBB-condition [11, 21, 27]:

$$(6.6) \quad \begin{cases} \text{There exists } \alpha_h > 0 \text{ such that, for all } q^h \in Q_h, \\ \qquad \|q^h\|_{0/\ker B_h^*} \leqq \sup\limits_{\boldsymbol{v}^h \in V_h} \dfrac{I(q^h, \operatorname{div} \boldsymbol{v}^h)}{\|\boldsymbol{v}^h\|_1} \\ \text{(with } \ker B_h^* = \{q^h \in Q_h \mid I(q^h \operatorname{div} \boldsymbol{v}^h) = 0\ \forall \boldsymbol{v}^h \in V_h\}). \end{cases}$$

The approximate hydrostatic pressure $p_\varepsilon^h \in Q_h$ is uniquely defined by

$$(6.7) \qquad p_\varepsilon^h(\boldsymbol{\xi}_j^e) = -\varepsilon^{-1} \operatorname{div} \boldsymbol{u}_\varepsilon^h(\boldsymbol{\xi}_j^e), \qquad 1 \leqq e \leqq E, \qquad 1 \leqq j \leqq G.$$

Condition (6.6) has been studied for certain types of elements by Oden, Kikuchi and Song (see [21, 27]).

Properties and conditions similar to (6.4), (6.6) and (6.7) hold for RIP methods for other types of boundary-value problems with constraints. We comment on some special cases in the next section.

7. Discrete LBB-condition and numerical example

As an example of a numerical scheme based on the RIP-ideas discussed in the previous section, we consider here plane-strain equilibrium problems in incompressible elasticity.

7.1. *Discrete* LBB-*condition*

Consider first the case of infinitesimal deformations of linearly elastic material for which the total potential energy is given by (5.23) with a fixed boundary $\partial\Omega$. We shall assume that Ω is a rectangular domain in $\mathbb{R}^2$ on which there is constructed a uniform mesh of 9-node, C^0-biquadratic elements:

$$(7.1)\qquad V_h = \{\boldsymbol{v}^h = (v_1^h, v_2^h) \in (C^0(\Omega))^2 \,|\, v_i^h|_{\Omega_e} \in Q_2(\bar{\Omega}_e),\ 1 \leqq e \leqq E\} \subset (H_0^1(\Omega))^2.$$

Here Ω_e is a finite element in the mesh ($\bar{\Omega} = \bigcup_{e=1}^{E} \bar{\Omega}_e$, $\Omega_e \cap \Omega_f = \emptyset$, $e \neq f$) and $Q_2(\bar{\Omega}_e)$ is the space of tensor products of quadratics in x_1 and x_2 on the rectangle $\bar{\Omega}_e$. In view of (6.1), the spaces (7.1) have the interpolation property: $\forall \boldsymbol{v} \in V$,

$$(7.2)\qquad \|\boldsymbol{v} - \boldsymbol{\phi}^h\|_1 \leqq Ch^2 \|\boldsymbol{v}\|_3,$$

where $\|\cdot\|_m = \|\cdot\|_{m,2,\Omega}$, $m = 1,3$.

The RIP-approximation of the minimization problem characterized by (5.23) consists of seeking $\boldsymbol{u}_\varepsilon^h \in V_h$ such that

$$(7.3)\qquad \mu \int_\Omega \varepsilon_{ij}^D(\boldsymbol{u}_\varepsilon^h)\varepsilon_{ij}(\boldsymbol{v}^h)dx + \varepsilon^{-1} I(\operatorname{div} \boldsymbol{u}_\varepsilon^h \operatorname{div} \boldsymbol{v}^h) = f(\boldsymbol{v}^h), \qquad \forall \boldsymbol{v}^h \in V_h,$$

where $I(\cdot)$ is an approximation of the penalty functional $2^{-1}\int_\Omega (\operatorname{div} \boldsymbol{v})^2 dx$ by an appropriate numerical integration formula. Equivalently $I(\operatorname{div} \boldsymbol{u}_\varepsilon^h \operatorname{div} \boldsymbol{v}^h)$ may correspond to the approximate penalty functional produced by a perturbed Lagrangian scheme (mixed finite element method) in which discontinuous piecewise polynomial approximations of the pressures are employed (recall (3.11)). In view of this latter property, we shall take

$$(7.4)\qquad \begin{cases} Q_h = \{q^h \in L^2(\Omega) \,|\, q^h|_{\Omega_e} \in P_1(\Omega_e),\ 1 \leqq e \leqq E\}, \\ (q^h, \operatorname{div} \boldsymbol{v}^h) = I(q^h \operatorname{div} \boldsymbol{v}^h), \qquad \boldsymbol{v}^h \in V_h, \qquad q^h \in Q_h, \end{cases}$$

where $P_1(\Omega_e)$ is the space of linear functions in x_1 and x_2 defined on element Ω_e.

The scheme (7.3), (7.4) is stable and convergent if the discrete LBB-condition (6.5) holds (with $\ker B_h^* = \{\text{constant}\}$ in this case). The fact that this condition does indeed hold for these choices of spaces has been proved by the author and the method performs very well in numerical tests.

In the case of finite deformations of incompressible elastic bodies, the spaces V_h and Q_h of (7.1) and (7.4) can also be used as a basis for the construction of perturbed mixed finite element methods which are equivalent to certain RIP methods. Again the LBB-condition (6.4ii) is a crucial requirement. However, if the approximation motion $\boldsymbol{u}_\varepsilon^h$ is a diffeomorphism from the reference configuration Ω of the body onto the current configuration, then it is sufficient to show that the linear LBB-condition (6.6) holds for the spaces V_h and Q_h. Since this condition does hold for the spaces in (7.1) and (7.4), these appear to be excellent candidates for use in solving problems in finite elasticity. We next describe results of a numerical experiment designed to test this assertion.

7.2. *Finite deformation of an elastic cylinder*

We consider as an example problem the analysis of finite deformations of an infinite pressurized hollow cylinder composed of a Mooney–Rivlin material with material constants $E_1 = 80$ psi and $E_2 = 20$ psi. The outside radius of the undeformed cylinder is $r_0 = 18.625$ in. and the initial inside radius is $r_i = 7.000$ in. An analysis of this particular problem can be found in the book by Oden [16].

We consider two finite element models of a strip of the cylinder, each consisting of five equal elements. In the first model, we use a mixed method which employs 9-node C^0-biquadratic elements for the displacements and conforming C^0-bilinear approximations for the pressures. The second model employs a non-conforming perturbed Lagrangian element (RIP) described earlier in which displacements are again approximated by 9-node biquadratics but discontinuous linear pressures are used.

The calculated hydrostatic pressure distribution corresponding to an applied internal pressure of 150 psi is shown in Figure 1. Note that the nonconforming element performs exceptionally well and yields results of equal accuracy to the more complicated conforming model. Even more remarkable is that the computing time required for the conforming model was 32.3 sec on a CDC Cyber 175 whereas the nonconforming model required 28.0 sec, a difference of 15 percent.

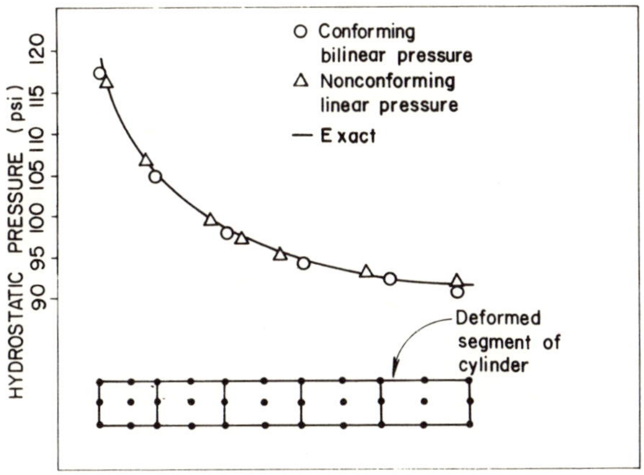

Figure 1. Hydrostatic pressure distributions in a finitely deformed elastic cylinder of Mooney–Rivlin material. The displacement field is modelled by five 9-node biquadratic elements and pressures are approximated using equal conforming bilinear and nonconforming linear basis functions.

Acknowledgement

The results presented in this work were developed during the course of research supported by the National Science Foundation under Grant ENG 75–08746 and the Air Force Office of Scientific Research under Contract F-49620-78-C-0083.

I wish to register a special note of thanks to Professor N. Kikuchi who has worked with me on various penalty methods over the last three years and whose numerous discussions with me on this subject were extremely helpful. I also wish to thank Mr Amin Aly for performing the numerical experiments given in Section 7.

References

1. I. Babuska, *Error bounds for finite element method*, Numeriche Mathematik **16** (1971), 322–333.

2. I. Babuska, *The finite element method with Lagrange multipliers*, Numeriche Mathematik **20** (1973).

3. I. Babuska, *The finite element method with penalty*, Mathematics of Computation **27** (1973), 221–228.

4. J. M. Ball, *Convexity conditions and existence theorems in nonlinear elasticity*, Archives for Rational Mechanics and Analysis **63** (1977), 337–403.

5. M. Bercovier, *Perturbation of mixed variational problems — applications to mixed finite element methods*, R.A.I.R.O. **12** (1978).

6. F. Brezzi, *On the existence, uniqueness, and approximation of saddle-point problems arising from Lagrange multipliers*, R.A.I.R.O. **8** (1974).

7. P. G. Ciarlet, *The Finite Element Method for Elliptic Problems*, North-Holland, Amsterdam, 1978.

8. T. J. R. Hughes, *Equivalence of finite elements for near incompressibility*, Journal of Applied Mechanics **44** (1977), 181–183.

9. A. Isihara, N. Hashitsume and M. Tatibana, *Statistical theory of rubber-like elasticity. IV. Two-dimensional stretching*, Journal of Chem. Physics **19** (1951), 1508–1512.

10. N. Kikuchi, *Convergence of a penalty method for variational inequalities*, TICOM Report 79–16, The University of Texas at Austin, October, 1979.

11. N. Kikuchi and J. T. Oden, *Contact problems in elasticity*, SIAM Studies in Applied Mathematics, Philadelphia, Pa., to appear.

12. O. A. Ladyszhenskaya, *The Mathematical Theory of Viscous Incompressible Flow*, Gordon and Breach, N.Y., 1969.

13. P. Le Tallec and J. T. Oden, *Existence and characterization of hydrostatic pressures in finite deformations of incompressible bodies*, Journal of Elasticity, to appear.

14. D. S. Malkus, *Finite element analysis of incompressible solids*, PhD Dissertation, Department of Mathematics, Boston University, 1975.

15. D. S. Malkus and T. J. R. Hughes, *Mixed finite element methods–reduced and selective integration technique – a unification of concepts*, Computer Methods in Applied Mechanics and Engineering **15** (1978), 63–81.

16. J. T. Oden, *Finite Elements of Nonlinear Continua*, McGraw-Hill, New York, 1972.

17. J. T. Oden, *A theory of penalty methods for finite element approximations of highly nonlinear problems in continuum mechanics*, Computers and Structures **8** (1978), 445–449.

18. J. T. Oden, *Recent developments in the theory of finite element approximations of boundary-value problems in nonlinear elasticity*, Computer Methods in Applied Mechanics and Engineering **17/18** (1979), 183–202.

19. J. T. Oden, *RIP Methods for Stokesian Flows*, in *Finite Elements in Fluids*, Vol. IV (R. H. Gallagher et al., eds.), John Wiley and Sons, Ltd., London, to appear.

20. J. T. Oden and J. N. Reddy, *An Introduction to the Mathematical Theory of Finite Elements*, Wiley Interscience, New York, 1976.

21. J. T. Oden, N. Kikuchi and Y. J. Song, *Reduced integration and exterior penalty methods for finite element approximations of contact problems in incompressible elasticity*, TICOM Report 80–2, The University of Texas at Austin, 1980.

22. J. T. Oden and N. Kikuchi, *Finite element methods for constrained problems in elasticity*, International Journal for Numerical Methods in Engineering, to appear.

23. K. Ohtake, J. T. Oden and N. Kikuchi, *Analysis of certain unilateral problems in Von Karman plate theory by a penalty method. Part I, A variational principle with penalty*, Computer Methods in Applied Mechanics and Engineering, to appear.

24. K. Ohtake, J. T. Oden and N. Kikuchi, *Analysis of certain unilateral problems in Von Karman plate theory by a penalty method. Part II, Approximation and numerical analysis*, Computer Methods in Applied Mechanics and Engineering, to appear.

25. Y. G. Reshetnyak, *On the stability of conformal mappings in multi-dimensional spaces*, Sibirskii Math. **8** (1967), 91–114.

26. R. Rostamian, *Internal constraints in boundary value problems of continuum mechanics*, Indiana University Mathematics Journal **27** (1978), 637–656.

27. Y. J. Song, J. T. Oden and N. Kikuchi, *Discrete LBB-conditions for RIP-finite element methods*, TICOM Report 80–3, The University of Texas at Austin, 1980.

28. O. C. Zienkiewicz, R. L. Taylor and J. M. Too, *Reduced integration technique in general analysis of plates and shells*, International Journal of Numerical Methods in Engineering **3** (1971).

DEFORMATION AND VIBRATION OF ROTATING ELASTIC CYLINDERS

R. W. OGDEN[†] AND D. M. HAUGHTON[‡]
School of Mathematics, University of Bath, Bath BA2 7AY, UK

(Received September 12, 1980)

1. The circular cylindrical configuration

This paper summarizes briefly the work described in references [1] and [2]. Firstly, we consider the circular cylindrical deformation of an initially circular elastic cylinder due to axial loading and rotation about its axis. In its undeformed configuration the cylinder is defined by

$$0 \leqq R \leqq A, \qquad 0 \leqq \Theta \leqq 2\pi, \qquad 0 \leqq Z \leqq L$$

in cylindrical polar coordinates (R, Θ, Z). Let ω be the angular speed, F the resultant axial load, and λ the azimuthal principal stretch. With cylindrical polar coordinates (r, θ, z) defined in the current configuration and the material assumed incompressible the (homogeneous) deformation is defined by

$$r = \lambda R, \qquad \dot{\theta} = \omega, \qquad z = \lambda^{-2} Z.$$

The current radius and length of the cylinder are respectively $a = \lambda A$ and $l = \lambda^{-2} L$.

For an (incompressible) isotropic elastic solid the strain-energy function $W(\lambda_1, \lambda_2, \lambda_3)$ per unit volume is replaced by

$$\tilde{W}(\lambda) = W(\lambda, \lambda^{-2}, \lambda)$$

for the deformation in question, where the indices 1, 2, 3, correspond respectively to the principal directions (of strain) defined by the cylindrical polar coordinates θ, z, r.

[†] Now at: Department of Mathematics, Brunel University, Uxbridge, Middlesex.
[‡] Now at: Department of Mathematics, University of Glasgow.

Proceedings of the IUTAM Symposium on Finite Elasticity, Lehigh University, August 10–15, 1980. Invited paper presented by R. W. Ogden.

Carlson, D.E. and Shield, R.T. (eds.)
Proceedings of the IUTAM Symposium on Finite Elasticity
ISBN 90 247 2629 8

The corresponding principal Cauchy stresses σ_1, σ_2, σ_3 are such that

$$\sigma_1 = \sigma_3 = \tfrac{1}{2}\rho\omega^2(a^2 - r^2), \qquad 0 \leqq r \leqq a,$$

and

$$\sigma_3 - \sigma_2 = \tfrac{1}{2}\lambda \tilde{W}'(\lambda),$$

where ρ is the density of the material and a prime denotes differentiation with respect to λ. The lateral boundary of the cylinder is assumed free of traction.

The axial load on the ends of the cylinder is given by

$$F = \tfrac{1}{4}\pi a^2\{\rho\omega^2 a^2 - 2\lambda \tilde{W}'(\lambda)\}, \tag{1}$$

and the inequalities

$$\tilde{W}'(\lambda) \gtreqless 0 \text{ according as } \lambda \gtreqless 1 \tag{2}$$

are imposed. The conditions (2) are sensible since they imply (a) for $\omega = 0$, tension (compression) is accompanied by extension (contraction), and (b) for $F = 0$, rotation is accompanied by shortening of the cylinder.

When $F = 0$ the existence of a value of λ for each value of ω implies that the growth condition

$$\tilde{W}'(\lambda)/\lambda \to \infty \qquad \text{as } \lambda \to \infty \tag{3}$$

should be satisfied. We note that the neo-Hookean solid does not conform to this requirement.

For realistic constitutive laws the possibility of ω having turning points with respect to λ must be considered. Such points are characterized by

$$\lambda \tilde{W}''(\lambda) - \tilde{W}'(\lambda) = 0 \tag{4}$$

when $F = 0$. For $F > 0$ turning points are also evident, but not for $F < 0$ (see [1] for details).

2. Bifurcations

Bifurcation into a non-circular-cylindrical mode of deformation is considered next. Linearization of the general equations and use of separable solutions leads to coupled non-linear ordinary differential equations with r as the independent variable. Application of the boundary conditions then yields the bifurcation criteria for different modes of deformation.

For prismatic modes, in which the cross-section of the cylinder is independent of z, a simple bifurcation criterion is obtained in the form

$$\lambda \tilde{W}'(\lambda) + 2F/\pi a^2 = (m - m^{-1})\mathcal{B}_{3131}, \tag{5}$$

where m is the angular mode number and $\mathcal{B}_{3131}$, which depends on λ in general, is an elastic modulus [1]. Given F/A^2, equation (5) determines the values of λ at which bifurcation can occur. Note that this formula is valid for an arbitrary form of strain-energy function. For specific forms of strain-energy function satisfying the growth condition (3) calculations in [1] show that two distinct bifurcation points are associated with each value of m, the λ's for $m+1$ being between those for m. More particularly, only a finite number of modes can be activated for the strain-energy functions detailed in [1].

For axisymmetric and asymmetric bifurcation modes rotation has an effect which is comparable to that of axial compression, but in tension the behaviour is complicated by the existence of turning points of ω. Depending on the ratio A/L bifurcation may occur under pure axial compression, but for $F>0$ rotation is required to effect bifurcation. In particular, for cylinders with A/L *greater* than some critical value bifurcation into an axisymmetric mode can occur under pure axial compression, but for asymmetric modes, on the other hand, A/L is required to be *less* than some critical value (which depends on the magnitude of the compression). Explicit bifurcation criteria for these modes are given in [1] but they are not repeated here because of their algebraic complexity.

For long cylinders the asymmetric (or bending) mode is clearly the most important and the results described in [1] are thereby consistent with the classical theory of whirling shafts which predicts a critical value of ω proportional to L^{-2}. The longer the cylinder the smaller the value of ω required to produce whirling. In fact, the asymptotic form of ω for large values of L/A is given by

$$\rho\omega^2 a^2/\mu = \tfrac{3}{4}\pi^4(a/l)^4$$

when $\lambda = 1$.

3. Vibrations

Finally, we examine the effect of axial loading and rotation on infinitesimal waves propagating along the cylinder. For $\omega = 0$, torsional vibrations and longitudinal vibrations (axisymmetric with no azimuthal displacement) decouple, but when $\omega \neq 0$ there is no decoupling. The dispersion relation for these waves [2] is an equation for c^2, where c is the wave speed. Thus, for any wave propagating in the positive z-direction there is a corresponding wave travelling in the opposite direction with the *same* speed. The sign of the azimuthal displacement is different for the two waves, however, and is reversed by a change in the sense of rotation.

For flexural (or asymmetric) waves, on the other hand, the situation is different since the product ωc occurs in the dispersion relation [2], and waves travel in the positive and negative z-directions with different speeds. A change in the sense of rotation reverses this effect.

The bifurcation results discussed in Section 2 above correspond to points where $c = 0$ on the graphs given in [2] in which c^2 is plotted as a function of ω^2, λ or A/L. It is consistent with these results that for short cylinders the axisymmetric waves have the lowest wave speeds while for long cylinders the flexural waves are slowest.

Acknowledgement

The work of D. M. Haughton was supported by an SRC Research Grant. We are grateful to Professor W. T. Koiter for a helpful comment.

References

1. D. M. Haughton and R. W. Ogden, *Bifurcation of finitely-deformed rotating elastic cylinders*, Quart. J. Mech. Appl. Maths. **33** (1980), 251–265.

2. D. M. Haughton, *Wave speeds in rotating elastic cylinders at finite deformation*, Quart. J. Mech. Appl. Maths.

BOUNDARY INTEGRAL EQUATIONS FOR INEXTENSIBLE MATERIALS

A. C. PIPKIN

Division of Applied Mathematics, Brown University, Providence, Rhode Island 02912, USA

(Received September 11, 1980)

ABSTRACT

Various theories of materials composed of inextensible fibers are reviewed. Two-dimensional pure traction boundary-value problems are discussed. In each theory, equations relating the kinematic variables to the prescribed boundary tractions are derived, and the approximate solution of these equations by iteration is discussed.

1. Introduction

Rivlin's work [1–4] on various theories of materials reinforced with inextensible fibers has been followed by a large body of work in that area. Different theories are used, depending on the intended application: the pure network theory [2] applies most directly to materials like fish-nets; the theory of networks with shear resistance (Adkins [5]) applies to closely-woven or coated fabrics; the theory of one family of fibers embedded in a matrix of compliant material [6, 7] models some aspects of the behavior of reinforced rubber; the various small-deformation theories [8] are intended for use in connection with highly anisotropic strong composite materials. All of these theories treat the material as a continuum, with the fibers spaced infinitesimally close together, and all treat the fibers as incapable of extension or contraction.

The constraint of inextensibility allows much of the analysis of deformation to be carried out purely kinematically, without reference to stress–strain relations or equilibrium conditions. For this reason, problems with prescribed boundary displacements are usually very easy to solve. From a mathematical point of view, pure traction boundary-value problems are more difficult and more interesting.

Proceedings of the IUTAM Symposium on Finite Elasticity, Lehigh University, August 10–15, 1980. Invited paper.

Carlson, D.E. and Shield, R.T. (eds.)
Proceedings of the IUTAM Symposium on Finite Elasticity
ISBN 90 247 2629 8

In the present paper we disuss dead-loading traction boundary-value problems, in plane stress or plane strain, for several theories. In these theories the resultant force on the part of the boundary to one side of a given fiber can be related to the values of the kinematic variables along that fiber. In two dimensions, with two families of fibers, these relations give a pair of integral or differential-difference equations for the kinematic variables. We examine several of the systems of equations set up in this way.

In one of the theories (Section 5), one of the relevant equations is derived from a J-integral rather than from equilibrium conditions in the more usual sense. This is the only new result given in the present paper. Our main object is to point out common features and analogies among various theories that are superficially quite distinct.

2. Infinitesimal deformations

We first consider infinitesimal plane deformations of a material composed of inextensible fibers that initially lie along the x- and y-directions of a system of Cartesian coordinates. The x-displacements of particles on a fiber $y =$ constant must all be the same in order to maintain the length of each segment of that fiber, so the displacement parallel to the x-axis is a function of y alone, $u(y)$. Similarly, the displacement parallel to the y-axis is $v(x)$.

The shearing stress σ_{xy} is related to the strain in the usual way,

$$\sigma_{xy} = G\,[u'(y) + v'(x)], \tag{2.1}$$

but the normal stress components σ_{xx} and σ_{yy} are reactions to the constraint conditions, to be determined from the equilibrium equations and boundary conditions. An expression for σ_{xx} is found by using (2.1) in the x-component of the equilibrium equation and integrating with respect to x. Use of the traction boundary conditions at both ends of the fiber $y =$ constant then yields a relation between the boundary tractions and the displacements. A similar relation is obtained by integrating the y-component of the equilibrium equation over the length of the fiber $x =$ constant. Each of these two equations can then be integrated once formally.

The resulting pair of equations can be derived more directly from global analysis of equilibrium (England [9]). Let $F_x(y)$ be the x-component of the resultant force on the part of the boundary above the fiber $y =$ constant. For simplicity here and elsewhere, we suppose that this line and the part of the boundary above it bound a connected region. Equilibrium of this region requires $F_x(y)$ to be equal to the resultant shearing force on the line $y =$ constant. Evaluating the latter from (2.1), we get

$$u'(y)\Delta x(y)+\Delta v(y)=F_x(y)/G. \tag{2.2}$$

Here $\Delta x(y)$ is the length of the fiber y, and $\Delta v(y)$ is the difference between the values of $v(x)$ at the two ends of the fiber y. Thus $\Delta v(y)$ depends in a complicated way on $v(x)$ and the boundary shape. Similarly, if $F_y(x)$ is the y-component of the resultant force on the part of the boundary to the right of the line x = constant, we obtain

$$v'(x)\Delta y(x)+\Delta u(x)=F_y(x)/G. \tag{2.3}$$

The system of differential-difference equations (2.2)–(2.3) can be solved analytically when the boundary of the region is simple enough, but here we will discuss only the approximate solution of these equations by iteration. The iterative procedure is used to find the difference quotients defined by

$$Q(x)=\Delta u(x)/\Delta y(x) \quad \text{and} \quad R(y)=\Delta v(y)/\Delta x(y). \tag{2.4}$$

When these quotients are known, so that Δu and Δv are known, u and v can be found by integrating (2.2) and (2.3).

In the absence of boundary forces, an infinitesimal rotation is a solution, $u = Cy$, $v = -Cx$. The difference quotients for this solution are constants, $Q = -R = C$. In the cases in which it has been possible to solve (2.2)–(2.3) analytically, it has been possible to write them as differential-difference equations for Q and R, and this particular solution of the homogeneous equations has been used to lower the order of the system.

Integral equations for Q and R can always be obtained. By dividing (2.2) by $\Delta x(y)$ and then integrating over the length of a fiber x = constant, one obtains Q in terms of an integral over R, and similarly (2.3) yields R as an integral over Q:

$$\begin{aligned} Q(x)\Delta y(x) &= \int [f_x(y)-R(y)]U(x,y)dy, \\ R(y)\Delta x(y) &= \int [f_y(x)-Q(x)]U(x,y)dx. \end{aligned} \tag{2.5}$$

Here f_x and f_y come from the inhomogeneous terms in (2.2)–(2.3). $U(x, y)$ is unity inside the region and zero outside, so the limits of integration can be taken to be any fixed limits that include the whole body.

Iteration generates a convergent sequence of approximate solutions Q_n, R_n. R_n is used under the integral sign in (2.5a) to give Q_n, and Q_n is then used in (2.5b) to evaluate R_{n+1}.

The convergence of this iterative procedure is most easily proved by first combining the two equations in (2.5) to obtain a single Fredholm equation for $Q(x)$,

$$Q(x)\Delta y(x) = f(x) + \int K(x, x')Q(x')dx', \tag{2.6}$$

where

$$K(x, x') = \int U(x, y)U(x', y)\Delta x(y)^{-1}dy. \tag{2.7}$$

The kernel K is symmetric and positive semi-definite, at least. Also, since

$$\int U(x', y)dx' = \Delta x(y) \quad \text{and} \quad \int U(x, y)dy = \Delta y(x), \tag{2.8}$$

then

$$\int K(x, x')dx' = \Delta y(x), \tag{2.9}$$

so Q = constant is a weighted eigenfunction with eigenvalue unity. The corresponding Fredholm condition on $f(x)$, that its integral must vanish, turns out to be equivalent to the condition that the resultant moment of the boundary tractions must be zero. It is easy to show [10] that all other eigenfunctions have eigenvalues less than unity, so the solution of (2.6) by iteration will converge; the amount of rotation in the solution obtained by iteration is non-unique, being equal to the rotational component of the initial guess.

For the more difficult theory involving only one family of inextensible fibers, $u = u(y)$ still, but $v(x, y)$ is harmonic. Morland [11] has proved existence of solutions by using Green's functions to obtain an integral equation analogous to (2.6).

3. Finite deformations of pure networks

We now turn to a finite-deformation theory which, surprisingly, is even simpler than the preceding small-deformation theory. Rivlin's network theory [2] concerns inextensible fibers knotted together or held together by friction where they cross, with no resistance to shearing deformations. The fibers initially lie along the X and Y directions of a system of Cartesian coordinates. It is convenient to express all variables as functions of the initial position X, Y of the particle at which the variable is evaluated.

In a plane deformation the fibers initially in the $\boldsymbol{i}$ and $\boldsymbol{j}$ directions through a particle turn toward the directions $\boldsymbol{a}$ and $\boldsymbol{b}$, where $\boldsymbol{a}$ and $\boldsymbol{b}$ are unit vectors. If the position of the particle in the deformed state is $\boldsymbol{x}(X, Y)$, then inextensibility requires that $\boldsymbol{x}_{,X} = \boldsymbol{a}$ and $\boldsymbol{x}_{,Y} = \boldsymbol{b}$, so

$$dx = \boldsymbol{a}dX + \boldsymbol{b}dY, \qquad |\boldsymbol{a}| = |\boldsymbol{b}| = 1. \tag{3.1}$$

The integrability condition $\boldsymbol{a}_{,Y} = \boldsymbol{b}_{,X}$ on the unit vectors $\boldsymbol{a}$ and $\boldsymbol{b}$ implies that wherever they are not parallel, each is a function of only one variable:

$$\boldsymbol{a} = \boldsymbol{a}(X), \qquad \boldsymbol{b} = \boldsymbol{b}(Y). \tag{3.2}$$

This implies that the fibers in one family lie along congruent curves after the deformation. There are also exceptional cases in which the mapping $\boldsymbol{x}(X, Y)$ is degenerate, mapping a finite area onto a curve [12]. A less severe singularity that arises very easily is a *fold*, a curve across which $\operatorname{sgn} J$ is discontinuous, where

$$J = \boldsymbol{k} \cdot \boldsymbol{a} \times \boldsymbol{b}. \tag{3.3}$$

The forms of $\boldsymbol{a}$ and $\boldsymbol{b}$ are generally different on the two sides of a fold.

The analysis of equilibrium is most easily carried out in terms of Rivlin's stress function $\boldsymbol{F}(X, Y)$. The increase of $\boldsymbol{F}$ along a directed arc represents the force exerted across the arc by the material on the right on the material on the left (orientations refer to the regions occupied by the same pieces of material in the undeformed state). The forces per unit length exerted across infinitesimal arcs in the Y- and X-directions are

$$\boldsymbol{F}_{,Y} = T_a\boldsymbol{a} + S\boldsymbol{b} \quad \text{and} \quad -\boldsymbol{F}_{,X} = T_b\boldsymbol{b} + S\boldsymbol{a}. \tag{3.4}$$

Here we have introduced notation for the components with respect to the local basis $\boldsymbol{a}$, $\boldsymbol{b}$. T_a and T_b are in the nature of fiber tensions, which are reactions to the constraints of fiber inextensibility. S is a shearing stress, and for Rivlin's theory we take $S = 0$ as a constitutive equation.

Equilibrium equations in a more usual form can be obtained by eliminating $\boldsymbol{F}$ from (3.4) by cross-differentiation, but this is working backward. Instead, (3.4) is to be integrated. When $S = 0$, (3.4) yields

$$(\boldsymbol{k} \times \boldsymbol{b} \cdot \boldsymbol{F})_{,X} = (\boldsymbol{k} \times \boldsymbol{a} \cdot \boldsymbol{F})_{,Y} = 0, \tag{3.5}$$

since $\boldsymbol{b}$ and $\boldsymbol{a}$ are independent of X and Y, respectively. Then $\boldsymbol{k} \times \boldsymbol{b} \cdot \boldsymbol{F}$ depends only on Y, and $\boldsymbol{k} \times \boldsymbol{a} \cdot \boldsymbol{F}$ only on X, so

$$\boldsymbol{F} = J^{-1}[N(Y)\boldsymbol{a}(X) + M(X)\boldsymbol{b}(Y)], \tag{3.6}$$

where J is defined in (3.3). This integration implicitly assumes the absence of folds.

Dead-loading traction boundary-value problems are posed by prescribing the values of $\boldsymbol{F}$ around the boundary of the sheet; we take the sheet to be simply-connected. The boundary tractions implied by given boundary values of $\boldsymbol{F}$ are then automatically in translational equilibrium. No rotational equilibrium condition arises because in finite deformations, rotational equilibrium is a property of the unknown deformed shape.

With $\boldsymbol{F}$ prescribed on the boundary, the unknown functions M, N, $\boldsymbol{a}$ and $\boldsymbol{b}$ in (3.6) can be deduced algebraically. So far as $\boldsymbol{a}$ and $\boldsymbol{b}$ are concerned, the algebraic result can be obtained more easily from a direct physical argument like that used in Section 2 to derive the system (2.2)–(2.3). Let $\Delta\boldsymbol{F}(X)$ be the difference between the boundary values of $\boldsymbol{F}$ at the upper and lower ends of the fiber X. This represents the resultant force on the part of the boundary initially to the right of that fiber. In the deformed state, the fiber X is curved, but all of the fibers crossing it have the same direction, $\boldsymbol{a}(X)$. Since forces are transmitted only by the tensions in these fibers, the resultant force across the fiber X must be parallel to $\boldsymbol{a}(X)$. Hence,

$$K\boldsymbol{a}(X) = \Delta\boldsymbol{F}(X), \tag{3.7}$$

and a similar argument yields

$$L\boldsymbol{b}(Y) = -\Delta\boldsymbol{F}(Y), \tag{3.8}$$

where $\Delta\boldsymbol{F}(Y)$ is the difference between the boundary values of $\boldsymbol{F}$ at the right and left ends of the fiber Y. K is the magnitude of $\Delta F(X)$ and L is the magnitude of $\Delta\boldsymbol{F}(Y)$.

In the algebraic derivation of (3.7) and (3.8) from (3.6), the signs of K and L are indeterminate and may even vary with position. Taking K and L to be positive everywhere comes from the assumption that all fibers are under tension rather than compression. Solutions are obtained with any prescription of signs for K and L, but solutions in which fibers carry compressive loads are not likely to be physically meaningful.

The results (3.7) and (3.8) give $\boldsymbol{a}$ and $\boldsymbol{b}$ directly in terms of the prescribed boundary forces, and integration of (3.1) then yields the deformation. Here, for once, is a theory in which traction boundary-value problems are relatively trivial.

4. Coated networks

Adkins [5] generalized Rivlin's network theory by allowing the fibers to be intially curved and non-parallel, and letting the shearing stress S in (3.4) be any function of the local deformation (see also Green and Adkins [13]). Here we consider only a slight generalization of Rivlin's theory. With other things as in Section 3, we suppose that S is proportional to the tangent of the angle of shear:

$$S = G\boldsymbol{a}\cdot\boldsymbol{b}/|J|. \tag{4.1}$$

This form of S allows integration of (3.4) much as Rivlin did for the case $S=0$. In place of (3.5) we find that

$$(4.2)\qquad (\boldsymbol{k}\times\boldsymbol{b}\cdot\boldsymbol{F})_{,X}=(\boldsymbol{k}\times\boldsymbol{a}\cdot\boldsymbol{F})_{,Y}=G\boldsymbol{a}\cdot\boldsymbol{b}\operatorname{sgn}J.$$

The integration to find $\boldsymbol{F}$ can be carried out separately in each region in which sgn J is constant, yielding [12]

$$(4.3)\qquad \boldsymbol{F}=\boldsymbol{F}_0(X,Y)+G\operatorname{sgn}J\boldsymbol{k}\times\boldsymbol{x}(X,Y).$$

Here $\boldsymbol{F}_0$ has the form (3.6), but need not be the same function in adjacent regions with different values of sgn J. For simplicity, let us consider deformations with sgn $J=1$ everywhere, so that the sheet has no folds and has not turned over bodily.

In dead-loading traction boundary-value problems, $\boldsymbol{a}$ and $\boldsymbol{b}$ satisfy integral equations analogous to the differential-difference system (2.2)–(2.3) satisfied by $v(x)$ and $u(y)$ in the infinitesimal case. The terms analogous to the displacement differences $\Delta v(y)$ and $\Delta u(x)$ are the end-to-end vectors for the deformed fibers,

$$(4.4)\qquad \Delta\boldsymbol{x}(Y)=\int U(X,Y)\boldsymbol{a}(X)dX,$$

and

$$(4.5)\qquad \Delta\boldsymbol{x}(X)=\int U(X,Y)\boldsymbol{b}(Y)dY.$$

Here U is the characteristic function for the initial region, as in Section 2, and the integrals are over all X or Y.

From either (4.2) or (3.6) and (4.3),

$$(4.6)\qquad \boldsymbol{k}\times\boldsymbol{a}\cdot\boldsymbol{F}=M(X)+G\boldsymbol{a}(X)\cdot\boldsymbol{x}(X,Y).$$

On eliminating M from the two expressions obtained by evaluating (4.6) at the two ends of the fiber X, we obtain

$$(4.7)\qquad \boldsymbol{k}\times\boldsymbol{a}\cdot[\Delta\boldsymbol{F}(X)-G\boldsymbol{k}\times\Delta\boldsymbol{x}(X)]=0.$$

Thus, $\boldsymbol{a}$ is parallel to the bracketed quantity, so we obtain

$$(4.8)\qquad K\boldsymbol{a}(X)=\Delta\boldsymbol{F}(X)+G\Delta\boldsymbol{x}(X)\times\boldsymbol{k}.$$

A similar development leads to the second equation

$$(4.9)\qquad L\boldsymbol{b}(Y)=-\Delta\boldsymbol{F}(Y)+G\boldsymbol{k}\times\Delta\boldsymbol{x}(Y).$$

These equations generalize (3.7) and (3.8) to the case $G\neq 0$, and are analogous to (2.2) and (2.3).

K and L are the magnitudes of the right-hand members of (4.8) and (4.9), respectively. The signs of K and L are not determined by the preceding derivation, and taking them to be positive is not directly justified by the argument in Section 3 for a pure network, because now forces are not transmitted entirely by fiber tensions. However, (4.8) and (4.9) can also be derived as the Euler equations for an energy-minimization problem [14], and this derivation shows that K and L must be positive for minimum energy. Solutions with K or L negative can be obtained, but they are unstable.

It should be possible to solve the system (4.8) and (4.9) by iteration, and energy minimization gives an interesting and useful interpretation of the iterative procedure. For any given field $\boldsymbol{b}(Y)$, with $\Delta\boldsymbol{x}(X)$ defined in terms of it by (4.5), (4.8) gives the field $\boldsymbol{a}(X)$ that minimizes the energy with respect to variations of $\boldsymbol{a}$ alone. Then with that $\boldsymbol{a}$, (4.9) gives the field $\boldsymbol{b}$ that minimizes the energy with respect to variations of $\boldsymbol{b}$ alone. Thus sequences obtained by iteration are lowering the energy as fast as is possible by changing only one field at a time. Uniform convergence of the sequences of fields $\boldsymbol{a}$ and $\boldsymbol{b}$ is easy to prove directly if the boundary loads are sufficiently large [14].

In setting up (4.8) and (4.9) it was assumed that the solution involves no folds, but the fields $\boldsymbol{a}$ and $\boldsymbol{b}$ that satisfy (4.8) and (4.9) may be such that J $(= \boldsymbol{k} \cdot \boldsymbol{a} \times \boldsymbol{b})$ changes sign, contradicting the hypothesis. More complicated equations replace (4.8) and (4.9) when folds may be present. By using these more complicated relations it is possible to derive conditions sufficient to ensure that the solution has no folds [14], namely that the corresponding pure network solution (3.7)–(3.8) has no folds and that the boundary loads are uniformly large in a certain sense.

The interpretation of iteration as an energy-minimization procedure suggests that iteration must always converge, but this has not been proved. It appears that when the boundary loads are small in comparison to G, a different iterative procedure should be used, in which the first step is to rotate the sheet rigidly until the boundary loads are in rotational equilibrium. The first-order equations for small displacements superposed on this large rotation are the equations (2.2)–(2.3) of the infinitesimal theory. A difficulty arises if the loading system is astatic when the sheet is treated as rigid, because then the initial large rotation is not uniquely determined. We intend to go into this more thoroughly in a forthcoming paper [17].

5. Incompressible materials with one family of fibers

Some of the ideas used in the preceding theories arise in more complicated form in the theory of plane strain of incompressible materials reinforced with one family of inextensible fibers [6, 7]. We suppose that the fibers are initially straight, and parallel to the X-direction. In plane strain, the constraints of fiber inextensibility and bulk incompressibility imply that the spacing between fibers cannot change in a deformation, so the deformed fibers lie along parallel curves. Their orthogonal trajectories are then straight lines, which we call *normal lines*. The fibers and normal lines are characteristics of the governing equations. Integral equations can be set up which relate the values of the kinematic variables along a characteristic to the load on the part of the boundary cut off by that characteristic.

Let $\theta(X, Y)$ be the angle of the fiber through X, Y after the deformation, measured from the X-axis. Let $\boldsymbol{a}(\theta)$ be a unit vector in this direction, and let $\boldsymbol{n} = \boldsymbol{k} \times \boldsymbol{a}$ be a perpendicular unit vector. The constraint conditions imply that the deformation $\boldsymbol{x}(X, Y)$ has the local form

$$d\boldsymbol{x} = \boldsymbol{a}dX + (\boldsymbol{n} + k\boldsymbol{a})dY, \tag{5.1}$$

where we omit the irrelevant dependence on Z. Here $k(X, Y)$ is the amount of shear at the particle considered.

The integrability condition for (5.1), $\boldsymbol{a}_{,Y} = (\boldsymbol{n} + k\boldsymbol{a})_{,X}$, can be separated into components in the $\boldsymbol{a}$ and $\boldsymbol{n}$ directions after using the facts that $\boldsymbol{a}'(\theta) = \boldsymbol{n}(\theta)$ and $\boldsymbol{n}'(\theta) = -\boldsymbol{a}(\theta)$. The component equations are

$$(k - \theta)_{,X} = 0 \quad \text{and} \quad \theta_{,Y} = k\theta_{,X}. \tag{5.2}$$

The first of these compatibility conditions gives

$$k(X, Y) = \theta(X, Y) + D'(Y), \tag{5.3}$$

where D' is arbitrary. The second can be put into vector form as $\boldsymbol{n} \cdot \nabla\theta = 0$, which states that θ and thus $\boldsymbol{n}(\theta)$ are constant along any trajectory of the field $\boldsymbol{n}$, so these trajectories, the normal lines, are straight.

In kinematic boundary-value problems it is possible to use the straightness of normal lines and inextensibility of fibers to construct solutions geometrically [15]. In some mixed problems it is possible to determine the deformation by equating the shearing force on a normal line to the parallel component of the external force on the part of the boundary cut off by that line, as in the infinitesimal theory (Section 2). But pure traction boundary-value problems are much more difficult. Our discussion of such problems here is patterned after the work of Craig and Hart [16].

The pre-images of the normal lines in the undeformed body are seen from (5.2b) to be the curves with slopes $dY/dX = -1/k$. With k given by (5.3) and θ constant along each normal line, we find that these characteristics are given by

$$X + Y\theta(X_0) + D(Y) = X_0 \text{ (say)}, \qquad \theta = \theta(X_0). \tag{5.4}$$

If we take $D(0) = 0$, as we may, the parameter X_0 that numbers the normal line is the value of X at the place where the normal line crosses the axis $Y = 0$.

Later we derive a generalization of Craig and Hart's equations by using path-independent integrals. One such integral can be derived by noting that (5.2) implies that $\theta_{,Y} = kk_{,X}$, whence

$$\oint (\theta dX + \tfrac{1}{2}k^2 dY) = 0, \tag{5.5}$$

where the path of integration is the boundary of any simply-connected part of the body.

The stress has the form [15]

$$\boldsymbol{\sigma} = T\boldsymbol{aa} - P\boldsymbol{nn} + S(\boldsymbol{an} + \boldsymbol{na}), \tag{5.6}$$

in dyadic notation. We omit the irrelevant component $\sigma_{zz}\boldsymbol{kk}$. T and P are reactions, not determined by constitutive equations. The shearing stress S is some function of the amount of shear, $S(k)$, related to the strain energy density $W(k)$ by $S = W'(k)$.

The equilibrium equations yield two equations in characteristic form, governing the variation of T along fibers and P along normal lines [15]. The use of traction boundary conditions at both ends of each characteristic leads to integral equations governing θ and k. However, as in the theories discussed previouly, an equivalent system can be derived more easily in another way.

Let $\Delta\boldsymbol{F}(X_0)$ be the resultant force on the part of the boundary originally to the right of the normal line X_0. In the deformed body this normal line is straight, with tangent $\boldsymbol{n}(\theta)$. The tangential component of the boundary force, $\boldsymbol{n} \cdot \Delta\boldsymbol{F}(X_0)$, is equal to the resultant shearing stress S on the normal line:

$$\int S[\theta(X_0) + D'(Y)]dY = \boldsymbol{n}(\theta) \cdot \Delta\boldsymbol{F}(X_0). \tag{5.7}$$

The limits of integration are the values of Y at the places where the normal line crosses the boundary, which are not known. These values also enter into the evaluation of $\Delta\boldsymbol{F}(X_0)$.

Because fibers are generally curved, the integral of S along a fiber is not necessarily equal to any component of the boundary load. However, departing from the path followed by Craig and Hart [16], we can find an equilibrium condition involving the integral of S along a fiber by using a J-integral. If the material is homogeneous in its undeformed state (requiring in particular that the fibers be straight and parallel initially), then the equilibrium equation implies that

$$\oint \boldsymbol{a} \cdot (\boldsymbol{\sigma} - W\boldsymbol{I}) \cdot \boldsymbol{\nu} ds = 0, \tag{5.8}$$

where the path of integration is any closed curve in the deformed body. $\boldsymbol{I}$ is the identity, ds is arc length along the curve, and $\boldsymbol{\nu}$ is the unit outward normal.

For the path of integration we use the fiber Y = constant (denoting this part of the path by F) and the part of the external boundary initially below that fiber (denoted by B). On B, $\boldsymbol{\sigma} \cdot \boldsymbol{\nu} ds = d\boldsymbol{F}$ is the increment of the boundary traction, which is prescribed (as in Section 3). On F, $\boldsymbol{\nu} = \boldsymbol{n}$ and thus $\boldsymbol{a} \cdot \boldsymbol{\sigma} \cdot \boldsymbol{\nu} = S$. On both F and B, $\boldsymbol{a} \cdot \boldsymbol{\nu} ds = dY$. Hence, we obtain

$$\int_F S dX + \int_B \boldsymbol{a} \cdot d\boldsymbol{F} = \oint W dY. \tag{5.9}$$

The relations (5.7) and (5.9) are two equations to be satisfied by $\theta(X_0)$ and $D(Y)$. As in the theories discussed previously, each unknown is constant along characteristics of one family or the other. The points where the normal line X_0 intersects the boundary, needed in (5.7), are additional unknowns to be determined algebraically from (5.4) and the equations of the undeformed boundary. Faith in counting equations and unknowns suggests that the problem might be well-set, but it is so formidably complicated that this is far from clear.

Craig and Hart [16] studied the special case in which $S = Gk$ and $W = Gk^2/2$. In this case (5.7) can be integrated to yield

$$\theta(X_0)\Delta Y(X_0) + \Delta D(X_0) = \boldsymbol{n}(\theta) \cdot \Delta \boldsymbol{F}(X_0)/G, \tag{5.10}$$

where the differences $\Delta f(X_0)$ for $f = Y$, D, and $\boldsymbol{F}$ are the differences between the values of f at the two ends of the normal line X_0. With (5.4), (5.10) can be written more briefly as

$$-\Delta X(X_0) = \boldsymbol{n}(\theta) \cdot \Delta \boldsymbol{F}(X_0)/G. \tag{5.11}$$

Craig and Hart's second equation can be derived from (5.9). We first note that with $S = Gk$ and k given by (5.3), the integral of S along a fiber can be expressed as

$$\int_F S dX = GD'(Y)\Delta X(Y) + \int_B G\theta dX - \oint G\theta dX. \tag{5.12}$$

The integral of θ along the fiber has been written as the difference between the integral along B and the loop integral. $\Delta X(Y)$ is the difference between the boundary values of X at the two ends of the line $Y = \text{constant}$. From (5.5), with $W = Gk^2/2$, the loop integral in (5.12) can be written as

$$\oint G\theta dX = -\oint W dY. \tag{5.13}$$

Collecting (5.12) and (5.13) in (5.9), we obtain the equation derived in a different way by Craig and Hart [16]:

$$\int_B (\boldsymbol{a} \cdot d\boldsymbol{F} + G\theta dX) + GD'(Y)\Delta X(Y) = 0. \tag{5.14}$$

Craig and Hart [16] studied the system (5.4), (5.10) (or (5.11)), and (5.14). They first showed that when there is no traction on the boundary, these equations imply that $k = 0$ and that θ is an arbitrary constant, the expected result (rigid rotation). This result gives some assurance that the system actually determines k and θ, which is not obvious by inspection. They also discussed ordinary perturbation methods, expanding all unknowns in powers of a loading parameter, and solved some particular problems by this method.

The relations (5.14) and (5.10) are analogous to (2.2) and (2.3), respectively. $\theta(X_0)$ is the analog of $v'(x)$ and $D'(Y)$ is the counterpart of $u'(y)$. This suggests that (5.14) and (5.10) might be solved by iteration. With any first guess for θ, say $\theta = 0$, $D'(Y)$ is found from (5.14) and integrated to give $D(Y)$. These values of θ and D are used in (5.4) to determine the points at which the normal line X_0 crosses the boundary. Then the terms in (5.10) can be evaluated, and (5.10) can be used to give an improved approximation for θ. The integral of this θ is then used in (5.14) and the process is repeated.

6. Discussion

In each of the theories that we have considered, the governing partial differential equations form a hyperbolic system. In the theories with two families of fibers, the fibers are the characteristics; in Section 5, with only one family of fibers, the fibers and their orthogonal trajectories are characteristics. Each kinematic variable is a function of only one characteristic coordinate: $u(y)$ and $v(x)$, $\boldsymbol{a}(X)$ and $\boldsymbol{b}(Y)$, $\theta(X_0)$ and $D(Y)$. In each case, two of the stress components are reactions to the constraint conditions, and when the equilibrium equations are written in characteristic form, each equation is a first-order equation governing the variation of one of the reactions along a characteristic. Specifying boundary tractions around the whole boundary means that values are specified at two points on each characteristic. The kinematic variables must adjust themselves to make the two specified values compatible with one another. The conditions on the kinematic variables required in order to make the boundary data compatible become the basic equations to be solved.

It was not necessary to use the equilibrium equations explicitly, in the way just described, because in each case the basic data-compatibility equations could be derived directly from global equilibrium considerations. In one case (Section 5), equilibrium was expressed more conveniently by a J-integral rather than in terms of forces.

It appears that the differential-difference or integral equations for the kinematic variables can generally be solved by iteration. Convergence of iteration has been proved for the infinitesimal theory (Section 2) and in special cases for the coated network theory (Section 4), and in the latter theory there is reason to suspect that iteration always converges. In the theory discussed in Section 5, the iterative technique outlined there is based merely on analogy with the techniques used in the other theories.

Acknowledgement

This paper was prepared under a grant MCS79–03392 from the National Science Foundation. We gratefully acknowledge this support.

References

1. J. E. Adkins and R. S. Rivlin, *Large elastic deformations of isotropic materials X. Reinforcement by inextensible cords*, Phil. Trans. Roy. Soc. London **A248** (1955), 201–223.

2. R. S. Rivlin, *Plane strain of a net formed by inextensible cords*, J. Rational Mech. Anal. **4** (1955), 951–974.

3. R. S. Rivlin, *The deformation of a membrane formed by inextensible cords*, Arch. Rational Mech. Anal. **2** (1959), 447–476.

4. R. S. Rivlin, *Networks of inextensible cords*, in *Nonlinear Problems of Engineering*, Academic Press, New York, 1964.

5. J. E. Adkins, *Finite plane deformation of thin elastic sheets reinforced with inextensible cords*, Phil. Trans. Roy. Soc. London **A249** (1956), 125–150.

6. J. F. Mulhern, T. G. Rogers and A. J. M. Spencer, *A continuum model for fibre-reinforced plastic materials*, Proc. Roy. Soc. London **A301** (1967), 473–492.

7. A. C. Pipkin, *Finite deformations of ideal fiber-reinforced composites*, in *Composite Materials, Vol.* 2: *Mechanics of Composite Materials* (G. P. Sendeckyj, ed.), Academic Press, New York, 1974.

8. A. C. Pipkin, *Stress analysis for fiber-reinforced materials*, Adv. Appl. Mech. **19** (1979), 1–51.

9. A. H. England, *The stress boundary value problem for plane strain deformations of an ideal fibre-reinforced materal*, J. Inst. Math. Appl. **9** (1972), 310–322.

10. A. C. Pipkin and V. Sanchez, *Existence of solutions of plane traction problems for ideal composites*, SIAM J. Appl. Math. **26** (1974), 213–220.

11. L. W. Morland, *Existence of solutions of plane traction problems for inextensible transversely isotropic elastic solids*, Austral. Math. Soc. **B19** (1975), 40–54.

12. A. C. Pipkin, *Some developments in the theory of inextensible networks*, Q. Appl. Math., **38** (1980), 343–355.

13. A. E. Green and J. E. Adkins, *Large Elastic Deformations*, Chap. VII, Oxford, London, 1960.

14. A. C. Pipkin, *Plane traction problems for inextensible networks*, Q. J. Mech. Appl. Math., **34** (1981), 415–429.

15. A. C. Pipkin and T. G. Rogers, *Plane deformations of incompressible fiber-reinforced materials*, J. Appl. Mech. **38** (1971), 634–640.

16. M. S. Craig and V. G. Hart, *The stress boundary-value problem for finite plane deformations of a fibre-reinforced material*, Q. J. Mech. Appl. Math. **32** (1979), 473–498.

17. A. C. Pipkin, *Finite plane stress of stiff fiber-reinforced sheets*, J. Inst. Math. Appl. **27** (1981), 195–209.

STABILITY OF A THICK NEO-HOOKEAN PLATE

KENNETH N. SAWYERS

Center for the Application of Mathematics, Lehigh University, Bethlehem, PA 18015, USA

(Received September 15, 1980)

ABSTRACT

An account is given of some recent results on the stability of critical states of a thick plate subjected to compressive loading. Flexural and barreling types of bifurcations are considered, and stability results for both are compared on a common basis. The phenomenon of surface wrinkling is shown to stem from barreling bifurcations.

1. Introduction

We consider a rectangular plate of isotropic incompressible neo-Hookean elastic material to be situated with its edges parallel to the axes of a fixed coordinate system. The faces of the plate which are perpendicular to the 1- and 3-axes are acted on by normal forces applied as dead loads. These faces are free of tangential tractions and are constrained to remain parallel to their original orientations. The faces initially perpendicular to the 2-axis are free of all tractions. The load in the 1-direction is compressive. Under these conditions, one possible equilibrium configuration of the plate is a pure homogeneous deformation, with principal axes parallel to the coordinate axes, having extension ratios λ_1, λ_2, λ_3, say. At certain critical values of the ratio $\lambda = \lambda_2/\lambda_1$, bifurcations are possible, these being infinitesimal non-homogeneous deformations in the 12-plane.

This bifurcation problem has been studied by a number of workers including Biot [1], Levinson [2], Nowinski [3] and Sawyers and Rivlin [4].[†] They have found that the possible bifurcations are of the flexural or barreling type, accordingly as the critical value of λ lies in the range $1 < \lambda < 3.383$ or

Proceedings of the IUTAM Symposium on Finite Elasticity, Lehigh University, August 10–15, 1980. Invited paper.

[†] A closely related problem, that of a thick circular plate compressed by radial loading, was studied by Guo Zhong-Heng [5].

Carlson, D.E. and Shield, R.T. (eds.)
Proceedings of the IUTAM Symposium on Finite Elasticity
ISBN 90 247 2629 8

$\lambda > 3.383$, respectively. However, no consideration was given to the question of whether the underlying critical states of pure homogeneous deformation are stable.

The stability question has recently been investigated by Sawyers and Rivlin [6], whose analysis is based on the procedure of Koiter [7–9]. An expression is given for the total potential energy, G, of the system, and stability is governed by the condition that the second variation of G be positive definite, the variation being taken with respect to kinematically admissible deformations which lie in a small neighborhood of the critical state.

The present paper gives an account of the relevant calculations, which are carried out without restriction on the magnitude of the aspect ratio (l_2/l_1), where $2l_1$ and $2l_2$, respectively, are the undeformed 1- and 2-dimensions of the plate. A normalization of the bifurcation displacements is introduced which allows a direct comparison to be made of the stability results for both the flexural and barreling cases. Results for the barreling case are interpreted as applying to the phenomenon of wrinkling of a free surface.

2. Critical equilibrium states

Let $\boldsymbol{\xi}$ be the vector position of a generic particle of the plate, whose bounding surfaces in the initial, undeformed, state are the planes $\xi_A = \pm l_A$ $(A = 1, 2, 3)$. Let $\boldsymbol{X}$ and $\boldsymbol{x}$ be the vector positions of this particle in the state of pure homogeneous deformation characterized by λ_1, λ_2, λ_3 (state I) and in a neighboring state (state II), respectively. The latter is taken to be a state of plane strain in the 12-plane, superposed on an additional small uniform stretch in the 3-direction having extension ratio $\lambda_3(1+\bar{E})$, where $\bar{E}$ is spatially constant. Then

$$X_A = \lambda_A \xi_A \qquad (A = 1, 2, 3), \qquad \lambda_1\lambda_2\lambda_3 = 1, \tag{2.1}$$

and, upon writing $\boldsymbol{x} = \boldsymbol{X} + \bar{\boldsymbol{u}}$, we have[†]

$$\bar{u}_\alpha = \bar{u}_\alpha(\xi_1, \xi_2), \qquad \bar{u}_3 = \lambda_3 \bar{E} \xi_3. \tag{2.2}$$

Incompressibility of the material requires that

$$(1+\bar{E})(\lambda_2 \bar{u}_{1,1} + \lambda_1 \bar{u}_{2,2} + \bar{u}_{1,1}\bar{u}_{2,2} - \bar{u}_{1,2}\bar{u}_{2,1}) + \lambda_1\lambda_2\bar{E} = 0. \tag{2.3}$$

Let ΔW be the increase in strain-energy density at the generic particle in the deformation from state I to state II. For a neo-Hookean material this may be written, in appropriate units, as

[†] We adopt the convention that lower-case Greek subscripts take on the values 1, 2. The usual summation convention applies only to repeated lower-case subscripts. Differentiation with respect to ξ_i is denoted by a comma.

$$\begin{aligned}\Delta W &= \tfrac{1}{2}(x_{\alpha,\beta}x_{\alpha,\beta} + x_{3,3}^2) - \tfrac{1}{2}\lambda_j\lambda_j \\ &= \lambda_1\bar{u}_{1,1} + \lambda_2\bar{u}_{2,2} + \lambda_3^2\bar{E} + \tfrac{1}{2}(\bar{u}_{\alpha,\beta}\bar{u}_{\alpha,\beta} + \lambda_3^2\bar{E}^2).\end{aligned} \tag{2.4}$$

In state I only two components of the Piola–Kirchhoff stress tensor are non-zero, viz.,

$$\Pi_{11} = \lambda_1 - \lambda_2^2/\lambda_1, \qquad \Pi_{33} = \lambda_3 - \lambda_2^2/\lambda_3. \tag{2.5}$$

The kinematic condition that the faces of the plate given by $\xi_1 = \pm l_1$ remain parallel to their initial orientation can be expressed as $\bar{u}_{1,2}(\pm l_1, \xi_2) = 0$ for $-l_2 \leqq \xi_2 \leqq l_2$, or as

$$\bar{u}_1(l_1, \xi_2) = -\bar{u}_1(-l_1, \xi_2) = \lambda_1\bar{e}l_1, \qquad \text{say,} \tag{2.6}$$

where $\bar{e}$ is spatially constant. It follows from $(2.2)_3$, (2.5) and (2.6) that the work done by the surface forces in passing from state I to state II is

$$\mathcal{P} = 8l_1l_2l_3\{(\lambda_1^2 - \lambda_2^2)\bar{e} + (\lambda_3^2 - \lambda_2^2)\bar{E}\}. \tag{2.7}$$

We employ (2.3), (2.4) and (2.7) to express the total potential energy of the system as†

$$G[\bar{\boldsymbol{u}}] = 2l_3\iint \Delta W d\xi_1 d\xi_2 - \mathcal{P} = 2l_3\iint \{\gamma[\bar{\boldsymbol{u}}] - \lambda_2^2\bar{E}^3/(1+\bar{E})\}d\xi_1 d\xi_2, \tag{2.8}$$

where the functional γ is defined by

$$\gamma[\bar{\boldsymbol{u}}] = \tfrac{1}{2}\{\bar{u}_{\alpha,\beta}\bar{u}_{\alpha,\beta} + 2\lambda(\bar{u}_{1,2}\bar{u}_{2,1} - \bar{u}_{1,1}\bar{u}_{2,2}) + (\lambda_3^2 + 2\lambda_2^2)\bar{E}^2\}. \tag{2.9}$$

Upon regarding $\bar{\boldsymbol{u}}$ as a small displacement field, it follows immediately from (2.8) that the second variation of G is given by

$$G_2[\bar{\boldsymbol{u}}] = 2l_3\iint \gamma[\bar{\boldsymbol{u}}]d\xi_1 d\xi_2. \tag{2.10}$$

The first step in the stability analysis consists of establishing the condition under which state I is a critical state. Here we bear in mind that state I can be stable only if G_2 is non-negative for all sufficiently small, kinematically admissible values of $\bar{\boldsymbol{u}}$. In particular, this must be so when $\bar{u}$ satisfies the linearized incompressibility condition obtained from (2.3), viz.,

$$\bar{u}_{2,2}/\lambda_2 + \bar{u}_{1,1}/\lambda_1 + \bar{E} = 0. \tag{2.11}$$

Moreover, since state I is stable if G_2 is positive definite, the condition that state I be *at* the stability limit is for G_2 to have a zero minimum for some

† The domain of integration for double integrals is the rectangle $[-l_1, l_1] \times [-l_2, l_2]$. That for single integrals is the interval $[-l_2, l_2]$.

non-trivial $\bar{\boldsymbol{u}}$. This condition leads to a search for the field $\bar{\boldsymbol{u}}$ which renders G_2 stationary, i.e., for which $\delta G_2 = 0$.

Let $-2l_3 p(\xi_1, \xi_2)$ be the Lagrange multiplier that is introduced to account for the constraint (2.11), and let u_α, E be the solution of the equilibrium and boundary conditions that stem from the variational problem $\delta G_2 = 0$. This field is given by

$$\begin{gathered} u_1 = \Phi'(\xi_1)U'(\xi_2)/\lambda\,\Omega^2, \qquad u_2 = \Phi(\xi_1)U(\xi_2), \qquad E = 0, \\ p = \lambda_1 \Phi(\xi_1)\beta'(\xi_2)/\lambda\,\Omega^2, \qquad [\beta = U'' - \Omega^2 U], \end{gathered} \tag{2.12}$$

where

$$\Phi(\xi_1) = \begin{Bmatrix} \cos\Omega\xi_1 \\ \\ \sin\Omega\xi_1 \end{Bmatrix}, \qquad \Omega = n\pi/2l_1 \qquad (n = 1, 2, 3, \cdots). \tag{2.13}$$

In (2.12) the primes denote differentiation with respect to the indicated arguments, and, in (2.13), the upper (lower) line in Φ corresponds to n even (odd), where n is the number of half -wavelengths parallel to the 1-direction in the mode considered. The function U in (2.12) has the form

$$U(\xi_2) = \begin{cases} M(\cosh\lambda\,\Omega\xi_2 - m\cosh\Omega\xi_2) & [\text{flexure}], \\ \\ M(\sinh\lambda\,\Omega\xi_2 - m\sinh\Omega\xi_2) & [\text{barreling}], \end{cases} \tag{2.14}$$

where M is an arbitrary constant and

$$m = (\lambda\sinh 2\lambda\eta/\sinh 2\eta)^{1/2}, \qquad \eta = \Omega l_2. \tag{2.15}$$

The relationship between η and λ, i.e., the critical bifurcation condition, is defined by the equation

$$\frac{\sinh(\lambda+1)\eta}{\sinh(\lambda-1)\eta} = \nu\frac{(\lambda+1)\{\lambda(\lambda+1)^2 + (\lambda-1)^2\}}{(\lambda-1)\{(\lambda+1)^2 - \lambda(\lambda-1)^2\}}, \tag{2.16}$$

where the numerical factor ν is $+1$ (-1) for the flexural (barreling) case. Expressions equivalent to (2.16) have been derived in [1–4] for the flexural case and in [3, 4] for the barreling case. It is of interest to note that (2.16) is independent of the value of λ_3. Figure 1 shows the dependence of η on λ, curve I pertaining to flexure and curve II to barreling. Both curves have the vertical line $\lambda \sim 3.383$ as an asymptote.

The substitution from (2.12) into (2.10), with (2.13)–(2.16), yields the result

$$G_2[\boldsymbol{u}] = 2l_3 \iint \gamma[\boldsymbol{u}]\,d\xi_1 d\xi_2 \equiv 0. \tag{2.17}$$

Thus, a critical state is a state of neutral equilibrium.

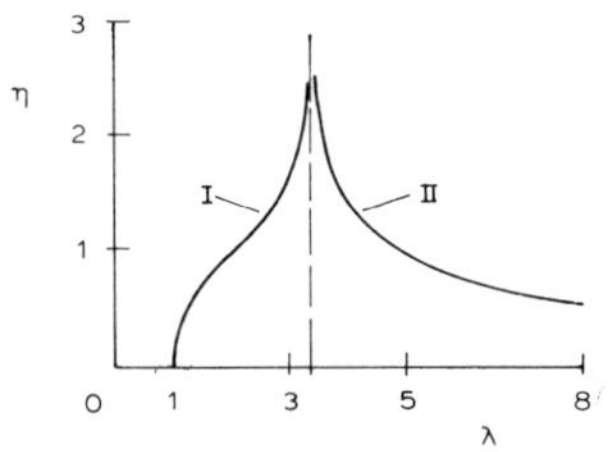

Figure 1. Relation between η and critical values of λ: curve I, flexure; curve II, barreling.

3. Stability of the critical states

To proceed, we write the displacement $\bar{\boldsymbol{u}}$ of (2.2) as the sum of a small constant multiple of the bifurcation field (2.12) and a field orthogonal to it. That is, $\bar{\boldsymbol{u}} = a\boldsymbol{u} + \boldsymbol{v}$, where

$$(3.1)\qquad \begin{gathered} \bar{u}_\alpha = au_\alpha + v_\alpha(\xi_1, \xi_2), \qquad \bar{u}_3 = v_3 = \lambda_3 \bar{E} \xi_3, \\ \iint u_{\alpha,\beta} v_{\alpha,\beta} d\xi_1 d\xi_2 = 0. \end{gathered}$$

The substitution from $(3.1)_{1,2}$ into (2.3) and (2.9) yields

$$(3.2)\qquad \begin{aligned} &v_{1,1}/\lambda_1 + v_{2,2}/\lambda_2 + \bar{E}/(1+\bar{E}) \\ &\quad = -\lambda_3\{a^2(u_{1,1}u_{2,2} - u_{1,2}u_{2,1}) + a\Lambda[\boldsymbol{u}, \boldsymbol{v}] + v_{1,1}v_{2,2} - v_{1,2}v_{2,1}\}, \\ &\gamma[\bar{\boldsymbol{u}}] = a^2\gamma[\boldsymbol{u}] - a\lambda\Lambda[\boldsymbol{u}, \boldsymbol{v}] + au_{\alpha,\beta}v_{\alpha,\beta} + \gamma[\boldsymbol{v}], \\ &\Lambda[\boldsymbol{u}, \boldsymbol{v}] \stackrel{\text{def}}{=} u_{1,1}v_{2,2} + u_{2,2}v_{1,1} - u_{1,2}v_{2,1} - u_{2,1}v_{1,2}. \end{aligned}$$

The result of employing (3.1) and (3.2) in (2.8), with (2.9)–(2.17), is

$$(3.3)\qquad \begin{aligned} &G[\bar{\boldsymbol{u}}] = G_2[\bar{\boldsymbol{u}}] - 8l_1l_2l_3\lambda_2^2\bar{E}^3/(1+\bar{E}), \\ &G_2[\bar{\boldsymbol{u}}] = 2l_3\iint\{\gamma[\boldsymbol{v}] - a^2\lambda_3 p\Lambda[\boldsymbol{u}, \boldsymbol{v}] - a\lambda_3 p(v_{1,1}v_{2,2} - v_{1,2}v_{2,1})\}d\xi_1 d\xi_2. \end{aligned}$$

With a held fixed at some small value, we contend, following Koiter, that state I is stable if G is positive for every nontrivial, kinematically admissible field $\boldsymbol{v}$ of order a^2, and is unstable if G is negative for any such field. We write

$$(3.4)\qquad v_\alpha = a^2 u^*_\alpha, \qquad \bar{E} = a^2 E^*,$$

and substitute into $(3.2)_1$ and (3.3). With the neglect of terms of orders a^3 and a^5, respectively, this yields

$$u^*_{1,1}/\lambda_1 + u^*_{2,2}/\lambda_2 + E^* + \lambda_3(u_{1,1}u_{2,2} - u_{1,2}u_{2,1}) = 0,$$

(3.5)
$$G[\bar{u}] = G_2[\bar{u}] = a^4\bar{G}[u^*],$$

$$\bar{G}[u^*] = 2l_3 \iint \{\gamma[u^*] - \lambda_3 p \Lambda[u, u^*]\} d\xi_1 \xi_2.$$

We now determine the field u^* which renders $\bar{G}$ stationary. Let $-2l_3 p^*(\xi_1, \xi_2)$ be the Lagrange multiplier that is introduced to account for the constraint $(3.5)_1$. (No restriction is imposed by the orthogonality condition for this problem.) The solution to the variational problem $\delta\bar{G} = 0$ is found to have the following form:

$$u^*_1 = U_1(\xi_2)\sin 2\Omega\xi_1 + \lambda_1 e^* \xi_1,$$

(3.6)
$$u^*_2 = V(\xi_2)\cos 2\Omega\xi_1 + \tfrac{1}{2}UU'/\lambda_2 - \lambda_2(e^* + E^*)\xi_2,$$

$$p^* = P(\xi_2)\cos 2\Omega\xi_1 + \tfrac{1}{2}(UU')' + \tfrac{1}{2}\beta' U'/\lambda^2\Omega^2 - 2\lambda_2^2(e^* + E^*),$$

where

$$e^* = -\lambda\lambda_3 H(2\lambda + \lambda_3^3)/\kappa, \qquad E^* = \lambda_3 H(\lambda^2 - 1)/2\kappa,$$

$$H = (2\lambda^2\Omega^2 l_2)^{-1} \int (U''^2 + \Omega^2 U'^2) d\xi_2, \qquad \kappa = 5\lambda^3 + 3\lambda(\lambda\lambda_3^3 + 1) + \lambda_3^3,$$

(3.7)
$$U_1 = -\{V' + (-1)^n \alpha/2\lambda_2\}/(2\lambda\Omega), \qquad \alpha = UU'' - U'^2,$$

$$P = \lambda_1\{V''' - 4\Omega^2 V' + (-1)^n[\alpha'' - 4\Omega^2\alpha + 2(\beta'' U - \beta' U')]/2\lambda_2\}/(4\lambda\Omega^2),$$

$$V(\xi_2) = M_1 \sinh 2\lambda\Omega\xi_2 + M_2 \sinh 2\Omega\xi_2 - M_3 \sinh(\lambda + 1)\Omega\xi_2 - M_4 \sinh(\lambda - 1)\Omega\xi_2.$$

Expressions for the constants $M_1, \cdots, M_4$ are shown in the Appendix.

The substitution from (2.12) and (3.6) into $(3.5)_3$ yields an expression for $\bar{G}$ which may be written in the form

(3.8)
$$\bar{G} = (l_1 l_3 \lambda_3 M^4 \eta^3/4l_2^3)(\bar{g}_2 - K\bar{g}_1),$$

where

(3.9)
$$K = (7\lambda^2 + 4\lambda\lambda_3^3 + 1)/\kappa,$$
$$\bar{g}_1 = \eta(\lambda^2 - 1)^2\{\nu + (\lambda^2 - 3)\sinh 2\lambda\eta/[2\lambda\eta(\lambda^2 + 1)]\}^2,$$

and $\bar{g}_2$ is given in the Appendix. Thus, from $(3.5)_2$, we see that state I is stable or unstable accordingly as the sign of $(\bar{g}_2 - K\bar{g}_1)$ is positive or negative.

4. Conclusions

Stability results for flexural bifurcations, in the limiting case of a thin plate (i.e., $\eta \to 0$), have been derived in [6]. The relevant expression for $\bar{G}$ may be cast in the form

$$\bar{G} = (l_1 l_3 \lambda_3 / 6 l_2^3)\eta^6 \{G_0^* + O(\eta^4)\},$$
$$G_0^* = 1 - (\lambda - 1)[16 + 1/(2 + \lambda_3^3)], \tag{4.1}$$

which is based on the normalizing condition, $U(0) = 1$. Accordingly, from $(2.14)_1$, it follows that an optimum comparison between (4.1) and (3.8) can be made by taking

$$M = (1 - m)^{-1}, \qquad \bar{G} = (l_1 l_3 \lambda_3 / 6 l_2^3)\eta^6 G^*,$$
$$G^* = 3(\bar{g}_2 - K\bar{g}_1)/(2\eta^3[1 - m]^4). \tag{4.2}$$

Figure 2 shows plots of G_0^* and G^* as functions of $\lambda - 1$, with $\lambda_3 = 1$, for the range $0 \leqq \lambda - 1 < 0.2$. Good agreement between them is noted at the smaller values of $\lambda - 1$.

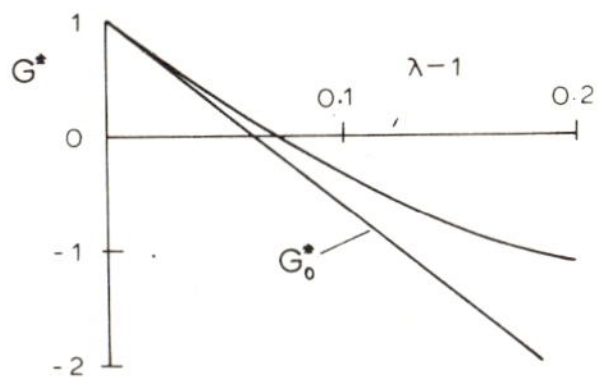

Figure 2. Comparison between the asymptotic calculation (G_0^*) and the exact calculation of G^* for small values of $\lambda - 1$, with $\lambda_3 = 1$.

Another choice for the normalization of U may be based on the requirement that $\max|U(\xi_2)| = 1$. From $(2.14)_1$, with (2.15) and (2.16), it is found that this maximum occurs at the point $\xi_2 = 0$ provided that $1 < \lambda < \sim 2.05$. As λ increases from 2.05 to 3.383, the points at which the maxima occur move steadily from $\xi_2 = 0$ towards the outer edges, $\xi_2 = \pm l_2$. (Recall that U is an even function of ξ_2.) For the barreling case, where $(2.14)_2$ applies, it is found that the points at which the maximum values of $|U(\xi_2)|$ occur are $\xi_2 = \pm l_2$ if $\lambda = 3.383$. As λ increases from 3.383, these points move inward from the outer edges. (Here, U is an odd function of ξ_2.)

Let $\bar{\xi}_2$ denote the point in the interval $[0, l_2]$ where $\max|U(\xi_2)|$ occurs, with U given by either $(2.14)_1$ or $(2.14)_2$, and let $\bar{t} = \bar{\xi}_2 / l_2$. A plot of $\bar{t}$ is shown in Figure 3 as a function of $\lambda - 1$. Based on the discussion of the preceding section, we may choose the value of M as follows:

$$M = \begin{array}{ll} |1 - m|^{-1}, & 1 < \lambda < 2.05, \\ |\cosh \lambda \Omega \bar{\xi}_2 - m \cosh \Omega \bar{\xi}_2|^{-1}, & 2.05 < \lambda < 3.383, \\ |\sinh \lambda \Omega \bar{\xi}_2 - m \sinh \Omega \bar{\xi}_2|^{-1}, & 3.383 < \lambda. \end{array} \tag{4.3}$$

This choice allows a direct comparison to be made between the results for flexural and barreling bifurcations.

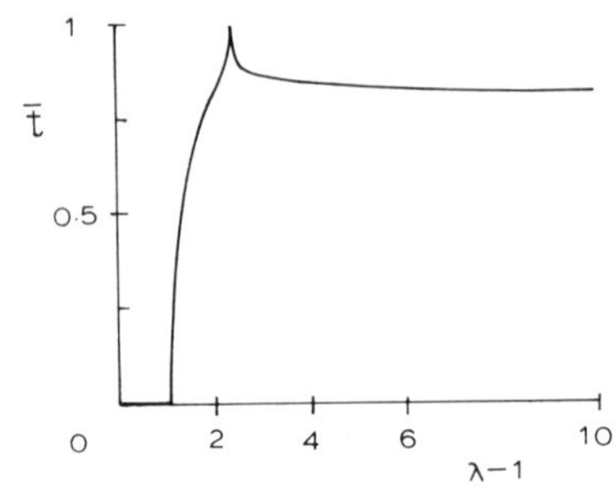

Figure 3. Relation between the dimensionless thickness coordinate $\bar{t}$, where $\max|U|$ occurs, and $\lambda - 1$.

Following $(4.2)_3$, we rewrite (3.8) in the form shown in $(4.2)_2$ with

$$G^* = 3M^4(\bar{g}_2 - K\bar{g}_1)/(2\eta^3), \tag{4.4}$$

where M is given by (4.3). The behavior of G^* is shown in Figure 4.

In Figures 2 and 4 we see that G^* is positive for values of λ between 1 and about 1.072, and negative for larger values of λ (up to 3.383). Although this conclusion is based on the condition $\lambda_3 = 1$, the results are extremely insensitive to the value assigned to λ_3. Indeed, a study of the limiting cases $\lambda_3 \to 0$ and $\lambda_3 \to \infty$ shows that $G^* = 0$ for values of λ about 1.071 and 1.073, respectively. [This insensitivity is reflected also in the asymptotic expression for G^* in $(4.1)_2$.] Thus, the *ultimate* stability limit for flexural bifurcations is reached for $\lambda \sim 1.072$. The corresponding value of η is about 0.32. From $(2.13)_2$ and $(2.15)_2$, we obtain $\eta = n\pi l_2/2l_1$. Therefore, with respect to the lowest-order flexural mode ($n = 1$), the ultimate stability limit can be reached for a plate having an aspect ratio (l_2/l_1) about 0.2. For a thicker plate the critical state is unstable, indicating that buckling must occur by some means other than a smooth bifurcation from a homogeneously deformed state. For a thinner plate (i.e., for $l_2/l_1 < 0.1$), the possible stability of critical states corresponding to certain higher-order flexural modes is not ruled out, provided that lower-order ones be suppressed by the imposition of appropriate passive (kinematic) constraints.

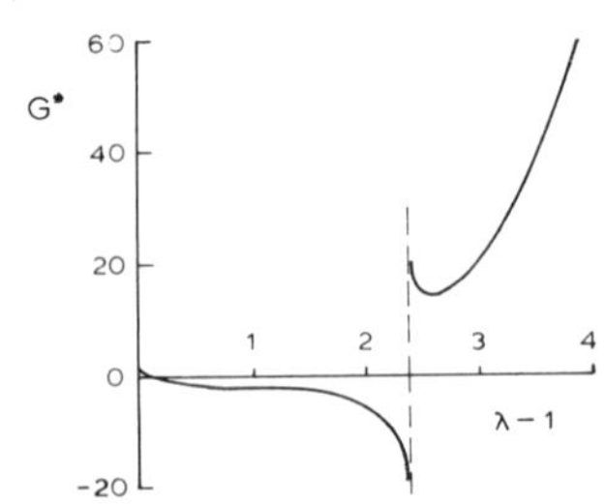

Figure 4. Relation between G^* and $\lambda - 1$, with $\lambda_3 = 1$.

In Figure 4 we see that G^* is positive for all values of $\lambda > 3.383$, which indicates that the underlying state of pure homogeneous deformation, corresponding to any barreling bifurcation, is stable. However, as the compressive load on the faces $\xi_1 = \pm l_1$ is increased from zero, values of λ corresponding to unstable critical states for flexure will be realized before any corresponding to a barreling bifurcation is attained. Nevertheless, it is possible to consider a variation of the stated problem which does permit the barreling results to have (at least theoretical) validity.†

We imagine the plate to be cut in half along the plane $\xi_2 = 0$. This surface is then constrained to remain in contact with a rigid frictionless platen, so that

$$\bar{u}_2(\xi_1, 0) = 0, \qquad -l_1 \leqq \xi_1 \leqq l_1. \tag{4.5}$$

As before, the surface $\xi_2 = l_2$ remains traction free. The condition (4.5) rules out the possibility of any flexural bifurcations. Now, as the compressive load is increased, a barreling-type bifurcation of high (infinite) order is reached at $\lambda = 3.383$, which corresponds to a wrinkling of the free surface, $\xi_2 = l_2$. This conclusion, that surface wrinkling obtains as a limiting case of barreling bifurcations, was also reached by Usmani [10], whose investigation is based on a study of surface waves.

5. Appendix

We employ the following notation:

$$\begin{aligned}
\Delta &= 4\lambda^3 \sinh 2\lambda\eta \cosh 2\eta - (\lambda^2+1)^2 \cosh 2\lambda\eta \sinh 2\eta,\\
f(\lambda) &= (\lambda+1)\{3(\lambda^2-1)(5\lambda^2+2\lambda+1)/[(3\lambda+1)(\lambda+3)] + 7\lambda^2+1\},\\
g(\lambda) &= 3(\lambda^2-1)(\lambda+1)^2(7\lambda^2-2\lambda+3)/[(3\lambda+1)(\lambda+3)] + 8\lambda(2\lambda^2-1),\\
f_1 &= m\{f(\lambda)\sinh(\lambda+1)\eta - \nu f(-\lambda)\sinh(\lambda-1)\eta\}\\
&\quad - (7\lambda^2+1)(\lambda \sinh 2\lambda\eta + m^2 \sinh 2\eta),\\
f_2 &= m\{g(\lambda)\cosh(\lambda+1)\eta + \nu g(-\lambda)\cosh(\lambda-1)\eta\}\\
&\quad - 4(2\lambda^2-1)\{\lambda^2\cosh 2\lambda\eta + m^2\cosh 2\eta - \nu(\lambda^2+m^2)\},\\
F_1 &= \{4\lambda^2 f_1 \cosh 2\eta - (\lambda^2+1) f_2 \sinh 2\eta\}/\Delta,\\
F_2 &= 2\lambda\{\lambda f_2 \sinh 2\lambda\eta - (\lambda^2+1) f_1 \cosh 2\lambda\eta\}/\Delta,\\
F(\lambda) &= 24\lambda(\lambda^2-1)(\lambda+1)/[(3\lambda+1)(\lambda+3)].
\end{aligned} \tag{5.1}$$

† The question of whether the assumed incompressible neo-Hookean behavior and idealized boundary conditions might be unrealistic for states of such severe compression necessarily lies outside the scope of this investigation.

Then

$$
(5.2)\qquad \begin{aligned}
M_1 &= (-1)^n \Omega M^2 F_1/(32\lambda\lambda_2),\\
M_2 &= (-1)^n \Omega M^2 F_2/(32\lambda\lambda_2),\\
M_3 &= (-1)^n \Omega m M^2 F(\lambda)/(32\lambda\lambda_2),\\
M_4 &= (-1)^n \Omega \nu m M^2 F(-\lambda)/(32\lambda\lambda_2).
\end{aligned}
$$

With the further notation,

$$
(5.3)\qquad \begin{aligned}
\gamma_1 &= F_2 \sinh 2\lambda\eta + F_2 \sinh 2\eta\\
&\quad - m\{F(\lambda)\sinh(\lambda+1)\eta + \nu F(-\lambda)\sinh(\lambda-1)\eta\},\\
\gamma_2 &= 2\lambda F_1 \cosh 2\lambda\eta + 2F_2 \cosh 2\eta\\
&\quad - m\{(\lambda+1)F(\lambda)\cosh(\lambda+1)\eta + \nu(\lambda-1)F(-\lambda)\cosh(\lambda-1)\eta\},\\
\gamma_3 &= \lambda^2 \cosh 2\lambda\eta + m^2 \cosh 2\eta - \nu(\lambda^2+m^2)\\
&\quad - 2\lambda m\{\cosh(\lambda+1)\eta - \nu \cosh(\lambda-1)\eta\},
\end{aligned}
$$

the expression for $\bar{g}_2$ reads

$$
(5.4)\qquad \begin{aligned}
2\lambda^2\bar{g}_2 &= (\lambda^2-1)^2 \sinh 2\lambda\eta\,[\nu(\lambda^2-1)m^2 + 2\lambda^2\cosh 2\lambda\eta + (\lambda^2+1)m^2\cosh 2\eta\\
&\quad - 2\lambda m\{(\lambda+1)\cosh(\lambda+1)\eta + \nu(\lambda-1)\cosh(\lambda-1)\eta\}]/(\lambda^2+1)\\
&\quad - [8\lambda(\lambda^2+1)^2\gamma_1\gamma_3 + (7\lambda^2+1)(\lambda^2-1)^2\gamma_2 \sinh 2\lambda\eta]/[64\lambda^3(\lambda^2+1)]\\
&\quad - 3(\lambda^2-1)\left[\eta\sigma + \sum_{k=1}^{3}\{\bar{\sigma}_{k,4-k} + \bar{\sigma}_{k,k-4}\} + \sum_{k=1}^{2}\{\bar{\sigma}_{k-1,3-k} + \bar{\sigma}_{k,k-2}\}\right]\Big/(32\lambda^2)\\
&\quad + \left[\eta\tau + \sum_{k=0}^{3}\{\bar{\tau}_{k,4-k} + \bar{\tau}_{k+1,k-3}\} + \sum_{k=1}^{2}\{\bar{\tau}_{k-1,3-k} + \bar{\tau}_{k,k-2}\}\right]\Big/(16\lambda)
\end{aligned}
$$

where

$$
(5.5)\qquad \begin{aligned}
\sigma &= \tfrac{1}{2}m^2(\lambda^2-1)\{(\lambda-1)F(\lambda) + (\lambda+1)F(-\lambda)\},\\
\tau &= 4[16\{\lambda^2(\lambda^2-1)m^2 + \lambda^2(\lambda^4+m^4) + (\lambda^2+m^2)^2\}\\
&\quad - 2\lambda m^2(\lambda^2-1)\{(\lambda+1)^3 + (\lambda-1)^3\}\\
&\quad + 2\{(\lambda-1)^4 + (\lambda+1)^4\}(2\lambda^2+1)m^2 - (\lambda^2-1)^2(\lambda^2+1)m^2],
\end{aligned}
$$

and, for integers j, k,

$$
(5.6)\qquad \begin{aligned}
\bar{\sigma}_{j,k} &= \sigma_{j,k}\sinh(j\lambda+k)\eta/(j\lambda+k),\\
\bar{\tau}_{j,k} &= \tau_{j,k}\sinh(j\lambda+k)\eta/(j\lambda+k),
\end{aligned}
$$

with

$$\sigma_{3,1} = -\lambda(\lambda-1)^2 mF_1, \qquad \sigma_{3,-1} = -\nu\lambda(\lambda+1)^2 mF_1,$$

$$\sigma_{1,3} = -(\lambda-1)^2 mF_2, \qquad \sigma_{1,-3} = -\nu(\lambda+1)^2 mF_2,$$

$$\sigma_{2,2} = \tfrac{1}{2}(\lambda+1)(\lambda-1)^2 m^2 F(\lambda), \qquad \sigma_{2,-2} = \tfrac{1}{2}(\lambda-1)(\lambda+1)^2 m^2 F(-\lambda),$$

$$\sigma_{2,0} = \tfrac{1}{2}\nu\{16\lambda^3 F_1 + m^2[(\lambda+1)^3 F(\lambda) + (\lambda-1)^3 F(-\lambda)]\}, \tag{5.7}$$

$$\sigma_{0,2} = \tfrac{1}{2}\nu\{16\lambda^2 F_2 + m^2[(\lambda+1)^3 F(\lambda) + (\lambda-1)^3 F(-\lambda)]\},$$

$$\sigma_{1,1} = -\nu(\lambda+1)m\{4\lambda^2 F(\lambda) + (\lambda+1)(\lambda F_1 + F_2)\},$$

$$\sigma_{1,-1} = -(\lambda-1)m\{4\lambda^2 F(-\lambda) + (\lambda-1)(\lambda F_1 + F_2)\},$$

and

$$\tau_{4,0} = 64\lambda^4(2\lambda^2-1), \qquad \tau_{0,4} = 64\lambda^2 m^4,$$

$$\tau_{3,1} = -32\lambda^2(\lambda+1)m\{(\lambda+1)(3\lambda^2-1) + 2\lambda(\lambda-1)\},$$

$$\tau_{3,-1} = -32\nu\lambda^2(\lambda-1)m\{(\lambda-1)(3\lambda^2-1) - 2\lambda(\lambda+1)\},$$

$$\tau_{1,3} = -64\lambda(\lambda+1)m^3(\lambda^2+2\lambda-1),$$

$$\tau_{1,-3} = -64\nu\lambda(\lambda-1)m^3(\lambda^2-2\lambda-1),$$

$$\tau_{2,2} = 2m^2\{(\lambda-1)^4[4-(\lambda+1)(3\lambda-1)] + 8\lambda(\lambda+1)^3(\lambda^2+3\lambda-2) + 32\lambda^2(3\lambda^2-1)\},$$

$$\tau_{2,-2} = 2m^2\{(\lambda+1)^4[4-(\lambda-1)(3\lambda+1)] + 8\lambda(\lambda-1)^3(\lambda^2-3\lambda-2) + 3\lambda^2(3\lambda^2-1)\}, \tag{5.8}$$

$$\tau_{2,0} = 4\nu(\lambda^2-1)\{m^2(\lambda^2-1)(5\lambda^2+3) - 32\lambda^2(\lambda^2+m^2)\},$$

$$\tau_{0,2} = 4\nu(\lambda^2-1)m^2\{(\lambda^2-1)(3\lambda^2+5) - 64\lambda^2\},$$

$$\tau_{1,1} = -32\nu(\lambda-1)m\{(\lambda^2+m^2)(\lambda-1)(\lambda^2-\lambda+2) - 2\lambda(\lambda+1)[m^2+\lambda(\lambda^2+3\lambda+1)]\},$$

$$\tau_{1,-1} = -32(\lambda+1)m\{(\lambda^2+m^2)(\lambda+1)(\lambda^2+\lambda+2) + 2\lambda(\lambda-1)[m^2-\lambda(\lambda^2-3\lambda+1)]\}.$$

Acknowledgement

This work was carried out with the support of the National Science Foundation by a grant to Lehigh University. The author is grateful to Professors W. T. Koiter and R. S. Rivlin for their helpful comments.

References

1. M. A. Biot, *Exact theory of buckling of a thick slab*, Appl. Sci. Res. **A12** (1963), 183–198.

2. M. Levinson, *Stability of a compressed neo-Hookean rectangular parallelepiped*, J. Mech. Phys. Solids **16** (1968), 403–415.

3. J. L. Nowinski, *On the elastic stability of thick columns*, Acta Meccanica **7** (1969), 279–286.

4. K. N. Sawyers and R. S. Rivlin, *Bifurcation conditions for a thick elastic plate under thrust*, Int. J. Solids Structures **10** (1974), 483–501.

5. Zhong-Heng, Guo, *Vibration and stability of a cylinder subjected to finite deformation*, Arch. Mech. Stos. **14** (1962), 757–768.

6. K. N. Sawyers and R. S. Rivlin, *Stability of a thick elastic plate under thrust*, J. Elasticity, to appear.

7. W. T. Koiter, *On the stability of elastic equilibrium*, Thesis, Delft, 1945 (in Dutch). English translation NASA TT-F-10 (1967), 833.

8. W. T. Koiter, *Elastic stability and post-buckling behaviour*, Proc. Symp. Nonlinear Problems (R. E. Langer, ed.), Univ. Wisconsin Press, Madison, 1963, pp. 257–275.

9. W. T. Koiter, *Current trends in the theory of buckling*, Proc. IUTAM Symp. Buckling of Structures (B. Budiansky, ed.), Springer-Verlag, Berlin, 1976, pp. 1–16.

10. S. A. Usmani, *Some stability problems in finite elasticity*, Thesis, University of Kentucky, Lexington, 1973.

FINITE EXTENSION AND TORSION OF THIN ELASTIC STRIPS

R. T. SHIELD

Department of Theoretical and Applied Mechanics, University of Illinois at Urbana-Champaign, Urbana, Illinois 61801, USA

(Received September 11, 1980)

ABSTRACT

The large extension and torsion of highly elastic bars of slender section is examined by an approximate method. The approach treats the cylinder of elongated section as a membrane of non-uniform thickness and the problem reduces to a one-dimensional problem when the deformation is the same at each section. The solution for a particular strain-energy function can be obtained by numerical integration. Results are presented for a Mooney material and for a material with an empirical strain-energy function for a bar with an elongated rectangular section. When the section is not symmetrical about the centroid, the location of the axis of torsion varies with the amount of twist, and results showing the migration of the axis of torsion for a tapered section are given.

1. Introduction

The problem of the finite extension and torsion of a circular cylinder of incompressible isotropic elastic material was solved by Rivlin [1] using a general form for the strain-energy function. Green and Wilkes [2] treated the same problem for a compressible isotropic material and solutions can be obtained by numerical integration of an ordinary differential equation when the form of the strain-energy function is known. Rivlin's treatment [1] included the finite extension and torsion of a circular cylindrical tube of incompressible isotropic material, and Green and Adkins [3–6] extended the analysis to compressible and to non-isotropic materials.

The torsion of finitely extended cylinders with cross sections other than circular or a circular annulus leads to two-dimensional nonlinear partial

Proceedings of the IUTAM Symposium on Finite Elasticity, Lehigh University, August 10–15, 1980, Invited paper.

Carlson, D.E. and Shield, R.T. (eds.)
Proceedings of the IUTAM Symposium on Finite Elasticity
ISBN 90 247 2629 8

differential equations under the assumption that the state of strain is the same at each section. No exact solutions have been obtained but the twisting moment has been determined to first order in the twist [7] and second-order effects have been considered by various writers (see, for example, [7–10]). Additional references and a historical survey can be found in [11]. The problem has been formulated by Lee and Shield [12] and it was shown in general terms how estimates for the axial force and twisting moment could be obtained through the use of variational principles [13]. As an illustration, an elliptical cylinder of neo-Hookean material was considered in [12] and accurate estimates for the end loads were obtained for cylinders with axes in the ratios 2:1 and 4:1 for a wide range of extension and twist.

Here we consider a cylinder of elongated section and assume that the section is slender enough so that the membrane approximation can be used. The elastic material of the thin strip is orthotropic with preferred directions in the axial direction, along the section and transverse. Within the membrane assumption, in Section 2 we show that for a given material and known section thickness distribution exact numerical solutions for large extension and twist can be derived by integration of a non-linear ordinary differential equation with two-point boundary conditions.

Section 3 considers small twist on finite extension and the twisting moment M is found correct to third order in the twist per unit length ψ for an arbitrary section shape and a general form for the strain energy. The axial load L is given correct to second order in ψ. With the approach of [10], L can be found to $O(\psi^4)$ from the third order form for M but the details are omitted. When the strip is unextended, the initial axis of twist is not necessarily the line of centroids (symmetry of the section is not assumed), and it is shown that the initial axis is determined by vanishing of the third moment of the section about the axis.

Section 4 specializes the results to incompressible isotropic materials. For a Mooney material and for a material with an empirical strain-energy function, numerical results are presented showing the variation of the end loads M and L with the angle of twist for various extensions for an elongated rectangular section. A trapezoidal section with thickness varying in the ratio 1:7 was also treated numerically in order to illustrate the migration of the axis of torsion as the twist increases for a non-symmetrical section.

2. Extension and torsion of a thin elastic strip

We consider a thin elastic strip of length ℓ and width 2a composed of highly elastic material. The strip is flat in the unstressed state but a thin strip with an initial twist could be treated in a similar manner. We use a rectangular cartesian coordinate system to denote particle locations and use x_i and y_i $(i = 1, 2, 3)$ to denote coordinates in the undeformed and deformed states, respectively. Initially the middle surface of the strip lies in the plane $x_2 = 0$ and we use x, z for the coordinates x_1, x_3 of particles on the middle surface. The strip occupies the region

$$-\tfrac{1}{2}h(x) \leqq x_2 \leqq \tfrac{1}{2}h(x), \qquad -a \leqq x \leqq a, \qquad 0 \leqq z \leqq \ell$$

in the undeformed state and we assume that $h \ll a$ so that the membrane approximation, which ignores strain variations in the transverse direction, can be used. The cross section is not necessarily symmetric about $x_1 = 0$ so that the thickness $h(x)$ is not restricted to be an even function. We will treat strips of homogeneous material but the analysis can be extended straightforwardly to a strip with material properties dependent on x.

The strip is extended and twisted by forces applied to the end sections $z = 0, \ell$ with the major surfaces and the edges $x = \pm a$ traction free. We assume that the strain is the same at each section and the deformation is indicated in Figure 1. A particle initially on the middle surface at the point $(x, 0, z)$ moves to the location

$$y_i = (r \cos \psi z, r \sin \psi z, \lambda z),$$

where $r = r(x)$ and the constants ψ, λ denote the angle of twist per unit initial length and the overall axial extension ratio. The material line P_0P parallel to the x_3-axis becomes a circular helix $P_0'P'$ of radius r and angle $\tan^{-1}(\psi r/\lambda)$. For a section symmetric about $x = 0$, the axis of torsion is $x = 0$ so that $r(0)$ is zero, but for other sections the location x_A of the axis of torsion will vary with λ and ψ and will be determined from $r(x_A) = 0$. With the membrane approximation material lines normal to the middle surface remain normal to the middle surface. The principal directions of strain at P' are in the radial direction $A'P'$, tangential to the helix $P_0'P'$ and in the transverse direction. The corresponding principal extension ratios will be denoted by λ_1, λ_2 and λ_3, with λ_1 radial and λ_3 transverse, and we have

$$\lambda_1 = \frac{dr}{dx} = r', \qquad \lambda_2 = \frac{ds}{dz} = (\psi^2 r^2 + \lambda^2)^{1/2}, \tag{2.1}$$

where $s(z)$ denotes arc length of the helix $P_0'P'$.

We suppose that the material of the strip is orthotropic with the preferred directions in the x_i-directions. For the deformed strip, with principal axes of strain along the preferred directions, the strain energy W per unit initial volume is then a function of λ_1, λ_2, λ_3,

$$W = W(\lambda_1, \lambda_2, \lambda_3).$$

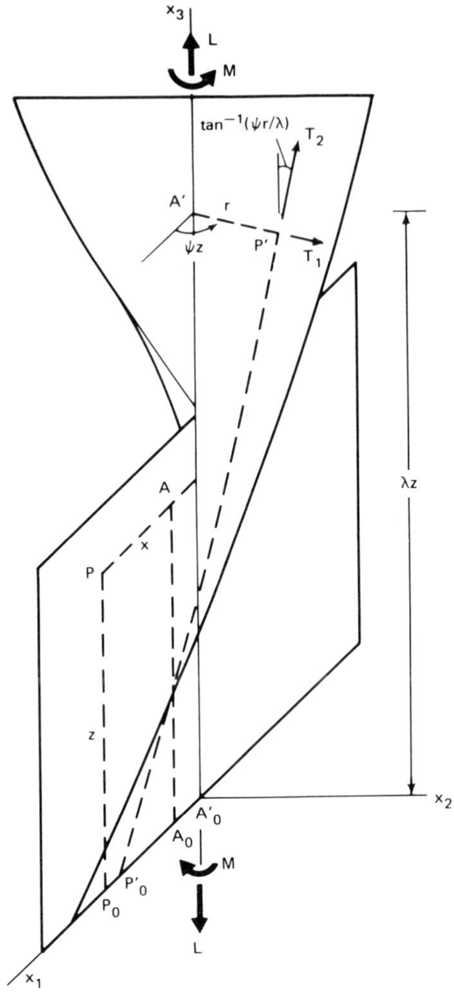

Figure 1. Extension and torsion of a thin flat strip with axial extension ratio λ and twist ψ per unit initial length.

For a compressible material, zero transverse stress requires

$$\frac{\partial W}{\partial \lambda_3} = 0, \tag{2.2}$$

and we can view this equation as serving to determine λ_3 in terms of λ_1 and λ_2. For an incompressible material, $\lambda_3 = 1/\lambda_1\lambda_2$ so that in either case we can assume that W is determined by λ_1 and λ_2,

$$W = w(\lambda_1, \lambda_2) = w^*(r, r').$$

The strain energy per unit area of the initial middle surface is hW and the principal stress resultants per unit length of the deformed strip are

$$T_1 = \frac{h(x)}{\lambda_2} \frac{\partial w}{\partial \lambda_1}, \qquad T_2 = \frac{h(x)}{\lambda_1} \frac{\partial w}{\partial \lambda_2} \tag{2.3}$$

in the radial direction and tangential to the helix $P_0'P'$, respectively. Note that for a compressible material

$$\frac{\partial W}{\partial \lambda_\alpha} = \frac{\partial w}{\partial \lambda_\alpha} \qquad (\alpha = 1, 2)$$

when (2.2) holds.

The curvatures of the deformed middle surface in the directions of the principal stress resultants are both zero and T_1, T_2 are independent of z. It follows that for equilibrium of the strip we need only consider balance of force in the radial direction which leads to

$$\frac{d}{dx}(\lambda_2 T_1) - \lambda_1 T_2 k \lambda_2 = 0 \qquad |x| \leqq a, \tag{2.4}$$

where k is the curvature of the helix $P_0'P'$ at P',

$$k = r\psi^2/(\psi^2 r^2 + \lambda^2).$$

If we use (2.3) in (2.4), equilibrium requires

$$\frac{d}{dx}\left(h(x)\frac{\partial w}{\partial \lambda_1}\right) - \frac{r\psi^2}{\lambda_2} h(x) \frac{\partial w}{\partial \lambda_2} = 0 \qquad |x| \leqq a. \tag{2.5}$$

Alternatively, the equation for $r(x)$ can be obtained from the Euler differential equation associated with the total strain energy of the strip given by

$$U = \ell \int_{-a}^{a} h(x) w(\lambda_1, \lambda_2) dx,$$

and viewed as a functional of r. We then have

$$\frac{d}{dx}\left(h(x)\frac{\partial w^*}{\partial r'}\right) - h(x)\frac{\partial w^*}{\partial r} = 0 \qquad |x| \leqq a, \tag{2.6}$$

and this agrees with (2.5) when (2.1) is used. Zero traction on the edges requires

$$T_1 = 0 \qquad \text{at } x = \pm a, \tag{2.7}$$

which leads to

$$\frac{\partial w}{\partial \lambda_1} = \frac{\partial w^*}{\partial r'} = 0 \qquad \text{at } x = \pm a \tag{2.8}$$

when h is non-zero at the edges. For a section with h zero at the edge $x = a$, such as an elongated elliptical section, the point $x = a$ becomes a singular point of the differential equation for $r(x)$ and we require

$$h(x)\frac{\partial w^*}{\partial r'} \to 0 \qquad \text{as } x \to a.$$

For tensile T_2, the sign of the second term in (2.4) depends upon the sign of k and therefore of r, and with (2.7) it follows that T_1 is compressive and the compression is a maximum at the axis of torsion $r = 0$. For a stretched strip, as the twist ψ is increased a stage will be reached when the compressive stress in the radial direction will lead to buckling with sections folding over on themselves.

Equation (2.5) or (2.6) is a second-order differential equation for $r(x)$ which is linear in the second derivative r'' and so will usually be amenable to standard methods of numerical integration. When h is constant, the functional U has an integrand which does not depend explicitly on x and the Euler differential equation then has the first integral

$$r'\frac{\partial w^*}{\partial r'} - w^* = \text{constant}. \tag{2.9}$$

Because this is, in general, a non-linear equation for r', it will usually be simpler to integrate (2.5) or (2.6) directly rather than to use (2.9).

The resultant force and moment on a section z = constant arise from the stress resultant T_2 which is inclined at an angle $\tan^{-1}(\psi r/\lambda)$ to the x_3-axis. The twisting moment M and the axial force L are given by

$$\begin{aligned} M &= \int_{-a}^{a} \frac{\psi r T_2}{(\psi^2 r^2 + \lambda^2)^{1/2}}\, r r' dx = \int_{-a}^{a} \frac{\psi r^2}{\lambda_2} h(x) \frac{\partial w}{\partial \lambda_2}\, dx, \\ L &= \int_{-a}^{a} \frac{\lambda T_2}{(\psi^2 r^2 + \lambda^2)^{1/2}}\, r' dx = \int_{-a}^{a} \frac{\lambda}{\lambda_2} h(x) \frac{\partial w}{\partial \lambda_2}\, dx. \end{aligned} \tag{2.10}$$

The expressions for M and L can also be derived from the relations (see also [10])

$$M = \frac{1}{\ell}\frac{\partial U}{\partial \psi}, \qquad L = \frac{1}{\ell}\frac{\partial U}{\partial \lambda}. \tag{2.11}$$

The total strain energy U depends on ψ, λ through the explicit dependence of λ_2 on ψ and λ but also because the solution $r(x)$ of (2.6) and (2.8) varies with ψ, λ. If we use the equilibrium equation (2.6) and the boundary condition (2.8) we can show that

$$M = \frac{1}{\ell}\frac{\partial U}{\partial \psi} = \int_{-a}^{a} h(x) \left(\frac{\partial w}{\partial \psi}\right)_{\text{exp}} dx, \tag{2.12}$$

where $(\partial w/\partial\psi)_{\exp}$ indicates an explicit differentiation of w with respect to ψ (holding r and r' constant). The relation (2.12) is a special case of a relation obtained by Lee [14] for a cylinder of arbitrary cross section (see also [12]). Because the explicit dependence of w on ψ and r occurs only through the product ψr, we also have

$$M = \frac{1}{\psi}\int_{-a}^{a} h(x) r \frac{\partial w}{\partial r}\, dx,$$

again a special case of a relation [12] for an arbitrary cross section. Similarly we can show that

$$L = \int_{-a}^{a} h(x)\left(\frac{\partial w}{\partial \lambda}\right)_{\exp} dx = \frac{\lambda}{\psi^2}\int_{-a}^{a} \frac{h(x)}{r}\frac{\partial w}{\partial r}\, dx. \tag{2.13}$$

Expression (2.12) for M and the first of (2.13) for L can easily be seen to agree with (2.10).

The resultant axial force L acts along the axis of torsion, as it must if there is no bending of the strip. The condition that L acts along the axis of torsion $r = 0$ is

$$\int_{-a}^{a} \frac{rT_2}{(\psi^2 r^2 + \lambda^2)^{1/2}}\, r'dx = 0, \tag{2.14}$$

and this is also the condition that there be no resultant load perpendicular to the x_3-axis. With the equilibrium equation (2.4), condition (2.14) becomes

$$\int_{-a}^{a} \frac{d}{dx}(\lambda_2 T_1)dx = 0,$$

which is satisfied when the edge boundary conditions (2.7) hold. Note that ordinary discontinuities in the thickness $h(x)$ are allowable but T_1 must be continuous for equilibrium.

3. Small twist on finite extension

For simple extension of the strip we will have $\partial W/\partial\lambda_1$ zero in addition to (2.2) for a compressible material or in addition to $\lambda_3 = 1/\lambda_1\lambda_2$ for an incompressible material. In either case λ_1 and λ_3 will be determinable in terms of $\lambda_2 = \lambda$, and we shall denote the value of λ_1 for simple extension of amount λ in the x_3-direction by $\mu(\lambda)$. The axial force $L_0(\lambda)$ in simple extension is given by

$$L_0(\lambda) = \left(\frac{\partial w}{\partial \lambda_2}\right)_{(\mu,\lambda)}\int_{-a}^{a} h(x)dx = A_0 t(\lambda), \tag{3.1}$$

where A_0 is the initial cross-sectional area and $t(\lambda)$ is the nominal stress for simple extension in the x_3-direction. For small ψ and non-zero L_0, the strip will twist about the centroid x_c of the section, the line of action of L_0, and we will have

$$r = \mu(x - x_c) + O(\psi^2). \tag{3.2}$$

This can also be seen by substituting

$$r = \mu(x - x_A), \qquad \psi = 0$$

into condition (2.14) to give

$$\frac{\mu}{\lambda} t(\lambda) \int_{-a}^{a} (x - x_A) h(x) dx = 0,$$

which requires x_A to coincide with x_c when $t(\lambda)$ is non-zero.

The value of M for small ψ is, from (2.10),

$$M = \psi \frac{\mu^2}{\lambda} t(\lambda) I_0 + O(\psi^3), \tag{3.3}$$

where I_0 is the moment of inertia of the initial section about the centroid,

$$I_0 = \int_{-a}^{a} (x - x_c)^2 h(x) dx.$$

Expression (3.3) agrees with that found previously [7] for small twist superposed on simple extension for cylinders of isotropic material if the geometrical torsional rigidity S_0 is neglected in comparison with I_0.

It was shown in [10] that the relations (2.11) can be used to determine L to second order in ψ when M is known to first order, and with (3.3) we get

$$L = A_0 t(\lambda) + \tfrac{1}{2}\psi^2 I_0 \frac{d}{d\lambda}\left(\frac{\mu^2}{\lambda} t(\lambda)\right) + O(\psi^4). \tag{3.4}$$

This value for L agrees with that given by Shield [10] for cylinders of arbitrary section when S_0 is set equal to zero.

For small ψ the second term in (2.4) or (2.5) is $O(\psi^2)$, and setting $\lambda_2 = \lambda$ and $r = \mu(x - x_c)$ in this term we obtain

$$\lambda_2 T_1 = h(x) \frac{\partial w}{\partial \lambda_1} = \psi^2 \frac{\mu}{\lambda} t(\lambda) \int_{-a}^{x} (\xi - x_c) h(\xi) d\xi + O(\psi^4). \tag{3.5}$$

If we set

$$r = \mu(x - x_c) + \psi^2 u(x) + O(\psi^4)$$

and expand $\partial w / \partial \lambda_1$ in (3.5) about (μ, λ) in powers of ψ^2, we find that

$$u = u_0 + \frac{\mu}{\lambda w_{11}} \left\{ t(\lambda) \int_{x_c}^{x} \frac{d\eta}{h(\eta)} \int_{-a}^{\eta} (\xi - x_c) h(\xi) d\xi - \frac{\mu}{6} w_{12} (x - x_c)^3 \right\}, \tag{3.6}$$

where u_0 is a constant. We are using the notation

$$w_{\alpha\beta} = \frac{\partial^2 w}{\partial \lambda_\alpha \partial \lambda_\beta} \qquad (\alpha, \beta = 1, 2)$$

and these derivatives are evaluated for the state of simple extension in the x_3-direction. For a section symmetric about $x = 0$, both x_c and u_0 are zero. For other sections the constant u_0 must be chosen so that (2.14) is satisfied to $O(\psi^2)$ and this leads to

$$t(\lambda) \int_{-a}^{a} uhdx + \mu w_{12} \int_{-a}^{a} (x - x_c) u' hdx = \frac{1}{2} \frac{\mu^3}{\lambda} \left[\frac{t(\lambda)}{\lambda} - w_{22} \right] \int_{-a}^{a} (x - x_c)^3 hdx.$$

With u known, r is known to $O(\psi^2)$ and M can be found to $O(\psi^3)$ from (2.10), which gives, on expanding in powers of ψ,

$$\begin{aligned} M = \psi \frac{\mu^2}{\lambda} t(\lambda) I_0 + \psi^3 \frac{\mu}{\lambda} \Big\{ 2t(\lambda) \int_{-a}^{a} (x - x_c) uhdx + \mu w_{12} \int_{-a}^{a} (x - x_c)^2 u' hdx \\ + \frac{1}{2} \frac{\mu^3}{\lambda} \left[w_{22} - \frac{t(\lambda)}{\lambda} \right] \int_{-a}^{a} (x - x_c)^4 hdx \Big\} + O(\psi^5). \end{aligned} \tag{3.7}$$

If we write (3.7) as

$$M = \psi m_1(\lambda) + \psi^3 m_3(\lambda) + O(\psi^5),$$

then the relations (2.11) imply [10] that L has the form

$$L = L_0(\lambda) + \tfrac{1}{2}\psi^2 \frac{dm_1}{d\lambda} + \tfrac{1}{4}\psi^4 \frac{dm_3}{d\lambda} + O(\psi^6). \tag{3.8}$$

For a rectangular section (h constant), expression (3.7) for M gives

$$\begin{aligned} M/ha^2 = \tfrac{2}{3}\psi a \mu^2 t(\lambda)/\lambda + \tfrac{1}{5}(\psi a)^3 (\mu/\lambda)^2 \{ \mu^2 [w_{22} - t(\lambda)/\lambda] \\ - \mu^2 w_{12}^2 / w_{11} - \tfrac{4}{3} t(\lambda)[2t(\lambda) + \mu w_{12}]/w_{11} \} + O(\psi^5). \end{aligned} \tag{3.9}$$

A separate treatment is needed for $\lambda = 1$ because with L_0 zero the initial axis of twist is not necessarily the line of centroids. In this case the first term in the expansion of the stress resultant T_2 for small ψ results from the extension of longitudinal fibers due to the twisting and we have

$$T_2 = \tfrac{1}{2} E \psi^2 r^2 + O(\psi^4), \tag{3.10}$$

where E is Young's modulus in the axial direction of the (orthotropic) strip. If we use (3.10) and $r = x - x_A$ in (2.14), we find that the initial location of the axis of torsion is determined by

$$\int_{-a}^{a} (x - x_A)^3 h(x)dx = 0. \tag{3.11}$$

The twisting moment M is, from (2.10),

$$M = \tfrac{1}{2}\psi^3 E \int_{-a}^{a} (x - x_A)^4 h(x)dx + O(\psi^5), \tag{3.12}$$

and we also have

$$L = \tfrac{1}{2}\psi^2 E \int_{-a}^{a} (x - x_A)^2 h(x)dx + O(\psi^4). \tag{3.13}$$

Note that (3.4) and (3.7) were derived on the assumption that x_A coincides with x_c when twisting begins and so (3.4) and (3.7) for $\lambda = 1$ do not agree with (3.13) and (3.12) unless $x_A = x_c$ satisfies (3.11). The derivative $dt/d\lambda$ is E when $\lambda = 1$ and

$$E = (w_{22} - w_{12}^2/w_{11})_{\lambda_1=\lambda_2=1}.$$

Weber [15] has given an intuitive theory for isotropic materials and small strains which estimates the contribution to M due to the extension of longitudinal fibers. The theory is incorrect except for cylinders of elongated section (see [16] for a discussion of the Weber theory). For elongated sections, the Weber theory adds the cubic term in (3.12) to the linear St. Venant moment, but it is assumed that $x_A = x_c$. For a rectangular section, the two contributions are approximately equal when ψa is h/a. Goodier and Shaw [16] report good agreement of the Weber theory with experiment for steel bars of rectangular section.

4. Results for rubber-like materials

In this section we assume that the material is incompressible and isotropic so that $\lambda_3 = 1/\lambda_1\lambda_2$ and W is a function $W(I_1, I_2)$ of the principal strain invariants

$$I_1 = \lambda_1^2 + \lambda_2^2 + \frac{1}{\lambda_1^2\lambda_2^2}, \qquad I_2 = \lambda_1^2\lambda_2^2 + \frac{1}{\lambda_1^2} + \frac{1}{\lambda_2^2}, \tag{4.1}$$

in which λ_1 and λ_2 have the values (2.1). The derivatives $\partial w/\partial\lambda_\alpha$ are given by

$$\begin{aligned} \frac{\partial w}{\partial \lambda_1} &= 2\left(\lambda_1 - \frac{1}{\lambda_1^3\lambda_2^2}\right)(W_1 + \lambda_2^2 W_2), \\ \frac{\partial w}{\partial \lambda_2} &= 2\left(\lambda_2 - \frac{1}{\lambda_1^2\lambda_2^3}\right)(W_1 + \lambda_1^2 W_2), \end{aligned} \tag{4.2}$$

and the equilibrium equation (2.5) becomes

$$r''\left\{(3+r'^4\lambda_2^2)(W_1+\lambda_2^2W_2)+\frac{2}{r'^2\lambda_2^2}(r'^4\lambda_2^2-1)^2(W_{11}+2\lambda_2^2W_{12}+\lambda_2^4W_{22})\right\}$$

$$(4.3)\qquad = -\psi^2\frac{r}{\lambda_2^2}\left\{r'^2(3-r'^2\lambda_2^4)W_1+r'^4(r'^2\lambda_2^4+1)W_2 + \frac{2}{\lambda_2^2}(r'^4\lambda_2^2-1)(r'^2\lambda_2^4-1)[W_{11}+(r'^2+\lambda_2^2)W_{12}+r'^2\lambda_2^2W_{22}]\right\} - \frac{h'}{h}r'(r'^4\lambda_2^2-1)(W_1+\lambda_2^2W_2),$$

where we have used the notation

$$W_\alpha = \frac{\partial W}{\partial I_\alpha}, \qquad W_{\alpha\beta} = \frac{\partial^2 W}{\partial I_\alpha \partial I_\beta} \qquad (\alpha, \beta = 1, 2).$$

In order for T_1 to vanish at $x = \pm a$ we need $\lambda_1^2\lambda_2 = 1$ at these points or

$$(4.4)\qquad r'^4(\psi^2 r^2 + \lambda^2) = 1 \qquad \text{at } x = \pm a.$$

For a given section shape $h(x)$ and given values of ψa and λ, (4.3) and (4.4) provide a two-point boundary value problem for $r(x)$. The numerical approach used assumed a trial value for r at $x = -a$, determined r' at $x = -a$ from (4.4), and then integrated (4.3) from $x = -a$ to $x = a$ using a fourth-order Runge–Kutta method. The condition (4.4) at $x = a$ was tested and iterations were made with adjustments to the value of r at $x = -a$ until the boundary condition at $x = a$ was satisfied to the desired order of accuracy. (For symmetric sections an alternate test is $r = 0$ at $x = 0$.) Expressions (2.10) for the moment and axial force were evaluated numerically. For a non-symmetric section the location x_A of the axis of torsion ($r = 0$) was found by identifying the interval of integration in which r changed sign and then using the known values of r, r' and r'' at the ends of the interval to obtain an accurate estimate for x_A.

For a Mooney material, W_1 and W_2 are constants,

$$W_1 = C_1, \qquad W_2 = C_2, \qquad W_{\alpha\beta} = 0.$$

Figures 2 and 3 show the twisting moment and axial force for a strip with a rectangular section for a neo-Hookean material ($C_1 = 0$) and for a Mooney material with $C_2/C_1 = 0.25$. The shear modulus G for small strains,

$$G = 2(W_1 + W_2) \qquad \text{for } I_1 = I_2 = 3,$$

is used to non-dimensionalize M and L, and the variations of M and L with $\psi a/\lambda$, the twist per unit length of the extended strip, are shown for values of λ

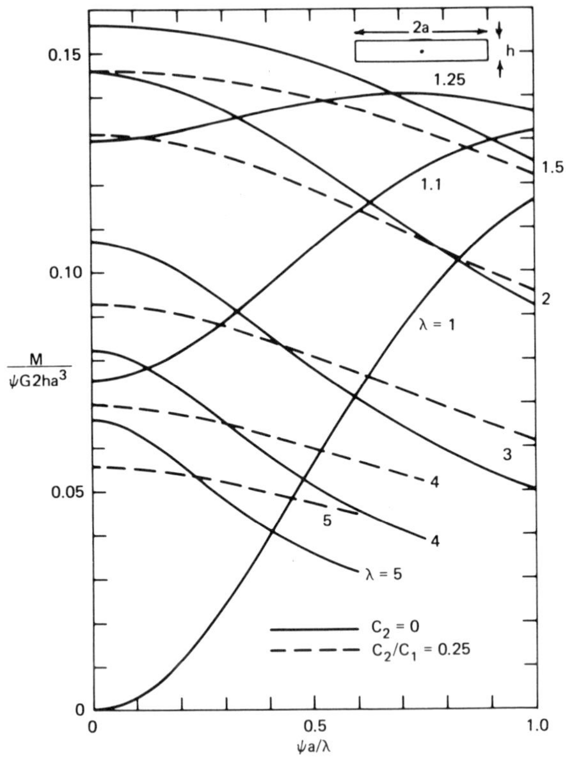

Figure 2. Twisting moment for torsion of an extended rectangular strip of neo-Hookean material ($C_2 = 0$) and of Mooney material with $C_2/C_1 = 0.25$.

from 1 to 5. Estimates for M and L for an elliptical cylinder of neo-Hookean material with semi-axis ratios of $b/a = 0.5$ and 0.25 were obtained by Lee and Shield [12] through variational principles. Figures 2 and 3 show strong similarities with the corresponding results for the elliptical cylinder with $b/a = 0.25$. In contrast to Figure 2, M/ψ is constant for fixed λ for a circular cylinder of Mooney material.

The tendency of an extended cylinder with the axial force held constant to elongate or shorten when twisted a small amount was discussed in [10] and the results in [10] for S_0/I_0 zero apply to the flat strip. For larger values of ψa, the tendency to elongate or shorten as ψa is increased can be estimated from Figure 3; alternatively, Figure 2 and the relation

$$\frac{\partial L}{\partial \psi} = \frac{\partial M}{\partial \lambda},$$

which follows from (2.11), can be used.

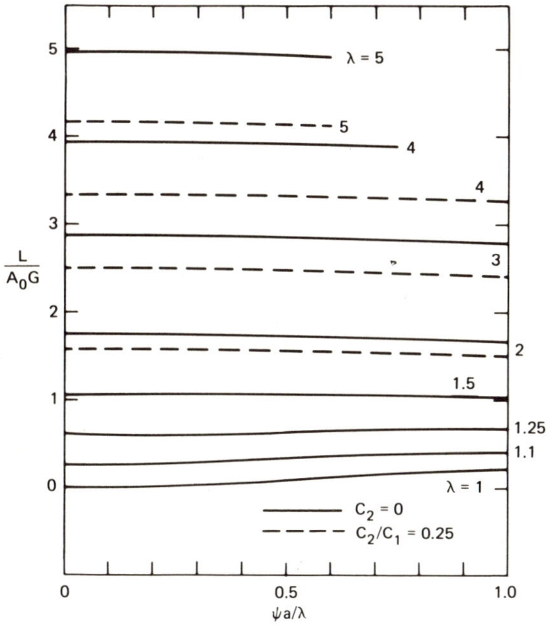

Figure 3. Axial force for torsion of an extended rectangular strip of neo-Hookean material ($C_2 = 0$) and of Mooney material with $C_2/C_1 = 0.25$.

Empirical values for W_1 and W_2 were determined by Shield [17] from results of Treloar for a latex rubber. It was assumed in [17] that W_1 depends only on I_1 and W_2 only on I_2, and the values

$$W_1/G = 0.2946 + 0.7342 \times 10^{-7}(I_1 - 3)^4 \qquad 3 \leqq I_1 \leqq 56$$

$$(4.5) \qquad W_2/G = \begin{cases} 0.1107 + 0.3885 \times 10^{-2}(4.928 - I_2)^{4.864} & 3 \leqq I_2 \leqq 4 \\ 0.2268/I_2^{0.5} & 4 < I_2 \leqq 200 \end{cases}$$

provide a good fit with the empirical values of [17]. The system (4.3) and (4.4) was integrated numerically for a rectangular section using the empirical forms (4.5). Variations of M and L with $\psi a/\lambda$ are shown in Figures 4 and 5 for this case. The figures show appreciable qualitative differences with those given in [10] for M and L for a circular cylinder with the empirical values (4.5). Figure 6 indicates the shape of the deformed section for $\lambda = 2$ and $\psi a = 0, 1, 2, 3$. The curves in Figure 6 are essentially plots of λ_3 versus r, and the width and thickness for $\psi = 0$ are $a/\sqrt{2}$ and $h/\sqrt{2}$. The compression in the radial direction which develops on twisting thickens the section towards the center. It should be remembered that the deformed section is not flat but is inclined at the varying angle $\tan^{-1}(\psi r/\lambda)$ to the plane $x_3 =$ constant, and this warping of the section is not shown in Figure 6.

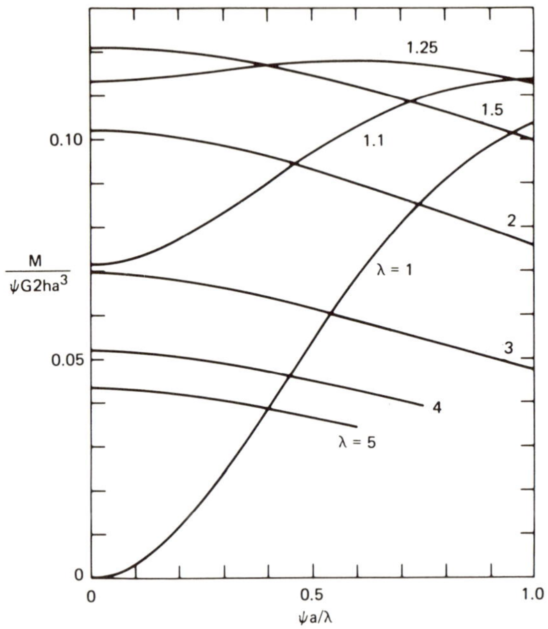

Figure 4. Twisting moment for torsion of an extended rectangular strip with empirical strain energy.

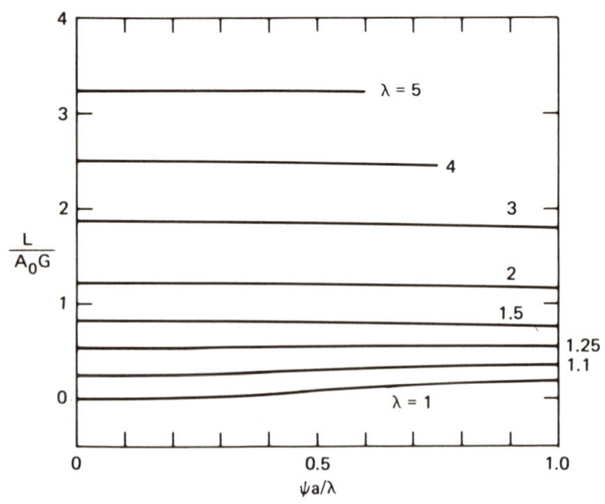

Figure 5. Axial force for torsion of an extended rectangular strip with empirical strain-energy.

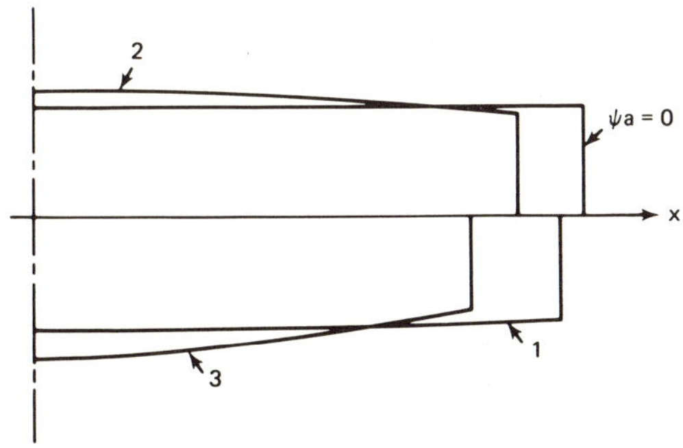

Figure 6. Thickness of the deformed section for $\lambda = 2$ and $\psi a = 0, 1, 2, 3$ for an initially rectangular strip with empirical strain energy.

A strip with the tapered section

$$h = h_0(1 + 0.75x/a)$$

was also treated numerically for a material defined by the empirical values (4.5). For an extended strip the initial axis of twist passes through the centroid $x_c = 0.25a$, but the initial axis for the unextended strip ($\lambda = 1$) is found to be $x_A = 0.1661a$ from (3.11). Figure 7 shows the migration of the axis of torsion as the amount of twist varies for values of λ from 1 to 2. The curves for $\lambda = 3, 4, 5$ are not shown but they are similar to that for $\lambda = 2$ with less movement of the axis from the centroid.

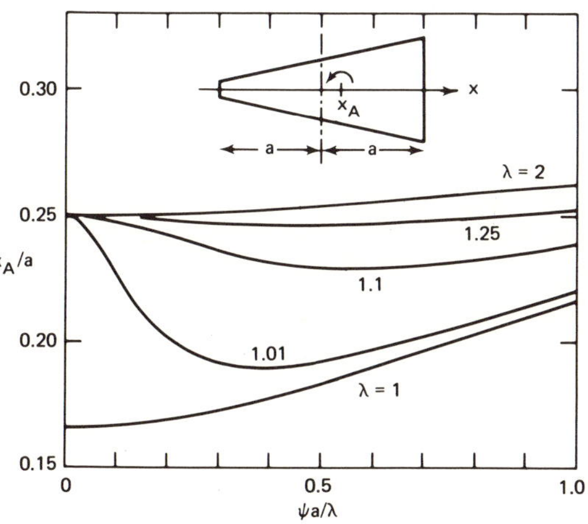

Figure 7. Migration of the axis of torsion for a tapered section $h/h_0 = 1 + 0.75x/a$ with empirical strain-energy.

References

1. R. S. Rivlin, *Large elastic deformations of isotropic materials VI. Further results in the theory of torsion, shear and flexure*, Phil. Trans. Roy. Soc. **A242** (1949), 173–195; reprinted in *Problems in Nonlinear Elasticity*, Gordon and Breach, New York, 1965.

2. A. E. Green and E. W. Wilkes, *A note on the extension and torsion of a circular cylinder of a compressible elastic isotropic material*, Quart. J. Mech. Appl. Math. **6** (1953), 240–249.

3. A. E. Green, *Finite elastic deformation of compressible isotropic bodies*, Proc. Roy. Soc. **A227** (1955), 271–278.

4. J. E. Adkins, *Finite deformation of materials exhibiting curvilinear aeolotropy*, Proc. Roy. Soc. **A229** (1955), 119–134.

5. J. E. Adkins, *Some general results in the theory of large elastic deformations*, Proc. Roy. Soc. **A231** (1955), 75–90.

6. A. E. Green and J. E. Adkins, *Large Elastic Deformation*, Oxford University, 2nd ed., 1970.

7. A. E. Green and R. T. Shield, *Finite extension and torsion of cylinders*, Phil. Trans. Roy. Soc. **A244** (1951), 47–86; reprinted in *Problems in Nonlinear Elasticity*, Gordon and Breach, New York, 1965.

8. R. S. Rivlin, *The solution of problems in second-order elasticity theory*, J. Rat. Mech. Anal. **2** (1953), 53–81.

9. A. E. Green, *A note on second-order effects in the torsion of incompressible cylinders*, Proc. Camb. Phil. Soc. **50** (1954), 488–490.

10. R. T. Shield, *An energy method for certain second-order effects with application to torsion of elastic bars under tension*, J. Appl. Mech. **47** (1980), 75–81.

11. C. Truesdell and W. Noll, *The non-linear field theories of mechanics*, Handbuch der Physik **III/3**, Springer, Berlin, 1965.

12. S. J. Lee and R. T. Shield, *Applications of variational principles in finite elasticity*, J. Appl. Math. Phys. (ZAMP) **31** (1980), 454–472.

13. S. J. Lee and R. T. Shield, *Variational principles in finite elastostatics*, J. Appl. Math. Phys. (ZAMP) **31** (1980), 437–453.

14. S. J. Lee, *Variational principles in finite elasticity with applications*, Ph.D. Thesis, University of Illinois at Urbana-Champaign, December, 1979; T. & A. M. Report No. 437.

15. C. Weber, *Die Lehre der Drehungsfestigkeit*, VDI Forschungsarbeiten **249** (1921).

16. J. N. Goodier and W. A. Shaw, *Nonlinear effects in elastic torsion of bars of slender section*, J. Mech. Phys. Solids, **10** (1962), 35–52.

17. R. T. Shield, *On the stability of finitely deformed elastic membranes*, Part I, J. Appl. Math. Phys. (ZAMP) **22** (1971), 1016–1028.

GROWTH AS A FINITE DISPLACEMENT FIELD

RICHARD SKALAK

Department of Civil Engineering and Engineering Mechanics, Bioengineering Institute, Columbia University, New York, N.Y. 10027, USA

(Received September 11, 1980)

ABSTRACT

The growth of an animal or plant may be described by a time-dependent displacement field. When cells proliferate throughout a tissue, growth may be regarded as a spatial distribution of mass sources. To allow for anisotropic growth, a growth rate tensor may be defined which is similar to a strain rate tensor. Some tissues, notably bones, horns, and seashells, grow by accretion on certain surfaces which move as new tissue is produced. Such growth is modeled by a surface distribution of mass sources. At each point of the surface, a vector is assigned whose magnitude and direction specify the rate and direction of growth. Growth can lead to internal stresses that may be relieved by stress relaxation or additional growth. The vocabulary of distributed dislocations may be adapted to describe growth which produces internal stresses. A psychological aspect of growth concerns the identification of invariant aspects of a growth displacement field which are recognized as maturation of form.

Introduction

The purpose of the present paper is to suggest that some of the vocabulary of finite elasticity can be adapted to the description of growth. At the present time, some use is being made of finite elasticity in biomechanics for such problems as finite deformations of the skin in plastic surgery or deformation of red blood cells in capillaries. These applications pertain to short time periods and do not take growth into account. There is, on the other hand, a large literature in anatomy, zoology, botany, and evolutionary theory which describes changes of form from early stages of life to adulthood or from one species to another over long periods of time (see, for example, D'Arcy Thompson, 1969). This literature is largely unconnected to modern continuum mechanics although the concepts and vocabulary of mechanics may

Proceedings of the IUTAM Symposium on Finite Elasticity, Lehigh University, August 10–15, 1980. Invited paper.

Carlson, D.E. and Shield, R.T. (eds.)
Proceedings of the IUTAM Symposium on Finite Elasticity
ISBN 90 247 2629 8

lead to improved insights. The main ideas of the present paper have been recently submitted to a biological journal (Skalak et al., 1980) but are herein expressed in more general notation and extended to include use of distributed dislocation theory.

Volumetric growth

Consider an animal or organ which grows by proliferation of its cells throughout the tissue involved. As a first approximation, neglecting any stresses present, the kinematics of finite elasticity may be adapted directly. Only the equation of conservation of mass needs to be modified. Due to the creation of new tissue, it appears that mass sources are distributed throughout the space. Further, growth may be anisotropic at any point, producing greater extension in one direction than another. If the successive states are stress-free, then there is a compatible displacement field which carries any mass point present at time t_0 to its position at time t.

Let x^i be fixed cartesian coordinates of material points at the reference time to when the body occupies volume V_0 and let θ^i by any curvilinear coordinates in V_0 so

$$x^i = x^i(\theta^j) \qquad i, j = 1, 2, 3. \tag{1}$$

At any later time, t, let the cartesian coordinates of the same mass points be y^i:

$$y^i = y^i(\theta^j, t), \tag{2}$$

where θ^i are regarded as convected coordinates of the material points. Due to the growth, a line element ds_0 at t_0 extends to ds at t and

$$d^2s - d^2s_0 = 2\gamma_{ij} d\theta^i d\theta^j, \tag{3}$$

where γ_{ij} is the apparent strain tensor which will be here called the growth tensor. It is defined by

$$\gamma_{ij} = (G_{ij} - g_{ij})/2, \tag{4}$$

where G_{ij} and g_{ij} are the metric tensors associated with the deformed and original states respectively.

It is useful to define also a rate of growth tensor at any time t. The velocity components in cartesian coordinates are v^i:

$$v^i = \frac{\partial y^i}{\partial t} = \dot{y}^i. \tag{5}$$

The contravariant velocity components with respect to θ^i at time t are

$$u^i = v^j \frac{\partial \theta^i}{\partial y^j} \quad (6)$$

The rate of deformation tensor usually used in fluid mechanics is here defined as the growth rate tensor, d_{ij}:

$$d_{ij} = (\nabla_j u_i + \nabla_i u_j)/2, \quad (7)$$

where $\nabla_j u_i$ indicates covariant differentiation in θ^i using the metric G_{ij}. One aspect of physical significance of d_{ij} is that

$$\frac{1}{ds}\frac{d}{dt}(ds) = d_{ij} n^i n^j, \quad (8)$$

where n^i are contravariant components of a unit vector in the direction of the line element ds. In biology, the left-hand side of (8) would be called the specific growth rate of ds.

Growth usually results in volumetric expansion and addition of mass, although decrease of some volumes and of mass may also be part of a growth process. Mass may even be added without a change of volume by increasing density. For a complete description of growth it is necessary to know the mass density ρ:

$$\rho = \rho(\theta^i, t). \quad (9)$$

The rate at which mass must be added per unit volume is

$$h(\theta^i, t) = \frac{\partial \rho}{\partial t} + \rho \nabla_i u_i, \quad (10)$$

where $\partial\rho/\partial t$ is the material derivative (θ^i = constant) and $\nabla_i u_i$ is the covariant derivative using the metric G_{ij}. In continuum mechanics in which no mass is added, $h = 0$ and equation (10) becomes the continuity equation. In the biological context h will usually be positive, but may be negative corresponding to atrophy or loss of weight.

Surface growth

Bones, horns and seashells grow primarily by accretion of new tissue on certain parts of their external surfaces. Where two bones meet, as at the knee or at a suture of the skull, there is, in effect, a surface S_0 on which new bone is created and extruded on each side of S_0. Suppose surface growth starts on S_0 at t_0 (Figure 1A) so that at a later time, t, two new regions R_3 and R_4 (Figure 1B) have been created and the original regions R_1, R_2 are separated as shown in Figure 1B. In this case, the displacement is not continuous since points P

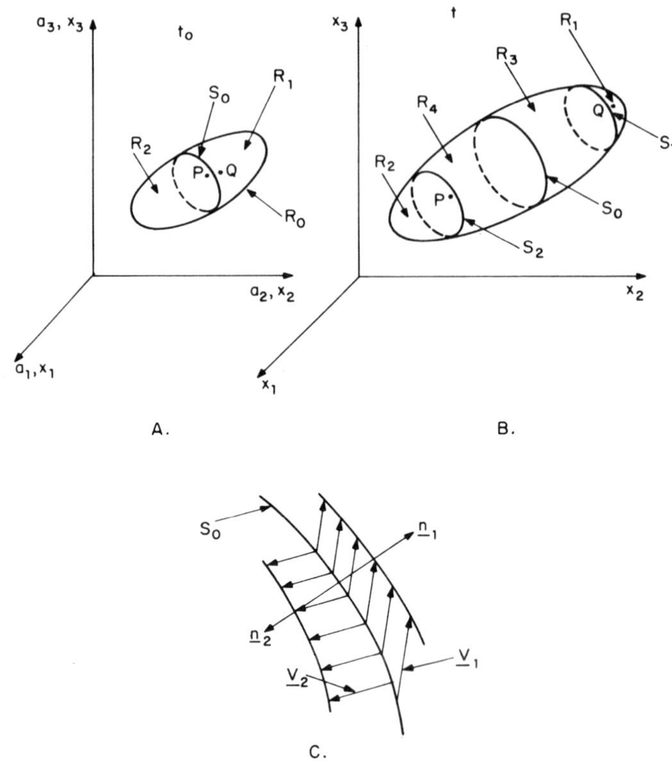

Figure 1. Schematic representation of surface growth.

and Q originally on either side of S_0 are located on S_1 and S_2, Figure 1B, a finite distance apart at time t. Further, the points in R_3 and R_4 have no material coordinates at time t_0 because they did not yet exist. The definition of convected coordinates can be readily amended to accommodate these points as follows. Let θ^i be chosen so that the surface S_0 is given by

$$\theta^3 = C^3(t), \tag{11}$$

where C^3 is an increasing function of time. The coordinates θ^1, θ^2 on S_0 are chosen so that θ^1, $\theta^2 = \text{constant}$, $\theta^3 = C^3(t)$, is a material point in the sense that it locates the same cell which extrudes the new tissue at each time. The line θ^1, $\theta^2 = \text{constant}$, $C^3(t_0) \leqq \theta^3 \leqq C^3(t)$ locates all tissue that was extruded by the same cell. The surface S_0 may move and change size and shape. It is given parametrically by y_0^i:

$$y_0^i = y_0^i(\theta_1, \theta_2, t), \tag{12}$$

where y_0^i is formed by substituting (11) into (2). The velocity of the generating cells on S_0 is given by

$$v_0^i = \dot{y}_0^i, \tag{13}$$

where the dot means holding θ_1, θ_2 constant in (12). The velocity v^i of the new material being generated on S_0 is given by (5) in which θ_1, θ_2, θ_3 are constant. The relative velocity of the new tissue generated with respect to the generating cells on S_0 is V^i:

$$V^i = v^i - v_0^i = V^i(\theta_1, \theta_2, t), \tag{14}$$

where v^i is evaluated setting $\theta^3 = C^3(t)$. The vector V^i is the growth velocity on S_0 and may have different values V_1^i and V_2^i on the two sides of S_0 as shown in Figure 1C. The normal component $(V^i n_i)$ will be positive for accretion and negative for resorption of bone which does occur to some extent during growth. The tangential components of $\boldsymbol{V}_1$ and $\boldsymbol{V}_2$ will be equal if there is no slip of the two bones, but may be unequal to allow for slip.

When a bone is in contact with a soft tissue which grows by volumetrically distributed sources, one growth rate on S_0, say $\boldsymbol{V}_2$ is zero. The bone is then on the $\boldsymbol{V}_1$ side. If there is no slip, the material velocity of the soft tissue adjacent to the bone must equal the surface velocity v_0^i (equation (13)). If slip is permitted, then only the normal components of the soft tissue velocity and the surface are equal.

In many situations, the bony tissue in a region like R_4 (Figure 1B) moves essentially as a rigid body once it is formed. In this case, the distribution of $\boldsymbol{V}_1$ and $\boldsymbol{V}_2$ is constrained to be such that it is commensurate with a rigid body motion plus the motion of S_0. There could be volumetric growth in a region like R_4 with growth rate given by (7). However, some modification of the growth tensor, equation (4), is required because g_{ij} does not exist at t_0 for points in R_4. The strain can be at most computed from the time of generation, say τ, of any material originating on S_0. Then the growth tensor within R_4 from time τ to t is

$$\gamma_{ij} = (G_{ij}(t) - G_{ij}(\tau))/2. \tag{15}$$

Growth on a surface is specified when the surface S_0, the growth velocities $\boldsymbol{V}_1$ and $\boldsymbol{V}_2$ and the densities ρ_1 and ρ_2 on each side of S_0 are known. The rates at which mass is created on each side of S_0 are

$$H_1 = \rho_1 V_1^i n_1^i, \qquad H_2 = \rho_2 V_2^i n_2^i,$$

where $\boldsymbol{n}_1$ and $\boldsymbol{n}_2$ are the normals shown in Figure 1C. Negative H corresponds to resorption.

The fact that $\boldsymbol{V}_1$ and $\boldsymbol{V}_2$ are vectors is essential because the direction as well as the magnitude of mass production is important to controlling the end result. For example, if S_0 is a plane stationary disk with a linear spatial distribution of $|\boldsymbol{V}_1|$ on it and $\boldsymbol{V}_1$ is parallel to the normal, then a horn is produced which is part of a torus. But if $|\boldsymbol{V}_1|$ is linear and $\boldsymbol{V}_1$ is at a constant angle to the normal, then a logarithmic spiral horn will be generated which is more realistic in many cases. (See Skalak et al., 1980.)

Superposition of growth and stress

In the analysis outlined above, the animal was considered to be stress-free at all times and the growth described was assumed to take place over long times. The growth strains involved may be very large, but the growth velocities will normally be small, typically measured in centimeters per year. If at any time a rapid motion such as running is to be analyzed, the current position of the tissues in the growth analysis above may be used as the initial coordinates for the computations of strains and displacements due to applied stresses or accelerations. During such computation, the stress effects are of short duration although the strains may again be quite large. The current coordinates of the growth analysis can be regarded as fixed during the short time interval of the stress analysis without taking into account changes due to growth.

The scheme discussed above where growth is neglected during stressing and stressing is neglected during growth may be appropriate for many purposes, but may be inadequate where there are stresses continually present. For example, the arteries are under a considerable tension at all times during their growth history. There are also important cases, especially in bone growth, where the stresses and growth are believed to interact and affect each other. (See Cowin and Buskirk, 1979.) Internal stresses and growth interact in trees and have been discussed by Archer and Byrnes (1974) and Archer (1976).

Interaction of stresses and growth

In the analysis discussed above, the growth strain field is assumed to be compatible at any time and therefore stress-free. If instead a growth tensor is considered to be arbitrarily specified by the growth processes, the first question to consider is whether or not any internal stresses will be generated. The answer is, or course, that no stresses will be necessary if the specified tensor growth strains are compatible. If they are not compatible, then internal stresses are required in the final state to maintain continuity of the body. The situation is similar to that encountered in the theory of distributed dislocations in crystals in which the plastic deformation (or dislocation density) may be specified and an elastic reponse is invoked in order to maintain integrity of the crystal. Another analogous situation is that of thermal stresses. If a temperature field is specified, the free thermal expansion of each element may or may not yield a compatible strain field. If not, thermal stresses are required to maintain continuity.

Consider first the linear case, in which both the growth strains and the elastic strains due to internal stresses are small. The specified growth strains are first applied to each element of the body assuming no stresses are present. This corresponds to the so-called natural state which is the relaxed (stress-free) state. The degree to which the growth field e_{ij} is not compatible may be measured by the incompatibility tensor η_{ij} given by:

$$\eta_{ij} = \varepsilon_{ilm}\varepsilon_{jkn}\nabla_l\nabla_k e_{mn}, \tag{17}$$

where cartesian tensors are used, ε_{ilm} is the alternating tensor and e_{mn} is the strain. On the other hand, the incompatibility may be expressed in terms of a dislocation density, α_{ij}, which is related to the Burgers vector, b_i, for any closed curve C by

$$b_i = \iint_A \alpha_{ij} dA_j, \tag{18}$$

where A is a surface capping C. The Burgers vector measures the dislocation observed in traversing the curve C. The incompatibility tensor is given in terms of the dislocation density by

$$\eta_{ij} = \mathrm{sym}(\varepsilon_{ilk}\nabla_k\alpha_{jl}) \tag{19}$$

where sym stands for the symmetrical part.

When internal stresses are invoked to restore the continuity of the body, the elastic stress strains must be such as to produce an equal and opposite incompatibility tensor to that of the growth field in the natural state. This condition is the basis of methods of finding the elastic stress due to the specified dislocations (e.g., Kröner, 1958 and Nabarro, 1979). The relation of the incompatibility tensor and the dislocation density may be written in direct notation as

$$\mathrm{Inc}\,\boldsymbol{e} = \eta, \tag{20}$$

where $\mathrm{Inc}\,\boldsymbol{e}$ stands for the right-hand side of (17). Equation (20) is called the fundamental geometric equation by Kröner (1958).

Returning to the case of large strains, the nonlinear generalization of equation (20) is given by the form (see Gairola, 1979):

(21) $$\mathrm{Inc}\,\boldsymbol{e} = \eta + Q,$$

where now all quantities are referred to the curvilinear coordinates, θ^i, of the final position and the strains involved are those required to pass from the final deformed position to the natural state. Then

(22) $$(\mathrm{Inc}\,\boldsymbol{e})^{qr} = -\varepsilon^{qjm}\varepsilon^{rkl}\nabla_j\nabla_k e_{ml},$$

where ε^{ijk} is the generalized alternating tensor and ∇_j stands for covariant differentiation and

(23) $$\eta^{qr} = -\mathrm{sym}[\varepsilon^{qjm}\nabla_j\alpha^{r}_{m}],$$

(24) $$Q = \mathrm{sym}\,\tfrac{1}{2}[\varepsilon^{qjm}\varepsilon^{rkl}h^{np}(-2e_{jlp} + k_{jlp})(-2e_{mkn} + k_{mkn})],$$

where

(25) $$k_{jkl} = \tfrac{1}{2}(\varepsilon_{jkm}\alpha^{m}_{l} + \varepsilon_{ljm}\alpha^{m}_{k} - \varepsilon_{klm}\alpha^{m}_{j}),$$

and

(26) $$e_{jkl} = \tfrac{1}{2}(\nabla_j e_{kl} + \nabla_k e_{jl} - \nabla_l e_{jk}).$$

In principle, these equations form a basis for the determination of stresses due to given growth strains. In practice, it is not likely that the growth strains corresponding to the natural state will be known in advance. Only the final deformed state is directly available for observation. Further, some hypothesis of the effect of stresses on the rate of growth will be necessary to develop a complete description of the growth process. The present paper is only intended to discuss the kinematics involved.

Applications

The original motivation for the present study was the possible application to clinical orthodontic and cranio-facial growth problems. Present practice does not usually include any description in terms of growth tensors. The long range goal is to see if more fundamental descriptions of growth may be useful in clinical diagnosis and treatment.

Application to the description of plant as well as animal growth may be useful to give a more detailed picture of when and where growth takes place. Examples may be found in Silk and Erickson (1979).

The kind of description discussed above may also be useful in phylogenetic studies (e.g., Dullemeijer, 1974 and Gould, 1977). The connection to phylogenetic developments may be made by seeking closely related species which provide sufficient points in series of changing shapes so a continuum model may be used to describe the transformation. Even with only two species, the description of the differences in terms of strain-like tensors may be useful. Some examples of two-dimensional analyses of this type are given by Bookstein (1978).

An interesting psychological aspect that may develop out of the proposed description is the identification of the geometric invariants by which people judge age, for example. A recent paper by Pittinger et al. (1979) shows that nonuniform growth of the skull is an important factor in age judgements. This kind of information may be stated more precisely in terms of growth tensors.

Acknowledgements

Discussion with colleagues and collaborators including Drs. M. L. Moss, P. Dullemeijer, H. Vilmann, M. Shinozuka, E. Otten and F. Bookstein have been essential. I am indebted to Professor J. H. Weiner for the suggestion to utilize distributed dislocation theory. This research was partially supported by NIH Grant 1-RO1-HD-14371.

References

R. R. Archer and F. E. Byrnes, *On the distribution of three growth stresses*, Part I: *An anisotropic plane strain theory*, Wood Sci. and Tech. **8** (1974), 184–196.

R. R. Archer, *On the distribution of tree growth stresses*, Part II: *Stresses due to asymmetric growth strains*, Wood Sci. and Tech. **10** (1976), 293–309.

F. L. Bookstein, *The Measurement of Biological Shape and Shape Change*, Lecture Notes in Biomathematics **24**, Springer-Verlag, New York, 1978.

S. C. Cowin and W. C. Van Buskirk, *Surface remodeling induced by a medullary pin*, J. Biomechanics **12** (1979), 269–276.

P. Dullemeijer, *Concepts and Approaches in Animal Morphology*, Van Gorcum and Co., Assen, Netherlands, 1974.

B. K. D. Gairola, *Nonlinear elastic problems*, in *Dislocations in Solids*, Vol. 1: *The Elastic Theory* (F. R. N. Nabarro, ed.), North-Holland Publishing Co., Amsterdam, 1979, pp. 223–342.

S. J. Gould, *Ontogeny and Phylogeny*, Harvard Univ. Press, Cambridge, Mass, 1977.

E. Kröner, *Continuum theory of dislocations and self-stresses*, Ergebnisse der Angervanandten Mathematik **5**, Springer-Verlag, Berlin, 1958. (English translation by I. Raasch and C. S. Hartley.)

F. R. N. Nabarro (ed.), *Dislocations in Solids*, Vol. 1: *The Elastic Theory*, North-Holland Publishing Co., Amsterdam, 1979.

J. B. Pittenger, R. E. Shaw and L. S. Mark, *Perceptual information for the age level of faces as a higher order invariant of growth*, J. of Experimental Psychology **5** (1979), 478–493.

W. K. Silk and R. O. Erickson, *Kinematics of plant growth*, J. Theor. Biol. **76** (1979), 481–501.

R. Skalak, G. Dasgupta, M. Moss, E. Otten, P. Dullemeijer and H. Vilman, *Analytical description of growth*, Submitted to J. Theor. Biol., 1980.

D. W. Thompson, *On Growth and Form* (abridged edition), Cambridge University Press, Cambridge, 1969.

STRESS CONCENTRATION LAYERS IN FINITE DEFORMATION OF FIBRE-REINFORCED ELASTIC MATERIALS

A. J. M. SPENCER

Department of Theoretical Mechanics, University of Nottingham, Nottingham, England

(Received September 26, 1980)

ABSTRACT

We consider finite plane strain of incompressible elastic materials reinforced by inextensible fibres. Deformations of inextensible materials often give rise to stress concentration layers, which are sheets of fibres which carry infinite direct stress but finite force, and across which the shear stress may be discontinuous. For linear elastic materials reinforced by straight parallel fibres these layers are well understood, and asymptotic methods of analysis have been developed for linear elastic materials with small but finite extensibility. When finite deformations occur, a qualitatively new feature arises because, in general, the fibres become curved and then the normal stress across them may also be discontinuous. We consider two examples, namely simple shear and shear bending of a rectangular block. In both of these examples the solution for inextensible material involves surface stress concentration layers. For 'almost inextensible' material we obtain approximate solutions for the displacement and stress in the neighbourhood of these surfaces, and show that the solution for an ideally inextensible material can be interpreted as the limit of the solution for material with small but finite extensibility as the ratio of shear modulus to fibre extension modulus tends to zero.

1. Introduction

A simple model of a fibre-reinforced material can be constructed by assuming the material to be inextensible in the fibre-direction. The theory of such materials was initiated by Adkins and Rivlin [1] in the context of finite elasticity theory. Pipkin and Rogers [2] developed the plane strain theory of finite deformations of ideal fibre-reinforced materials, and showed that a substantial part of this theory does not depend on the particular mechanical

Proceedings of the IUTAM Symposium on Finite Elasticity, Lehigh University, August 10–15, 1980. Invited paper.

Carlson, D.E. and Shield, R.T. (eds.)
Proceedings of the IUTAM Symposium on Finite Elasticity
ISBN 90 247 2629 8

response of the material. Further developments, including the constitutive theory for finite elastic ideal fibre-reinforced materials, are given in Spencer [3].

A feature of solutions using the theory of materials reinforced by inextensible fibres is the occurrence of singular sheets of fibres, or stress concentration layers. These are fibre sheets which carry infinite direct stress in the fibre direction, although the force they carry is finite. They arise when the kinematic constraints and boundary conditions impose a discontinuity in shear strain, and hence in shear stress, across the fibres; such a shear stress discontinuity can only be equilibriated by an infinite direct stress.

In plane linear elasticity, with the fibres aligned in parallel straight lines, these stress concentration layers are now well understood. No material is truly inextensible. In some illuminating examples in plane linear elasticity, Everstine and Pipkin [4] showed that in anisotropic linear elasticity the important parameter is $\varepsilon = (\mu/E)^{1/2}$, where μ is a shear modulus and E the extension modulus in the fibre direction. The inextensible material corresponds to the limit $\varepsilon \rightarrow 0$. For materials in which ε is small but finite, Everstine and Pipkin showed that the stress concentration layers correspond to thin layers in which the direct stress is of order σ/ε (where σ is a characteristic stress) and across which the shear stress changes rapidly. These layers have thickness of order εL (where L is a characteristic length) and along them the direct stress decays in a length of order L/ε. Subsequently Everstine and Pipkin [5] and Spencer [6] developed boundary layer theories for the analysis of the stress within these layers.

The main purpose of this paper is to begin to seek a similar understanding of the stress concentration layers for a material which has finite elastic response. For materials reinforced by initially straight fibres, there is a qualitative difference between linear and finite deformation theories, because under finite deformation the fibres will generally become curved, and then an inextensible fibre can sustain discontinuous normal stress as well as shear stress. We do not at this stage attempt a general theory of boundary layers in almost inextensible finite elastic material. Instead we try to gain insight into the phenomenon by analysing two particular problems.

In Section 2 we outline the theory of transversely isotropic elastic materials. In Section 3 this is specialized to the case of plane strain. In Section 4 we describe how the theory is modified when the material is inextensible in the fibre direction, and in Section 5 we consider the case of a material which is 'almost inextensible' in the fibre direction. A regular perturbation scheme is outlined in Section 6, but this breaks down in the neighbourhood of a stress concentration layer. In Section 7 we consider the problem of simple shear of a block of 'almost inextensible' material, and interpret the stress concentration layers by means of an asymptotic analysis. A similar analysis is performed in Section 8 for the problem of shear bending of a rectangular block.

2. Constitutive equations

Deformations are referred to a rectangular cartesian coordinate system, and in this section all vector and tensor components are components in this system. Suppose that a body undergoes a deformation in which a typical particle which initially has position vector $\boldsymbol{X}$, with components X_R ($R = 1,2,3$), moves to the point with position vector $\boldsymbol{x}$ and coordinates x_i ($i = 1,2,3$). Then the deformation is described by equations of the form

$$\boldsymbol{x} = \boldsymbol{x}(\boldsymbol{X}), \qquad \text{or} \qquad x_i = x_i(X_R). \tag{2.1}$$

We consider transversely isotropic elastic materials. The direction of the axis of transverse isotropy in the initial configuration is defined by a unit vector $\boldsymbol{A}$, with components A_R. For convenience, and with a view to applications of the theory to fibre-reinforced materials, we call the direction of $\boldsymbol{A}$ the *fibre direction* and the trajectories of $\boldsymbol{A}$ the *fibres*; however, the theory is applicable to any transversely isotropic material, and it is not necessary for this anisotropy to be introduced through the presence of physical fibres. The vector $\boldsymbol{A}$ is in general a function of $\boldsymbol{X}$, although in the examples we consider in Sections 7 and 8, $\boldsymbol{A}$ will be taken to be constant.

It is assumed that the material has a strain-energy function W per unit volume which is a function of the deformation-gradient tensor $\boldsymbol{F}$, with components $F_{iR} = \partial x_i / \partial X_R$. It is also assumed that the only preferred direction in the initial configuration (which is assumed to be stress-free) is the direction of $\boldsymbol{A}$, so that the mechanical properties of the material are completely specified by the dependence of W on $\boldsymbol{F}$ and $\boldsymbol{A}$. Then it was shown in [3] that W can be expressed as a function of the five invariants

$$\begin{gathered} I_1 = \operatorname{tr} \boldsymbol{C}, \qquad I_2 = \tfrac{1}{2}\{(\operatorname{tr} \boldsymbol{C})^2 - \operatorname{tr} \boldsymbol{C}^2\}, \qquad I_3 = \det \boldsymbol{C}, \\ I_4 = \boldsymbol{A}\,.\,\boldsymbol{C}\,.\,\boldsymbol{A}, \qquad I_5 = \boldsymbol{A}\,.\,\boldsymbol{C}^2\,.\,\boldsymbol{A}, \end{gathered} \tag{2.2}$$

where the tensor $\boldsymbol{C}$ and its cartesian components C_{RS} are given by

$$\boldsymbol{C} = \boldsymbol{F}^T\boldsymbol{F}, \qquad C_{RS} = \frac{\partial x_i}{\partial X_R}\frac{\partial x_i}{\partial X_S}, \tag{2.3}$$

and, here and subsequently, the usual repeated index summation convention is employed. Results equivalent to (2.2) seem to have been given first by Ericksen and Rivlin [8].

The cartesian components σ_{ij} of the Cauchy stress $\boldsymbol{\sigma}$ are given in the usual way as

$$\sigma_{ij} = \frac{1}{\sqrt{I_3}} \frac{\partial x_i}{\partial X_R} \frac{\partial x_j}{\partial X_S} \left(\frac{\partial W}{\partial C_{RS}} + \frac{\partial W}{\partial C_{SR}} \right). \tag{2.4}$$

3. Plane strain

We now consider plane strain deformations in planes normal to the x_3-axis, so that (2.1) reduce to

$$x_1 = x_1(X_1, X_2), \qquad x_2 = x_2(X_1, X_2), \qquad x_3 = X_3. \tag{3.1}$$

We also consider that the fibres lie in the deformation planes, so that

$$A_3 = 0. \tag{3.2}$$

Then

$$C_{13} = C_{23} = 0, \qquad C_{33} = 1, \tag{3.3}$$

and it can be verified from (2.2), (3.2) and (3.3) that, in the case of plane strain,

$$I_2 = I_1 + I_3 - 1, \qquad I_5 = (I_1 - 1)I_4 - I_3. \tag{3.4}$$

Hence, for plane strain, W can be expressed as a function of three invariants, say I, I_3 and I_4, where

$$\begin{aligned} I &= I_1 - 1 = C_{11} + C_{22}, \\ I_3 &= \det \boldsymbol{C}, \\ I_4 &= A_\alpha A_\beta C_{\alpha\beta} \qquad (\alpha, \beta = 1, 2). \end{aligned} \tag{3.5}$$

We note that $\sqrt{I_3}$ is the ratio of the volume of a material volume element after the deformation to its volume before the deformation, and $\sqrt{I_4}$ is the extension ratio of a line element which lies along a fibre direction. The invariant I is closely associated with the only remaining possible plane strain deformation, namely shear in the deformation planes; in fact it is shown in [3] that for deformations in which $I_3 = 1$, $I_4 = 1$ (in which case the deformation is one of shear along the fibre directions) then

$$\gamma^2 = I - 2, \tag{3.6}$$

where γ is the tangent of the angle of shear.

For plane strain, when W is regarded as a function of I, I_3 and I_4, the stress is determined by (2.4) and (3.5) to be

$$\boldsymbol{\sigma} = \frac{2}{\sqrt{I_3}} \boldsymbol{F} \left\{ \frac{\partial W}{\partial I} \boldsymbol{I} + \frac{\partial W}{\partial I_3} I_3 \boldsymbol{C}^{-1} + \frac{\partial W}{\partial I_4} \boldsymbol{A} \otimes \boldsymbol{A} \right\} \boldsymbol{F}^T.$$

This may be expressed more conveniently as

$$\boldsymbol{\sigma} = \frac{2}{\sqrt{I_3}}\left\{I_3 \frac{\partial W}{\partial I_3}\boldsymbol{I} + \frac{\partial W}{\partial I}\boldsymbol{B} + I_4 \frac{\partial W}{\partial I_4}\boldsymbol{a}\otimes\boldsymbol{a}\right\},$$

where the tensor $\boldsymbol{B}$, with cartesian components B_{ij}, is given by

$$\boldsymbol{B} = \boldsymbol{F}\boldsymbol{F}^T, \tag{3.7}$$

and $\boldsymbol{a}$ is a unit vector in the fibre direction in the deformed configuration (that is, in the direction of a material line element which has direction $\boldsymbol{A}$ in the reference configuration), so that

$$\sqrt{I_4}\boldsymbol{a} = \boldsymbol{F}.\boldsymbol{A} \quad \text{or} \quad \sqrt{I_4}a_i = A_R \frac{\partial x_i}{\partial X_R}. \tag{3.8}$$

If the material is incompressible, then $I_3 = 1$ and W depends only on I and I_4. Then we have

$$\boldsymbol{\sigma} = -p\boldsymbol{I} + 2\frac{\partial W}{\partial I}\boldsymbol{B} + 2I_4\frac{\partial W}{\partial I_4}\boldsymbol{a}\otimes\boldsymbol{a}, \tag{3.9}$$

where p is an arbitrary hydrostatic pressure. Henceforth we shall assume for simplicity that the material is incompressible, although this leads to a slight logical inconsistency later. This point is discussed in Section 9.

We now suppose that in the initial configuration the fibres are straight and parallel to the X_1-axis, so that $A_1 = 1$, $A_2 = 0$.

It is useful to consider two simple deformations. In a shearing deformation in the fibre direction, defined by

$$x_1 = X_1 + \gamma X_2, \qquad x_2 = X_2, \qquad x_3 = X_3, \tag{3.10}$$

it follows from (3.9) that

$$\sigma_{12} = 2\frac{\partial W}{\partial I}\gamma. \tag{3.11}$$

Hence $2\partial W/\partial I$ may be interpreted as a shear modulus. We denote

$$2\partial W/\partial I = \mu(I, I_4) \quad \text{and} \quad \mu_0 = \mu(2, 1), \tag{3.12}$$

so that μ_0 is the shear modulus at zero strain.

Secondly, we consider the incompressible plane strain extension in the X_1-direction, defined by

$$x_1 = \lambda X_1, \qquad x_2 = \lambda^{-1}X_2, \qquad x_3 = X_3. \tag{3.13}$$

Then if p is chosen so that $\sigma_{22} = 0$, (3.9) gives

$$\sigma_{11} = 2\frac{\partial W}{\partial I}(\lambda^2 - \lambda^{-2}) + 2\lambda^2\frac{\partial W}{\partial I_4} = \mu(\lambda^2 - \lambda^{-2}) + 2\lambda^2\frac{\partial W}{\partial I_4}. \tag{3.14}$$

We see that if $\sigma_{11} = 0$ when $\lambda = 1$, it is necessary that $\partial W/\partial I_4 = 0$ when $I = 2$ and $I_4 = 1$. The extension modulus at zero strain is denoted by E_0, and is given by $d\sigma_{11}/d\lambda$, evaluated at $\lambda = 1$. Hence, from (3.14),

$$E_0 = 8\frac{\partial W}{\partial I} + 4\frac{\partial^2 W}{\partial I_4^2} = 4\left(\mu_0 + \frac{\partial^2 W}{\partial I_4^2}\right), \tag{3.15}$$

where the derivatives of W are evaluated at $I = 2$, $I_4 = 1$.

4. Ideal fibre-reinforced material

We define an 'ideal' fibre-reinforced material to be a material which is inextensible in the fibre direction. For any deformation of such a material, $I_4 = 1$. Thus, in the case of finite elastic plane strain of an incompressible material, W becomes a function of I only. Since I is related to the amount of shear γ by (3.6), the fact that W only depends on I reflects the property that for this class of deformations the only possible deformation mechanism is shear on the fibres. The inextensibility constraint gives rise to an arbitrary tension T in the fibre direction. For an ideal fibre-reinforced material, the constitutive equation (3.9) is therefore replaced by

$$\boldsymbol{\sigma} = -p\boldsymbol{I} + T\boldsymbol{a} \otimes \boldsymbol{a} + \mu(I)\boldsymbol{B}. \tag{4.1}$$

A feature of solutions for plane strain of ideal fibre-reinforced materials [2, 3] is that the shear strain may be discontinuous across the fibres (and also, in the case of an incompressible material, across the normal trajectories of the fibres, although such normal line discontinuities do not feature in the examples considered in this paper). These discontinuities in shear strain give rise to discontinuities in shear stress, which can only be equilibriated by a singular tension in the fibre across which the shear stress is discontinuous; such fibres carry infinite stress, but finite force. If the singular fibre is curved, equilibrium also demands that the normal stress across it is discontinuous. Fibres which form a boundary surface may also carry singular tension, if the shear stress on the interior side of the boundary, which is determined by the deformation, does not match the applied boundary shear traction.

Suppose a singular fibre carries a tension F (see Figure 1), and has radius of curvature ρ, and let σ_+ and τ_+ be the normal and tangential components of traction acting on the convex side of the fibre, and σ_- and τ_- the corresponding components acting on the concave side. Then the equilibrium of a thin element which includes a segment of fibre requires that

$$\begin{aligned} \tau_+ - \tau_- &= -dF/ds, \\ \sigma_+ - \sigma_- &= F/\rho, \end{aligned} \tag{4.2}$$

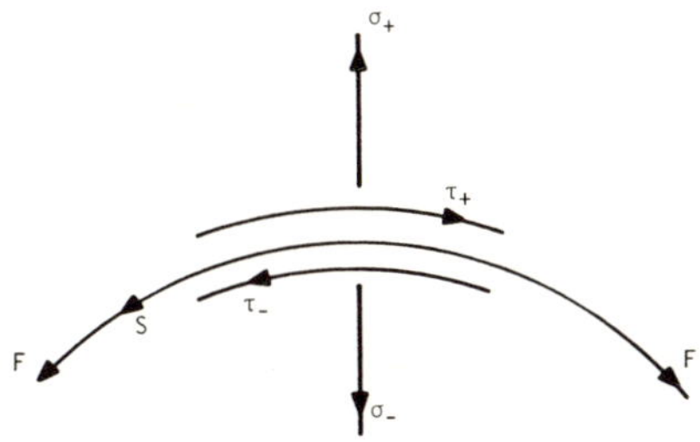

Figure 1. Tractions on a singular fibre.

where dF/ds is the rate of change of F along the fibre, in the sense indicated in Figure 1.

5. Almost inextensible fibres

We are interested in the description of composite materials which consist of a matrix which is capable of undergoing large elastic deformations reinforced by relatively high-modulus elastic fibres. For such materials the shear response is dominated by the shear behaviour of the matrix, and the extension in the fibre direction is mainly determined by the extension modulus of the fibres. Consequently the ratio μ_0/E_0 is often small; for example, for rubber reinforced by nylon fibres this ratio may be of order 4×10^{-3} (depending, of course, on the fibre concentration) and for rubber reinforced by steel fibres it may be of order 10^{-4}. Hence it is often legitimate to treat μ_0/E_0 as a small parameter.

It is supposed, therefore, that

$$\mu_0/E_0 = \varepsilon^2, \qquad \varepsilon^2 \ll 1, \tag{5.1}$$

and we consider the implications which this has for the strain-energy function. We assume that $\mu/\mu_0 = O(1)$, so that a shear strain of order one produces a shear stress of order μ_0. From (5.1), a fibre extension of order ε^2 is required in order to produce a fibre tension of order μ_0. Hence we consider the response to deformations in which $I-2$ is of order one and $I_4 - 1$ is of order ε^2.

Since $E_0 = \varepsilon^{-2}\mu_0$, it follows from (3.15) that $\partial^2 W/\partial I_4^2$ is of order $\varepsilon^{-2}\mu_0$. Hence we express W in the form

$$W(I, I_4) = W_0(I) + (I_4 - 1)\Phi(I) + \tfrac{1}{2}\varepsilon^{-2}(I_4 - 1)^2\Psi(I, I_4), \tag{5.2}$$

where $W_0(I) = W(I, 1)$, $\Phi(I) = \partial W(I, 1)/\partial I_4$ and, since $\partial W(2, 1)/\partial I_4 = 0$, we have $\Phi(2) = 0$. Then, from (3.12) and (5.2),

$$\mu = 2\frac{\partial W_0}{\partial I} + 2(I_4 - 1)\frac{\partial \Phi}{\partial I} + \varepsilon^{-2}(I_4 - 1)^2\frac{\partial \Psi}{\partial I}. \tag{5.3}$$

We do not expect μ to be strongly influenced by small fibre extensions of order ε^2, and so we assume that Φ and Ψ and their derivatives with respect to I are at most of order μ_0.

When W is expressed in the form (5.2), the stress–strain relation (3.9) takes the form

$$\begin{aligned}\boldsymbol{\sigma} = &-p\boldsymbol{I} + 2\left\{\frac{\partial W_0}{\partial I} + (I_4-1)\frac{\partial \Phi}{\partial I} + \tfrac{1}{2}\varepsilon^{-2}(I_4-1)^2\frac{\partial \Psi}{\partial I}\right\}\boldsymbol{B} \\ &+ 2I_4\left\{\Phi + \varepsilon^{-2}(I_4-1)\Psi + \tfrac{1}{2}\varepsilon^{-2}(I_4-1)^2\frac{\partial \Psi}{\partial I_4}\right\}\boldsymbol{a}\otimes\boldsymbol{a}.\end{aligned} \tag{5.4}$$

In a fibre-reinforced elastic material it may happen that the fibre tension is not strongly influenced by shear in the matrix and is approximately linear in the fibre extension, and that the shear response of the matrix is not strongly influenced by the fibre tension. For such materials we may write, approximately,

$$W = W_0(I) + \tfrac{1}{8}\varepsilon^{-2}\mu_0(I_4-1)^2. \tag{5.5}$$

This in effect assumes that the strain energy of the matrix arises only from shear, and that of the fibres arises only from their extension. For the strain-energy function (5.5) $2\partial W/\partial I = 2\partial W_0/\partial I = \mu(I)$, and the stress–strain relation (5.4) reduces to

$$\boldsymbol{\sigma} = -p\boldsymbol{I} + \mu(I)\boldsymbol{B} + \tfrac{1}{2}\varepsilon^{-2}\mu_0 I_4(I_4-1)\boldsymbol{a}\otimes\boldsymbol{a}. \tag{5.6}$$

A further simplification arises if μ is taken to have the constant value μ_0. This is a good approximation for some rubber-like materials over quite a wide range of values of I.

6. Regular perturbations

It is natural to attempt to solve problems for almost inextensible fibre-reinforced materials by a perturbation procedure based on the theory for ideal fibre-reinforced materials as a first approximation. Consider an ideal fibre-reinforced material with strain-energy function $W_0(I)$. Then, from (4.1), the stress $\boldsymbol{\sigma}_0$ is given by

$$\boldsymbol{\sigma}_0 = -p\boldsymbol{I} + T\boldsymbol{a}\otimes\boldsymbol{a} + \mu(I)\boldsymbol{B}. \tag{6.1}$$

Now suppose that in an almost inextensible material the deformation differs from (3.1) by a displacement of order $\varepsilon^2 L$, where L is a typical length. Then the particle initially at (X_1, X_2, X_3) moves to (x_1', x_2', X_3) where

$$x_1' = x_1 + \varepsilon^2 u_1(x_1, x_2), \qquad x_2' = x_2 + \varepsilon^2 u_2(x_1, x_2), \tag{6.2}$$

and u_1 and u_2 are of order L. The theory closely resembles the theory of small deformations superposed on large deformations, due to Green, Rivlin and Shield [9] and also described by Green and Zerna [10], and the theory of almost incompressible materials discussed by Oldroyd [11], Spencer [12, 13], and others. Suppose the deformation (3.1) gives rise to the tensor $\boldsymbol{B}$, the vector $\boldsymbol{a}$ and the invariants I, $I_4 = 1$. Then the deformation (6.2) gives rise to the tensor $\boldsymbol{B} + \varepsilon^2\boldsymbol{B}'$, the vector $\boldsymbol{a} + \varepsilon^2\boldsymbol{a}'$, and invariants $I + \varepsilon^2 I'$, $1 + \varepsilon^2 I_4'$. By substituting these expressions into (5.4), and discarding terms of order ε^2 and higher, we obtain

(6.3) $$\boldsymbol{\sigma} = -p\boldsymbol{I} + 2\frac{\partial W_0}{\partial I}\boldsymbol{B} + 2(\Phi + I_4'\Psi)\boldsymbol{a} \otimes \boldsymbol{a},$$

where Φ and Ψ are evaluated at $I_4 = 1$. By comparing (6.1) and (6.3) we see that in the limit $\varepsilon \to 0$, T is identified with $\Phi + I_4'\Psi$. Thus

(6.4) $$T = \Phi + I_4'\Psi.$$

In particular, if W_4 is of the form (5.5)

(6.5) $$T = \tfrac{1}{2}\mu_0 I_4',$$

and this, together with the incompressibility condition, determines the incremental displacement $\varepsilon^2\boldsymbol{u}$. In order to find the first-order correction to the stress, it is neccessary to introduce either the next higher order displacement correction or, preferably, a first order fibre tension $\varepsilon^2 T'$ which is treated as arbitrary at this stage.

We note that this procedure must fail if the theory for ideal fibre-reinforced materials predicts the presence of singular fibres, because T is infinite on such a fibre, and so a finite value of I_4' cannot be obtained. Experience with problems in linear elasticity [4–6] suggests that these singular fibres, or stress-concentration layers, correspond in almost inextensible materials to thin zones of high stress and that a different asymptotic analysis is needed within these zones. We do not attempt a general theory of these regions, but we shall discuss them in the context of two specific examples.

7. Simple shear

As the first example we consider simple shear. In order to make the boundary conditions as simple as possible we suppose that the deformed body is the rectangular block $0 \leqq x_1 \leqq L$, $0 \leqq x_2 \leqq h$, and the undeformed body has the parallelogram cross-section shown in Figure 2. Then the deformation is described by

(7.1) $$x_1 = X_1 + \gamma X_2, \qquad x_2 = X_2, \qquad x_3 = X_3.$$

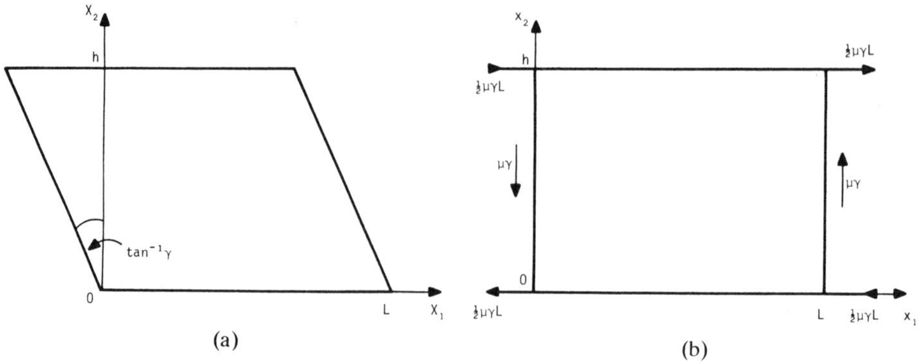

Figure 2. Shear into a rectangular block.

We seek solutions in which the surfaces $x_2 = 0$ and $x_2 = h$ are free from traction. The fibre direction $\boldsymbol{A}$ is taken to be the X_1-direction. Then

$$\text{(7.2)} \qquad \boldsymbol{F} = \begin{bmatrix} 0 & \gamma & 0 \\ 0 & 1 & 0 \\ 0 & 0 & 1 \end{bmatrix}, \quad \boldsymbol{B} = \begin{bmatrix} 1+\gamma^2 & \gamma & 0 \\ \gamma & 1 & 0 \\ 0 & 0 & 1 \end{bmatrix}, \quad \boldsymbol{a} = \begin{bmatrix} 1 \\ 0 \\ 0 \end{bmatrix},$$

$$\text{(7.3)} \qquad I = 2 + \gamma^2, \qquad I_3 = 1, \qquad I_4 = 1,$$

and so the deformation is consistent with the constraints of incompressibility and fibre inextensibility.

Ideal fibre-reinforced material

For an ideal fibre-reinforced material, (4.1) and (7.2) and the boundary conditions

$$\text{(7.4)} \qquad \sigma_{22} = 0, \qquad \sigma_{12} = 0, \qquad \text{at } x_2 = 0,\ x_2 = h,$$

give the solution

$$\text{(7.5)} \qquad \begin{aligned} \sigma_{11} &= \mu\gamma(x_1 - \tfrac{1}{2}L)\{\delta(x_2 - h) - \delta(x_2)\} + f(x_2), \\ \sigma_{22} &= 0, \qquad \sigma_{12} = \mu\gamma\{H(x_2) - H(x_2 - h)\}, \end{aligned}$$

where $H(x_2)$ denotes the Heaviside step function, and $\delta(x_2)$ the Dirac delta function. The function $f(x_2)$ represents an arbitrary fibre tension, which may be singular. We assume the boundary conditions to be such that $f(x_2) = 0$. Then the surfaces $x_1 = 0$, $x_1 = L$ are free from normal traction, except that normal concentrated loads of magnitude $\frac{1}{2}L\mu\gamma$ act at the four corners. Shear traction of magnitude $\mu\gamma$ is applied to the surfaces $x_1 = 0$, $x_1 = L$. The fibres $x_2 = 0$ and $x_2 = h$ are singular, and carry forces which vary linearly with x_1.

Regular perturbation

Now suppose that the material is almost inextensible in the sense described in Section 5, and that the boundary conditions are unchanged. Suppose that the result of this change is to superpose a small displacement $\varepsilon^2 \boldsymbol{u}(x_1, x_2)$ on the deformation (7.1). Then the incompressibility condition gives

$$\frac{\partial u_1}{\partial x_1} + \frac{\partial u_2}{\partial x_2} = 0. \tag{7.6}$$

The fibre tension T is given by (4.1) and (7.2) as

$$T = \sigma_{11} - \sigma_{22} - \gamma\sigma_{12} = -\mu\gamma^2 \tag{7.7}$$

and, for this deformation,

$$I_4' = 2\frac{\partial u_1}{\partial x_1}.$$

We adopt the strain-energy function (5.5), so that T and I_4' are related by (6.5), which gives

$$\mu_0 \frac{\partial u_1}{\partial x_1} = -\mu\gamma^2. \tag{7.8}$$

Hence, from (7.6) and (7.8),

$$u_1 = -\frac{\mu}{\mu_0}\gamma^2 x_1 + f_1(x_2), \qquad u_2 = \frac{\mu}{\mu_0}\gamma^2 x_2 + f_2(x_1), \tag{7.9}$$

where $f_1(x_2)$ and $f_2(x_1)$ are determined by boundary conditions.

However, T is given by (7.7) only in the interior of the block, and T is singular on $x_2 = 0$ and $x_2 = h$. Hence (7.9) does not apply in the neighbourhood of the upper and lower surfaces.

Boundary layers

On the basis of known exact and asymptotic solutions in linear elasticity [4–6] we anticipate that the singular fibres of the theory for an ideal fibre-reinforced material will, for almost inextensible fibres, become thin layers in which the direct stress is large and across which the shear stress varies rapidly. Suppose that in the neighbourhood of $x_2 = 0$ the displacement superimposed on (7.1) has components $U_1(x_1, x_2)$, $U_2(x_1, x_2)$. Since the material is incompressible and almost inextensible, we expect U_1 and U_2 to be small compared to L. However, at the surface the shear stress, and hence the shear strain, is zero, and therefore certain derivatives of U_1 and U_2 must be of order one.

Suppose the boundary layer thickness is of order δ. Then across the boundary layer σ_{12} changes from zero to a value of order μ_0. Hence $\partial\sigma_{12}/\partial x_2 = O(\mu_0/\delta)$. From the equilibrium equation

$$\frac{\partial\sigma_{11}}{\partial x_1} + \frac{\partial\sigma_{12}}{\partial x_2} = 0, \tag{7.10}$$

it follows that $\partial\sigma_{11}/\partial x_1 = O(\mu_0/\delta)$, and hence

$$\sigma_{11} = O(\mu_0 L/\delta). \tag{7.11}$$

Also vanishing shear at the surface demands that $\partial U_1/\partial x_2$ is of order one there. Hence U_1 is of order δ, and $\partial U_1/\partial x_1$ is of order δ/L. But σ_{11} is of order $E_0\partial U_1/\partial x_1 = \varepsilon^{-2}\mu_0\partial U_1/\partial x_1$, and so

$$\sigma_{11} = O(\mu_0\delta/L\varepsilon^2), \tag{7.12}$$

and it follows from (7.11) and (7.12) that δ is of order εL. Also from the incompressibility condition we have $\partial U_2/\partial x_2 = -\partial U_1/\partial x_1 = O(\delta/L)$, and so $U_2 = O(\delta^2/L)$. These considerations suggest the scalings:

$$U_1 = \varepsilon\bar{u}_1(x_1, x_2), \qquad U_2 = \varepsilon^2\bar{u}_2(x_1, x_2), \qquad x_2 = \varepsilon\xi, \tag{7.13}$$

where $\bar{u}_1$, $\bar{u}_2$, ξ are of order L.

Now, for the deformation with U_1, U_2 superposed on (7.1)

$$\begin{bmatrix} F_{11} & F_{12} \\ F_{21} & F_{22} \end{bmatrix} = \begin{bmatrix} 1+\dfrac{\partial U_1}{\partial x_1} & \gamma + \dfrac{\partial U_1}{\partial X_2} \\ \dfrac{\partial U_2}{\partial X_1} & 1 + \dfrac{\partial U_2}{\partial X_2} \end{bmatrix} = \begin{bmatrix} 1+\dfrac{\partial U_1}{\partial x_1} & \gamma\left(1+\dfrac{\partial U_1}{\partial x_1}\right) + \dfrac{\partial U_1}{\partial x_2} \\ \dfrac{\partial U_2}{\partial x_1} & 1 + \gamma\dfrac{\partial U_2}{\partial x_1} + \dfrac{\partial U_2}{\partial x_2} \end{bmatrix} \tag{7.14}$$

By substituting (7.13) into (7.14) and discarding terms of order ε^2, we obtain

$$\begin{bmatrix} F_{11} & F_{12} \\ F_{21} & F_{22} \end{bmatrix} = \begin{bmatrix} 1+\varepsilon\dfrac{\partial\bar{u}_1}{\partial x_1} & 1+\varepsilon\dfrac{\partial\bar{u}_1}{\partial x_1} + \dfrac{\partial\bar{u}_1}{\partial\xi} \\ 0 & 1+\varepsilon\dfrac{\partial\bar{u}_2}{\partial\xi} \end{bmatrix}. \tag{7.15}$$

Hence the incompressibility condition gives, to this order,

$$\frac{\partial\bar{u}_1}{\partial x_1} + \frac{\partial\bar{u}_2}{\partial\xi} = 0. \tag{7.16}$$

Also

$$\begin{gathered} I_4 = 1 + 2\varepsilon\frac{\partial\bar{u}_1}{\partial x_1} + O(\varepsilon^2), \qquad I = B_{11} + B_{22}, \\ B_{11} = 1 + \left(\gamma + \frac{\partial\bar{u}_1}{\partial\xi}\right)^2 + O(\varepsilon), \qquad B_{22} = 1 + O(\varepsilon), \qquad B_{12} = \gamma + \frac{\partial\bar{u}_1}{\partial\xi} + O(\varepsilon). \end{gathered} \tag{7.17}$$

It follows from (5.6) that the leading terms in the expressions for σ_{11} and σ_{12} are

$$\frac{\sigma_{11}}{\mu_0} = \frac{1}{\varepsilon}\frac{\partial \bar{u}_1}{\partial x_1}, \qquad \frac{\sigma_{12}}{\mu_0} = \frac{\mu}{\mu_0}\left(\gamma + \frac{\partial \bar{u}_1}{\partial \xi}\right). \tag{7.18}$$

In general μ is a function of I and so $(7.18)_2$ is highly non-linear. However, the solution is greatly simplified if μ has the constant value μ_0, and we consider this case only. Then (7.18) and the equilibrium equation (7.10) give Laplace's equation

$$\frac{\partial^2 \bar{u}_1}{\partial x_1^2} + \frac{\partial^2 \bar{u}_1}{\partial \xi^2} = 0, \tag{7.19}$$

and this has to be solved subject to the conditions

$$\begin{aligned} &\frac{\partial \bar{u}_1}{\partial \xi} + \gamma = 0, \qquad \xi = 0,\ 0 < x_1 < L, \\ &\bar{u}_1 \to 0, \quad \xi \to \infty,\ 0 < x_1 < L. \end{aligned} \tag{7.20}$$

With the aid of the Fourier series representation

$$\gamma = \frac{2\gamma}{\pi}\sum_{n=1}^{\infty}\frac{1}{n}\{1-(-1)^n\}\sin\frac{n\pi x_1}{L}, \tag{7.21}$$

a solution for $\bar{u}_1$ is readily obtained as

$$\bar{u}_1 = \frac{2L\gamma}{\pi^2}\sum_{n=1}^{\infty}\frac{1}{n^2}\{1-(-1)^n\}\sin\frac{n\pi x_1}{L}\exp\left(-\frac{n\pi\xi}{L}\right). \tag{7.22}$$

Then from (7.6)

$$\bar{u}_2 = \frac{2L\gamma}{\pi^2}\sum_{n=1}^{\infty}\frac{1}{n^2}\{1-(-1)^n\}\cos\frac{n\pi x_1}{L}\exp\left(-\frac{n\pi\xi}{L}\right) + g(x_1), \tag{7.23}$$

and the leading terms in σ_{11} and σ_{12} are, from (7.18),

$$\begin{aligned} \frac{\sigma_{11}}{\mu_0} &= \frac{2\gamma}{\varepsilon\pi}\sum_{n=1}^{\infty}\frac{1}{n}\{1-(-1)^n\}\cos\frac{n\pi x_1}{L}\exp\left(-\frac{n\pi\xi}{L}\right), \\ \frac{\sigma_{12}}{\mu_0} &= \frac{2\gamma}{\pi}\sum_{n=1}^{\infty}\frac{1}{n}\{1-(-1)^n\}\sin\frac{n\pi x_1}{L}\left\{1-\exp\left(-\frac{n\pi\xi}{L}\right)\right\}. \end{aligned} \tag{7.24}$$

The second equilibrium equation, with the boundary condition $\sigma_{22} = 0$ when $x_2 = 0$, shows that σ_{22}/μ_0 is of order ε in the boundary layer.

Since $\xi = x_2/\varepsilon$, it is evident that the leading terms in the expressions for $\bar{u}_1$ and σ_{11} decay rapidly and exponentially as x_2 increases. The maximum value of σ_{11} occurs at the surface and is of order μ_0/ε. The tensile force carried in the boundary layer is

$$F = \varepsilon \int_0^\infty \sigma_{11} d\xi = \frac{2\mu_0 \gamma L}{\pi^2} \sum_{n=1}^{\infty} \frac{1}{n^2} \{1 - (-1)^n\} \cos \frac{n\pi x_1}{L} = \mu_0 \gamma (\tfrac{1}{2}L - x_1). \tag{7.25}$$

Thus, to leading order, the tensile force transmitted by the boundary layer is independent of ε and is equal to the force (7.5) in the singular fibre $x_2 = 0$ in the inextensible theory.

By introducing the limits

$$\varepsilon^{-1} \exp(-\varepsilon^{-1} x) \to \delta(x), \qquad \exp(-\varepsilon^{-1} x) \to 1 - H(x), \qquad \text{as } \varepsilon \to 0 \tag{7.26}$$

for $x \geqq 0$, it follows that in the limit $\varepsilon \to 0$, (7.24) reduce to

$$\sigma_{11} = -\mu_0 \gamma (x_1 - \tfrac{1}{2}L) \delta(x_2), \qquad \sigma_{12} = \mu \gamma H(x_2),$$

which agrees with (7.5) in the neighbourhood of $x_2 = 0$. Hence the singular stress in the ideal fibre-reinforced material can be interpreted as the limit, as $\varepsilon \to 0$, of the stress in the almost inextensible material.

A similar solution applies in the neighbourhood of the surface $x_2 = h$. The solution breaks down at the corners of the block, where there are discontinuities in σ_{12}.

8. Shear bending of a rectangular block

The problem of simple shear considered in Section 7 is rather special in that (a) the deformation is homogeneous, and (b) the fibres remain straight during the deformation. We now consider a problem which does not have these simplifying features.

Consider a body which in its stress-free state is a rectangular block bounded by $X_1 = \pm L$, $X_2 = a$ and $X_2 = b$, and is reinforced by fibres which are parallel to the X_1-axis. The deformation (which is illustrated in Figure 3) defined by

$$r = X_2, \qquad \theta = -X_1/X_2, \qquad x_3 = X_3, \tag{8.1}$$

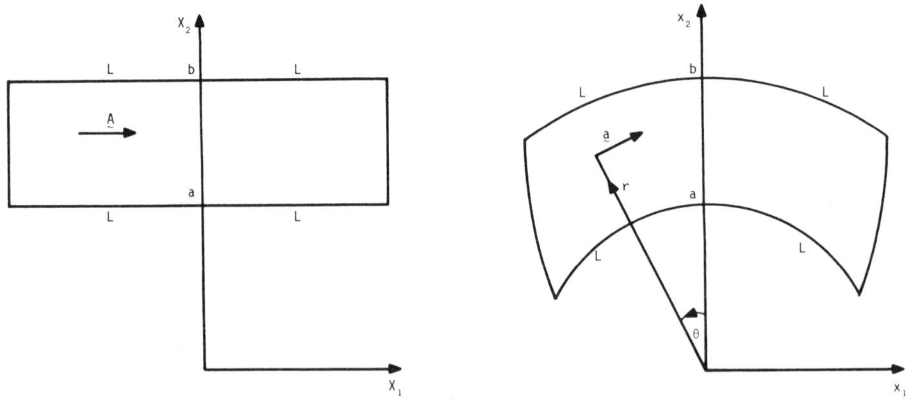

Figure 3. Shear bending of a rectangular block.

where

$$x_1 = -r \sin\theta, \qquad x_2 = r\cos\theta, \tag{8.2}$$

deforms the block into part of the cylinder $a < r < b$, with the surfaces $X_1 = \pm L$ deforming into the surfaces $r = \pm L\theta$. For this deformation

$$\boldsymbol{F} = \boldsymbol{F}_0 = \begin{pmatrix} 0 & 1 & 0 \\ -1 & \theta & 0 \\ 0 & 0 & 1 \end{pmatrix}, \qquad \boldsymbol{B} = \boldsymbol{B}_0 = \begin{pmatrix} 1 & \theta & 0 \\ \theta & 1+\theta^2 & 0 \\ 0 & 0 & 1 \end{pmatrix}, \tag{8.3}$$

where $\boldsymbol{F}_0$ is the matrix of components of the deformation gradient tensor referred to coordinates r, θ, x_3 in the deformed body and X_1, X_2, X_3 in the undeformed body, and the components of $\boldsymbol{B}_0$ are referred to r, θ, x_3 coordinates. Also

$$I = 2 + \theta^2, \qquad I_3 = 1, \qquad I_4 = 1, \tag{8.4}$$

and so the deformation is consistent with the constraints of incompressibility and fibre inextensibility.

Ideal fibre-reinforced material

For an ideal fibre-reinforced material the stress corresponding to the deformation (8.1) is given by (4.1) as

$$\sigma_{rr} = -p + \mu, \qquad \sigma_{\theta\theta} = -p + T + \mu(1+\theta^2), \qquad \sigma_{r\theta} = \mu\theta, \tag{8.5}$$

where, from (8.4), $\mu = \mu(\theta)$. The equilibrium equations are

$$\begin{aligned} &\frac{\partial \sigma_{rr}}{\partial r} + \frac{1}{r}\frac{\partial \sigma_{r\theta}}{\partial \theta} + \frac{\sigma_{rr} - \sigma_{\theta\theta}}{r} = 0, \\ &\frac{\partial \sigma_{r\theta}}{\partial r} + \frac{1}{r}\frac{\partial \sigma_{\theta\theta}}{\partial \theta} + \frac{2\sigma_{r\theta}}{r} = 0. \end{aligned} \tag{8.6}$$

We impose the boundary conditions

$$\begin{aligned} &\sigma_{rr} = \sigma_{r\theta} = 0, \qquad r = b, \ -L/b < \theta < L/b, \\ &\sigma_{r\theta} = 0, \qquad r = a, \ -L/a < \theta < L/a, \\ &\sigma_{\theta\theta} = 0, \qquad \theta = 0, \ a < r < b. \end{aligned} \tag{8.7}$$

The last condition is rather artificial, but is adequate for illustrative purposes. Then the solution of (8.5) and (8.6) which satisfies (8.7) and the jump conditions (4.2) at $r = a$ and $r = b$ is

$$\sigma_{rr} = \left(\frac{b}{r} - 1\right)\left[2\int_0^\theta \mu(\theta)\theta d\theta + \frac{d}{d\theta}\{\mu(\theta)\theta\}\right]$$

$$-\frac{1}{r}\{1 - H(r-b)\}\left\{b\int_0^\theta \mu(\theta)\theta d\theta + F_b\right\}$$

$$+\frac{1}{r}\{1 - H(r-a)\}\left\{a\int_0^\theta \mu(\theta)\theta d\theta - F_a\right\},$$

(8.8) $$\sigma_{\theta\theta} = -2\int_0^\theta \mu(\theta)\theta d\theta + \delta(r-b)\left\{b\int_0^\theta \mu(\theta)\theta d\theta + F_b\right\}$$

$$-\delta(r-a)\left\{a\int_0^\theta \mu(\theta)\theta d\theta - F_a\right\},$$

$$\sigma_{r\theta} = \mu(\theta)\theta\{H(r-a) - H(r-b)\}.$$

The fibres $r = a$ and $r = b$ are singular, there are jumps in σ_{rr} as well as in $\sigma_{r\theta}$ across $r = a$ and $r = b$, and a normal traction component must be applied to the surface $r = a$ to maintain the deformation. F_a and F_b represent constant tensile forces in the singular fibres.

Regular perturbations

Suppose now that the fibres have slight extensibility and that in consequence small radial and tangential displacements $U(r, \theta)$ and $V(r, \theta)$ are superposed on the deformation (8.1). The deformation gradient for the superposed displacement (in polar coordinates) is

(8.9) $$\boldsymbol{F}_1 = \begin{bmatrix} 1 + \dfrac{\partial U}{\partial r} & \dfrac{1}{r}\dfrac{\partial U}{\partial \theta} & 0 \\ \dfrac{\partial V}{\partial r} - \dfrac{V}{r} & 1 + \dfrac{1}{r}\left(U + \dfrac{\partial V}{\partial \theta}\right) & 0 \\ 0 & 0 & 1 \end{bmatrix}.$$

The overall deformation gradient from the initial state is $\boldsymbol{F} = \boldsymbol{F}_1\boldsymbol{F}_0$, which from (8.3) and (8.9) gives

(8.10) $$\boldsymbol{F} = \begin{bmatrix} -\dfrac{1}{r}\dfrac{\partial V}{\partial \theta} & 1 + \dfrac{\partial U}{\partial r} + \dfrac{\theta}{r}\dfrac{\partial V}{\partial \theta} & 0 \\ -1 - \dfrac{1}{r}\left(U + \dfrac{\partial V}{\partial \theta}\right) & \theta + \dfrac{\partial V}{\partial r} - \dfrac{V}{r} + \dfrac{\theta}{r}\left(U + \dfrac{\partial V}{\partial \theta}\right) & 0 \\ 0 & 0 & 1 \end{bmatrix}$$

If we assume that $U = \varepsilon^2 u$, $V = \varepsilon^2 v$, where u and v are of order L, then to leading order the incompressibility condition is

$$\frac{\partial u}{\partial r} + \frac{u}{r} + \frac{1}{r}\frac{\partial v}{\partial \theta} = 0, \tag{8.11}$$

and

$$I_4 = 1 + \frac{2\varepsilon^2}{r}\left(u + \frac{\partial v}{\partial \theta}\right). \tag{8.12}$$

Hence (6.5) and (8.5) give

$$T = \sigma_{\theta\theta} - \sigma_{rr} - \theta\sigma_{r\theta} = \frac{\mu_0}{r}\left(u + \frac{\partial v}{\partial \theta}\right). \tag{8.13}$$

With (8.8), equations (8.11) and (8.13) determine u and v. However this approach breaks down in the neighbourhood of the stress-concentration layers.

Boundary layers

We now consider the stress and deformation near the surface $r = b$. By arguments similar to those used in Section 7 we introduce the following scalings in the region adjacent to $r = b$:

$$U = \varepsilon^2 \bar{u}, \qquad V = \varepsilon \bar{v}, \qquad r = b + \varepsilon\xi. \tag{8.14}$$

Then, to order ε, (8.10) becomes

$$\boldsymbol{F} = \begin{bmatrix} 0 & 1 + \varepsilon\dfrac{\partial \bar{u}}{\partial \xi} & 0 \\ -1 - \dfrac{\varepsilon}{b}\dfrac{\partial \bar{v}}{\partial \theta} & \theta + \dfrac{\partial \bar{v}}{\partial \xi} - \dfrac{\varepsilon}{b}\left(\bar{v} - \theta\dfrac{\partial \bar{v}}{\partial \theta}\right) & 0 \\ 0 & 0 & 1 \end{bmatrix}. \tag{8.15}$$

Hence the incompressibility condition is, to this order,

$$\frac{\partial \bar{u}}{\partial \xi} + \frac{1}{b}\frac{\partial \bar{v}}{\partial \theta} = 0, \tag{8.16}$$

and

$$\begin{gathered} I_4 = 1 + \frac{2\varepsilon}{b}\frac{\partial \bar{v}}{\partial \theta} + O(\varepsilon^2), \qquad I = B_{rr} + B_{\theta\theta}, \\ B_{rr} = 1 + O(\varepsilon), \qquad B_{\theta\theta} = 1 + \left(\theta + \frac{\partial \bar{v}}{\partial \xi}\right)^2 + O(\varepsilon), \qquad B_{r\theta} = \theta + \frac{\partial \bar{v}}{\partial \xi} + O(\varepsilon). \end{gathered} \tag{8.17}$$

It follows from (5.6) that the leading terms in the expressions for $\sigma_{\theta\theta}$ and $\sigma_{r\theta}$ are

$$\frac{\sigma_{\theta\theta}}{\mu_0}=\frac{1}{\varepsilon b}\frac{\partial \bar{v}}{\partial \theta}, \qquad \frac{\sigma_{r\theta}}{\mu_0}=\frac{\mu}{\mu_0}\left(\theta+\frac{\partial \bar{v}}{\partial \xi}\right). \tag{8.18}$$

By substituting (8.18) into the equilibrium equation $(8.6)_2$ we obtain a differential equation for $\bar{v}$. In general μ is a function of I, and so this equation is highly non-linear, and cannot be solved without specifying the dependence of μ on I. However, if μ has the constant value μ_0, $(8.6)_2$ and (8.18) give, to leading order in ε, Laplace's equation

$$\frac{1}{b^2}\frac{\partial^2 \bar{v}}{\partial \theta^2}+\frac{\partial^2 \bar{v}}{\partial \xi^2}=0, \tag{8.19}$$

and we consider this case only. The boundary conditions are

$$\begin{aligned} &\frac{\partial \bar{v}}{\partial \xi}+\theta=0, \qquad \xi=0,\\ &\bar{v}\to 0, \qquad \xi\to-\infty. \end{aligned} \tag{8.20}$$

The solution of (8.19) subject to (8.20) is

$$\bar{v}=\frac{2L^2}{\pi^2 b}\sum_{n=1}^{\infty}\frac{(-1)^n}{n^2}\sin\frac{n\pi b\theta}{L}\exp\frac{n\pi\xi}{L}. \tag{8.21}$$

It follows from (8.16) that

$$\bar{u}=-\frac{2L^2}{\pi^2 b}\sum_{n=1}^{\infty}\frac{(-1)^n}{n^2}\cos\frac{n\pi b\theta}{L}\exp\frac{n\pi\xi}{L}+h(\theta), \tag{8.22}$$

and the leading terms in $\sigma_{\theta\theta}$ and $\sigma_{r\theta}$ are, from (8.18),

$$\begin{aligned} &\frac{\sigma_{\theta\theta}}{\mu_0}=\frac{2L}{\varepsilon\pi b}\sum_{n=1}^{\infty}\frac{(-1)^n}{n}\cos\frac{n\pi b\theta}{L}\exp\frac{n\pi\xi}{L},\\ &\frac{\sigma_{r\theta}}{\mu_0}=-\frac{2L}{\pi b}\sum_{n=1}^{\infty}\frac{(-1)^n}{n}\sin\frac{n\pi b\theta}{L}\left(1-\exp\frac{n\pi\xi}{L}\right). \end{aligned} \tag{8.23}$$

From $(8.6)_1$, with the condition $\sigma_{rr}=0$ when $\xi=0$, it follows that

$$\frac{\sigma_{rr}}{\mu_0}=\frac{2L^2}{\pi^2 b^2}\sum_{n=1}^{\infty}\frac{(-1)^n}{n^2}\cos\frac{n\pi b\theta}{L}\left(\exp\frac{n\pi\xi}{L}-1\right). \tag{8.24}$$

The tensile force in the boundary layer is

$$F=\varepsilon\int_{-\infty}^{0}\sigma_{\theta\theta}d\xi=\frac{2\mu_0 L^2}{\pi^2 b}\sum_{n=1}^{\infty}\frac{(-1)^n}{n^2}\cos\frac{n\pi b\theta}{L}=\frac{1}{2}\mu_0 b\left(\theta^2-\frac{1}{3}\frac{L^2}{b^2}\right), \tag{8.25}$$

and, as $\xi \to -\infty$,

$$\frac{\sigma_{rr}}{\mu_0} \to -\frac{2L^2}{\pi^2 b^2} \sum_{n=1}^{\infty} \frac{(-1)^n}{n^2} \cos \frac{n\pi b\theta}{L} = -\frac{1}{2}\left(\theta^2 - \frac{1}{3}\frac{L^2}{b^2}\right). \tag{8.26}$$

These results are in agreement with values given by (8.8) (with $\mu = \mu_0$) for the fibre force and jump in σ_{rr} across $r = b$ in the solution for the ideal fibre-reinforced material, provided that the arbitrary constant fibre force F_b has the value $-\frac{1}{3}\mu_0 L^2/b$. In the neighbourhood of $r = b$, the above solution for the almost inextensible fibre-reinforced material reduces to that for the ideal fibre-reinforced material in the limit $\varepsilon \to 0$, with $F_b = -\frac{1}{3}\mu_0 L^2/b$.

If F_b does not take this special value, then in order that the boundary-layer solution should reduce to the ideal fibre-reinforced material solution in the limit $\varepsilon \to 0$, it is necessary to superimpose a further solution of (8.19) onto (8.21). The appropriate solution to be superposed is of the form

$$\bar{v} = \frac{A}{2\pi} \log \left\{\frac{\xi^2 + (L + b\theta)^2}{\xi^2 + (L - b\theta)^2}\right\}. \tag{8.27}$$

The corresponding superposed values of $\sigma_{\theta\theta}$, $\sigma_{r\theta}$ and σ_{rr} are, from $(8.6)_1$ and (8.18),

$$\sigma_{\theta\theta} = \frac{\mu_0 A}{\varepsilon\pi}\left\{\frac{L + b\theta}{\xi^2 + (L + b\theta)^2} + \frac{L - b\theta}{\xi^2 + (L - b\theta)^2}\right\},$$

$$\sigma_{r\theta} = \frac{\mu_0 A\xi}{\pi}\left\{\frac{1}{\xi^2 + (L + b\theta)^2} - \frac{1}{\xi^2 + (L - b\theta)^2}\right\},$$

$$\sigma_{rr} = \frac{\mu_0 A}{\pi}\left\{\tan^{-1}\left(\frac{\xi}{L + b\theta}\right) + \tan^{-1}\left(\frac{\xi}{L - b\theta}\right)\right\},$$

and the superposed tensile force in the boundary layer is

$$\varepsilon \int_{-\infty}^{0} \sigma_{\theta\theta} d\xi = \mu_0 A.$$

Hence, in the limit $\varepsilon \to 0$, the stress corresponding to (8.27) becomes a constant fibre force $\mu_0 A$ in the fibre $r = b$.

A similar boundary-layer solution obtains in the neighbourhood of $r = a$, except that at this surface the boundary condition $\sigma_{rr} = 0$ no longer applies. However the jump in σ_{rr} across the boundary layer is the same as the jump in σ_{rr} across the singular fibre $r = a$. Therefore, if it is assumed that the theory for ideal fibre-reinforced materials gives a valid first approximation except in the neighbourhood of the stress-concentration layers, the boundary-layer solution once more reduces to the solution for the ideal fibre-reinforced material in the limit $\varepsilon \to 0$.

9. Discussion

For the examples discussed in Sections 7 and 8, the boundary-layer theory seems to give an adequate interpretation of the singular fibres which are predicted by the theory for ideal fibre-reinforced materials. Just as in the linear theory, in slightly extensible materials the singular fibres correspond to layers of thickness of order εL. In these, for shear strains of order γ, the tensile stress is of order $\mu_0\gamma/\varepsilon$, and across them the shear stress changes by an amount of order $\mu_0\gamma$. If the singular fibres are curved with radius of curvature b, then the normal stress also changes by an amount of order $L\mu_0\gamma/b$. The solutions for the ideal fibre-reinforced material may be interpreted as the limits of the solutions for the almost inextensible material as $\varepsilon \to 0$.

Throughout this paper we have assumed the material to be incompressible. There is some inconsistency in requiring perfect incompressibility while permitting slight extensibility. It should be possible to develop a theory for materials which are slightly compressible and slightly extensible, but such a theory would be more complicated than the one described here. We would not expect slight compressibility to affect substantially the stress and deformation in the boundary layers discussed in this paper. However, it should be noted [2] that in plane strain of materials which are incompressible and inextensible in the fibre direction, singular stress may develop along the normal trajectories of the fibres as well as along the fibres, so that in almost incompressible, almost inextensible materials we might expect boundary layers to occur along these curves also. Singularities of this kind do not arise in the examples considered in this paper.

Finally, we observe that Parker [14, 15] has suggested an alternative approach to 'almost constrained' materials; in the case of a fibre-reinforced material this involves regarding the fibre tension as an independent variable and introducing the Legendre transform of W.

References

1. J. E. Adkins and R. S. Rivlin, *Large elastic deformations of isotropic materials X. Reinforcement by inextensible cords*, Phil. Trans. R. Soc. **A248** (1955), 201–223.

2. A. C. Pipkin and T. G. Rogers, *Plane deformations of incompressible fibre-reinforced materials*, J. Appl. Mech. **38** (1971), 634–640.

3. A. J. M. Spencer, *Deformations of Fibre-Reinforced Materials*, Clarendon Press, Oxford, 1972.

4. G. C. Everstine and A. C. Pipkin, *Stress channelling in transversely isotropic elastic composites*, Z. angew. Math. Phys. **22** (1971), 825–834.

5. G. C. Everstine and A. C. Pipkin, *Boundary layers in fiber-reinforced materials*, J. Appl. Mech. **40** (1973), 518–522.

6. A. J. M. Spencer, *Boundary layers in highly anisotropic plane elasticity*, Int. J. Solid Structures **10** (1974), 1103–1123.

7. A. J. M. Spencer, *Theory of invariants* in *Continuum Physics*, Vol. 1 (A. C. Eringen, ed.), Academic Press, New York, 1971.

8. J. E. Ericksen and R. S. Rivlin, *Large elastic deformations of homogeneous anisotropic materials*, J. Rat. Mech. Anal. **3** (1954), 281–301.

9. A. E. Green, R. S. Rivlin and R. T. Shield, *General theory of small elastic deformations superposed on finite elastic deformations*, Proc. R. Soc. **A211** (1951), 128–154.

10. A. E. Green and W. Zerna, *Theoretical Elasticity*, Clarendon Press, Oxford, 1954.

11. J. G. Oldroyd, *Finite strains in an anisotropic elastic continuum*, Proc. R. Soc. **A202** (1950), 345–358.

12. A. J. M. Spencer, *Finite deformations of an almost incompressible elastic solid*, in *Proceedings of the International Symposium on Second-Order Effects in Elasticity, Plasticity and Fluid Mechanics* (M. Reiner and D. Abir, eds.), Pergamon Press, Oxford and Jerusalem Academic Press, Jerusalem, 19??.

13. A. J. M. Spencer, *The static theory of finite elasticity*, J. Inst. Maths. Applics. **6** (197?), 164–200.

14. D. F. Parker, Discussion of *Some research directions in finite elasticity theory* by R. S. Rivlin, Rheologica Acta **16** (1977), 112.

15. D. F. Parker, *Finite-amplitude oscillations of a spherical cavity in a nearly-incompressible elastic material*, Archives of Mechanics **30** (1978), 777–790.

INCREMENTAL METHODS IN FINITE ELASTICITY, ESPECIALLY FOR RODS

ERWIN STEIN

Lehrstuhl für Baumechanik, Technische Universität Hannover, BR Deutschland

(Received June 17, 1981)

1. Introduction

In the course of the last decade we investigated incremental deformation processes of engineering structures like stiffened bridge girders, shells and large spatial rod systems, e.g., high cranes, in order to get insight into critical, postcritical and ultimate load behaviour. The discretization is realized by finite element methods using standard functionals and element models or appropriate modifications. According to the situation, the material behaviour may be described as elastic, elasto-plastic or elasto-viscoplastic. These nonlinear calculations are necessary in light of new safety-concepts where deterministic and stochastic loadings of different probabilities are to be considered and not — as before — admissible stresses for idealized working loads.

In this paper the geometrically nonlinear theory of prismatic thin rods without warping is presented within consistent approximations for infinitesimally small, moderate and large rotations of the rod axis with respect to a small parameter $\theta = \max((D/L)\sqrt{\varepsilon})$; D characteristic measure of cross section, L length of rod, ε maximum strain. The material is assumed to be linear elastic. Looking at the polar decomposition of the finite material deformation gradient, only the rotations are to be finite but the strains shall be infinitesimally small within a load increment. For practical calculations, the incremental application of moderate rotation theory proved to be efficient looking at the size of load steps and the number of iterations within one load step, i.e., evaluating the whole computation effort.

Proceedings of the IUTAM Symposium on Finite Elasticity, Lehigh University, August 10–15, 1980. Invited paper.

Carlson, D.E. and Shield, R.T. (eds.)
Proceedings of the IUTAM Symposium on Finite Elasticity
ISBN 90 247 2629 8

Beam theory has a very long tradition in mechanics. As a starting point for this paper the rigorous treatise of Kappus [1] is essential. The lectures of my colleague F. H. Schroeder on beam theory gave some important stimulations for the doctoral thesis of my student Kessel [2] in which a comprehensive theoretical study and finite element calculations of ultimate loads for high cranes were presented.

The concept of consistent approximations using power expansions in a three-dimensional theory was developed by Hay [3] for very thin rods and by John [4] and [4] Pietraszkiewicz [5] for thin shells.

2. Incremental deformations of elastic bodies

2.1. *Material description*

We assume isotropic elastic material, linear with respect to spatial strain rate tensor $\boldsymbol{D}$ and Jaumann stress flow $\mathring{\boldsymbol{T}}$

$$\begin{aligned} \mathring{\boldsymbol{T}} &= 2G\left[\boldsymbol{D} + \frac{\nu}{1-2\nu}\operatorname{tr}(\boldsymbol{D})\boldsymbol{1}\right], \\ \mathring{\boldsymbol{T}} &= \mathbb{L}[\boldsymbol{D}]. \end{aligned} \tag{1}$$

For small strains and large rotations equation (1) can be transformed into a linear relation between the rates of Green–Lagrange strain tensor and 2nd P. K. stress tensor

$$\begin{aligned} \dot{\boldsymbol{S}} &= 2G\left[\dot{\boldsymbol{E}} + \frac{\nu}{1-2\nu}\operatorname{tr}(\dot{\boldsymbol{E}})\boldsymbol{1}\right], \\ \dot{\boldsymbol{S}} &= \mathbb{L}[\dot{\boldsymbol{E}}], \end{aligned} \tag{2}$$

with same operator $\mathbb{L}$ as in equation (1).

PROOF. To show this result we write the material deformation gradient as

$$\boldsymbol{F} = \frac{\partial \boldsymbol{x}}{\partial \boldsymbol{X}} = \frac{\partial x_i}{\partial X_k}\, \boldsymbol{e}_i \otimes \boldsymbol{e}_k. \tag{3}$$

Polar decomposition gives

$$\boldsymbol{F} = \boldsymbol{R}\boldsymbol{U}, \tag{4}$$

and we assume that

$$\boldsymbol{U} = \boldsymbol{1} + \varepsilon \boldsymbol{A}, \tag{5}$$

with

$$|\boldsymbol{U}| \approx |1| \approx |\boldsymbol{A}| \quad \text{and} \quad \varepsilon \to 0. \tag{6}$$

Then

$$F \approx R = (R^T)^{-1} = (F^T)^{-1}; \qquad \det F = 1. \tag{7}$$

The relation between Truesdell stress flow and 2nd P. K. stress tensor is

$$\overset{\Delta}{T} := (\det F)^{-1} F \dot{S} F^T, \tag{8}$$

and the relation with Jaumann stress flow is

$$\overset{\Delta}{T} = \overset{\circ}{T} - TD - DT + \mathrm{tr}(D)T. \tag{9}$$

Introducing the material law, equation (1), into equation (9),

$$\overset{\Delta}{T} = (G\mathbf{1} - T)D + D(G\mathbf{1} - T) + \frac{2G\nu}{1-2\nu}\,\mathrm{tr}(D)\mathbf{1} + T\,\mathrm{tr}(D) \tag{10}$$

and due to assumption, equation (5),

$$T \ll G\mathbf{1}, \tag{11}$$

we get

$$\overset{\Delta}{T} = 2G\left[D + \frac{\nu}{1-2\nu}\,\mathrm{tr}(D)\mathbf{1}\right] = \overset{\circ}{T}. \tag{12}$$

Again with the assumption, equation (5), we get from equation (8),

$$\overset{\Delta}{T} = R\dot{S}R^T, \tag{13}$$

and with the known relation

$$\dot{E} = F^T D F,$$

and regarding equation (7) we obtain

$$D = R\dot{E}R^T. \tag{14}$$

Introducing equations (13) and (14) into equation (12), we arrive at

$$\dot{S} = 2G\left[\dot{E} + \frac{\nu}{1-2\nu}\,\mathrm{tr}(\dot{E})\mathbf{1}\right], \tag{15}$$

which was to be proved.

2.2. *Incremental form of principle of virtual work from state u^N to $u^N + \Delta u$*

The total Lagrangian representation of the incremental principle of virtual work can be written for a finite element discretization in the form

$$\text{(16)} \qquad \int_{V_0} \operatorname{tr}[\delta\Delta \boldsymbol{E}(\boldsymbol{S}^N + \Delta\boldsymbol{S})]dV_0 - \delta\Delta\boldsymbol{V}^T(\bar{\boldsymbol{R}}^N + \Delta\bar{\boldsymbol{R}}) = 0,$$

where $\Delta\boldsymbol{V}$ is the column matrix of incremental nodal displacements in FEM; $\Delta\bar{\boldsymbol{R}}$ is the given incremental nodal loads,

$$\text{(17)} \qquad \Delta\boldsymbol{E} = \operatorname{sym}\{\Delta\boldsymbol{F}^T\boldsymbol{F}^N + \tfrac{1}{2}\Delta\boldsymbol{F}^T\Delta\boldsymbol{F}\},$$

$\Delta\boldsymbol{E}$ is the incremental Green strain tensor in each point of the elements,

$$\text{(18)} \qquad \delta\Delta E = \operatorname{sym}\{\delta\Delta\boldsymbol{F}^T\boldsymbol{F}^N\} + \operatorname{sym}\{\delta\Delta\boldsymbol{F}^T\Delta\boldsymbol{F}\}.$$

With the symmetry of $\boldsymbol{S}^N$ and $\Delta\boldsymbol{S}$ we get from equation (16),

$$\text{(19)} \qquad \begin{aligned} &\int_{V_0} \operatorname{tr}(\delta\Delta\boldsymbol{E})\Delta\boldsymbol{S}dV_0 + \int_{V_0} \operatorname{tr}(\delta\Delta\boldsymbol{F}^T\Delta\boldsymbol{F}\boldsymbol{S}^N)dV_0 \\ &= \left[\delta\Delta\boldsymbol{V}^T\bar{\boldsymbol{R}}^N - \int_{V_0} \operatorname{tr}(\delta\Delta\boldsymbol{F}^T\boldsymbol{F}^N\boldsymbol{S}^N)dV_0\right] + \delta\Delta\boldsymbol{V}^T\Delta\bar{\boldsymbol{R}}. \end{aligned}$$

That the term in square brackets is zero can be shown by using the principle of virtual work for state N:

$$\text{(20)} \qquad \int_{V_0} \operatorname{tr}(\delta\boldsymbol{E}^N\boldsymbol{S}^N)dV_0 - \delta\boldsymbol{V}^{NT}\bar{\boldsymbol{R}}^N = 0.$$

Introducing incremental quantities instead of those for state N, i.e.,

$$\delta\boldsymbol{E}^N = \operatorname{sym}\{\delta\boldsymbol{F}^{NT}\boldsymbol{F}^N\}; \qquad \delta\boldsymbol{V}^N$$

changing into

$$\delta\Delta\boldsymbol{E} = \operatorname{sym}\{\delta\Delta\boldsymbol{F}^T\boldsymbol{F}^N\}; \qquad \delta\Delta\boldsymbol{V},$$

the prediction is confirmed. So, finally we get the wanted incremental form

$$\text{(21)} \qquad \int_{V_0} \operatorname{tr}(\delta\Delta\boldsymbol{E}\Delta\boldsymbol{S})dV_0 + \int_{V_0} \operatorname{tr}(\delta\Delta\boldsymbol{F}^T\Delta\boldsymbol{F}\boldsymbol{S}^N)dV_0 = \delta\Delta\bar{\boldsymbol{V}}^T\Delta\bar{\boldsymbol{R}}.$$

3. Theory of prismatic rods with finite rotations

3.1. *Configurations of rod space*

The configurations of rod space are defined by the following mappings of the topological body

$$\text{(22)} \quad \mathring{\chi}: \begin{cases} \mathfrak{B} \to \mathfrak{P} \\ B \to \mathring{\chi}(B) := P \end{cases} \qquad \text{initial configuration,}$$

(23) $\chi : \begin{cases} \mathfrak{B} \rightarrow \mathfrak{P} \\ B \rightarrow \mathring{\chi}(B) := p \end{cases}$ current configuration,

(24) $\chi \circ \mathring{\chi}^{-1} : \begin{cases} \mathring{\mathfrak{P}} \rightarrow \mathfrak{P} \\ p \rightarrow \chi \circ \mathring{\chi}^{-1}(P) = p \end{cases}$ deformation,

(25) $\boldsymbol{u} : \begin{cases} \mathring{\mathfrak{P}} \rightarrow \mathfrak{P} \\ \boldsymbol{X} \rightarrow \boldsymbol{X} + \boldsymbol{u} \end{cases}$ displacement vector of point P.

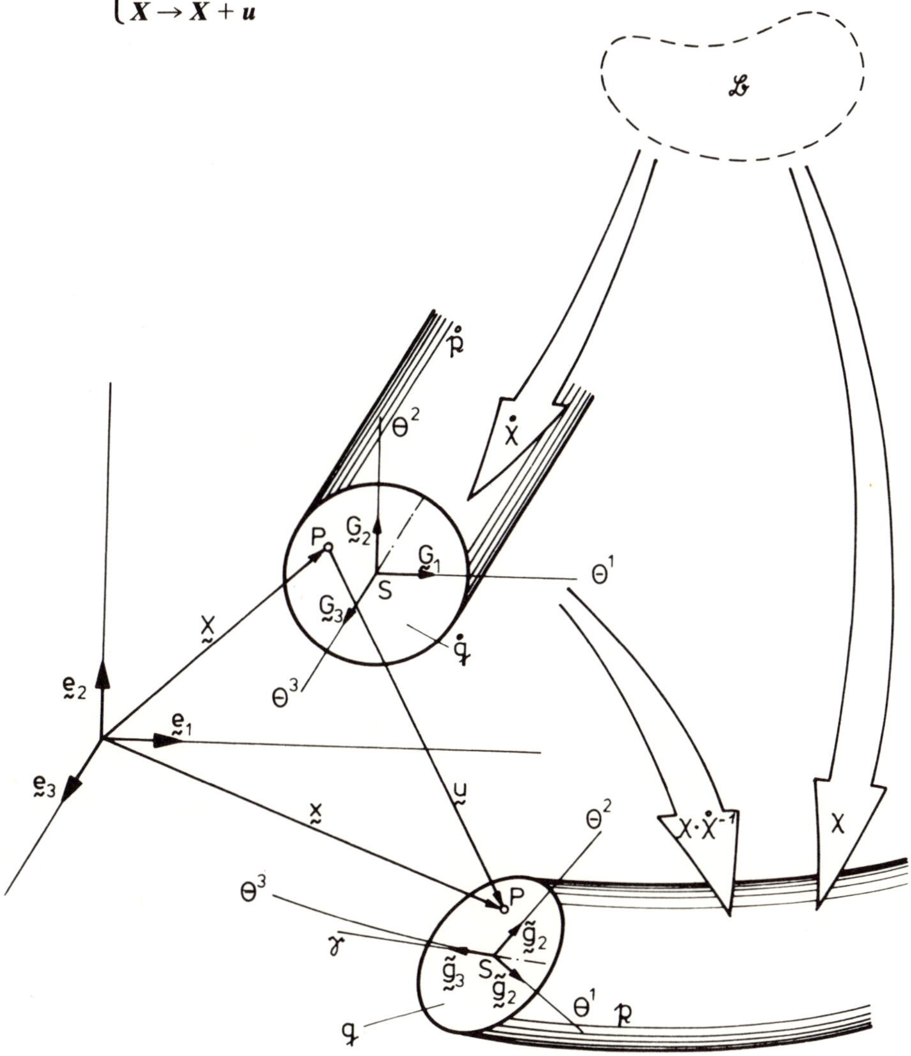

Figure 1. Deformation of a straight rod.

3.2. *Reduction of 3-dimension into 1-dimension problem*

The following well-known hypotheses are introduced:

A1: Bernoulli. The plane cross-section $\mathfrak{Q}^\circ$ is mapped by $\chi \circ \mathring{\chi}^{-1}$ into a plane section $\mathfrak{Q}$.

A2: de St. Venant. A straight line $\mathfrak{G}^\circ \subset \mathfrak{Q}^\circ$ through the common shear and gravity centre S is mapped by $\chi \circ \mathring{\chi}^{-1}$ into a straight line $\mathfrak{G} \subset \mathfrak{Q}$ through s.
The shape of the cross-section remains unchanged during deformation. For convected coordinates θ^i the displacement vector becomes

$$u = \tilde{u} + \theta^1(\tilde{g}_1 - G_1) + \theta^2(\tilde{g}_2 - G_2). \tag{26}$$

3.3. *Kinematics of rod axis with finite rotations*

3.3.1. *Directors of rod axis and their transformation.* The directors of the oriented points $S(\theta^3)$ are $G_1(\theta^3)$, $G_2(\theta^3)$ with $G_i \cdot G_j = \delta_{ij}$. The deformations are described by

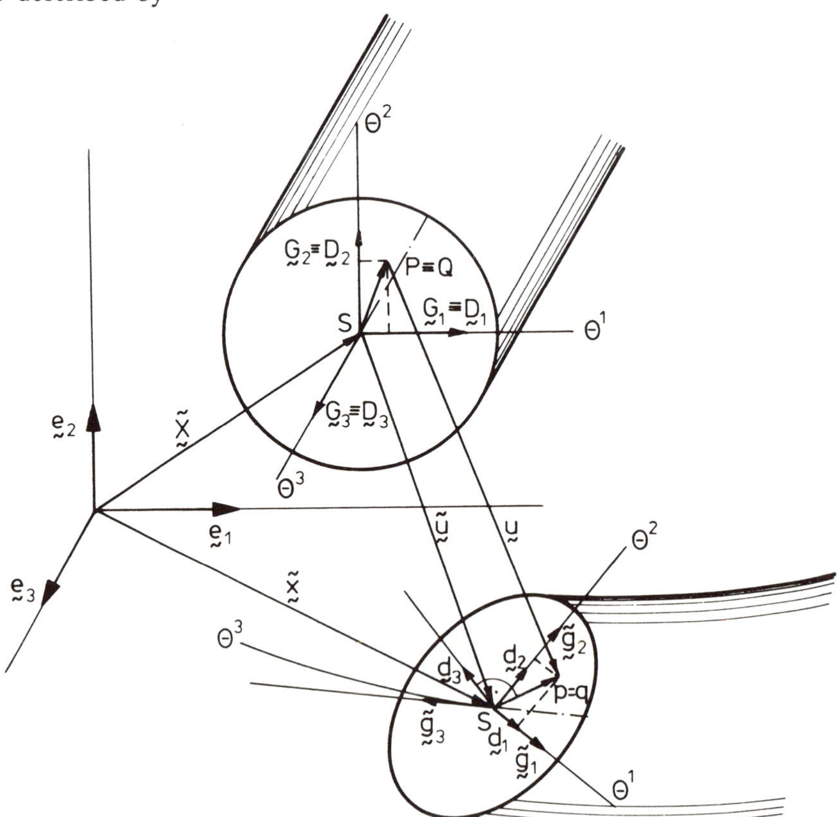

Figure 2. Base vectors and directors of rod axis.

$$\overset{\circ}{\vartheta} = \{(S, D)/S \in \overset{\circ}{\mathscr{S}},\ D = (\boldsymbol{D}_1, \boldsymbol{D}_2) = (\boldsymbol{G}_1, \boldsymbol{G}_2)\},$$

(27) $$\vartheta = \{(s, d)/s \in \mathscr{S},\ d = (\boldsymbol{d}_1, \boldsymbol{d}_2) \in \mathbb{E}^3 \times \mathbb{E}^3,\ \boldsymbol{d}_1 \perp \boldsymbol{d}_2\},$$

$$\chi \circ \overset{\circ}{\chi}{}^{-1}: \begin{cases} \overset{\circ}{\vartheta} \to \vartheta, \\ (S, D) \to (s, d). \end{cases}$$

With the displacement vector of point S

(27a) $$\tilde{\boldsymbol{u}}(\theta^3): \begin{cases} \overset{\circ}{\mathscr{S}} \to \mathscr{S} \\ S \to \tilde{\boldsymbol{u}} + S. \end{cases}$$

3.3.2. *Finite rotation vector.* As a measure for finite rotations we choose the vector

(28) $$\boldsymbol{\Omega} = \varphi \boldsymbol{\phi} = \omega^m \boldsymbol{G}_m \qquad \text{with } |\boldsymbol{\Omega}| = \varphi,$$

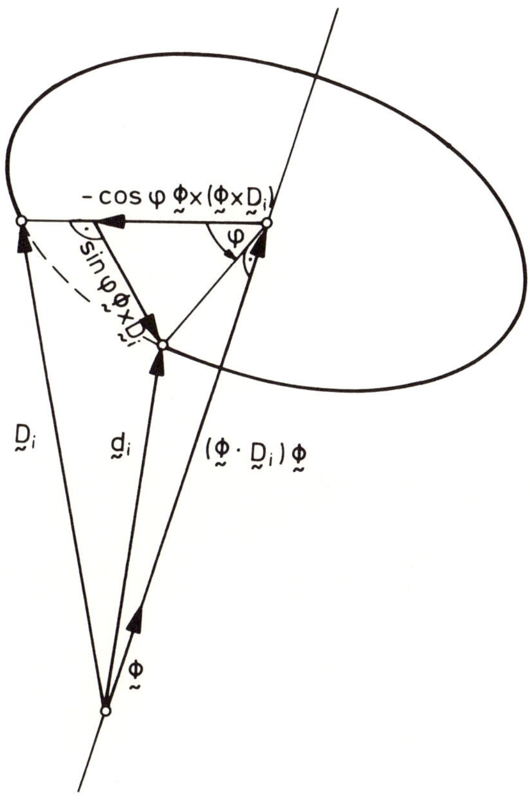

Figure 3. Representation of finite rotations.

which has the direction of rotation axis. φ is the angle. The Lagally representation of finite rotation angle φ is

$$\begin{aligned} \boldsymbol{d}_i &= \boldsymbol{\phi}(\boldsymbol{\phi}\cdot\boldsymbol{D}_i) - \cos\varphi\boldsymbol{\phi}\times(\boldsymbol{\phi}\times\boldsymbol{D}_i) + \sin\varphi\boldsymbol{\phi}\times\boldsymbol{D}_i, \\ \boldsymbol{d}_i &= \boldsymbol{\Phi}\boldsymbol{D}_i, \end{aligned} \tag{29}$$

with the versor (pure rotation)

$$\text{a)}\ \boldsymbol{\Phi} = \boldsymbol{\phi}\otimes\boldsymbol{\phi} + \cos\varphi(\boldsymbol{I} - \boldsymbol{\phi}\otimes\boldsymbol{\phi}) + \sin\varphi\boldsymbol{\phi}\times\boldsymbol{I}, \tag{30}$$

or in another form according to Pietraszkiewicz [5]

$$\hat{\boldsymbol{\Omega}} = \boldsymbol{\phi}\sin\varphi, \tag{31}$$

with the versor

$$\begin{aligned} \text{b)}\ \boldsymbol{\Phi} &= \frac{1}{\sin^2\varphi}\hat{\boldsymbol{\Omega}}\otimes\hat{\boldsymbol{\Omega}} + \cos\varphi\left(\boldsymbol{I} - \frac{1}{\sin^2\varphi}\hat{\boldsymbol{\Omega}}\otimes\hat{\boldsymbol{\Omega}}\right) + \hat{\boldsymbol{\Omega}}\times\boldsymbol{I} \\ &= \cos\varphi\boldsymbol{I} + \hat{\boldsymbol{\Omega}}\times\boldsymbol{I} + \frac{1-\cos\varphi}{\sin^2\varphi}\hat{\boldsymbol{\Omega}}\otimes\hat{\boldsymbol{\Omega}}, \end{aligned} \tag{32}$$

and the third form

$$\text{c)}\ \bar{\boldsymbol{\Omega}} = \boldsymbol{\phi}\tan\varphi/2, \tag{33}$$

in order to represent the components of versor $\boldsymbol{\Phi}$ in the components of rotation vector $\boldsymbol{\Omega}$.

With

$$\begin{aligned} &\bar{\omega} = \tan\varphi/2, \\ &\sin\varphi = \frac{2\bar{\omega}}{1+\bar{\omega}^2}, \qquad \cos\varphi = \frac{1-\bar{\omega}^2}{1+\bar{\omega}^2}, \end{aligned} \tag{34}$$

we get the versor from (30)

$$\boldsymbol{\phi} = \frac{1-(\bar{\boldsymbol{\Omega}}\cdot\bar{\boldsymbol{\Omega}})I + 2\bar{\boldsymbol{\Omega}}\otimes\bar{\boldsymbol{\Omega}} + 2\bar{\boldsymbol{\Omega}}\times\boldsymbol{I}}{1+\bar{\boldsymbol{\Omega}}\cdot\bar{\boldsymbol{\Omega}}}. \tag{35}$$

The component representation of $\bar{\boldsymbol{\Omega}}$ is

$$\bar{\boldsymbol{\Omega}} = \bar{\omega}^m\boldsymbol{D}_m. \tag{36}$$

From $\boldsymbol{d}_i = \boldsymbol{\Phi}\boldsymbol{D}_i$ we get Cayley's formulas

$$\begin{aligned} &(1+\bar{\omega}^{12}+\bar{\omega}^{22}+\bar{\omega}^{32})\begin{bmatrix}\boldsymbol{d}_1\\ \boldsymbol{d}_2\\ \boldsymbol{d}_3\end{bmatrix} \\ &= \begin{Bmatrix} 1+\bar{\omega}^{12}-\bar{\omega}^{22}-\bar{\omega}^{32} & 2(\bar{\omega}^1\bar{\omega}^2-\bar{\omega}^3) & 2(\bar{\omega}^1\bar{\omega}^3+\bar{\omega}^2) \\ 2(\bar{\omega}^1\bar{\omega}^2+\bar{\omega}^3) & 1-\bar{\omega}^{12}+\bar{\omega}^{22}-\bar{\omega}^{32} & 2(\bar{\omega}^2\bar{\omega}^3-\bar{\omega}^1) \\ 2(\bar{\omega}^1\bar{\omega}^3-\bar{\omega}^2) & 2(\bar{\omega}^2\bar{\omega}^3+\bar{\omega}^1) & 1-\bar{\omega}^{12}-\bar{\omega}^{22}+\bar{\omega}^{32} \end{Bmatrix}\begin{bmatrix}\boldsymbol{D}_1\\ \boldsymbol{D}_2\\ \boldsymbol{D}_3\end{bmatrix}. \end{aligned} \tag{37}$$

This matrix representation can be condensed as

$$
\begin{aligned}
&\boldsymbol{d} = \boldsymbol{R}\boldsymbol{D}, \\
&\boldsymbol{d} = [d_i]; \qquad \boldsymbol{D} = [D_j]; \qquad \boldsymbol{R} = [R^j_i],
\end{aligned} \tag{38}
$$

with the proper orthogonal rotation matrix $\boldsymbol{R}$, i.e.,

$$
\boldsymbol{R}^T = \boldsymbol{R}^{-1}. \tag{38a}
$$

Matrix $\boldsymbol{R}$ represents a linear unitarian mapping. With $\boldsymbol{R}$ one can refer directly to the rotation angle φ instead of $\tan\varphi/2$. The versor is

$$
\boldsymbol{\Phi} = \frac{\boldsymbol{\Omega}\otimes\boldsymbol{\Omega} + \cos\psi(\psi^2\boldsymbol{I} - \boldsymbol{\Omega}\otimes\boldsymbol{\Omega})}{\psi^2} + \frac{\sin\psi\,\boldsymbol{\Omega}\times\boldsymbol{I}}{\psi}, \tag{39}
$$

with $\psi = |\boldsymbol{\Omega}|$; $\psi^2 = \boldsymbol{\Omega}\cdot\boldsymbol{\Omega}$ and the versor components

$$
\begin{aligned}
\boldsymbol{\Phi} = {} & \frac{1-\cos\psi}{\psi^2}(\omega_i\omega^j\boldsymbol{D}_j\otimes\boldsymbol{D}^i) + \cos\psi\boldsymbol{D}_i\otimes\boldsymbol{D}^i \\
& + \frac{\sin\psi}{\psi}\,\omega^k\varepsilon^{\,j}_{ki}\boldsymbol{D}_j\otimes\boldsymbol{D}^i.
\end{aligned} \tag{40}
$$

The components of the third dyade in (40) can be written in matrix notation as

$$
\frac{\sin\psi}{\psi}[R^j_{i_{(1)}}] = \frac{\sin\psi}{\psi}\underbrace{\begin{bmatrix} 0 & \omega^3 & -\omega^2 \\ -\omega^3 & 0 & \omega^1 \\ \omega^2 & -\omega^1 & 0 \end{bmatrix}}_{[R^j_{i_{(1)}}]}, \tag{41}
$$

and finally the rotation matrix becomes

$$
\boldsymbol{R} = [R^j_i]; \qquad R^j_i = \delta^j_i + \frac{1-\cos\psi}{\psi^2}R^k_{i_{(1)}}\boldsymbol{R}^j_{k_{(1)}} + \frac{\sin\psi}{\psi}R^j_{i_{(1)}}. \tag{42}
$$

3.3.3. *Taylor expansion of rotation matrix and convected base vectors of rod axis.* From $d_i = R^j_iD_j$, the expansion

$$
\begin{aligned}
&\boldsymbol{d}_i = R^j_i\boldsymbol{D}_j, \\
&R^j_i = \delta^j_i + \left(1 - \frac{1}{3!}\psi^2 + \cdots\right)R^j_{i_{(1)}} + \left(\frac{1}{2} - \frac{1}{4!}\psi^2 + \cdots\right)R^k_{i_{(1)}}\boldsymbol{R}^j_{k_{(1)}},
\end{aligned} \tag{43}
$$

is introduced. The displacement vector of point P into point p is given by

$$
\boldsymbol{u} = \tilde{\boldsymbol{u}} + \theta^\alpha(cR^j_\alpha - \delta^j_\alpha)\boldsymbol{G}_j; \qquad \alpha = 1,2. \tag{44}
$$

From (26),

$$u = \tilde{u} + \theta^{\alpha}(\tilde{g}_{\alpha} - G_{\alpha}),$$

with

$$\tilde{g}_{\alpha} = cR^{j}_{\alpha} G_j = cR^{j}_{\alpha} D_j = cd_{\alpha}. \tag{45}$$

With the additional tangential base vector $\tilde{g}_3$,

$$\tilde{g}_3 = \tilde{x}_{,3} = (\delta^{j}_{3} + \tilde{u}^{j}_{,3})G_j, \tag{46}$$

we get the complete set of base vectors

$$\tilde{g}_i = \tilde{F}G_i\,; \qquad \tilde{F} = cd_{\alpha} \otimes G^{\alpha} + \tilde{x}_{,3} \otimes G^3. \tag{47}$$

The matrix of material deformation gradient components of points of the rod axis is

$$\tilde{F} = [\tilde{F}^{k}_{i}] = \begin{bmatrix} 1 - \frac{1}{2}(\omega^2\omega^2 + \omega^3\omega^3) + \cdots & \omega^3 + \frac{1}{2}\omega^1\omega^2 + \cdots & -\omega^2 + \frac{1}{2}\omega^1\omega^3 + \cdots \\ -\omega^3 + \frac{1}{2}\omega^1\omega^2 + \cdots & 1 - \frac{1}{2}(\omega^1\omega^1 + \omega^3\omega^3) + \cdots & \omega^1 + \frac{1}{2}\omega^2\omega^3 + \cdots \\ \tilde{u}^{1}_{,3} & \tilde{u}^{2}_{,3} & 1 + \tilde{u}^{3}_{,3} \end{bmatrix}. \tag{48}$$

3.3.4. *Convected base vectors in rod space*. The base vectors of points p in rod space are defined by

$$g_{\alpha} = x_{,\alpha} = cR^{j}_{\alpha} G_j, \tag{49}$$

$$g_3 = x_{,3} = (\delta^{j}_{3} + \tilde{u}^{j}_{,3} + \theta^{\beta} cR^{j}_{\beta,3})G_j. \tag{50}$$

This mapping can be represented using the deformation gradient F

$$g_i = FG_i\,; \qquad F = x_{,k} \otimes G^k. \tag{51}$$

Introducing directors yields

$$F = cd_{\alpha} \otimes G^{\alpha} + \tilde{x}_{,3} \otimes G^3 + \theta^{\alpha} cd_{\alpha,3} \otimes G^3. \tag{52}$$

For transformation of base vectors of the axis into those of rod space we use the shifter μ

$$g_i = \mu \tilde{g}_i\,; \qquad \mu = \tilde{g}_j \otimes \tilde{g}^j + \theta^{\alpha} cd_{\alpha,3} \otimes g^3 = I + \theta^{\alpha} \tilde{g}_{\alpha,3} \otimes \tilde{g}^3 \tag{53}$$

with $\mu^{k}_{\alpha} = \delta^{k}_{\alpha}$ and

$$\mu^{k}_{3} = \delta^{k}_{3} + \theta^{\alpha} R^{m}_{\alpha,3} R^{k}_{m}. \tag{54}$$

3.4. *Definition of strains and derived quantities*

3.4.1. *Introduction of Green–Lagrange strain tensor in rod space.* From

(55) $$\boldsymbol{E} \equiv \boldsymbol{\gamma} := \tfrac{1}{2}(\boldsymbol{F}^T\boldsymbol{F} - \boldsymbol{I}); \qquad \boldsymbol{I} = \boldsymbol{G}_i \otimes \boldsymbol{G}^i,$$

we have in component form

(56) $$\gamma_{ik} = \tfrac{1}{2}(g_{ik} - G_{ik}),$$

and for points S of rod axis

(57) $$\tilde{\gamma}_{ik} = \tfrac{1}{2}(\tilde{g}_{ik} - G_{ik}).$$

With the metric coefficients

(58) $$\begin{cases} G_{ik} = \delta_i^k \\ \tilde{g}_{\alpha\beta} = g_{\alpha\beta} = c^2\delta_\alpha^\beta \\ \tilde{g}_{\alpha 3} = cR_\alpha^m(\delta_3^m + \tilde{u}_{,3}^m) \\ \tilde{g}_{33} = (\delta_3^m + \tilde{u}_{,3}^m)^2 \\ g_{\alpha 3} = \tilde{g}_{\alpha 3} + \theta^\beta c^2 R_\alpha^m R_{\beta,3}^m \\ g_{33} = \tilde{g}_{33} + 2\theta^\beta c R_{\beta,3}^m(\delta_3^m + \tilde{u}_{,3}^m) + \theta^\beta\theta^\gamma c^2 R_{\beta,3}^m R_{\gamma,3}^m \end{cases}$$

we get the strain components

(59) $$\begin{cases} \gamma_{\alpha\beta} = (c^2 - 1)\delta_\alpha^\beta \\ \gamma_{\alpha 3} = \tilde{\gamma}_{\alpha 3} + \tfrac{1}{2}\theta^\beta c^2 R_\alpha^m R_{\beta,3}^m \\ \gamma_{33} = \tilde{\gamma}_{33} + \theta^\beta c R_{\beta,3}^m(\delta_3^m + \tilde{u}_{,3}^m) + \tfrac{1}{2}\theta^\beta\theta^\gamma c^2 R_{\beta,3}^m R_{\gamma,3}^m. \end{cases}$$

The additional normal hypothesis yields the conditions

(60) $$\begin{aligned} \frac{1-\cos\psi}{\psi^2}\,\omega^\alpha\omega^3 + \frac{\sin\psi}{\psi}\,e_{\alpha\beta}\omega^\beta &= \tilde{u}_{,3}^\alpha/\sqrt{\tilde{g}_{33}}; \\ \frac{\cos\psi - 1}{\psi^2}\,\omega^\lambda\omega^\lambda &= \tilde{u}_{,3}^3/\sqrt{\tilde{g}_{33}}. \end{aligned}$$

3.4.2. *Definition of curvatures of rod axis.* The change of curvature can be described by

(61a) $$\tilde{\kappa}_{ik} = \boldsymbol{d}_{i,3}\cdot\boldsymbol{d}_k - \boldsymbol{D}_{i,3}\cdot\boldsymbol{D}_k,$$

(61b) $$\tilde{\kappa}_{ik} = R_{i,3}^m\boldsymbol{D}_m \cdot R_k^1\boldsymbol{D}_1 = R_{i,3}^m R_k^m,$$

with the skew-symmetric matrix

$$(62)\qquad [\tilde{\kappa}_{ik}] = \begin{bmatrix} 0 & \omega^3_{,3} - \tfrac{1}{2}\omega^1\omega^2_{,3} + \tfrac{1}{2}\omega^1_{,3}\omega^2 + \cdots & -\omega^2_{,3} - \tfrac{1}{2}\omega^1\omega^3_{,3} + \tfrac{1}{2}\omega^1_{,3}\omega^3 + \cdots \\ \text{skew} & 0 & \omega^1_{,3} - \tfrac{1}{2}\omega^2\omega^3_{,3} + \tfrac{1}{2}\omega^2_{,3}\omega^3 + \cdots \\ \text{symmetric} & & 0 \end{bmatrix}.$$

Then the strains in rod space can be expressed by the strains of the rod axis and its changes of curvature

$$(63a)\qquad \gamma_{\alpha 3} = \tilde{\gamma}_{\alpha 3} + \tfrac{1}{2}c^2\theta^\beta\tilde{\kappa}_{\beta\alpha} \qquad \text{with } \tilde{\kappa}_{\beta\alpha} = 0 \text{ for } \alpha = \beta,$$

$$(63b)\qquad \gamma_{33} = \tilde{\gamma}_{33} + c\theta^\beta\tilde{\kappa}_{\beta 3}R^l_3F^l_3 + \tfrac{1}{2}c^2\theta^\beta\theta^\gamma\tilde{\kappa}_{\beta m}\tilde{\kappa}_{\gamma m}.$$

The determinant of the metric coefficients is

$$(64)\qquad g = |g_{ik}| = g_{33} - g_{\alpha 3}g_{\alpha 3} \qquad \text{for } c = 1.$$

Taylor expansions of g and $\sqrt{g}$ are

$$(65)\qquad g = (1 + \tilde{u}^3_{,3} + \theta^\alpha R^3_{\alpha,3})^2 - R^3_{\alpha(1)}R^3_{\alpha(1)} - 2R^3_{\alpha(1)}\tilde{u}_{,3} + \cdots,$$

$$(66)\qquad \sqrt{g} = 1 + \tilde{u}^3_{,3} + \theta^\alpha R^3_{\alpha,3} - \tfrac{1}{2}R^3_{\alpha(1)}R^3_{\alpha(1)} - R^3_{\alpha(1)}\tilde{u}^\alpha_{,3} + \cdots.$$

3.4.3. *Polar decomposition of **F***. The matrix representation of the decomposition of material deformation gradient is

$$(67)\qquad \boldsymbol{F} = \boldsymbol{R}\boldsymbol{U},$$

with the component matrix $\boldsymbol{U}$ of the stretch tensor. Then we get for the strain components

$$(68)\qquad \boldsymbol{\gamma} = \tfrac{1}{2}(\boldsymbol{U}^2 - \boldsymbol{I}); \qquad \boldsymbol{U} = \sqrt{\boldsymbol{I} + 2\boldsymbol{\gamma}}.$$

Corresponding to this decomposition we introduce the fictictious intermediate base vectors

$$(69)\qquad \hat{\boldsymbol{g}}_i = U^k_i\boldsymbol{G}_k,$$

which result from pure stretching.

The following rotation

$$(70)\qquad \boldsymbol{g}_i = \hat{R}^k_i\hat{\boldsymbol{g}}^k$$

leads to the convected base vectors $\boldsymbol{g}_i$ in the deformed rod. With the modified rotation measure $\hat{\boldsymbol{\Omega}}$ with $|\hat{\boldsymbol{\Omega}}| = \sin\hat{\varphi}$ we get the corresponding versor $\boldsymbol{\Phi}$

$$(71)\qquad \tilde{\boldsymbol{g}}_i = \boldsymbol{\Phi}\hat{\boldsymbol{g}}_i,$$

$$(72)\qquad \boldsymbol{\Phi} = \cos\hat{\varphi}\boldsymbol{I} + \hat{\boldsymbol{\Omega}} \times \boldsymbol{I} + \frac{1 - \cos\hat{\varphi}}{\sin^2\hat{\varphi}}\hat{\boldsymbol{\Omega}} \otimes \hat{\boldsymbol{\Omega}}.$$

The components of $\boldsymbol{\Phi}$ are rational functions so that equation (71) can be calculated explicitly.

From equation (71) we find

$$\hat{\boldsymbol{\Omega}}\hat{\boldsymbol{g}}^k = \tfrac{1}{2}\varepsilon^{ijk}\tilde{\boldsymbol{g}}_i\tilde{\boldsymbol{g}}_j\,; \qquad \varepsilon_{ijk} = \sqrt{g}\,e_{ijk}, \tag{73}$$

and finally

$$\hat{\boldsymbol{\Omega}} = \tfrac{1}{2}\varepsilon^{ijs}(\tilde{\boldsymbol{g}}_i\cdot\hat{\boldsymbol{g}}_j)\hat{\boldsymbol{g}}_s. \tag{74}$$

With respect to the original base vectors one gets

$$\hat{\boldsymbol{\Omega}} = \tfrac{1}{2}\varepsilon^{ijs}\tilde{F}_i^m U_j^m U_s^r \boldsymbol{G}_r. \tag{75}$$

The linear terms of the components of $\hat{\boldsymbol{\Omega}}$ are

$$\hat{\omega}^1 = \tfrac{1}{2}(\tilde{u}^2_{,3} - \omega^1), \qquad \hat{\omega}^2 = \tfrac{1}{2}(\tilde{u}^1_{,3} + \omega^2), \qquad \hat{\omega}^3 = \omega^3. \tag{76}$$

3.5. *Introduction of stresses and stress resultants*

3.5.1. *Stresses in convected coordinates.* A deformed area element in rod space, referred to convected coordinates θ^1; θ^2, is given by

$$d\boldsymbol{a}_i = d\theta^j\boldsymbol{g}_i \times d\theta^k\boldsymbol{g}_k = da_i\,\frac{1}{\sqrt{g^{(ii)}}}\,\boldsymbol{g}^i, \tag{77}$$

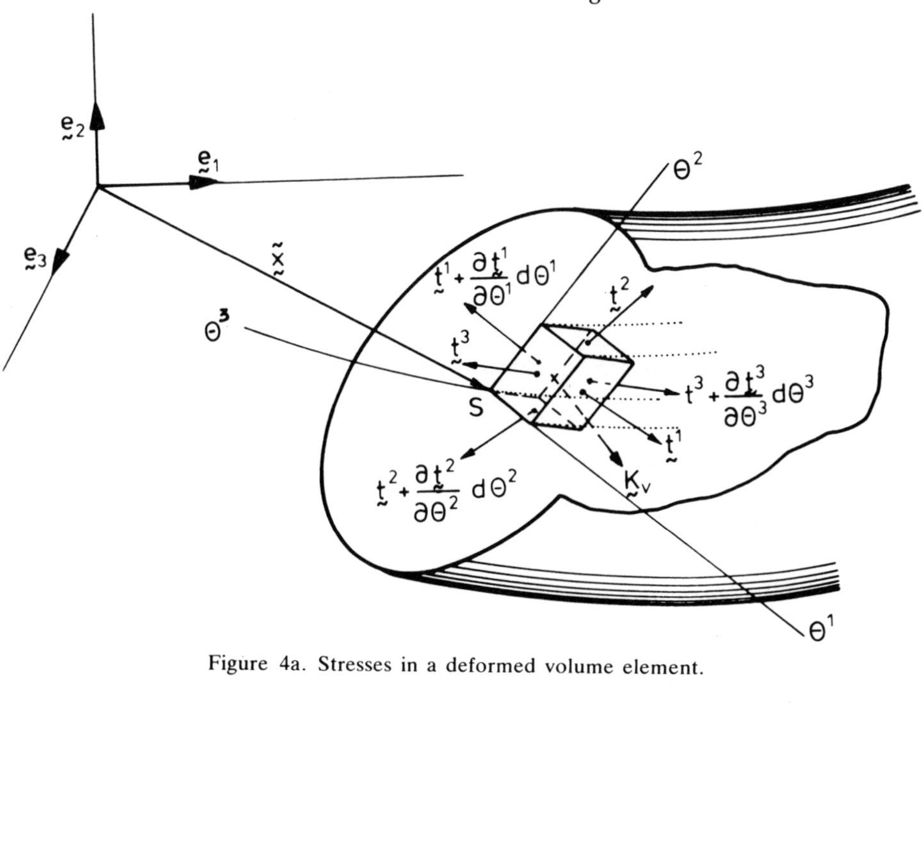

Figure 4a. Stresses in a deformed volume element.

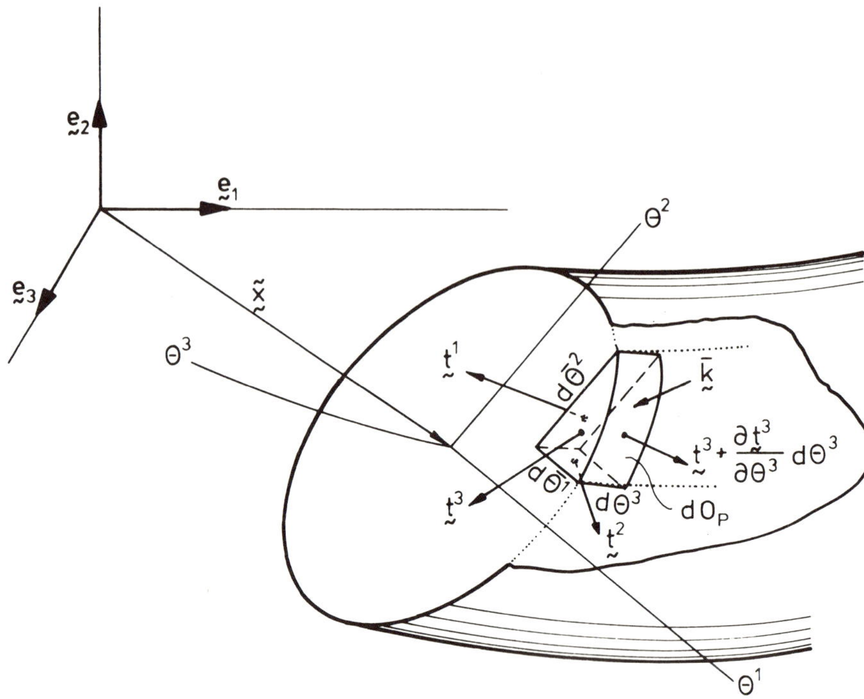

Figure 4b. Stresses in a deformed surface element.

with

$$da_i = \chi \circ \mathring{\chi}^{-1}(dA_i) = \frac{\sqrt{g}\sqrt{g^{(ii)}}}{\sqrt{G}\sqrt{G^{(ii)}}}\, dA_i .$$

The equilibrium conditions for a deformed tetrahedron are

(78) $$\boldsymbol{t}^{(n)} da_{(n)} = \boldsymbol{t}^i da_i = \boldsymbol{T}^i dA_i ,$$

with the Cauchy stress tensor $\boldsymbol{T} \equiv \boldsymbol{\tau}$

(79) $$\boldsymbol{t}_{(n)} = \boldsymbol{\tau}\boldsymbol{n} ; \qquad \boldsymbol{\tau} = \tau^{ij}\boldsymbol{g}_i \otimes \boldsymbol{g}_j$$

and the 2nd Piola–Kirchhoff stress tensor

(80) $$\boldsymbol{S} \equiv \boldsymbol{\sigma} = \sigma^{ij}\boldsymbol{G}_i \otimes \boldsymbol{G}_j ; \qquad \sigma^{ij} = \sqrt{\frac{g}{G}}\, \tau^{ij} .$$

3.5.2. *Stress resultants and equilibrium conditions.* The equilibrium conditions of a deformed rod element are

(81a) $$\boldsymbol{N}_{,3}+\bar{\boldsymbol{N}}=\boldsymbol{0},$$

(81b) $$\boldsymbol{M}_{,3}-\tilde{\boldsymbol{x}}_{,3}\times\boldsymbol{N}+\bar{\boldsymbol{M}}=\boldsymbol{0},$$

with the vector of section forces

(82a) $$\boldsymbol{N}:=\int_{(A)}\boldsymbol{T}^3 dA,$$

and the vector of section moments

(82b) $$\boldsymbol{M}:=\int_{(A)}\theta^\alpha\tilde{\boldsymbol{g}}_\alpha\times\boldsymbol{T}^3 dA.$$

The components of forces and moments are defined as

(83a) $$\boldsymbol{N}=n^k\tilde{\boldsymbol{g}}_k\,;\qquad n^k=\int_{(A)}\sigma^{i3}\mu_i^{\;k}\,dA,$$

and

(83b) $$\boldsymbol{M}=m^k\tilde{\boldsymbol{g}}_k\,;\qquad m^k=e_{i\alpha j}\int_{(A)}\theta^\alpha\sigma^{i3}g^{jm}\mu_m^{\;k}\sqrt{g}\,dA.$$

The given loads at the boundary can be expressed in a force vector

(84a) $$\bar{\boldsymbol{N}}:=\oint_{(s)}\bar{\boldsymbol{f}}ds,$$

and a moment vector

(84b) $$\bar{\boldsymbol{M}}:=-\oint_{(s)}\theta^\alpha\tilde{\boldsymbol{g}}_\alpha\times\bar{\boldsymbol{f}}ds,$$

with s as the circumferential coordinate of the cross-section. Volume loads are neglected for simplicity.

3.5.3. *Principle of virtual work and constitutive kinematic variables.* The principle of virtual work for the rod can be written in the form

(85) $$\int_{(L)}(\boldsymbol{N}d\boldsymbol{\varepsilon}+\boldsymbol{M}d\boldsymbol{\kappa})d\theta^3=\int_{(L)}(\bar{\boldsymbol{N}}\delta\tilde{\boldsymbol{u}}+\bar{\boldsymbol{M}}\delta\boldsymbol{r})d\theta^3+|\bar{\boldsymbol{N}}\delta\boldsymbol{u}|_0^L+|\bar{\boldsymbol{M}}\delta\boldsymbol{r}|_0^L,$$

with

(86a) $$\delta\boldsymbol{\varepsilon}=\delta\tilde{\boldsymbol{u}}_{,3}+\tilde{\boldsymbol{x}}_{,3}\times\delta\boldsymbol{r},$$

(86b) $$\delta\boldsymbol{\kappa}=\delta\boldsymbol{r}_{,3},$$

and

(87) $$d\boldsymbol{d}_i=d\boldsymbol{r}\times\boldsymbol{d}_i\,;\qquad d\boldsymbol{r}=\tfrac{1}{2}dR_j^{\;m}R_i^{\;m}e_{ijk}\boldsymbol{d}_k.$$

The constitutive kinematic variables $\boldsymbol{\varepsilon}$ and $\boldsymbol{\kappa}$ are derived as work-conforming quantities with $\boldsymbol{N}$ and $\boldsymbol{M}$.

Strains:

$$\delta\varepsilon_i = \delta\boldsymbol{\varepsilon}\tilde{\boldsymbol{g}}_i = (\delta\tilde{\boldsymbol{u}}_{,3}\tilde{\boldsymbol{x}}_{,3} \times \delta\boldsymbol{r})\tilde{\boldsymbol{g}}_i, \tag{88a}$$

$$\delta\varepsilon_\alpha = \delta\tilde{\boldsymbol{g}}_3\boldsymbol{d}_\alpha + \tilde{\boldsymbol{g}}_3\delta\boldsymbol{d}_\alpha = \delta(\tilde{\boldsymbol{g}}_3\boldsymbol{d}_\alpha), \tag{88b}$$

$$d\varepsilon_3 = \tilde{\boldsymbol{g}}_3\delta\tilde{\boldsymbol{g}}_3,$$

or

$$\delta\varepsilon_\alpha = \delta(\tilde{F}^k_3\tilde{F}^k_\alpha); \qquad \delta\varepsilon_3 = \tilde{F}^k_3\delta\tilde{F}^k_3. \tag{88c}$$

Thus we get

$$\varepsilon_\alpha = \tilde{F}^k_3\tilde{F}^k_\alpha = 2\tilde{\gamma}_{3\alpha}, \tag{89a}$$

$$\varepsilon_3 = \tfrac{1}{2}(\tilde{F}^k_3\tilde{F}^k_3 - \delta^3_3) = \tilde{\gamma}_{33}. \tag{89b}$$

Changes of curvature:

$$\delta\boldsymbol{\kappa} = \delta(\kappa_m\boldsymbol{d}_m) = \delta\boldsymbol{r}_{,3} = \delta(\tfrac{1}{2}R^{\,k}_{i,3}R^{\,k}_{j}e_{jim}\boldsymbol{d}_m), \tag{90}$$

$$\kappa_m = \tfrac{1}{2}R^{\,k}_{i,3}R^{\,k}_{j}e_{jim}. \tag{91a}$$

For base vectors $\tilde{\boldsymbol{g}}_\alpha$, $\tilde{\boldsymbol{g}}_3$ the components are

$$\tilde{\kappa}_\alpha = \kappa_\alpha\,; \qquad \tilde{\kappa}_3 = \kappa_m R^{\,k}_{m}\tilde{F}^k_3. \tag{91b,c}$$

3.6. *Constitutive equations*

We assume an elastic material

$$\boldsymbol{\sigma} = \frac{\partial \overset{(3)}{W}_s(\boldsymbol{\gamma})}{\partial\boldsymbol{\gamma}}, \tag{92}$$

and restrict to:

A3: W_s is given by a bilinear form

$$W_s = \tfrac{1}{2}\boldsymbol{E}\cdot(\boldsymbol{\gamma}\otimes\boldsymbol{\gamma}) = \tfrac{1}{2}E^{ijkl}\gamma_{ij}\gamma_{kl}, \tag{93}$$

so that we have the linear relation

$$\sigma^{ij} = E^{ijkl}\gamma_{kl}, \tag{94}$$

and for isotropic material

$$\sigma^{ij} = \frac{E}{1+\nu}\,\gamma_{ij} + \frac{E\nu}{(1+\nu)(1-2\nu)}\,\delta_{ij}\gamma_{ll}. \tag{95}$$

The one-dimensional theory of rods is represented by

(96a,b) $$N = \frac{\partial \overset{(1)}{W_s}(\boldsymbol{\varepsilon}, \boldsymbol{\kappa})}{\partial \boldsymbol{\varepsilon}}; \qquad M = \frac{\partial \overset{(1)}{W_s}}{\partial \boldsymbol{\kappa}},$$

with the elastic strain energy

(97) $$W = \Pi_{(i)} = \int_{(L)} \overset{(1)}{W_s} d\theta^3 = \int_{(L)} \int_{(A)} \overset{(3)}{W_s} dA d\theta^3.$$

3.7. *Taylor expansion of rod equations and estimation of stresses*

One can establish a system of 18 differential equations of 1st order for the displacements u^i, the rotations ω^i, the constitutive variables ε_i, κ_i and for the section forces N^i and section moments M^i.

Taylor expansions of these rod equations according to Hay [3] with respect to a small parameter $\overset{*}{\varepsilon}$ defined by

(98) $$\overset{*}{\varepsilon} := \max\left(\frac{D}{L}, \varepsilon\right); \qquad \overset{*}{\varepsilon} \in \mathbb{R}^+,$$

with D, L characteristic measure of cross-section and system length, ε strain in rod space, and the restrictive assumption

(98a) $$\text{A4:} \quad \varepsilon < 10^{-2},$$

lead to the following power series

(99) $$\begin{aligned} g_{ik} &= g_{ik_{(0)}} + g_{ik_{(1)}} + \overset{*}{\varepsilon}^2 g_{ik_{(2)}} + O(\overset{*}{\varepsilon}^3); \qquad \overset{*}{\varepsilon} > 0 \\ \gamma_{ik} &= \gamma_{i\gamma_{(0)}} + \cdots \end{aligned}$$

with

$$g_{ik_{(0)}} = \delta_{ik}; \qquad 2\gamma_{ik_{(1)}} = g_{ik_{(1)}}; \qquad 2\gamma_{ik_{(2)}} = g_{ik_{(2)}}; \qquad \cdots$$

The disadvantage of Hay's parameter $\overset{*}{\varepsilon}$ is that only very thin rods can be considered. So, following John [4] in his error estimations of shell theory, we define a measure θ

(100) $$\theta := \max\left(\frac{D}{L}, \sqrt{\varepsilon}\right), \qquad \theta \in \mathbb{R}^+,$$

which leads to the following estimations of order of magnitude

(101) $$\tau^{\alpha\beta} = O(E\varepsilon); \qquad \tau^{\alpha 3} = O(E\varepsilon\theta), \qquad \tau^{33} = O(E\varepsilon\theta^2).$$

Introducing furthermore the realistic assumption

(102) $$\text{A5:} \quad D \ll L \text{ and therefore } \frac{D/L}{\sqrt{\varepsilon}} = O(1), \text{ e.g., } \frac{D}{L} = 10^{-1},$$

it comes out that the $\tau^{\alpha\beta}$ are smaller by one order of magnitude in comparison with τ^{33}, and it follows that $\tau^{\alpha\beta}_{(1)}=0$; $\gamma_{(\alpha\alpha)}=-\nu\cdot\gamma_{33}$. The approximation for the specific strain energy of the rod then becomes

$$\overset{(1)}{W}_s=\frac{1}{2}\int_{(A)}[4G\gamma_{\alpha 3}\gamma_{\alpha 3}+E\gamma_{33}\gamma_{33}+O(E\nu\varepsilon^2\theta^2)]dA. \tag{103}$$

It seems to be an essential advantage of the theory and approximation given in this paper that the neglect of stresses $\tau^{\alpha\beta}$ against τ^{33} in the strain energy is a consequence of proper general assumptions and not an additional hypothesis.

3.8. *Classification of the magnitude of rotations*

With the value φ of the rotation vector,

$$|\mathbf{\Omega}|=\sqrt{\omega^m\omega^m}=\varphi, \tag{104}$$

we define orders of magnitude for the components, according to Pietraszkiewicz [5] who introduced these quantities for shells:

$$\begin{array}{lll} \text{(i)} & \max\omega^i<O(\theta^2), & \text{then } \omega^i \textit{ small} \text{ rotations} \\ \text{(ii)} & \max\omega^i<O(\theta), & \text{then } \omega^i \textit{ moderate} \text{ rotations} \\ \text{(iii)} & \max\omega^i<O(\sqrt{\theta}), & \text{then } \omega^i \textit{ large} \text{ rotations} \\ \text{(iv)} & \max\omega^i<O(1), & \text{then } \omega^i \textit{ finite} \text{ rotations.} \end{array} \tag{105}$$

3.8.1. *Small rotations.* For small rotations we get

$$\begin{gathered} \varphi\lesssim 10^{-2} \qquad \text{or} \qquad \varphi\lesssim 1^0, \\ R^j_i=\delta^j_i+R^j_{i(1)}+O(\theta^4), \end{gathered} \tag{106}$$

with metric coefficients

$$\begin{cases} g_{\alpha\beta}=\delta^\beta_\alpha+O(\varepsilon) \\ g_{\alpha 3}=\varepsilon g_{\alpha 3_{(1)}}+O(\varepsilon^2)=\tilde{u}^\alpha_{,3}+R^3_{\alpha(1)}+\theta^\beta R^\alpha_{\beta(1),3}+O(\varepsilon\theta^2) \\ g_{33}=1+\varepsilon g_{33_{(1)}}+O(\varepsilon^2)=1+2\tilde{u}^3_{,3}+2\theta^\alpha R_{\alpha(1),3}+O(\varepsilon\theta^2) \end{cases} \tag{107}$$

and the constitutive relations

$$\begin{cases} N^\alpha=GA[\tilde{u}^\alpha_{,3}+R^{(3)}_{\alpha(1)}]+GS^\beta R^\alpha_{\beta(1),3}+\displaystyle\int_{(A)}O(E\varepsilon\theta^2)dA \\ N^3=EA\tilde{u}^3_{,3}+ES^\beta R^3_{\beta(1),3}+\displaystyle\int_{(A)}O(E\varepsilon^2)dA \end{cases} \tag{108}$$

$$(109)\quad \begin{cases} M^{\alpha} = Ee_{\beta\alpha}S^{\beta}\tilde{u}^{3}_{,3} + Ee_{\beta\alpha}I^{\beta\gamma}R^{\ 3}_{\gamma(1),3} + \int_{(A)} O(E\varepsilon^2)dA \\ M^3 = Ge_{\alpha\beta}S^{\beta}\tilde{u}^{\alpha}_{,3} + R^{\ 3}_{\alpha(1)} + Ge_{\alpha\beta}I^{\beta\gamma}R^{\ \alpha}_{\gamma(1),3} + \int_{(A)} O(E\varepsilon\theta^2)dA \end{cases}$$

with

$$(110)\qquad S = \int_{(A)} \theta^{\beta}dA\,; \qquad I^{\beta\gamma} = \int_{(A)} \theta^{\beta}\theta^{\gamma}dA.$$

3.8.2. *Moderate rotations.* For moderate rotations

$$\varphi \lessapprox 0.1 \qquad \text{or} \qquad \varphi \lessapprox 10^0,$$

$$(111)\qquad R^{\ j}_{i} = \delta^{j}_{1} + R^{\ j}_{i(1)} + \tfrac{1}{2}R^{\ k}_{i(1)}R^{\ j}_{k(1)} - \tfrac{1}{6}\psi^2 R^{\ j}_{i(1)} + O(\theta^4)$$

we get the metric coefficients

$$\begin{aligned} g_{\alpha_3} &= \varepsilon g_{\alpha_{3(1)}} + O(\varepsilon^2) \\ &= \tilde{u}^{\alpha}_{,3} + R^{\ 3}_{\alpha(1)} + \theta^{\beta}R^{\ \alpha}_{\beta(1),3} + R^{\ m}_{\alpha(1)}\tilde{u}^{m}_{,3} + \tfrac{1}{2}R^{\ k}_{\alpha(1)}R^{\ 3}_{k(1)} + \theta^{\beta}R^{\ m}_{\alpha(1)}R^{\ m}_{\beta(1),3} \\ &\quad - \tfrac{1}{6}\Psi^2 R^{\ 3}_{\alpha(1)} + \tfrac{1}{2}R^{\ k}_{\alpha(1)}R^{\ m}_{k(1)}\tilde{u}^{m}_{,3} + \tfrac{1}{2}\theta^{\beta}(R^{\ k}_{\beta(1)}R^{\ \alpha}_{k(1)})_{,3} + O(\varepsilon\theta^2) \end{aligned}$$

and

$$(113)\qquad \begin{aligned} g_{33} &= 1 + \varepsilon g_{33_{(1)}} + O(\varepsilon^2) \\ &= 1 + 2\tilde{u}^{3}_{,3} + 2\theta^{\alpha}R^{\ 3}_{\alpha(1),3} \\ &\quad + \tilde{u}^{\alpha}_{,3}\tilde{u}^{\alpha}_{,3} + \theta^{\alpha}(R^{\ k}_{\alpha(1)}R^{\ 3}_{k(1)})_{,3} + 2\theta^{\alpha}R^{\ m}_{\alpha(1),3}R^{\ m}_{3(1)} + O(\varepsilon\theta^2). \end{aligned}$$

It is essential that there exist already cubic terms of ω^i in $g_{\alpha 3}$. The constitutive equations are

$$(114)\qquad \begin{aligned} n^{\alpha} &= GA\{\tilde{u}^{\alpha}_{,3} + R^{\ 3}_{\alpha(1)} + \tfrac{1}{2}R^{\ k}_{\alpha(1)}R^{\ 3}_{k(1)} + R^{\ m}_{\alpha(1)}\tilde{u}^{m}_{,3} - \tfrac{1}{6}\psi^2 R^{\ 3}_{\alpha(1)} \\ &\quad + \tfrac{1}{2}R^{\ k}_{\alpha(1)}R^{\ m}_{k(1)}\tilde{u}^{m}_{,3}\} + \int_{A} O(E\theta^4)dA, \\ n^3 &= EA\{\tilde{u}^{3}_{,3} + \tfrac{1}{2}\tilde{u}^{\alpha}_{,3}\tilde{u}^{\alpha}_{,3}\} + \int_{A} O(E\varepsilon^2)dA, \end{aligned}$$

$$(115)\qquad \begin{aligned} m^{\alpha} &= Ee_{\beta\alpha}I^{\beta\gamma}R^{\ 3}_{\gamma(1),3} + \int_{A} O(E\varepsilon\theta)dA, \\ m^3 &= Ge_{\alpha\beta}I^{\beta\gamma}\{R^{\ \alpha}_{\gamma(1),3} + \tfrac{1}{2}(R^{\ k}_{\gamma(1)}R^{\ \alpha}_{k(1)})_{,3} + R^{\ k}_{\gamma(1),3}R^{\ k}_{\alpha(1)}\} \\ &\quad + \int_{A} O(E\varepsilon\theta^2)dA. \end{aligned}$$

3.8.3. *The so-called theory of 2nd order.* With the theory of moderate rotations we have to introduce 3 additional approximations, namely:

A6: Bernoulli's normal hypothesis:

$$\sqrt{(1+\tilde{u}^{3}_{,3})^2+(\tilde{u}^{1}_{,3})^2}\sin\omega^2=\tilde{u}^{1}_{,3},$$
(116)
$$\sqrt{(1+\tilde{u}^{3}_{,3})^2+(\tilde{u}^{1}_{,3})^2}\cos\omega^2=1+\tilde{u}^{3}_{,3}.$$

A7: Strains in axis-direction are neglected:

(117) $$\varepsilon_3=\tilde{\gamma}_{33}=0.$$

A8: In the relations for ε_i, κ_i, only linear terms are considered.

From A7 it follows that

(118) $$\sin\omega=\omega-\tfrac{1}{6}\omega^3+\cdots=w',$$

or, see Kappus [1],

(119) $$\omega'=\frac{w''}{\sqrt{1-w'^2}}=w''(1+\tfrac{1}{2}w'^2+\tfrac{3}{8}w'^4+\cdots).$$

For moderate rotations the approximation

(120) $$\sin\omega=\omega+O(\theta^3)=w'$$

is admissible.

The constitutive equation then becomes for bending with respect to a main axis 1–1 of inertia

(121) $$M=-EI^{11}\omega'=-EI^{11}w''(1+O(\theta^2)),$$

and with A8

(122) $$M=-EI^{11}w''.$$

3.8.4. *Large rotations.* Here we can admit rotations of magnitude

$$\varphi\lesssim 0.5 \qquad \text{or} \qquad \varphi\lesssim 30^0.$$

The rotation matrix $\boldsymbol{R}$ contains terms with power 6 if the errors in g_{ij} are of $O(\varepsilon\theta^2)$.

For the plane problem and with assumption A6 we get the normal force

(123a) $$N=EA(\tilde{u}^{3}_{,3}+\tfrac{1}{2}\tilde{u}^{m}_{,3}\tilde{u}^{m}_{,3})+\int_A O(E\varepsilon^2)dA,$$

the bending moment

(123b) $$M = -EI^{11}\omega_{,3}^2 + \int_A O(E\varepsilon\theta)dA$$

and with A7 the rotation

(124) $$\sin\omega = \omega - \tfrac{1}{6}\omega^3 + O(\theta^2\sqrt{\theta}) = w',$$
$$\omega = w' + \tfrac{1}{6}w'^3 + O(\theta^2\sqrt{\theta}),$$

and the bending moment

(125) $$M = -EI^{11}\omega' = -EI^{11}w''(1 + w'^2 + O(\theta^2)).$$

Introducing A8 one gets

(126) $$M = -EI^{11}w''(1 + O(\theta)).$$

In Figure 5 comparisons of different approximations for the bending of a cantilever beam with a single force at the free end are given. The so-called exact solution is the solution of Bishopp and Drucker [6] for the equation

(127) $$M = -EI^{11}\frac{w''}{(1 + w'^2)^{3/2}}\,.$$

The straight line belongs to the completely linearized theory (small rotations). The curve in between is the solution of large rotation theory with the additional assumptions A6–A8.

Looking at the errors one can see that large rotation theory is valid up to $\varphi \sim 30°$.

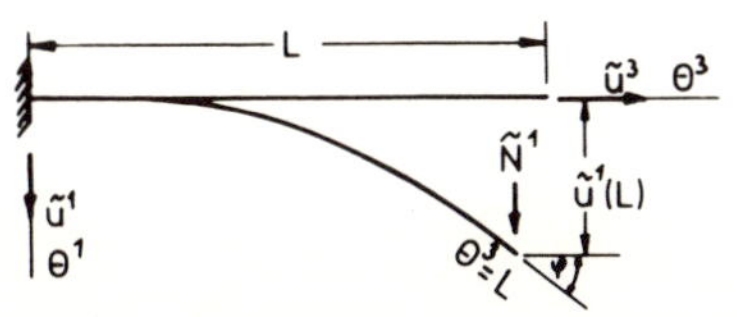

α^2	$\tilde{u}^1(L)/L$				
	linear		equ. 125		exact
1	0,333	10%	0,304	0%	0,30
2	0,666	36%	0,521	6%	0,49
3	1,0	67%	0,679	11%	0,60

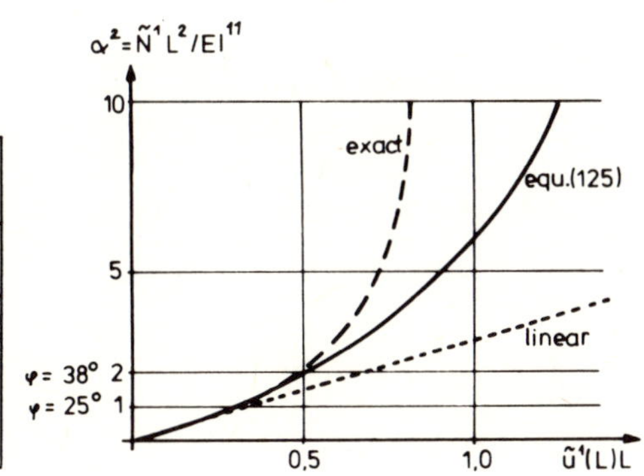

Figure 5. Comparison of deflections of a cantilever beam for finite, large and small rotation theory.

References

1. R. Kappus, Z. Angew. Math. Mech. **19** (5) (1939), 271.

2. M. H. Kessel, Forschungs- und Seminarberichte aus dem Bereich der Mechanik der Universität Hannover, Bericht Nr. F 80/1, Februar 1980.

3. G. E. Hay, Trans. Am. Math. Soc. **51** (1942), 65.

4. F. John, Com. Pure Appl. Math. **18** (1965), 235.

5. W. Pietraszkiewicz, Mitteilungen aus dem Institut für Mechanik, Ruhr-Universität Bochum, Nr. 10, 1977.

6. K. E. Bishopp and D. C. Drucker, Quart. Appl. Math. **3** (1945), 272.

LOCAL THEOREMS OF EXISTENCE AND UNIQUENESS IN FINITE ELASTOSTATICS

TULLIO VALENT

Seminario Matematico, Università di Padova, Via Belzoni, 35100 Padova, Italy

(Received September 11, 1980)

1. Introduction

We confine ourselves to dealing with the displacement problem in finite elastostatics (see problems ((P), (D)) and ((P_1), (D)) of Section 3), but part of the results can be used in treating other boundary-value problems. We present local theorems of existence and uniqueness (in fact, local diffeomorphism theorems) in Sobolev spaces and Schauder spaces. (See Theorems 8.1 and 8.2, Corollaries 8.1 and 8.2 and Theorems 10.1 and 10.2). Following an idea of Gurtin and Spector [8] a "semiglobal" result is also given (see Theorems 8.3, 8.4 and 8.5). These theorems are obtained via the Implicit Function Theorem in the case when the body force depends on the deformation, and via the Inverse Function Theorem when we consider only body forces independent of the deformation.

Note that the solutions, whose existence we prove, actually correspond to orientation-preserving and globally invertible deformations. The possibility for the solutions to satisfy such a requirement is founded on a basic topological property of the set of admissible displacements in suitable spaces. (See Section 6).

A large part of this work concerns an analysis of the choice of a Banach space for the displacements and of a Banach space for the forces in order to apply the Inverse Function Theorem or the Implicit Function Theorem. This leads to various problems, in particular to the study of the differentiability of superposition operators. In this study, to which we devote Sections 4, 5 and 9, the main result is a nondifferentiability theorem (Theorem 4.1) from which we deduce that the Inverse Function Theorem does not work in the (Sobolev)

Proceedings of the IUTAM Symposium on Finite Elasticity, Lehigh University, August 10–15, 1980. Invited paper.

Carlson, D.E. and Shield, R.T. (eds.)
Proceedings of the IUTAM Symposium on Finite Elasticity
ISBN 90 247 2629 8

spaces related to the $W^{1,p}$-formulation of the formally linearized operator of the finite elastostatics operator. (See Remark 4.1.)

The results obtained in this work, besides clarifying the position of a linearized theory within the nonlinear theory, may constitute the first step in a global approach to the nonlinear problem.

We recall that the first use of the Inverse Function Theorem to prove local existence and uniqueness results in finite elastostatics was made by Stoppelli [15] in treating the problem of dead traction (see also [16–18]). Results similar to those of Stoppelli for other boundary-value problems have been successively obtained by Van Buren [24]. For a summary and a discussion of Stoppelli's and Van Buren's works see Grioli [7], Truesdell and Noll [19], Wang and Truesdell [25] and Marsden and Hughes [10].

2. Notations and technical preliminaries

Throughout this work Ω denotes a nonempty, bounded, open subset of $\mathbb{R}^n$, $(n \geqq 1)$, m denotes a nonnegative integer, p denotes a real number > 1, and λ denotes a real number such that $0 < \lambda \leqq 1$. $\partial\Omega$ is the boundary of Ω and $\bar{\Omega}$ its closure. If E is a nonempty, closed subset of $\mathbb{R}^n$, by a C^m function on E we mean, as usual, a function defined on E (with values in a normed space) which is the restriction to E of some function defined on an open neighborhood U of E in $\mathbb{R}^n$ and having all continuous derivatives up to order m on U. $C^m(\bar{\Omega})$ denotes the (Banach) space of C^m real-valued functions v on $\bar{\Omega}$, with the norm

$$|v|_m = \sum_{|\alpha| \leqq m} \sup_{x \in \bar{\Omega}} |D^\alpha v(x)|.$$

$C^{m,\lambda}(\bar{\Omega})$ denotes the (Banach) space of functions $v \in C^m(\bar{\Omega})$ such that, for $|\alpha| = m$, $D^\alpha v$ satisfies on $\bar{\Omega}$ a Holder condition of exponent λ, with the norm

$$\|v\|_{m,\lambda} = |v|_m + \sum_{|\alpha| = m} \sup_{\substack{x', x'' \in \bar{\Omega} \\ x' \neq x''}} \frac{|D^\alpha v(x') - D^\alpha v(x'')|}{|x' - x''|^\lambda}.$$

$W^{m,p}(\Omega)$ denotes the (Banach) space of elements v of $L^p(\Omega)$ such that, for every α with $|\alpha| \leqq m$, the weak derivative $D^\alpha v$ belongs to $L^p(\Omega)$, equipped with the norm

$$\|v\|_{m,p} = \sum_{|\alpha| \leqq m} \|D^\alpha v\|_{0,p},$$

where $L^p(\Omega)$ is the space of (equivalence classes of) measurable functions $v : \Omega \to \mathbb{R}$ such that $|v|^p$ is Lebesgue-integrable, and $\|\cdot\|_{0,p}$ is the usual norm of $L^p(\Omega)$. The closure in $W^{m,p}(\Omega)$, (resp. in $C^m(\bar{\Omega})$), of the set, $\mathscr{D}(\Omega)$, of infinitely-differentiable real-valued functions defined on Ω with compact support contained in Ω, is denoted by $W_0^{m,p}(\Omega)$, (resp. by $C_0^m(\bar{\Omega})$). Note that $C_0^m(\bar{\Omega}) = \{v \in C^m(\bar{\Omega})\colon D^\alpha v\,|_{\partial\Omega} = 0 \text{ for } |\alpha| \leqq m\}$. The normed (strong) dual of $W_0^{m,p}(\Omega)$ will be denoted by $W^{-m,p'}(\Omega)$, where $p' \in \mathbb{R}$ is related to p by $(1/p)+(1/p') = 1$. We recall that $W^{-m,p'}(\Omega)$ is the set of (continuous extensions to $W_0^{m,p}(\Omega)$ of) distributions in Ω which are equal to a finite sum of derivatives of order $\leqq m$ of functions belonging to $L^{p'}(\Omega)$. As usual, we will write $W^m(\Omega)$, $W_0^m(\Omega)$, $W^{-m}(\Omega)$ instead of $W^{m,2}(\Omega)$, $W_0^{m,2}(\Omega)$, $W^{-m,2}(\Omega)$. If $v = (v_i)_{i=1,\cdots,n}$ belongs respectively to $(W^{m,p}(\Omega))^n$, to $(W^{-m,p}(\Omega))^n$, to $(C^{m,\lambda}(\Omega))^n$, we put

$$\|v\|_{m,p} = \sum_{i=1}^n \|v_i\|_{m,p}, \qquad \|v\|_{-m,p} = \sum_{i=1}^n \|v_i\|_{-m,p}, \qquad \|v\|_{m,\lambda} = \sum_{i=1}^n \|v_i\|_{m,\lambda}.$$

We will say that Ω has the *cone property* if there exist positive constants α, h such that for any $x \in \Omega$ one can construct a right spherical cone with vertex x, opening α, and height h such that it lies in Ω. We will say that Ω *is of class* C^m (where m is an integer > 1) if $\bar{\Omega}$ is a submanifold of $\mathbb{R}^n$ with boundary, of class C^m, i.e., if for each $x \in \partial\Omega$ there exists an open neighborhood U_x of x in $\mathbb{R}^n$ and a C^m-diffeomorphism τ of U_x onto the unit ball $\{\xi \in \mathbb{R}^n : |\xi| < 1\}$ of $\mathbb{R}^n$ such that $\tau(\bar{\Omega} \cap U_x) = \{\xi \in \mathbb{R}^n : |\xi| < 1,\ \xi_n \geqq 0\}$. It is easy to see that, if Ω is of class C^1, it has the cone property. It is well-known (from the Sobolev Imbedding Theorem) that, if Ω has the cone property and $mp > n$, then each $v \in W^{m,p}(\Omega)$ can be identified with a continuous, bounded function of Ω into $\mathbb{R}$, and there exists a positive constant c such that

$$\sup_{x \in \Omega} |v(x)| \leqq c\, \|v\|_{m,p}.$$

Moreover, if Ω is of class C^1 and $mp > n$, then each $v \in W^{m,p}(\Omega)$ can be identified with a continuous function of $\bar{\Omega}$ into $\mathbb{R}$.

By using the well-known density of $C^m(\Omega) \cap W^{m,p}(\Omega)$ in $W^{m,p}(\Omega)$ and the Sobolev Imbedding Theorem one can verify that (see Adams [1], Theorem 5.23), *if* Ω *has the cone property and* $mp > n$, *then* $W^{m,p}(\Omega)$ *is a Banach algebra*, i.e.,

$$u, v \in W^{m,p}(\Omega) \Rightarrow uv \in W^{m,p}(\Omega), \qquad \|uv\|_{m,p} \leqq c_{m,p} \|u\|_{m,p} \|v\|_{m,p},$$

where $c_{m,p}$ is a positive number independent of u and v.

One can prove (see Valent [23], Osservazione I) that, *if* Ω *is connected and of class* C^1, *then* $C^{m,\lambda}(\bar{\Omega})$ *is a Banach algebra*, i.e.

$$u, v \in C^{m,\lambda}(\bar{\Omega}) \Rightarrow uv \in C^{m,\lambda}(\bar{\Omega}), \qquad \|uv\|_{m,\lambda} \leqq c_{m,\lambda} \|u\|_{m,\lambda} \|v\|_{m,\lambda},$$

where $c_{m,\lambda}$ is a positive number independent of u and v.

We now recall the statements of the Inverse Function Theorem and of the Implicit Function Theorem in Banach spaces, which we will use later. For the proofs, see e.g. Cartan [4].

INVERSE FUNCTION THEOREM. *Let X, Y be Banach spaces, U an open subset of X, and* $x_0 \in U$. *If* $f : U \to Y$ *is a continuously differentiable function, and if the differential at* x_0 *of f is an isomorphism*[1] *of X onto Y, then there exists an open neighborhood* U_0 *of* x_0 *in X contained in U and an open neighborhood* V_0 *of* $f(x_0)$ *in Y such that f is a* C^1*-diffeomorphism of* U_0 *onto* V_0 (i.e. such that the restriction of f to U is a one-to-one function of U_0 onto V_0, whose inverse is continuously differentiable).

IMPLICIT FUNCTION THEOREM. *Let X, Y, Z be Banach spaces, U an open subset of* $X \times Y$ *and let* $f : U \to Z$ *be a continuously differentiable function. Let* $(x_0, y_0) \in U$ *be such that* $f(x_0, y_0) = 0$ *and that the differential,* $d_y f(x_0, y_0)$ *of the function* $y \mapsto f(x_0, y)$ *at* y_0 *is an isomorphism of Y onto Z. Then there exists an open neighborhood* U_0 *of* (x_0, y_0) *in* $X \times Y$ *contained in U, an open neighborhood* U_0 *of* x_0 *in X and a continuously differentiable function* $g : V_0 \to Y$ *such that*

$$\{(x, y) \in U_0 : f(x, y) = 0\} = \{(x, y) : x \in V_0, y = g(x)\}.$$

Moreover, the differential, $dg(x_0)$, *of g at* x_0 *is given by*

$$dg(x_0) = -(d_y f(x_0, y_0))^{-1} \circ d_x f(x_0, y_0),$$

where $d_x f(x_0, y_0)$ *is the differential at* x_0 *of the function* $x \mapsto f(x, y_0)$.

3. The displacement problem in finite elastostatics

Let the bounded, open, subset Ω of $\mathbb{R}^n$ be a fixed reference configuration of an elastic body.[2]

The gradient of a function $v : \Omega \to \mathbb{R}^n$ will be denoted by Dv, i.e.,

$$Dv = (D_j v_i)_{i,j=1,\cdots,n}, \qquad \text{with } D_j v_i = \frac{\partial v_i}{\partial x_j}.$$

If $\sigma = (\sigma_{ij})_{i,j=1,\cdots,n}$ is a function of Ω into $\mathbb{R}^{n^2}$, we put, as usual,[3] $\operatorname{div} \sigma = D_j \sigma_{ij}$.

A displacement $u : \bar{\Omega} \to \mathbb{R}^n$ is associated to any deformation $\varphi : \bar{\Omega} \to \mathbb{R}^n$ by $\varphi = I + u$, where I is the identity function of $\bar{\Omega}$ into itself. Let

[1] Throughout this work by an isomorphism between two Banach spaces X, Y we mean an isomorphism for the topological linear space structures of X and Y.

[2] Throughout this work we will write n instead of 3, because all results which we will state are true for every $n \geqq 1$. Of course, they have a physical meaning in the case $n = 3$.

[3] We assume throughout this work the summation convention, i.e., a summation must be understood when an index is repeated twice.

$(x, z) \mapsto a(x, z) = (a_{ij}(x, z))_{i,j=1,\cdots,n}$ be a given function of $\Omega \times \mathbb{R}^{n^2}$ into $\mathbb{R}^{n^2}$ defining the material response, in the sense that $a(x, Du(x))$ is the first Piola–Kirchhoff stress at the point $x \in \Omega$ when the body is deformed by $I + u$. Moreover let $y \mapsto b(y) = (b_i(y))_{i=1,\cdots,n}$ be a given function of $\mathbb{R}^n$ into $\mathbb{R}^n$ defining a body force, in the sense that $b(x + u(x))$ is the body force per unit mass at the point $x + u(x)$ in the deformed configuration corresponding to the displacement u.

If the functions a and u are suitably smooth, let $Au : \Omega \to \mathbb{R}^{n^2}$, $A_{ij}u : \Omega \to \mathbb{R}$, and $A_{ij,hk}u : \Omega \to \mathbb{R}$, $(i, j, h, k = 1, \cdots, n)$, be the functions defined by

$$Au(x) = a(x, Du(x)), \qquad A_{ij}u(x) = a_{ij}(x, Du(x)),$$

(3.1)

$$A_{ij,hk}u(x) = \frac{\partial a_{ij}}{\partial z_{hk}}(x, Du(x)).$$

Moreover, let $Bu : \Omega \to \mathbb{R}^n$ be the function defined by

(3.2)
$$Bu(x) = b(x + u(x)).$$

We deal with the displacement problem in finite elastostatics. For simplicity, we consider the boundary condition $u_{|\partial\Omega} = 0$ for the displacement u. We will study the problem of finding $u : \bar{\Omega} \to \mathbb{R}^n$ such that

(P)
$$\begin{cases} \operatorname{div} Au + \rho Bu = 0 \\ u_{|\partial\Omega} = 0 \end{cases}$$

(D) *$I + u$ is an orientation-preserving C^1-diffeomorphism*[4] *of $\bar{\Omega}$ onto $\bar{\Omega}$*,

where ρ is a given real-valued function defined on Ω (it is the density in the reference configuration).

Note that the requirement (D) for the solutions of the boundary-value problem (P) is very strong. Actually the set of functions where we must find the solutions of the problem (P) is neither a linear space nor a convex set.

We begin by dealing with the case when b is a constant function. That leads to the problem of finding $u : \bar{\Omega} \to \mathbb{R}^n$ such that

(P_1)
$$\begin{cases} \operatorname{div} Au + f = 0 \\ u_{|\partial\Omega} = 0 \end{cases}$$

[4] By a C^1-diffeomorphism of $\bar{\Omega}$ onto $\bar{\Omega}$ we mean a one-to-one C^1 function of $\bar{\Omega}$ onto $\bar{\Omega}$ such that its inverse is a C^1 function on $\bar{\Omega}$. As is well-known, if φ is a C^1-diffeomorphism of $\bar{\Omega}$ onto $\bar{\Omega}$, then $\det D\varphi(x) \neq 0$ $\forall x \in \Omega$ (where $\det D\varphi(x)$ is the determinant of the matrix $D\varphi(x)$). A C^1-diffeomorphism φ of $\bar{\Omega}$ onto $\bar{\Omega}$ is an orientation-preserving C^1-diffeomorphism if and only if $\det D\varphi(x) > 0$ $\forall x \in \Omega$.

(D) $\quad$ *$I+u$ is an orientation-preserving C^1-diffeomorphism of $\bar{\Omega}$ onto $\bar{\Omega}$*;

where $f=(f_i)_{i=1,\cdots,n}$ is a given $\mathbb{R}^n$-valued function (or distribution) defined on Ω.

Let us consider the finite elastostatics operator $u \mapsto Tu$, where

$$Tu = -\operatorname{div} Au. \tag{3.3}$$

Local existence and uniqueness theorems for the problem (P_1) may be found by applying the Inverse Function Theorem to the operator T, provided there exist two Banach spaces S_u and S_f such that: (a) each $u \in S_u$ is a function of $\bar{\Omega}$ onto $\mathbb{R}^n$ somehow verifying the condition $u_{|\partial\Omega}=0$, and T acts from an open neighborhood U of some $u_0 \in S_u$ into S_f; (b) $T: U \to S_f$ is continuously differentiable; (c) the differential of T at u_0, $dT(u_0)$, is an isomorphism of S_u onto S_f.

By observing that the formal differential of T at any point is a second order linear partial differential operator, a natural choice of spaces S_u and S_f might seem the following: $S_u=(W_0^1(\Omega))^n$, $S_f=(W^{-1}(\Omega))^n$, or, more generally, the following: $S_u=(W_0^{1,p}(\Omega))^n$, $S_f=(W^{-1,p}(\Omega))^n$. Unfortunately these choices are not possible, as we will find out in the next section. For conditions (a), (b), (c) to be satisfied it is necessary that S_u is a space of functions smoother than those of $(W_0^{1,p}(\Omega))^n$.

4. A non-differentiability theorem

Now let a satisfy the Caratheodory condition (i.e. let a be continuous in z for almost all $x \in \Omega$ and measurable in x for each fixed z) and the growth condition $|a(x,z)| \leqq h(x)+k|z|$, with $h \in L^p(\Omega)$ and k constant >0. For every $u \in (W_0^{1,p}(\Omega))^n$ let $Au: \Omega \to \mathbb{R}^{n^2}$ be the function defined by (3.1). Under these conditions one can prove (see, e.g., Krasnoselskij [9]) that $u \mapsto Au$ is a continuous operator of $(W_0^{1,p}(\Omega))^n$ into $(L^p(\Omega))^n$ and hence $u \mapsto Tu$ (where $Tu=-\operatorname{div} Au$) is a continuous operator of $(W_0^{1,p}(\Omega))^n$ into $(W^{-1,p}(\Omega))^n$.

PROPOSITION 4.1. *Assume that the reference configuration is natural* (i.e. $a(x,0)=0\ \forall x \in \Omega$), *that the derivatives* $(\partial a_{ij}/\partial z_{hk})(x,0)$ *exist* $\forall x \in \Omega$, *that the functions* $x \mapsto (\partial a_{ij}/\partial z_{hk})(x,0)$ *are continuous on* $\bar{\Omega}$ (in the standard sense), *that*[5]

$$\frac{\partial a_{ij}}{\partial z_{hk}}(x,0)=\frac{\partial a_{ij}}{\partial z_{kh}}(x,0), \qquad \frac{\partial a_{ij}}{\partial z_{hk}}(x,0)=\frac{\partial a_{ji}}{\partial z_{hk}}(x,0), \tag{4.1}$$

and that, for each $x \in \bar{\Omega}$,

[5] The symmetries (4.1) are a consequence of the principle of frame-indifference (the former) and of the symmetry of the Cauchy stress tensor (the latter).

$$(4.2)\qquad \frac{\partial a_{ij}}{\partial z_{hk}}(x,0)z_{ij}z_{hk} > 0 \qquad \forall z = (z_{ij})_{i,j=1,\cdots,n} \in \mathbb{R}^{n^2}\setminus\{0\} \textit{ such that } z_{ij} = z_{ji}.$$

Then, if Ω is of class C^1, the formally linearized operator of T at 0, namely the linear differential operator $u \mapsto T_L u$, where

$$T_L u = -\left(D_j\left(\frac{\partial a_{ij}}{\partial z_{hk}}(x,0)D_k u_h\right)\right)_{i=1,\cdots,n},$$

is an isomorphism of $(W_0^{1,p}(\Omega))^n$ onto $(W^{-1,p}(\Omega))^n$.

Proof. First note that $u \in (W_0^{1,p}(\Omega))^n \Rightarrow T_L u \in (W^{-1,p}(\Omega))^n$. Note also that, by continuity on $\bar{\Omega}$ of $x \mapsto (\partial a_{ij}/\partial z_{hk})(x,0)$, from (4.2) it follows that

$$\frac{\partial a_{ij}}{\partial z_{hk}}(x,0)z_{ij}z_{hk} \geqq c_1|z|^2 \qquad \forall (x,z) \in \bar{\Omega}\times\mathbb{R}^{n^2} \text{ such that } z_{ij} = z_{ji},$$

where c_1 is a constant >0, which implies, by (4.1) and the Korn and Poincaré inequalities (see Fichera [5]),

$$(4.3)\qquad \int_\Omega \frac{\partial a_{ij}}{\partial z_{hk}}(x,0)D_j u_i(x)D_k u_h(x)dx \geqq c_2\|u\|_1^2 \qquad \forall u \in (W_0^1(\Omega))^n,$$

where c_2 is a positive constant, and $\|\cdot\|_1$ is the norm of $(W^1(\Omega))^n$. If $p = 2$, from (4.3) it easily follows that

$$(4.4)\qquad \|T_L u\|_{-1} \geqq c_2\|u\|_1, \qquad \|T_L^* u\|_{-1} \geqq c_2\|u\|_1 \qquad \forall u \in (W_0^1(\Omega))^n,$$

where T_L^* is the transpose of T_L and $\|\cdot\|_{-1}$ is the norm of $(W^{-1}(\Omega))^n$, which implies (by standard arguments) that T_L is an isomorphism of $(W_0^1(\Omega))^n$ onto $(W^{-1}(\Omega))^n$. If Ω is of class C^1, (4.3) implies that the analogous of inequalities (4.4) hold for each real $p > 1$, namely

$$(4.5)\qquad \begin{aligned} &\|T_L u\|_{-1,p} \geqq c_3\|u\|_{1,p} \qquad \forall u \in (W_0^{1,p}(\Omega))^n,\\ &\|T_L^* u\|_{-1,p'} \geqq c_3'\|u\|_{1,p'} \qquad \forall u \in (W_0^{1,p'}(\Omega)),\end{aligned}$$

where c_3 and c_3' are positive constants. The inequalities (4.5) can be proved by generalizing Simader's methods (see [14], Theorem 6.1) from a single equation to a system of equations. From (4.5) it follows that T_L is an isomorphism of $(W_0^{1,p}(\Omega))^n$ onto $(W^{-1,p}(\Omega))^n$. □

Even if T_L is an isomorphism of $(W_0^{1,p}(\Omega))^n$ onto $(W^{-1,p}(\Omega))^n$, nevertheless *the Inverse Function Theorem does not apply to the nonlinear operator $T:(W_0^{1,p}(\Omega))^n \to (W^{-1,p}(\Omega))^n$* defined by (3.3). Indeed, the following remark holds.

Remark 4.1. *Let the assumptions made on a at the beginning of the section be satisfied and assume further that the partial derivatives $\partial a/\partial z_{hk}$ exist and* *that, for each* $z \in \mathbb{R}^{n^2}$, *the functions* $x \mapsto (\partial a_{ij}/\partial z_{hk})(x, z)$ *belong to* $L^p(\Omega)$. *Let* u^0 *be any point of* $(W_0^{1,p}(\Omega))^n$. *If there exists a pair* (U, V), *with* U *a neighborhood of* u^0 *in* $(W_0^{1,p}(\Omega))^n$ *and* V *a neighborhood of* Tu^0 *in* $(W^{-1,p}(\Omega))^n$, *such that* T *is a homeomorphism of* U *onto* V *differentiable at* u^0, *then the function* a *is affine in* z.

This remark is a consequence of the following theorem, which has been proved by Valent and Zampieri [20] (see also Valent [21], [22]).

THEOREM 4.1. *Let* p *and* q *be real numbers with* $1 < p \leqq q$. *Let* $(x, z) \mapsto a(x, z)$ *be a function of* $\Omega \times \mathbb{R}^{n^2}$ *into* $\mathbb{R}$ *satisfying the Caratheodory condition and the growth condition* $|a(x, z)| \leqq h(x) + k|z|^{p/q}$, *with* $h \in L^q(\Omega)$ *and* k *positive constant. Assume also that the derivatives* $(\partial a/\partial z_{hk})(x, z)$ *exist and that, for each* $z \in \mathbb{R}^{n^2}$, *the functions* $x \mapsto (\partial a/\partial z_{hk})(x, z)$ *belong to* $L^q(\Omega)$. *For each* $u \in (W_0^{1,p}(\Omega))^n$ *let* Au *be the real-valued function defined on* Ω *by* $Au(x) = a(x, Du(x))$. *Then the continuous operator*[6] $u \mapsto Au$ *of* $(W_0^{1,p}(\Omega))^n$ *into* $L^q(\Omega)$ *is differentiable at a point* $u^0 \in (W_0^{1,p}(\Omega))^n$ *(if and) only if, for almost all* $x \in \Omega$, *the function* $z \mapsto a(x, Du^0(x) + z) - a(x, Du^0(x))$ *is linear.*

The proof of this theorem is rather difficult and lengthy, and will not be given here. We only remark that the proof would be much easier if we had $W^{1,p}(\Omega)$ instead of $W_0^{1,p}(\Omega)$.

PROOF OF REMARK 4.1. Assume that there exists a pair (U, V) with the property expressed in Remark 4.1. Since $u \mapsto Au$ (where Au is defined by (3.1)) is continuous from $(W_0^{1,p}(\Omega))^n$ to $(L^p(\Omega))^{n^2}$ and T induces a homeomorphism between U and V, the map $Au \mapsto Tu$ is a homeomorphism of the subset $A(U)$ of $(L^p(\Omega))^{n^2}$ onto V, and hence its inverse is differentiable at Tu^0. Then, since $T : (W_0^{1,p}(\Omega))^n \to (W^{-1,p}(\Omega))^n$ is differentiable at u^0, the operator $A : (W_0^{1,p}(\Omega))^n \to (L^p(\Omega))^{n^2}$ is also differentiable at u^0. Therefore, by Theorem 4.1, the function a is affine in z. ❑

5. Two differentiability theorems

We now state two differentiability results (which have been proved by Valent [23]) for a superposition operator, from which we deduce two differentiability results for the finite elastostatics operator T. Let N be an integer $\geqq 1$ and let $(x, y) \mapsto h(x, y)$ be a function of $\Omega \times \mathbb{R}^N$ into $\mathbb{R}$. For each function $\sigma : \Omega \to \mathbb{R}^N$ let $H\sigma$ be the real-valued function defined on Ω by

$$H\sigma(x) = h(x, \sigma(x)).$$

[6] By the assumptions on the function a, it is well-known that $u \mapsto Au$ is a continuous operator of $W_0^{1,p}(\Omega)$ into $L^q(\Omega)$. See, e.g., Krasnoselskij [9].

THEOREM 5.1. *Assume Ω has the cone property (in particular, Ω is of class C^1). If*[7] $h \in C^{m+1}(\bar{\Omega} \times \mathbb{R}^N)$ *and* $p(m+1) > n$, *then* $\sigma \mapsto H\sigma$ *is a continuous operator of* $(W^{m+1,p}(\Omega))^N$ *into* $W^{m+1,p}(\Omega)$. *If* $h \in C^{m+2}(\bar{\Omega} \times \mathbb{R}^N)$ *and* $p(m+1) > n$, *then* $\sigma \mapsto H\sigma$ *is a continuously differentiable operator of* $(W^{m+1,p}(\Omega))^N$ *into* $W^{m+1,p}(\Omega)$, *and its differential at any point* σ *is the (continuous, linear) operator* $\tau \mapsto dH(\sigma)\tau$, *of* $(W^{m+1,p}(\Omega))^N$ *into* $W^{m+1,p}(\Omega)$, *where* $dH(\sigma)\tau$ *is defined by*

$$(dH(\sigma)\tau)(x) = \frac{\partial h}{\partial y_j}(x, \sigma(x))\tau_j(x). \tag{5.1}$$

THEOREM 5.2. *Assume Ω is connected and of class C^1. If* $h \in C^{m+2}(\bar{\Omega} \times \mathbb{R}^N)$ *then, for any* $\lambda \in]0,1]$, $\sigma \mapsto H\sigma$ *is a continuous operator of* $(C^{m+1,\lambda}(\bar{\Omega}))^N$ *into* $C^{m+1,\lambda}(\bar{\Omega})$. *If* $h \in C^{m+3}(\bar{\Omega} \times \mathbb{R}^N)$ *then, for any* $\lambda \in]0,1]$, $\sigma \mapsto H\sigma$ *is a continuously differentiable operator of* $(C^{m+1,\lambda}(\bar{\Omega}))^N$ *into* $C^{m+1,\lambda}(\bar{\Omega})$, *and its differential at any point* σ *is the (continuous, linear) operator* $\tau \mapsto dH(\sigma)\tau$, *of* $(C^{m+1,\lambda}(\bar{\Omega}))^N$ *into* $C^{m+1,\lambda}(\bar{\Omega})$, *where* $dH(\sigma)\tau$ *is defined by* (5.1).

For the proof of Theorems 5.1 and 5.2 we refer to [23]. Here, we only emphasize that an important role in proving these theorems is played by the fact that (because of the hypotheses on Ω and p, m) $W^{m+1,p}(\Omega)$ and $C^{m+1,\lambda}(\bar{\Omega})$ are Banach algebras (see Section 2).

We now consider the finite elastostatics operator $u \mapsto Tu$, where Tu is defined by (3.3). By using Theorems 5.1 and 5.2 it is easy to deduce the following Corollaries 5.1 and 5.2, respectively.

COROLLARY 5.1. *Assume Ω has the cone property (in particular, Ω is of class C^1). If* $a_{ij} \in C^{m+2}(\bar{\Omega} \times \mathbb{R}^{n^2})$ *and* $p(m+1) > n$, *then* $u \mapsto Tu$ *is a continuously differentiable operator of* $(W^{m+2,p}(\Omega))^n$ *into* $(W^{m,p}(\Omega))^n$, *and its differential at any point u is the (continuous, linear) operator* $v \mapsto dT(u)v$, *where*

$$dT(u)v = -(D_j(A_{ij,hk}uD_kv_h))_{i=1,\cdots,n}. \tag{5.2}$$

COROLLARY 5.2. *Assume Ω is connected and of class C^1. If* $a_{ij} \in C^{m+3}(\bar{\Omega} \times \mathbb{R}^{n^2})$ *then, for any* $\lambda \in]0,1]$, $u \mapsto Tu$ *is a continuously differentiable operator of* $(C^{m+2,\lambda}(\bar{\Omega}))^n$ *into* $(C^{m,\lambda}(\bar{\Omega}))^n$, *and its differential at any point u is the (continuous, linear) operator* $v \mapsto dT(u)v$, *of* $(C^{m+2,\lambda}(\bar{\Omega}))^n$ *into* $(C^{m,\lambda}(\bar{\Omega}))^n$, *where* $dT(u)v$ *is defined by* (5.2).[8]

[7] $C^{m+1}(\bar{\Omega} \times \mathbb{R}^N)$ denotes the set of real-valued functions defined on $\bar{\Omega} \times \mathbb{R}^N$ which are the restriction to $\bar{\Omega} \times \mathbb{R}^N$ of some real-valued function defined on an open neighborhood of $\bar{\Omega} \times \mathbb{R}^N$ in $\mathbb{R}^n \times \mathbb{R}^N$ and having all continuous derivatives up to order $m+1$.

[8] A result of this type was obtained by Van Buren [24], in the case $m = 0$ and Ω convex.

6. A topological property of sets of admissible displacements

A function $u : \bar{\Omega} \to \mathbb{R}^n$ will be called an *admissible displacement* (for the problems ((P), (D)) and ((P_1), (D))) if u satisfies the condition (D) of Section 3 and the condition $u_{|\partial\Omega} = 0$.

We denote by $\mathcal{A}_{m+2,p}$ and by $\mathcal{A}_{m+2,\lambda}$ the set of those admissible displacements which lie in $(W^{m+2,p}(\Omega))^n$ and in $(C^{m+2,\lambda}(\bar{\Omega}))^n$, respectively.

Our goal, in this section, is to prove Theorem 6.1. We begin by showing the following basic

LEMMA 6.1. *Let Ω be connected and of class C^1. A C^1 function $\varphi : \bar{\Omega} \to \mathbb{R}^n$ such that $\varphi(x) = x \ \forall x \in \partial\Omega$ is an orientation-preserving C^1-diffeomorphism of $\bar{\Omega}$ onto $\varphi(\bar{\Omega})$ if (and only if) $\det D\varphi(x) > 0 \ \forall x \in \bar{\Omega}$.*

PROOF. Let $\varphi : \bar{\Omega} \to \mathbb{R}^n$ be a C^1 function such that $\varphi(x) = x \ \forall x \in \partial\Omega$ and that $\det D\varphi(x) > 0 \ \forall x \in \bar{\Omega}$. That means (see Section 2) that there exists a continuously differentiable $\mathbb{R}^n$-valued function ψ defined on an open neighborhood of $\bar{\Omega}$ in $\mathbb{R}^n$ such that $\psi(x) = \varphi(x) \ \forall x \in \bar{\Omega}$ and that $\det D\psi(x) > 0 \ \forall x \in \bar{\Omega}$. Therefore (by the Inverse Function Theorem) for each $x \in \bar{\Omega}$ there exists an open neighborhood U_x of x in $\mathbb{R}^n$ and an open neighborhood V_x of $\varphi(x)$ in $\mathbb{R}^n$ such that ψ is a C^1-diffeomorphism of U_x onto V_x. Since, moreover, φ induces the identity on $\partial\Omega$ and Ω is connected and of class C^1, it follows, by a theorem of Meisters and Olech (see [11], Theorem 1),[9] that $\varphi : \bar{\Omega} \to \mathbb{R}^n$ is (globally) one-to-one. It is now easy to conclude that φ is an orientation-preserving C^1-diffeomorphism of $\bar{\Omega}$ onto the subset $\varphi(\bar{\Omega})$ of $\mathbb{R}^n$. □

THEOREM 6.1. *Let Ω be connected and of class C^1. Then, for each $m \geqq 0$ and $\lambda \in]0, 1]$, $\mathcal{A}_{m+2,\lambda}$ is an open subset of the subspace $(C_0^0(\bar{\Omega}) \cap C^{m+2,\lambda}(\bar{\Omega}))^n$ of $(C^{m+2,\lambda}(\bar{\Omega}))^n$, while, for $p(m+1) > n$, $\mathcal{A}_{m+2,p}$ is an open subset of the subspace $(W_0^{1,p}(\Omega) \cap W^{m+2,p}(\Omega))^n$ of $(W^{m+2,p}(\Omega))^n$.*

PROOF. First of all we note that (since Ω is of class C^1) if $p(m+1) > n$ the following continuous imbedding holds (see Section 2): $W^{m+2,p}(\Omega) \subseteq C^1(\bar{\Omega})$, and therefore

[9] The statement of this theorem is the following. "Let X be a compact subset of $\mathbb{R}^n$ $(n \geqq 2)$ such that $\mathbb{R}^n \setminus \partial X$ is not connected and such that for each proper closed subset Γ of ∂X, $X \setminus \Gamma$ is connected. Let $\mathring{X}$ be the interior of X. Assume that $f : X \to \mathbb{R}^n$ is continuous on X and locally one-to-one on $X \setminus Z$, where Z is a subset of X such that $Z \cap \mathring{X}$ is a discrete set and such that $Z \cap \partial X \neq \partial X$. Then if $f_{|\partial X}$ is one-to-one, f is a homeomorphism of X onto $f(X)$." Note that, if Ω is a connected, bounded, open subset of $\mathbb{R}^n$ of class C^1, then the compact subset $\bar{\Omega}$ of $\mathbb{R}^n$ is such that $\mathbb{R}^n \setminus \partial\Omega$ is not connected and such that, for each proper closed subset Γ of $\partial\Omega$, $\bar{\Omega} \setminus \Gamma$ is connected. (That is easily seen.)

$$u \in (W^{m+2,p}(\Omega))^n, \qquad u_{|\partial\Omega} = 0 \Leftrightarrow u \in (W_0^{1,p}(\Omega) \cap W^{m+2,p}(\Omega))^n,$$

because, by a well-known result about $W_0^{m,p}(\Omega)$ spaces (see, e.g. Něcas [12]), for $u \in C^1(\bar{\Omega})$ the following equivalence holds: $u \in W_0^{1,p}(\Omega) \Leftrightarrow u_{|\partial\Omega} = 0$. Thus, if $p(m+1) > n$,

$$\mathcal{A}_{m+2,p} \subseteq (W_0^{1,p}(\Omega) \cap W^{m+2,p}(\Omega))^n.$$

We assume $p(m+1) > n$ and we prove that $\mathcal{A}_{m+2,p}$ is open in $(W_0^{1,p}(\Omega) \cap W^{m+2,p}(\Omega))^n$. Accordingly, we fix $u^0 \in \mathcal{A}_{m+2,p}$ and we show that there is a neighborhood U_0 of u^0 in $(W_0^{1,p}(\Omega) \cap W^{m+2,p}(\Omega))^n$ such that $U_0 \subseteq \mathcal{A}_{m+2,p}$. Put

$$\mu = \inf_{x \in \bar{\Omega}} \det D(I + u^0)(x).$$

From the hypotheses we have $\mu > 0$. The map $u \mapsto \det D(I+u)$ is obviously continuous from $(C^1(\bar{\Omega}))^n$ to $C^0(\bar{\Omega})$, and therefore from $(W^{1,p}(\Omega) \cap W^{m+2,p}(\Omega))^n$ to $C^0(\bar{\Omega})$ because of the continuity of the imbedding $W^{m+2,p}(\Omega) \subseteq C^1(\bar{\Omega})$. Hence there exists an open, convex neighborhood U_0 of u^0 in $(W_0^{1,p}(\Omega) \cap W^{m+2,p}(\Omega))^n$ such that $u \in U_0$ implies

$$\sup_{x \in \bar{\Omega}} |\det D(I+u)(x) - \det D(I+u^0)(x)| < \mu.$$

We immediately deduce that $u \in U_0$ implies

$$\sup_{x \in \bar{\Omega}} D(I+u)(x) > 0$$

and therefore, from Lemma 6.1, we conclude that, for each $u \in U_0$, $I + u$ is an orientation-preserving C^1-diffeomorphism of $\bar{\Omega}$ onto $(I+u)(\bar{\Omega})$. To conclude that $U_0 \subseteq \mathcal{A}_{m+2,p}$, it only remains to prove that, if $u \in U_0$, then $(I+u)(\bar{\Omega}) = \bar{\Omega}$. Let $u \in U_0$. Since (under our hypotheses) Ω coincides with the interior of $\bar{\Omega}$, from Brouwer's Theorem on the Invariance of Domain (cf., e.g., [11], p. 64) it follows that $(I+u)(\Omega)$ is the interior of $(I+u)(\bar{\Omega})$ and $(I+u)(\partial\Omega)$ is the boundary of $(I+u)(\bar{\Omega})$, which coincides with the boundary of $(I+u)(\Omega)$. On the other hand we have $(I+u)(\partial\Omega) = \partial\Omega$ because $(I+u)_{|\partial\Omega} =$ identity. Thus the boundary of $(I+u)(\Omega)$ (in $\mathbb{R}^n$) coincides with $\partial\Omega$, and $(I+u)(\bar{\Omega}) = \bar{\Omega}$ if and only if $(I+u)(\Omega) = \Omega$. We now prove that

$$(I+u)(\Omega) \subseteq \Omega.$$

Since U is convex, if we put $u_t = tu + (1-t)u^0$ with $t \in [0,1]$, we have $I + u_t \in U_0$ $\forall t \in [0,1]$. Let $x \in \Omega$. If $(I+u)(x) \notin \Omega$, then the straight line segment $\{x + u_t(x) : t \in [0,1]\}$, joining $x + u^0(x)$ and $x + u(x)$, contains some point of $\partial\Omega$, namely there exists $\lambda \in [0,1]$ such that $(I + u_\lambda)(x) \in \partial\Omega$, which is contrary to the fact that $x \in \Omega$, because, since $I + u_\lambda \in U_0$, $(I + u_\lambda)(\Omega)$ is an open subset of $\mathbb{R}^n$ and $\partial\Omega$ is its boundary in $\mathbb{R}^n$. It is now immediate to see that $(I+u)(\Omega) = \Omega$, by observing that Ω is an open, connected subset of $\mathbb{R}^n$ containing $(I+u)(\Omega)$ and having the same boundary as $(I+u)(\Omega)$. This completes the proof of the second statement of the theorem.

Quite analogous arguments show that $\mathscr{A}_{m+2,\lambda}$ is an open subset of $(C_0^0(\bar{\Omega}) \cap C^{m+2,\lambda}(\bar{\Omega}))^n$. □

7. Some results about linear differential operators

The following two theorems are useful in order to obtain local results of existence and uniqueness for the problem $((\mathrm{P}_1), (\mathrm{D}))$.

THEOREM 7.1. *Assume Ω is connected and of class C^{m+2}, $a_{ij} \in C^{m+2}(\bar{\Omega} \times \mathbb{R}^{n^2})$, and $p > n$. Then for each $u \in (W^{m+2,p}(\Omega))^n$ such that*

$$(7.1) \qquad \int_\Omega \frac{\partial a_{ij}}{\partial z_{hk}}(x, Du(x)) D_j v_i(x) D_k v_h(x) dx \geqq c(u) \| v \|_1^2 \qquad \forall v \in (W_0^1(\Omega))^n,$$

where $c(u)$ is a suitable positive number independent of v, the linear operator $dT(u)$ defined by (5.2) is an isomorphism of the subspace $(W_0^{1,p}(\Omega) \cap W^{m+2,p}(\Omega))^n$ of $(W^{m+2,p}(\Omega))^n$ onto $(W^{m,p}(\Omega))^n$.

THEOREM 7.2. *Assume Ω is connected and of class C^{m+3}, $a_{ij} \in C^{m+3}(\bar{\Omega} \times \mathbb{R}^{n^2})$, and $0 < \lambda < 1$. Then for each $u \in (C^{m+2,\lambda}(\bar{\Omega}))^n$ verifying (7.1), the linear operator $dT(u)$ defined by (5.2) is an isomorphism of the subspace $(C_0^0(\bar{\Omega}) \cap C^{m+2,\lambda}(\bar{\Omega}))^n$ of $(C^{m+2,\lambda}(\bar{\Omega}))^n$ onto $(C^{m,\lambda}(\bar{\Omega}))^n$.*

Theorem 7.2 is a consequence of well-known results (see, e.g., Agmon, Douglis and Nirenberg [2] and its bibliography, and note that, by Theorem 5.2, the functions $A_{ij,hk}u$ belong to $C^{m+1,\lambda}(\bar{\Omega})$ if $u \in (C^{m+2,\lambda}(\bar{\Omega}))^n$). Also, Theorem 7.1 can be deduced from well-known results about elliptic operators. One need only remark that such results have been obtained with slightly different hypotheses on the coefficients of the operator (see, e.g., Simader [14], Theorem 9.14, and observe that, by Theorem 5.1, the functions $A_{ij,hk}u$ belong to $W^{m+1,p}(\Omega)$ if $u \in (W^{m+2,p}(\Omega))^n$). However it is not difficult to show that the following estimate

$$\| dT(u)v \|_{m,p} \geqq k(u) \| v \|_{m+2,p} \qquad \forall v \in (W_0^{1,p}(\Omega) \cap W^{m+2,p}(\Omega))^n$$

(where $k(u)$ is a number > 0 independent of v), which is well-known (under (7.1)) when $A_{ij,hk}u \in C^{m+1}(\bar{\Omega})$,[10] still holds when $A_{ij,hk}u \in W^{m+1,p}(\Omega)$ with $p > n$. With the aid of this estimate it is possible to prove Theorem 7.1.

[10] See Agmon, Douglis and Nirenberg [2], and note that (7.1) implies that $dT(u)$ is uniformly strongly elliptic (cf. Truesdell and Noll [19] and Fichera [5]).

8. Local and "semiglobal" theorems of existence and uniqueness for the problem $((P_1), (D))$

When the assumptions of Corollary 5.1 are satisfied we set

$$\mathcal{T}_{m+2,p} = \{u \in \mathcal{A}_{m+2,p} : dT(u) \text{ is an isomorphism of the subspace } (W_0^{1,p}(\Omega) \cap W^{m+2,p}(\Omega))^n \text{ of } (W^{m+2,p}(\Omega))^n \text{ onto } (W^{m,p}(\Omega))^n\},$$

where T is the operator defined by (3.3). Analogously, when the assumptions of Corollary 5.2 are satisfied we set

$$\mathcal{T}_{m+2,\lambda} = \{u \in \mathcal{A}_{m+2,\lambda} : dT(u) \text{ is an isomorphism of the subspace } (C_0^0(\bar{\Omega}) \cap C^{m+2,\lambda}(\bar{\Omega}))^n \text{ of } (C^{m+2,\lambda}(\bar{\Omega}))^n \text{ onto } (C^{m,\lambda}(\bar{\Omega}))^n\}.$$

Note that (by the Open Mapping Theorem) from (5.2) it follows that $\mathcal{T}_{m+2,p}$ is the set of admissible displacements $u \in (W^{m+2,p}(\Omega))^n$ such that the linearized problem of (P_1) at u, i.e., the linear problem

$$(8.1) \qquad \begin{cases} D_j(A_{ij,hk}uD_kv_h + A_{ij}u) + f_i = 0, & (i = 1, \cdots, n), \\ v_{|\partial\Omega} = 0, \end{cases}$$

has one and only one solution $v \in (W^{m+2,p}(\Omega))^n$ for each $f \in (W^{m,p}(\Omega))^n$. Analogously, $\mathcal{T}_{m+2,\lambda}$ is the set of admissible displacements $u \in (C^{m+2,\lambda}(\bar{\Omega}))^n$ such that the problem (8.1) has a unique solution v in $(C^{m+2,\lambda}(\bar{\Omega}))^n$ for each $f \in (C^{m,\lambda}(\bar{\Omega}))^n$.

Assume the derivatives $\partial a_{ij} / \partial z_{hk}$ exist and are continuous on $\bar{\Omega} \times \mathbb{R}^{n^2}$. *We denote by* $\mathcal{S}_{m+2,p}$ (resp. $\mathcal{S}_{m+2,\lambda}$) *the set of all* $u \in \mathcal{A}_{m+2,p}$ (resp. $u \in \mathcal{A}_{m+2,\lambda}$) *verifying* (7.1) for some positive number $c(u)$.[11]

Note that, from the proof of the Proposition 4.1, it follows that the null displacement belongs to $\mathcal{S}_{m+2,p}$ and to $\mathcal{S}_{m+2,\lambda}$ if the function a satisfies the conditions $a(x,0) = 0\ \forall x \in \Omega$ and (4.1), (4.2). Note also that, if the assumptions of Theorem 7.1 (resp. Theorem 7.2) are satisfied, then

$$\mathcal{S}_{m+2,p} \subseteq \mathcal{T}_{m+2,p}, \qquad (\text{resp. } \mathcal{S}_{m+2,\lambda} \subseteq \mathcal{T}_{m+2,\lambda}).$$

Now we recall that, if Ω is connected and of class C^1, if $a_{ij} \in C^{m+2}(\bar{\Omega} \times \mathbb{R}^{n^2})$ and if $(m+1)p > n$, then (by Corollary 5.1 and Theorem 6.1) $u \mapsto Tu$ is a continuously differentiable operator of $(W^{m+2,p}(\Omega))^n$ into $(W^{m,p}(\Omega))^n$ and $\mathcal{A}_{m+2,p}$ is an open subset of the subspace $(W_0^{1,p}(\Omega) \cap W^{m+2,p}(\Omega))^n$ of $(W^{m+2,p}(\Omega))^n$. Therefore (by the Inverse Function Theorem) for

[11] In other words, $\mathcal{S}_{m+2,p}$ (resp. $\mathcal{S}_{m+2,\lambda}$) is the set of admissible displacements u belonging to $(W^{m+2,p}(\Omega))^n$ (resp. $(C^{m+2,\lambda}(\Omega))^n$), and such that the deformation $I + u$ is uniformly Hadamard-stable (in the terminology of Gurtin and Spector [8]), or infinitesimally strongly stable (according to the definition given by Fichera [6]).

each $u^0 \in \mathcal{T}_{m+2,p}$ there exists an open neighborhood U_0 of u^0 in $(W_0^{1,p}(\Omega) \cap W^{m+2,p}(\Omega))^n$ contained in $\mathcal{A}_{m+2,p}$ and an open neighborhood V_0 of Tu^0 in $(W^{m,p}(\Omega))^n$ such that $u \mapsto Tu$ is a C^1-diffeomorphism of U_0 onto V_0. Consequently the following theorem holds.

THEOREM 8.1. *Assume Ω is connected and of class C^1, $a_{ij} \in C^{m+2}(\bar{\Omega} \times \mathbb{R}^{n^2})$ and $p(m+1) > n$. Then for each $u^0 \in \mathcal{T}_{m+2,p}$ there exist two positive numbers ξ, η such that for all $f \in (W^{m,p}(\Omega))^n$ with $\|f - Tu^0\|_{m,p} < \xi$ the problem $((\mathrm{P}_1), (\mathrm{D}))$ of Section 3 has one and only one solution $u(f)$ in $(W^{m+2,p}(\Omega))^n$ satisfying the condition $\|u(f) - u^0\|_{m+2,p} < \eta$. Moreover the operator $f \mapsto u(f)$ is continuously differentiable.*

Analogously, from Corollary 5.2 and Theorem 6.1 we get the following

THEOREM 8.2. *Assume Ω is connected and of class C^1 and $a_{ij} \in C^{m+3}(\bar{\Omega} \times \mathbb{R}^{n^2})$. Then for each $u^0 \in \mathcal{T}_{m+2,\lambda}$ there exist two positive numbers ξ, η such that for all $f \in (C^{m,\lambda}(\bar{\Omega}))^n$ with $\|f - Tu^0\|_{m,\lambda} < \xi$ the problem $((\mathrm{P}_1), (\mathrm{D}))$ of Section 3 has one and only one solution $u(f)$ in $(C^{m+2,\lambda}(\bar{\Omega}))^n$ satisfying the condition $\|u(f) - u^0\|_{m+2,\lambda} < \eta$. Moreover the operator $f \mapsto u(f)$ is continuously differentiable.*

Recalling Theorems 7.1 and 7.2, from Theorems 8.1 and 8.2 we immediately deduce the following two corollaries.

COROLLARY 8.1. *Assume Ω is connected and of class C^{m+2}, $a_{ij} \in C^{m+2}(\bar{\Omega} \times \mathbb{R}^{n^2})$, and $p > n$. Then for each $u^0 \in \mathcal{S}_{m+2,p}$ [in particular for $u^0 = 0$ if the function a satisfies the conditions $a(x, 0) = 0$ $\forall x \in \Omega$ and* (4.1), (4.2)*] there exist two positive numbers ξ, η such that for all $f \in (W^{m,p}(\Omega))^n$ with $\|f - Tu^0\|_{m,p} < \xi$ the problem $((\mathrm{P}_1), (\mathrm{D}))$ of Section 3 has one and only one solution $u(f)$ in $(W^{m+2,p}(\Omega))^n$ satisfying the condition $\|u(f) - u^0\|_{m+2,p} < \eta$. Moreover the operator $f \mapsto u(f)$ is continuously differentiable.*

COROLLARY 8.2. *Assume Ω is connected and of class C^{m+3}, $a_{ij} \in C^{m+3}(\bar{\Omega} \times \mathbb{R}^{n^2})$, and $0 < \lambda < 1$. Then for each $u^0 \in \mathcal{S}_{m+2,\lambda}$ [in particular for $u^0 = 0$ if $a(x, 0) = 0$ $\forall x \in \Omega$ and if* (4.1), (4.2) *hold] there exist positive numbers ξ, η such that for all $f \in (C^{m,\lambda}(\bar{\Omega}))$ with $\|f - Tu^0\|_{m,\lambda} < \xi$ the problem $((\mathrm{P}_1), (\mathrm{D}))$ of Section 3 has one and only one solution $u(f)$ in $(C^{m+2,\lambda}(\bar{\Omega}))^n$ satisfying the condition $\|u(f) - u^0\|_{m+2,\lambda} < \eta$. Moreover the operator $f \mapsto u(f)$ is continuously differentiable.*

We conclude this section with some results of "semiglobal" type.[12] Before we prove these results it is convenient to make the following

[12] Results like the ones stated in Remark 8.1 and in Theorem 8.3 have been proved by Gurtin and Spector [8].

REMARK 8.1. *Let Ω be connected and of class C^1 and assume the derivatives $\partial a_{ij}/\partial z_{hk}$ exist and are continuous on $\bar{\Omega}\times\mathbb{R}^{n^2}$. Then $\mathcal{S}_{m+2,\lambda}$ is an open subset of the subspace $(C_0^0(\bar{\Omega})\cap C^{m+2,\lambda}(\bar{\Omega}))^n$ of $(C^{m+2,\lambda}(\bar{\Omega}))^n$, and, for $p(m+1)>n$, $\mathcal{S}_{m+2,p}$ is an open subset of the subspace $(W_0^{1,p}(\Omega)\cap W^{m+2,p}(\Omega))^n$ of $(W^{m+2,p}(\Omega))^n$.*

PROOF. Let $u^0\in\mathcal{S}_{m+2,\lambda}$ (resp. $u^0\in\mathcal{S}_{m+2,p}$, with $p(m+1)>n$). By definition of $\mathcal{S}_{m+2,\lambda}$ (and of $\mathcal{S}_{m+2,p}$) there exists a positive constant $c(u^0)$ such that (see (7.1) and notations (3.1))

$$(8.2)\qquad \int_\Omega A_{ij,hk}u D_jv_iD_kv_h dx \geqq c(u^0)\|v\|_1^2 \qquad \forall v\in(W_0^1(\Omega))^n.$$

Since the functions $\partial a_{ij}/\partial z_{hk}$ $(i,j,h,k=1,\cdots,n)$ are continuous on $\bar{\Omega}\times\mathbb{R}^{n^2}$ it is easily seen that $u\mapsto A_{ij,hk}u$ is a continuous operator of $(C^1(\bar{\Omega}))^n$ into $C^0(\bar{\Omega})$; therefore it is a continuous operator of $(C^{m+2,\lambda}(\bar{\Omega}))^n$ into $C^0(\bar{\Omega})$, and, if $p(m+1)>n$, of $(W^{m+2,p}(\bar{\Omega}))^n$ into $C^0(\bar{\Omega})$. Then it is not difficult to see that, given a number ε such that $0<\varepsilon<c(u^0)$, there exists a neighborhood U_0 of u^0 in $(C^{m+2,\lambda}(\bar{\Omega}))^n$, (resp. in $(W^{m+2,p}(\Omega))^n$), such that $u\in U_0$ implies

$$\left|\int_\Omega (A_{ij,hk}u^0-A_{ij,hk}u)D_jv_iD_kv_h dx\right| < \varepsilon\|v\|_1^2 \qquad \forall v\in(W_0^1(\Omega))^n.$$

From this and from (8.2) it follows that $u\in U_0$ implies

$$\int_\Omega A_{ij,hk}uD_jv_iD_kv_h dx \geqq (c(u^0)-\varepsilon)\|v\|_1^2 \qquad \forall v\in(W_0^1(\Omega))^n.$$

Since, by Theorem 6.1, U_0 may be chosen such that $U_0\subseteq\mathcal{A}_{m+2,\lambda}$ (resp. $U_0\subseteq\mathcal{A}_{m+2,p}$), the proof is complete. □

THEOREM 8.3. *Assume Ω and the function a are as in Corollary 5.1 (resp. in Corollary 5.2). Then the (continuously differentiable) operator $u\mapsto Tu$ of $(W^{m+2,p}(\Omega))^n$ into $(W^{m,p}(\Omega))^n$ with $p(m+1)>n$ (resp. of $(C^{m+2,\lambda}(\bar{\Omega}))^n$ into $(C^{m,\lambda}(\bar{\Omega}))^n$), defined by (3.3) is one-to-one on each convex subset of $\mathcal{S}_{m+2,p}$ (resp. of $\mathcal{S}_{m+2,\lambda}$).*

PROOF. It suffices to prove that the result holds for T acting from $(W^{m+2,p}(\Omega))^n$ to $(W^{m,p}(\Omega))^n$, because the proof of the result when T is considered as an operator of $(C^{m+2,\lambda}(\bar{\Omega}))^n$ into $(C^{m,\lambda}(\bar{\Omega}))^n$ is quite analogous. If N is any integer $\geqq 1$ we denote by $(\ ,\ )_0$ the scalar product on $(L^2(\Omega))^N$ defined by $(f,g)=\int_\Omega f_i(x)g_i(x)dx$, for $f=(f_i)_{i=1,\cdots,N}$ and $g=(g_i)_{i=1,\cdots,N}$. For each $u\in(W^{m+2,p}(\Omega))^n$ let Au and $A_{ij,hk}$ be the functions defined by (3.1). From Theorem 5.1 it follows that $u\mapsto Au$ is a continuously differentiable operator of $(W^{m+2,p}(\Omega))^n$ into $(W^{m+1,p}(\Omega))^{n^2}$ and that

$$dA(u)v = (A_{ij,hk}uD_kv_h)_{i,j=1,\cdots,n}, \qquad \text{for } u, v \in (W^{m+2,p}(\Omega))^n.$$

Therefore (7.1) becomes

$$\int_\Omega (dA(u)v, Dv)_0 dx \geqq c(u)\|v\|_1^2 \qquad \forall v \in (W_0^1(\Omega))^n. \tag{8.3}$$

Let now $\mathscr{C}$ be a convex subset of $\mathscr{S}_{m+2,p}$ and let $u, v \in \mathscr{C}$ be such that $u \neq v$. Note that $Au - Av = \int_0^1 dA(v + t(u-v))(u-v)\,dt$ and that, for all $t \in [0,1]$, (8.3) holds with $v + t(u-v)$ instead of u because $v + t(u-v) \in \mathscr{C}$ $\forall t \in [0,1]$ and $\mathscr{C} \subseteq \mathscr{S}_{m+2,p}$. Then, since it is easy to see that

$$(Tu, v)_0 = (Au, Dv)_0 \qquad \forall u, v \in \mathscr{A}_{m+2,p} \text{ with } p(m+1) > n,$$

we get

$$\begin{aligned}(Tu - Tv, v - v)_0 &= (Au - Av, Du - Dv)_0 \\ &= \int_0^1 (dA(v + t(u-v))(u-v), Du - Dv)_0 dt \\ &\geqq \int_0^1 c(v + t(u-v))(Du - Dv, Du - Dv)_0 dt > 0,\end{aligned}$$

which implies $Tu \neq Tv$. Thus T is one-to-one on $\mathscr{C}$ and the proof of the theorem is complete. □

THEOREM 8.4. *Assume Ω is connected and of class C^{m+2}, $a_{ij} \in C^{m+2}(\bar{\Omega} \times \mathbb{R}^{n^2})$ and $p > n$. If $\mathscr{C}$ is a convex, open subset of the subspace $(W_0^{1,p}(\Omega) \cap W^{m+2,p}(\Omega))^n$ of $(W^{m+2,p}(\Omega))^n$ contained in $\mathscr{S}_{m+2,p}$,[13] then the operator $u \mapsto Tu$, with Tu defined by (3.3), is a C^1-diffeomorphism of $\mathscr{C}$ onto an open subset of $(W^{m,p}(\Omega))^n$.*

PROOF. We know that $T: \mathscr{C} \to (W^{m,p}(\Omega))^n$ is one-to-one (see Theorem 8.3) and continuously differentiable (see Corollary 5.1). Moreover, for each $u \in \mathscr{C}$, $dT(u)$ is an isomorphism of $(W_0^{1,p}(\Omega) \cap W^{m+2,p}(\Omega))^n$ onto $(W^{m,p}(\Omega))^n$ (see Theorem 7.1). Then (by the Inverse Function Theorem) $T: \mathscr{C} \to (W^{m,p}(\Omega))^n$ is an open operator, and therefore T is a homeomorphism of $\mathscr{C}$ onto an open subset of $(W^{m,p}(\Omega))^n$. Consequently[14] we can conclude that T is a C^1-diffeomorphism of $\mathscr{C}$ onto an open subset of $(W^{m,p}(\Omega))^n$. □

[13] Note that, since (by Remark 8.1) $\mathscr{S}_{m+2,p}$ is open in $(W_0^{1,p}(\Omega) \cap W^{m+2,p}(\Omega))^n$, each point $u^0 \in \mathscr{S}_{m+2,p}$ has a convex, open neighborhood in $(W_0^{1,p}(\Omega) \cap W^{m+2,p}(\Omega))^n$ contained in $\mathscr{S}_{m+2,p}$.

[14] We recall that, if U is an open subset of a Banach space X and V is an open subset of a Banach space Y, then a continuously differentiable homeomorphism f of U onto V is a C^1-diffeomorphism of U onto V if and only if, for all $x \in U$, $df(x)$ is an isomorphism of X onto Y (see [4], p. 57).

A quite analogous argument shows that from Theorems 8.3, 7.2 and from Corollary 5.2 one can obtain the following theorem.

THEOREM 8.5. *Assume* Ω *is connected and of class* C^{m+3}, $a_{ij} \in C^{m+3}(\bar{\Omega} \times \mathbb{R}^{n^2})$, and $0 < \lambda < 1$. *If* $\mathscr{C}$ *is a convex, open subset of the subspace* $(C_0^0(\bar{\Omega}) \cap C^{m+2,\lambda}(\bar{\Omega}))^n$ *of* $(C^{m+2,\lambda}(\bar{\Omega}))^n$ *contained in* $\mathscr{S}_{m+2,\lambda}$,[15] *then the operator* $u \mapsto Tu$, *with* Tu *defined by* (3.3), *is a* C^1-*diffeomorphism of* $\mathscr{C}$ *onto an open subset of* $(C^{m,\lambda}(\bar{\Omega}))^n$.

9. On differentiability of an operator connected with the problem ((P), (D))

We now deal with the problem ((P), (D)) of Section 3.

If the functions $a : \Omega \times \mathbb{R}^{n^2} \to \mathbb{R}^{n^2}$ and $u : \bar{\Omega} \to \mathbb{R}^n$ are smooth enough, and the $\mathbb{R}^n$-valued function b is defined on some subset of $\mathbb{R}^n$ containing $(I+u)(\bar{\Omega})$, we set

$$\Phi(b, u) = \operatorname{div} Au + \rho Bu, \tag{9.1}$$

where $Au : \Omega \to \mathbb{R}^{n^2}$ and $Bu : \Omega \to \mathbb{R}^n$ are the functions defined by (3.1) and (3.2). In order to make use of the Implicit Function Theorem for the equation $\Phi(b, u) = 0$, we shall prove the following differentiability theorems for Φ.

THEOREM 9.1. *Assume* Ω *is connected and of class* C^1, $a_{ij} \in C^{m+2}(\bar{\Omega} \times \mathbb{R}^{n^2})$, $\rho \in W^{m+1,p}(\Omega)$, *and* $p(m+1) > n$. *Let* G *be a convex, bounded, open subset of* $\mathbb{R}^n$ *containing* Ω. *Then* $(b, u) \mapsto \Phi(b, u)$ *is a continuously differentiable operator of the open*[16] *subset* $(C^{m+1}(\bar{G}))^n \times \mathscr{A}_{m+2,p}$ *of the Banach space* $(C^{m+1}(\bar{G}))^n \times (W_0^{1,p}(\Omega) \cap W^{m+2,p}(\Omega))^n$ *into* $(W^{m,p}(\Omega))^n$, *and its differential at any* $(b^0, u^0) \in (C^{m+1}(\bar{G})) \times \mathscr{A}_{m+2,p}$ *is the (continuous, linear) operator* $(b, u) \mapsto d\Phi(b^0, u^0)(b, u)$ *of* $(C^{m+1}(\bar{G}))^n \times (W_0^{1,p}(\Omega) \cap W^{m+2,p}(\Omega))^n$ *into* $(W^{m,p}(\Omega))^n$, *where*[17]

$$\begin{aligned} &d\Phi(b^0, u^0)(b, u) \\ &\quad = (D_j(A_{ij,hk}u^0 D_k u_h) + \rho u_j D_j b_i^0 \circ (I+u^0) + \rho b_i \circ (I+u^0))_{i=1,\cdots,n}. \end{aligned} \tag{9.2}$$

PROOF. Recall that $(I+u)(\bar{\Omega}) = \bar{\Omega}\ \forall u \in \mathscr{A}_{m+2,p}$. Using Corollary 5.1 and Sobolev Imbeddings one can verify (by easy computations) that $(b, u) \mapsto \Phi(b, u)$ acts from $(C^{m+1}(\bar{G}))^n \times \mathscr{A}_{m+2,p}$ to $(W^{m,p}(\Omega))^n$ and that $(b, u) \mapsto d\Phi(b^0, u^0)(b, u)$, where $d\Phi(b^0, u^0)(b, u)$ is defined by (9.2), acts from $(C^{m+1}(\bar{G}))^n \times (W_0^{1,p}(\Omega) \cap W^{m+2,p}(\Omega))^n$ to $(W^{m,p}(\Omega))^n$ and is separately con-

[15] A remark analogous to the one made in Footnote 13 holds here too.

[16] Recall that, by Theorem 6.1, $\mathscr{A}_{m+2,p}$ is an open subset of $(W_0^{1,p}(\Omega) \cap W^{m+2,p}(\Omega))^n$.

[17] As usual, $b_i^0 \circ (I+u^0)$ denotes the composition of b_i^0 and $(I+u^0)$, i.e. the function of $\bar{\Omega}$ into $\mathbb{R}$ defined by $(b_i \circ (I+u^0))(x) = b_i(x + u^0(x))$.

tinuous (on b and on u); therefore this operator is continuous by a well-known result of Bourbaki (see [3], §3, Section 6, Theorem 3).

Note that (by Corollary 5.1) to prove our theorem it suffices to show that the operator $\Psi : (C^{m+1}(\bar{G}))^n \times \mathcal{A}_{m+2,p} \to (W^{m,p}(\Omega))^n$ defined by

$$\Psi(b, u) = \rho(b \circ (I + u))$$

is continuously differentiable and that its differential at any point (b^0, u^0) is the (continuous, linear) operator $d\Psi(b^0, u^0)$, of $(C^{m+1}(\bar{G}))^n \times (W_0^{1,p}(\Omega) \cap W^{m+2,p}(\Omega))^n$ into $(W^{m,p}(\Omega))^n$, defined by

$$d\Psi(b^0, u^0)(b, u) = \rho(u_j D_j b^0 \circ (I + u^0) + b \circ (I + u^0)).$$

It is easy to see that Ψ is Gâteaux differentiable[18] and that its Gâteaux differential at (b^0, u^0), $d_G\Psi(b^0, u^0)$, is exactly the operator $(b, u) \mapsto \rho(u_j D_j b^0 \circ (I + u^0) + b \circ (I + u^0))$. Therefore we must only prove that $(b^0, u^0) \mapsto d_G\Psi(b^0, u^0)$ is a continuous operator from $(C^{m+1}(\bar{G}))^n \times \mathcal{A}_{m+2,p}$ to the space of continuous, linear operators of $(C^{m+1}(\bar{G}))^n \times (W_0^{1,p}(\Omega) \cap W^{m+2,p}(\Omega))^n$ into $(W^{m,p}(\Omega))^n$, equipped with the topology of bounded convergence.[19]

It is not difficult to see that $(b^0, u^0) \mapsto d_G\Psi(b^0, u^0)$ is continuous if the following two facts are true.

(i) The operators $(b, u) \mapsto D_j b \circ (I + u)$, $(j = 1, \cdots, n)$, are continuous from $(C^{m+1}(\bar{G}))^n \times \mathcal{A}_{m+2,p}$ to $(W^{m,p}(\Omega))^n$.

(ii) For any $u^0 \in \mathcal{A}_{m+2,p}$ there exists a positive number $c(u^0)$ such that

$$\|b \circ (I + u) - b \circ (I + u^0)\|_{m,p} \leqq c(u^0)|b|_{m+1}\|u - u^0\|_{m+2,p}$$

$$\forall (b, u) \in (C^{m+1}(\bar{G}))^n \times \mathcal{A}_{m+2,p},$$

where

$$|b|_{m+1} = \sum_{|\alpha| \leqq m+1} \sup_{y \in \bar{G}} |D^\alpha b(y)|.$$

Therefore it remains only to be shown that, by our assumptions, (i) and (ii) hold. To justify (i) first observe that, for $u \in \mathcal{A}_{m+2,p}$ the linear operators $b \mapsto D_j b \circ (I + u)$, $(j = 1, \cdots, n)$, are continuous from $(C^{m+1}(\bar{G}))^n$ to $(W^{m,p}(\Omega))^n$, since it is not difficult to recognize that for each $u \in \mathcal{A}_{m+2,p}$ there exists a positive number $c_1(u)$ such that

$$\|D_j b \circ (I + u)\|_{m,p} \leqq c_1(u)|b|_{m+1} \qquad \forall b \in (C^{m+1}(\bar{G}))^n.$$

[18] For the definition of Gâteaux differentiability see, for example, Schwartz [12], p. 11.

[19] In fact the following result holds (see, e.g., Schwartz [12], Lemma 1.15). Let X, Y be Banach spaces, let U be an open subset of X. If $f : U \mapsto Y$ is Gâteaux differentiable and the Gâteaux differential, $d_G f$, of f is a continuous operator of U into the space of continuous, linear operators of X into Y equipped with the bounded convergence topology, then f is differentiable and $df = d_G f$.

Moreover easy arguments prove that, for $b \in (C^{m+1}(\bar{G}))^n$, the operators $u \mapsto D_j b \circ (I+u)$, $(j = 1, \cdots, n)$, are continuous from $\mathcal{A}_{m+2,p}$ to $(W^{m,p}(\Omega))^n$. Then, by the above-mentioned result of Bourbaki, (i) holds.

Finally (ii) can be proved by iterated application of the inequality $|b(x+u(x)) - b(x+u^0(x))| \leqq |b|_1 |u(x) - u^0(x)|$ $\forall x \in \bar{\Omega}$, which holds (by Taylor expansion of b of order 1) for any $b \in (C^1(\bar{G}))^n$ and any u, $u^0 \in (C^0(\bar{\Omega}))^n$. This completes the proof of Theorem 9.1. □

THEOREM 9.2. *Assume* Ω *is connected and of class* C^1, $a_{ij} \in C^{m+3}(\bar{\Omega} \times \mathbb{R}^{n^2})$, *and* $\rho \in C^{m,\lambda}(\bar{\Omega})$. *Let* G *be a convex, bounded, open subset of* $\mathbb{R}^n$ *containing* Ω. *Then* $(b,u) \mapsto \Phi(b,u)$ *is a continuously differentiable operator of the open*[20] *subset* $(C^{m,\lambda}(\bar{G}))^n \times \mathcal{A}_{m+2,\lambda}$ *of* $(C^{m+2}(\bar{G}))^n \times (C_0^0(\bar{\Omega}) \cap C^{m+2,\lambda}(\bar{\Omega}))^n$ *into* $(C^{m,\lambda}(\bar{\Omega}))^n$, *and its differential at any* $(b^0, u^0) \in (C^{m+2}(\bar{G}))^n \times \mathcal{A}_{m+2,\lambda}$ *is the* (*continuous, linear*) *operator* $(b,u) \mapsto d\Phi(b^0,u^0)(b,u)$ *of* $(C^{m+2}(\bar{G}))^n \times (C_0^0(\bar{\Omega}) \cap C^{m+2,\lambda}(\bar{\Omega}))^n$ *into* $(C^{m,\lambda}(\bar{\Omega}))^n$, *where* $d\Phi(b^0,u^0)(b,u)$ *is defined by* (9.1).

PROOF. One can proceed essentially as in the proof of Theorem 9.1, by keeping in mind Theorem 5.2, Corollary 5.2, and that (by the assumptions on Ω) $C^{m+1}(\bar{\Omega})$ is continuously imbedded in $C^{m,\lambda}(\bar{\Omega})$ (see Valent [23]) and $C^{m,\lambda}(\bar{\Omega})$ is a Banach algebra (see Section 2). □

10. Local theorems of existence and uniqueness for the problem ((P), (D))

By Theorems 6.1 and 9.1 (resp. 9.2), a straightforward application of the Implicit Function Theorem to the equation $\Phi(b,u) = 0$ (where $\Phi(u,v)$ is defined by (9.1)), gives the following Theorem 10.1 (resp. 10.2).

THEOREM 10.1. *Assume* Ω *is connected and of class* C^1, $a_{ij} \in C^{m+2}(\bar{\Omega} \times \mathbb{R}^{n^2})$, $\rho \in W^{m+1,p}(\Omega)$, *and* $p(m+1) > n$. *Let* G *be a convex, bounded, open subset of* $\mathbb{R}^n$ *containing* Ω, *and let* $(b^0, u^0) \in (C^{m+1}(\bar{G}))^n \times \mathcal{A}_{m+2,p}$. *If* $\Phi(b^0,u^0) = 0$ *and if the linear system*

$$(10.1) \qquad D_j(A_{ij,hk} u^0 D_k u_h) + \rho u_j D_j b_i^0 \circ (I + u^0) = g_i, \qquad (i = 1, \cdots, n)$$

has one and only one solution $u \in (W_0^{1,p}(\Omega) \cap W^{m+2,p}(\Omega))^n$ *for any* $g = (g_i)_{i=1,\cdots,n} \in (W^{m,p}(\Omega))^n$, *then there exist positive numbers* ξ, η *such that for all* $b \in (C^{m+1}(\bar{G}))^n$ *with* $|b - b^0|_{m+1} < \xi$ *the problem* ((P), (D)) *has one and only one solution* $u(b)$ *in* $(W^{m+2,p}(\Omega))^n$ *satisfying the condition* $\|u(b) - u^0\|_{m+2,p} < \eta$. *Moreover* (using the Landau symbol $\mathcal{O}$)

$$(10.2) \qquad u(b) = u^0 + u_L(b^0, u^0)(b - b^0) + \mathcal{O}(|b - b^0|_{m+1}),$$

[20] Recall that, by Theorem 6.1, $\mathcal{A}_{m+2,\lambda}$ is an open subset of $(C_0^0(\bar{\Omega}) \cap C^{m+2,\lambda}(\bar{\Omega}))^n$.

where $u_L(b^0, u^0)(b - b^0)$ *denotes the solution in* $(W_0^{1,p}(\Omega) \cap W^{m+2,p}(\Omega))^n$ *of* (10.1) *when* $g = \rho(b \circ (I + u^0) - b^0 \circ (I + u^0))$.

THEOREM 10.2. *Assume* Ω *connected and of class* C^1, $a_{ij} \in C^{m+3}(\bar{\Omega} \times \mathbb{R}^{n^2})$, *and* $\rho \in C^{m,\lambda}(\bar{\Omega})$. *Let* G *be a convex, bounded, open subset of* $\mathbb{R}^n$ *containing* Ω, *and let* $(b^0, u^0) \in (C^{m+2}(\bar{G}))^n \times \mathcal{A}_{m+2,\lambda}$. *If* $\Phi(b^0, u^0) = 0$ *and if* (10.1) *has one and only one solution* u *in* $(C_0^0(\bar{\Omega}) \cap C^{m+2,\lambda}(\bar{\Omega}))^n$ *for any* $g = (g_i)_{i=1,\cdots,n} \in (C^{m,\lambda}(\bar{\Omega}))^n$, *then there exist positive numbers* ξ, η *such that for all* $b \in (C^{m+2}(\bar{G}))^n$ *with* $|b - b^0|_{m+2} < \xi$ *the problem* ((P), (D)) *has one and only one solution* $u(b)$ *in* $(C^{m+2,\lambda}(\bar{\Omega}))^n$ *satisfying the condition* $\|u(b) - u^0\|_{m+2,\lambda} < \eta$, *and* (10.2) *holds, where* $u_L(b^0, u^0)(b - b^0)$ *denotes the solution in* $(C_0^0(\bar{\Omega}) \cap C^{m+2,\lambda}(\bar{\Omega}))^n$ *of* (10.1) *when* $g = \rho(b \circ (I + u^0) - b^0 \circ (I + u^0))$.

By general results about strongly elliptic operators and by remarks like the ones made in Section 7, we can get the following two remarks.

REMARK 10.1. Let Ω be of class C^{m+2} and let $p > n$. Then, under the assumptions of Theorem 10.1, the linear system (10.1) has one and only one solution u in $(W_0^{1,p}(\Omega) \cap W^{m+2,p}(\Omega))^n$ for any $g = (g_i)_{i=1,\cdots,n} \in (W^{m,p}(\Omega))^n$ *if*

$$\int_\Omega \frac{\partial a_{ij}}{\partial z_{hk}}(x, Du^0(x))D_j u_i(x) D_k u_h(x)dx$$
$$(10.3) \qquad + \int_\Omega \rho(x) \frac{\partial b_i^0}{\partial y_j}(x + u^0(x))u_i(x)u_j(x)dx$$
$$\geqq c\|u\|_1^2 \qquad \forall u \in (W_0^1(\Omega))^n, \text{ where } c > 0 \text{ is a constant.}$$

REMARK 10.2. Let Ω be of class $C^{m+2,\lambda}$, and let $0 < \lambda < 1$. Then, under the assumptions of Theorem 10.2, the linear system (10.1) has one and only one solution u in $(C_0^0(\bar{\Omega}) \cap C^{m+2,\lambda}(\bar{\Omega}))^n$ for any $g = (g_i)_{i=1,\cdots,n} \in (C^{m,\lambda}(\bar{\Omega}))^n$ if (10.3) holds.

Note that (see Section 4) the inequality (10.3) holds with $u^0 = b^0 = 0$ if the function a satisfies the condition $a(x, 0) = 0$ $\forall x \in \Omega$ and the Conditions (4.1), and (4.2).

REFERENCES

1. A. Adams, *Sobolev Spaces*, Academic Press, New York, 1975.

2. S. Agmon, A. Douglis and L. Nirenberg, *Estimates near the boundary for solutions of elliptic partial differential equations satisfying general boundary conditions II*, Comm. Pure and Appl. Mathematics **17** (1964), 35–92.

3. N. Bourbaki, *Eléments de Mathématique*, Fasc. XVIII, Livre V, Hermann, Paris, 1967.

4. H. Cartan, *Calcul différentiel*, Hermann, Paris, 1967.

5. G. Fichera, *Existence theorems in elasticity*, in *Handbuch der Physik*, Vol. IVa/2, Springer, Berlin–Heidelberg–New York, 1972, pp. 347–389.

6. G. Fichera, *Sulla propagazione delle onde in un mezzo elastico*, in *Continuum Mechanics and Related Problems of Analysis*, Moscow, 1972, pp. 567–574.

7. G. Grioli, *Mathematical Theory of Elastic Equilibrium*, Ergebnisse der Ang. Math. No. 7, Springer, Berlin–Göttingen–Heidelberg, 1962.

8. M. E. Gurtin and S. J. Spector, *On stability and uniqueness in finite elasticity*, Arch. Rational Mech. Anal. **70** (1979), 153–165.

9. M. A. Krasnoselskij et al., *Integral Operators in Spaces of Summable Functions*, Noordhoff Int. Publ., Leyden, 1976.

10. J. E. Marsden and J. R. Hughes, *Topics in the mathematical foundations of elasticity*, in *Nonlinear Analysis and Mechanics: Heriot-Watt Symposium* (R. E. Knops, ed.), Vol. II, Pitman, San Francisco, 1978.

11. H. Meisters and C. Olech, *Locally one-to-one mappings and a classical theorem on schlicht functions*, Duke Math. J. **30** (1963), 63–80.

12. J. Necas, *Les Méthodes Directes en Théorie des Equations Elliptiques*, Masson, Paris, 1967.

13. J. T. Schwartz, *Nonlinear Functional Analysis*, Gordon and Breach Science Publ., New York, 1969.

14. C. G. Simader, *On Dirichlet's Boundary Value Problems*, Lecture Notes in Mathematics, Springer, Berlin–Heidelberg–New York, 1972.

15. F. Stoppelli, *Un teorema di esistenza e di unicità relativo alle equazioni dell'elastostatica isoterma per deformazioni finite*, Ricerche Matematiche **3** (1954), 247–267.

16. F. Stoppelli, *Sulla sviluppabilità in serie di potenze di un parametro delle soluzioni delle equazioni dell'elastostatica isotrerma*, Ricerche Mathematiche **4** (1955), 58–73.

17. F. Stoppelli, *Su un sistema di equazioni integrodifferenziali interessante l'elastostatica*, Ricerche Matematiche **6** (1957), 11–26.

18. F. Stoppelli, *Sull'esistenza di soluzioni delle equazioni dell'elastostatica isoterma nel caso di sollecitazioni dotate di assi di equilibrio*, I, II, III, Ricerche Matematiche **6** (1957), 241–287; **7** (1958), 71–101, 138–152.

19. C. Truesdell and W. Noll, *The Non-Linear Field Theories of Mechanics*, in *Handbuch der Physik*, Vol. III/3, Springer, Berlin–Heidelberg–New York, 1965.

20. T. Valent and G. Zampieri, *Sulla differenziabilità di un operatore legato a una classe di sistemi differenziali quasi-lineari*, Rend. Sem. Mat. Univ. Padova **57** (1977), 311–322.

21. T. Valent, *Sulla differenziabilità dell'operatore di Nemytsky*, Rend. Acc. Naz. Lincei **65** (luglio-agosto 1978), 15–26.

22 T. Valent, *Osservazioni sulla linearizzazione di un operatore differenziale*, Rend. Acc. Naz. Lincei **65** (luglio-agosto 1978), 27–37.

23. T. Valent, *Teoremi di esistenza e unicità in elastostatica finita*, Rend. Sem. Mat. Univ. Padova **60** (1979), 165–181.

24. W. Van Buren, *On the existence and uniqueness of solutions to boundary value problems in finite elasticity*, Thesis, Carnegie–Mellon Univ., 1968. Research Report 68-ID7-MEKMA-RI, Westinghouse Research Laboratories, Pittsburgh, Pa, 1968.

25. C. C. Wang and C. Truesdell, *Introduction to Rational Elasticity*, Noordhoff, Groningen, 1973.

NONLINEAR WAVES IN RODS

T. W. WRIGHT

US Army Ballistic Research Laboratory, Aberdeen Proving Ground, MD 21005, USA

(Received October 22, 1980)

ABSTRACT

Rods are simple structures that are often regarded as one-dimensional continua, but because of finite transverse dimensions, propagating waves are subject to dispersion, which may mask other effects. A one-dimensional continuum theory with one internal, scalar variable can be used to model a solid rod with longitudinal waves. In this paper the material is assumed to be homogeneous, isotropic, and hyperelastic. First, the linear theory is reviewed to exhibit clearly the multiple wave hierarchy that exists in a rod. Next, expansion and scaling techniques are used in the fully nonlinear case to examine the main pulse. Finally, steady nonlinear waves are examined and both solitary waves and periodic waves are found to exist. Results are compared to the simple wave solutions available from the most elementary one-dimensional nonlinear theory for rods.

Introduction

The simplest theory for nonlinear wave propagation in a straight, uniform rod may be written as follows:

$$S_Z = \rho v_t, \qquad v_Z = \varepsilon_t. \tag{1}$$

The axial stress, S, depends only on the axial strain, ε, the axial particle velocity is v, and the reference density, ρ, is a constant. The axial coordinate along the rod is Z, t is time, and subscripts denote partial differentiation. In this version of rod theory, which is formally the same as one-dimensional, inviscid gas dynamics, the characteristic speed of propagation, c, is given by $\rho c^2 = dS/d\varepsilon$, and discontinuities in stress, strain, or velocity all travel at the shock speed given by $\rho c^2 = [S]/[\varepsilon]$, where $[\cdot]$ denotes the jump in a quantity across the shock.

Proceedings of the IUTAM Symposium on Finite Elasticity, Lehigh University, August 10–15, 1980. Invited paper.

Carlson, D.E. and Shield, R.T. (eds.)
Proceedings of the IUTAM Symposium on Finite Elasticity
ISBN 90 247 2629 8

Equation (1) has been used to interpret wave propagation experiments in which the strain–time profile is measured at several stations in the rod, [1, 2]. If each level of strain propagates at a constant velocity, $c_p(\varepsilon)$, then both stress and particle velocity may be computed as a function of strain as follows.

$$(2)\qquad \begin{aligned} v &= v_0 - \int_{\varepsilon_0}^{\varepsilon(Z,t)} c_p d\varepsilon, \\ S &= S_0 + \int_{\varepsilon_0}^{\varepsilon(Z,t)} \rho c_p^2 d\varepsilon. \end{aligned}$$

The initial values ahead of the wave are v_0, S_0 and ε_0. Equation (2) is a simple wave solution to (1), exhibiting amplitude dispersion, but not geometric dispersion.

A rod, no matter how slender, is actually a three dimensional object, of course. The dispersive influence of a finite diameter is well known in linear theories, e.g. [3, 4]. It is less well known in nonlinear theories [5, 6] where it may be expected to play a role as well, whenever the length scale in a wave pulse is comparable in magnitude to the rod diameter. In this paper the effects of finite lateral dimensions will be considered by modeling a rod as a one dimensional elastic structure with one internal variable used to represent the transverse, axisymmetric motion. Such a structure is a special case of an intrinsic rod theory as described by Antman [7].

One-dimensional equations and relationship to three-dimensional elasticity

A straight cylindrical rod of radius a is assumed to have an elastic stored energy density per unit length

$$(3)\qquad \hat{W} = \pi a^2 W(w', u, u').$$

The axial displacement is w, u is a measure of radial strain, and the dash represents differentiation with respect to the axial coordinate, Z. The kinetic energy density per unit length is given by

$$(4)\qquad K = \pi a^2(\tfrac{1}{2}\rho_1 \dot{w}^2 + \tfrac{1}{2}\rho_2 \dot{u}^2),$$

where ρ_1 and ρ_2 are the appropriate mass densities for the radial and axial motions respectively, and the dot denotes differentiation with respect to time. The Euler–Lagrange equations corresponding to (3) and (4) are given by (5):

$$(5)\qquad \begin{aligned} S' &= \rho_1 \ddot{w}, \\ Q' - P &= \rho_2 \ddot{u}. \end{aligned}$$

The forces S, P, and Q are obtained from the stored energy as follows:

(6) $$S = \frac{\partial W}{\partial w'}, \qquad P = \frac{\partial W}{\partial u}, \qquad Q = \frac{\partial W}{\partial u'}.$$

Equations (5) and (6) may be interpreted in terms of the three-dimensional theory of nonlinear elasticity. For the axisymmetric motions considered here, the deformation from an unstressed reference configuration to the present configuration may be written

(7) $$r = r(R, Z, t), \qquad \theta = \Theta, \qquad z = z(R, Z, t),$$

and the corresponding equations of motion are given by

(8) $$\frac{\partial T^{rR}}{\partial R} + \frac{\partial T^{rZ}}{\partial Z} + \frac{T^{rR} - T^{\theta\Theta}}{R} = \rho \ddot{r},$$
$$\frac{\partial T^{zR}}{\partial R} + \frac{\partial T^{zZ}}{\partial Z} + \frac{T^{zR}}{R} = \rho \ddot{z}.$$

The stress components used here are obtained from the unsymmetric Piola-Kirchhoff tensor referred to unit basis vectors in cylindrical coordinates (r, θ, z) and (R, Θ, Z),

(9) $$\boldsymbol{T} = T^{i\alpha} \boldsymbol{e}_i \otimes \boldsymbol{E}_\alpha .$$

If $(8)_1$ is multiplied by R, and if the resulting equation and $(8)_2$ are both averaged over the cross-section, equations (10) are obtained for the case of stress free lateral surfaces:

(10) $$\frac{1}{\pi a^2} \frac{\partial}{\partial Z} \int T^{zZ} dA = \frac{\rho}{\pi a^2} \int \ddot{z} dA,$$
$$\frac{1}{\pi a^2} \frac{\partial}{\partial Z} \int R T^{rZ} dA - \frac{1}{\pi a^2} \int (T^{rR} + T^{\theta\Theta}) dA = \frac{\rho}{\pi a^2} \int R \ddot{r} dA.$$

Now if the following identifications are made:

(11) $$S \leftrightarrow \frac{1}{\pi a^2} \int T^{zZ} dA,$$
$$P \leftrightarrow \frac{1}{\pi a^2} \int (T^{rR} + T^{\theta\Theta}) dA,$$
$$Q \leftrightarrow \frac{1}{\pi a^2} \int R T^{rZ} dA,$$

and if a first approximation for the position functions r and z is assumed to be

(12) $$r = R[1 + u(Z, t)],$$
$$z = Z + w(Z, t),$$

then equations (5) are obtained with

$$\rho_1 = \rho, \qquad \rho_2 = \tfrac{1}{2}\rho a^2. \tag{13}$$

In all of the preceding discussion it has been tacitly assumed that the Z-axis is an axis of material symmetry so that no angular motion will occur anywhere in the cross section. Furthermore, since the strain energy must be invariant under reversal of the z-axis, and if it is assumed to be invariant under reversal of the Z-axis as well, W must have the property

$$W(w', u, u') = W(w', u, -u'). \tag{14}$$

It follows that Q is odd in u' and S and P are even in u'.

It should be remarked that equations (5) and (6) have been used to describe one dimensional waves in porous materials [8–10], and it has also been suggested that perhaps they could be used to describe waves in a layered composite [11]. There are undoubtedly other applications as well.

Static solutions

It is convenient to distinguish two types of static solutions. In the first type u' is identically zero, so equations (5) reduce to

$$\begin{aligned} S(w', u, 0) &= S_0, \\ P(w', u, 0) &= 0. \end{aligned} \tag{15}$$

Since the reference configuration has been assumed to be unstressed, it is required that when $w' = u = 0$, (15) is satisfied with $S_0 = 0$. It will be further assumed that the slope along the curve defined by $(15)_2$ is always negative and bounded:

$$\infty > -\frac{du}{dw'} = \frac{P_{w'}}{P_u} = \nu(w', u, 0) > 0. \tag{16}$$

This corresponds to the physically reasonable assumption that the rod contracts (expands) laterally as it is stretched (compressed). Finally it will be assumed that S increases monotonically as w' increases along the curve $P(w', u, 0) = 0$. That is

$$S_{w'} - \nu S_u = E(w', u, 0) > 0. \tag{17}$$

The bar modulus, E, and Poisson's ratio, ν, will be considered as defined by (17) and (16) with the inequalities holding as well for all values of their arguments, not just along the curve of $(15)_2$. Since W has a minimum at the unstressed reference configuration ($w' = u = u' = 0$), $P_u > 0$ there, so by (16)

$P_{w'} > 0$ and $P_u > 0$ everywhere. This also implies from (17) that $W_{w'w'}W_{uu} - W^2_{uw'} > 0$ so that necking instabilities have been ruled out (see Antman [12], p. 97).

In the second type of static solution u' is not identically zero. There are two integrals of (5):

$$S(w', u, u') = S_0,$$

$$u'Q(w', u, u') + w'S_0 - W(w', u, u') = B. \tag{18}$$

If these may be solved for u' and w' as functions of u and the constants S_0 and B, then u may be found by quadrature, and the axial strain w' is given parametrically through u. The study of equations (18) will not be pursued further here except to remark that the version arising from linear elasticity gives boundary layer solutions with Q decaying exponentially from the ends. Phase plane analysis indicates that the behavior of the fully nonlinear equations is similar.

Solutions of the linearized equations

Before considering the dynamic nonlinear equations, it will be useful to examine some of the properties of solutions of the linearized equations. The three-dimensional strain energy for an isotropic material with displacement field given by (12), when averaged over the cross-section, leads to (19) for the forces:

$$S = (\lambda + 2\mu)w' + 2\lambda u,$$

$$P = 4(\lambda + \mu)u + 2\lambda w', \tag{19}$$

$$Q = \tfrac{1}{2}a^2\mu u',$$

where λ and μ are the usual Lamé constants. With nondimensional variables $\zeta = Z/a$ and $\tau = ct/a$, where c is a speed that is unspecified for now, and with $\bar{w} = w/a$, equations (5) become

$$w_{\zeta\zeta} + 2\frac{\lambda}{\lambda + 2\mu}u_\zeta = \frac{c^2}{c_1^2}w_{\tau\tau},$$

$$u_{\zeta\zeta} - \left[8\frac{\lambda + \mu}{\mu}u + 4\frac{\lambda}{\mu}w_\zeta\right] = \frac{c^2}{c_2^2}u_{\tau\tau}. \tag{20}$$

Subscripts indicate partial differentiation, and the overbar has been dropped from $\bar{w}$. The longitudinal and shear wave speeds are $c_1 = \sqrt{(\lambda + 2\mu)/\rho}$ and $c_2 = \sqrt{\mu/\rho}$. These equations in dimensional form were first given by Mindlin and Herrmann [13].

Since all the coefficients are constants, either w or u may be eliminated from (20) to obtain (21):

$$\left[\left(\frac{\partial^2}{\partial\zeta^2}-\frac{c^2}{c_1^2}\frac{\partial^2}{\partial\tau^2}\right)\left(\frac{\partial^2}{\partial\zeta^2}-\frac{c^2}{c_2^2}\frac{\partial^2}{\partial\tau^2}\right) - 8\frac{\lambda+\mu}{\mu}\frac{c_b^2}{c_1^2}\left(\frac{\partial^2}{\partial\zeta^2}-\frac{c^2}{c_b^2}\frac{\partial^2}{\partial\tau^2}\right)\right](w,u)=0. \tag{21}$$

The one-dimensional bar speed is $c_b=\sqrt{E/\rho}$ where E is Young's modulus, $E=\mu(3\lambda+2\mu)/(\lambda+\mu)$. Equation (21) exhibits what Whitham has called a hierarchy of wave speeds [14, 15]. Whereas discontinuities can only propagate with the speeds c_1/c or c_2/c, it is well known that in a thin bar the main disturbance in a pulse travels with speed c_b/c. Whitham has discussed extensively the hierarchical case where the orders of the highest and lowest derivatives differ by one. Wu [16] has pointed out that the case where the orders differ by two, as here, also exhibits a stable hierarchical wave structure, so long as $c_2<c_b<c_1$, which is always the case for linear elastic rods.

In Whitham's case the higher order waves decay exponentially, and the lower order wave, which carries the main pulse, has a diffusing front. In the present case, discontinuities in the higher order waves do not decay with time, but the pulse decays rapidly behind the front. The main pulse travels with the lower order speed, and it too has a diffusing front.

To give substance to the preceding remarks, consider a boundary, initial value problem for (20):

$$\begin{aligned} &\text{B.V.: } u_\zeta(0,\tau)=0, \qquad w_\tau(0,\tau)=f(\tau),\\ &\text{I.V.: } u(\zeta,0)=w(\zeta,0)=0. \end{aligned} \tag{22}$$

A straightforward application of the Laplace transform on τ leads to the following representations for u and w:

$$\begin{aligned} w &= \frac{1}{2\pi i}\int_{Br} C(p)e^{p\tau+m_1\zeta}dp+\frac{1}{2\pi i}\int_{Br} D(p)e^{p\tau+m_2\zeta}dp,\\ -\frac{2\lambda}{\lambda+2\mu}\,u &= \frac{1}{2\pi i}\int_{Br}\frac{C(p)}{m_1}\left(m_1^2-p^2\frac{c^2}{c_1^2}\right)e^{p\tau+m_1\zeta}dp\\ &\quad+\frac{1}{2\pi i}\int_{Br}\frac{D(p)}{m_2}\left(m_2^2-p^2\frac{c^2}{c_1^2}\right)e^{p\tau+m_2\zeta}dp. \end{aligned} \tag{23}$$

$C(p)$ and $D(p)$ are given by (24):

$$C(p) = -\frac{\bar{f}}{p}\frac{m_2^2 - p^2\dfrac{c^2}{c_1^2}}{m_1^2 - m_2^2},$$

(24)

$$D(p) = \frac{\bar{f}}{p}\frac{m_1^2 - p^2\dfrac{c^2}{c_1^2}}{m_1^2 - m_2^2}.$$

$\bar{f}(p)$ is the transform of $f(\tau)$, and $m_1(p)$ and $m_2(p)$ are the roots with negative real parts (for outgoing waves) of

(25) $$\left(m^2 - p^2\frac{c^2}{c_1^2}\right)\left(m^2 - p^2\frac{c^2}{c_2^2}\right) - 8\,\frac{\lambda+\mu}{\mu}\frac{c_b^2}{c_1^2}\left(m^2 - p^2\frac{c^2}{c_b^2}\right) = 0.$$

In each equation of (23) the first integral represents a wave whose front travels at speed c_1 and the second integral represents a wave whose front travels at speed c_2.

To examine the behavior of the solution near the wavefronts, let p be large. Expansions for m_1 and m_2 are:

$$m_1 = -\frac{c}{c_1}p - 4\,\frac{\lambda+\mu}{\lambda+2\mu}\,\frac{c_1}{c}\,\frac{c_1^2 - c_b^2}{c_1^2 - c_2^2}\,\frac{1}{p} + O\left(\frac{1}{p^3}\right),$$

(26)

$$m_2 = -\frac{c}{c_2}p - 4\,\frac{\lambda+\mu}{\mu}\,\frac{c_2}{c}\,\frac{c_b^2 - c_2^2}{c_1^2 - c_2^2}\,\frac{1}{p} + O\left(\frac{1}{p^3}\right).$$

The leading terms in the integrals for w_τ may be written asymptotically as

$$w_\tau \sim \frac{1}{2\pi i}\int_{Br}\left[1 + O\left(\frac{1}{p^2}\right)\right]\bar{f}e^{p\tau - (c/c_1)(p^2+r_1^2)^{1/2}\zeta}dp$$

(27)

$$+\frac{1}{2\pi i}\int_{Br} O\left(\frac{1}{p^2}\right)\bar{f}e^{p\tau - (c/c_2)(p^2+r_2^2)^{1/2}\zeta}dp.$$

In (27) the square root terms are equivalent to m_1 or m_2 within $O(1/p^3)$ where r_1^2 and r_2^2 are given by

$$r_1^2 = 8\,\frac{c_1^2 - c_b^2}{c^2},$$

(28)

$$r_2^2 = 8\,\frac{c_b^2 - c_2^2}{c^2}.$$

The first integral in (27) vanishes for $\tau < (c/c_1)\zeta$ and the second vanishes for $\tau < (c/c_2)\zeta$. Thus, near the leading wavefront only the first integral is required, which may be written, after a little manipulation and use of the convolution theorem, as follows:

$$w_\tau \sim -\frac{c_1}{c}\frac{\partial}{\partial \zeta}\int_{(c/c_1)\zeta}^{\tau} f(\tau-\tau')J_0\left[r_1\sqrt{\tau'^2-\frac{c^2}{c_1^2}\zeta^2}\right]d\tau'$$

(29)

$$\sim f\left(\tau-\frac{c}{c_1}\zeta\right)-\frac{c}{c_1}r_1\zeta\int_{(c/c_1)\zeta}^{\tau} f(\tau-\tau')\frac{J_1\left[r_1\sqrt{\tau'^2-\frac{c^2}{c_1^2}\zeta^2}\right]}{\sqrt{\tau'^2-\frac{c^2}{c_1^2}\zeta^2}}d\tau'.$$

J_0 and J_1 are Bessel functions. Suppose that $f(\tau)=w_\tau^0 h(\tau)$ where $h(\tau)$ is the Heaviside step function and w_τ^0 is a constant. The change of variable $\gamma=r_1\sqrt{\tau'^2-(c^2/c_1^2)\zeta^2}$ and use of the identity (taken from [17, p. 64])

(30) $$\int_0^\infty \frac{J_1(\gamma)d\gamma}{\sqrt{\gamma^2+\eta^2}}=\frac{1-e^{-\eta}}{\eta},$$

now reduces (29) to

(31) $$\frac{w_\tau}{w_\tau^0}\sim e^{-r_1(c/c_1)\zeta}+r_1\frac{c}{c_1}\zeta\int_{r_1\sqrt{\tau^2-(c^2/c_1^2)\zeta^2}}^{\infty}\frac{J_1(\gamma)d\gamma}{\sqrt{\gamma^2+\left(r_1\frac{c}{c_1}\zeta\right)^2}}.$$

Equation (31) shows that at points removed from the boundary the velocity becomes exponentially small shortly after passage of the wave front. Equation (20) is plotted in Figure 1 with $f(\tau)=h(\tau)$.

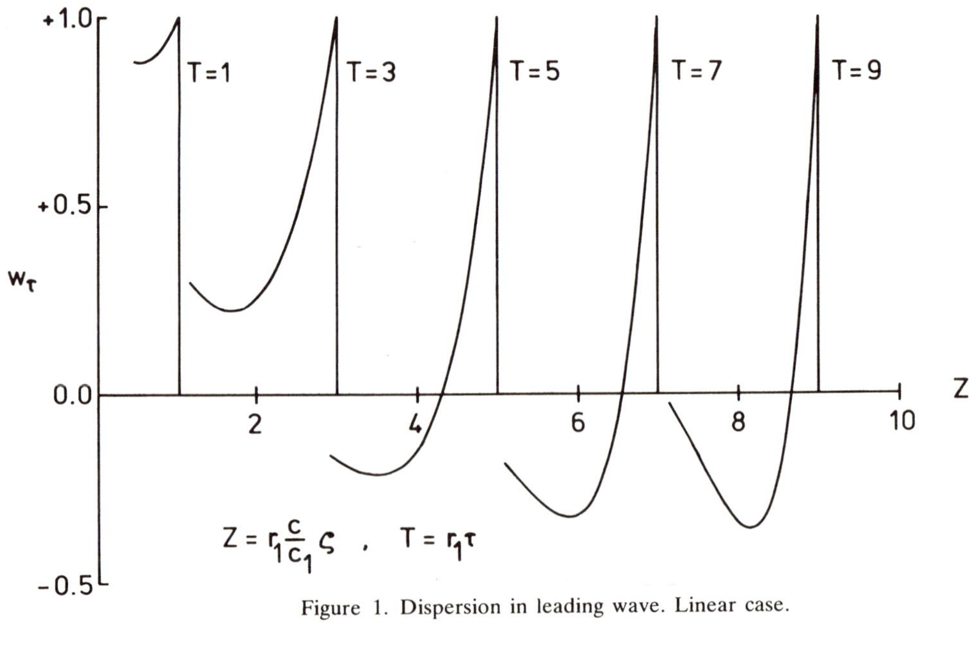

Figure 1. Dispersion in leading wave. Linear case.

Similar analysis may be applied to the other integrals in (23). The second integral for w_τ represents a wave whose front moves with speed c_2, and is similar in form to (29), but smoothed by a double integration in τ. u_ζ has a contribution at each wavefront that is similar to (29), but smoothed by a single integration in τ.

To examine the main pulse in (23) it is convenient to follow Whitham's approach to the problem [15, p. 345]. First, the exponent in one of the integrals is written as $\tau(p + m_1(p)\zeta/\tau)$, and then the saddle point method is used to find an asymptotic expansion valid at large τ. For a fixed value of ζ/τ saddles occur at

$$\frac{d}{dp}\left(p + m_1(p)\frac{\zeta}{\tau}\right) = 1 + m_1'(p)\frac{\zeta}{\tau} = 0. \tag{32}$$

Solution of (32) gives $p = \tilde{p}(\zeta/\tau)$, and the exponent is maximized with respect to ζ/τ when $m_1[\tilde{p}(\zeta/\tau)] = 0$ because of (32) and (33):

$$\frac{d}{d(\zeta/\tau)}\left[\tilde{p} + m_1(\tilde{p})\frac{\zeta}{\tau}\right] = m_1(\tilde{p}) + \left[1 + m_1'(\tilde{p})\frac{\zeta}{\tau}\right]\frac{d\tilde{p}}{d(\zeta/\tau)} = 0. \tag{33}$$

From (25) if $m_1 = 0$, either $p = 0$ or $p^2 = -8((\lambda + \mu)/\mu)(c_2^2/c^2)$. By differentiating (25) once with respect to p, it is clear that the correct choice is $p = 0$, and by differentiating (25) a second time with respect to p, it is found that

$$m_1'(0) = \pm\frac{c}{c_b}, \tag{34}$$

and so from (32)

$$\zeta/\tau = \pm\frac{c_b}{c}. \tag{35}$$

Equation (35) shows that the main disturbance travels with the bar speed, c_b. The correct asymptotic expansions will be obtained by expanding the integrands in (23) about $p = 0$. The following discussion will be simplified by taking $c = c_b$.

It has been shown that $m_1(0) = 0$, but is does not follow that $m_2(0) = 0$. Therefore, let $m^2 = a + bp^2 + dp^4 + \cdots$ in (25). By equating powers of p^2 it is found that

$$\begin{aligned} m_1^2 &= p^2 - \frac{1}{8}\frac{(c_1^2 - c_b^2)(c_b^2 - c_2^2)}{c_b^2(c_1^2 - c_2^2)}\,p^4 + \cdots, \\ m_2^2 &= 8\,\frac{\lambda + \mu}{\mu}\frac{c_b^2}{c_1^2} + \left(\frac{c_b^2}{c_1^2} + \frac{c_b^2}{c_2^2} - 1\right)p^2 + \cdots. \end{aligned} \tag{36}$$

Choosing roots to correspond to outgoing waves or bounded functions at infinity gives

$$m_1 = -p + \frac{1}{16}\frac{(c_1^2 - c_b^2)(c_b^2 - c_2^2)}{c_b^2(c_1^2 - c_2^2)}p^3 + \cdots, \tag{37}$$

$$m_2 = -\left\{a_2^{1/2} + \frac{1}{2}\frac{b_2}{a_2^{1/2}}p^2 + \cdots\right\}.$$

Since the second integrals for u and w in (23) give no contribution until $\tau > (c/c_2)\zeta$, near the head of the main pulse the velocity is given asymptotically as follows:

$$w_\tau \sim \frac{1}{2\pi i}\int_{Br} \bar{f} e^{p(\tau-\zeta)-\frac{1}{2}d_1p^3\zeta}(1 + O(p^2))dp. \tag{38}$$

where

$$d_1 = -\frac{1}{8}\frac{(c_1^2 - c_b^2)(c_b^2 - c_2^2)}{c_b^2(c_1^2 - c_2^2)}.$$

By the convolution theorem, (38) may be written

$$w_\tau = \frac{1}{(-\frac{3}{2}d_1\zeta)^{1/3}}\int_0^\tau f(\tau - \tau')\mathrm{Ai}\left[\frac{\zeta - \tau'}{(-\frac{3}{2}d_1\zeta)^{1/3}}\right]d\tau', \tag{39}$$

where Ai$(\cdot)$ is the Airy function with integral representation given in [18, Formula 10.4.32]. For a step function input, $f(\tau) = h(\tau)$, w_τ is simply the integral of the Airy function and is exponentially small for $\zeta > \tau$ and oscillates about an amplitude of 1 for $\tau > \zeta$. Figure 2 shows the waveform for a step input. Figure 3 shows the composite waveform schematically for both the leading wave and the main pulse. Note the progressive separation and decreasing importance of the leading fast wave. The initial development of this separation was shown numerically by Nunziato and Walsh in their consideration of small amplitude waves in porous soils [19], where the equations are formally the same as (20).

Nonlinear wave equations

The multiple wave structure, which the linearized equations show so clearly, has an exact counterpart in the fully nonlinear theory. The characteristic wave speeds of (5), which carry discontinuities in w'' and $\ddot{w}$ or u'' and $\ddot{u}$, are given by

$$c^2 = \frac{1}{2}\left[\frac{W_{u'u'}}{\rho_2} + \frac{W_{w'w'}}{\rho_1}\right] \pm \frac{1}{2}\left[\left(\frac{W_{u'u'}}{\rho_2} - \frac{W_{w'w'}}{\rho_1}\right)^2 + \frac{4W_{u'w'}^2}{\rho_1\rho_2}\right]^{1/2}. \tag{40}$$

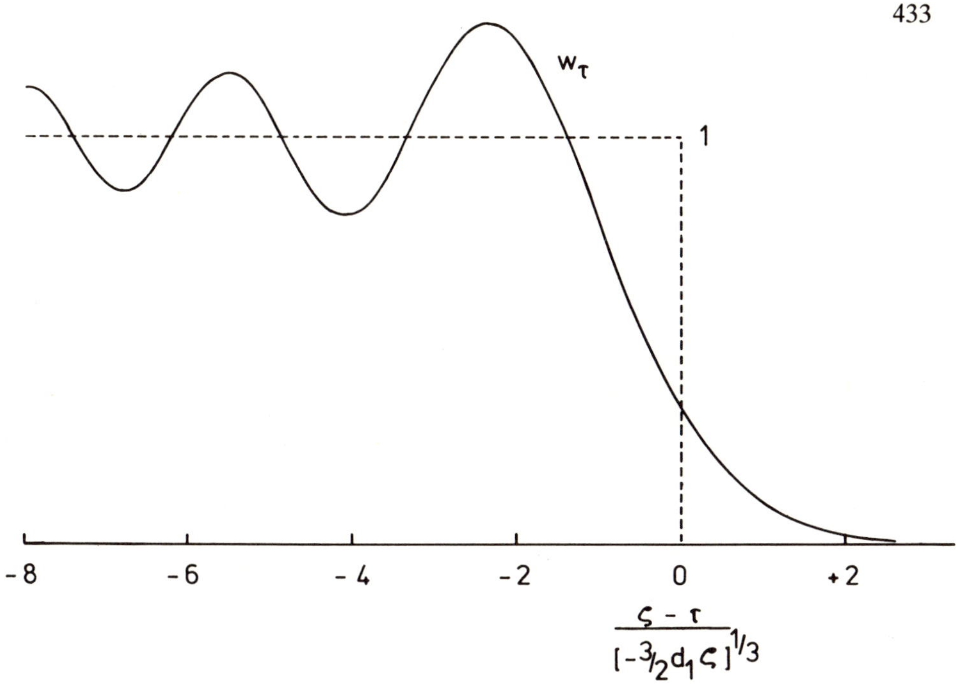

Figure 2. Dispersion in main pulse. Linear case.

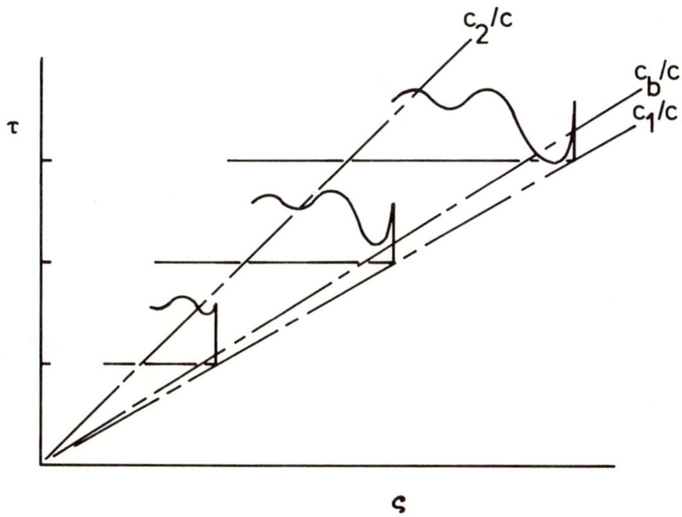

Figure 3. Composite sketch showing separation of leading wave and main pulse. Linear case.

On the other hand, the simplest one dimensional theory for a rod, which is given by (1), and which may be obtained from (5) by setting $u'=0$, $P(w', u, 0)=0$, has characteristic speeds given by

$$\rho c_b^2 = E, \tag{41}$$

where E is defined in (17). In analogy with the linear theory it might be expected that (41) represents a low order approximation for wave speeds in the main pulse, but that dispersive effects due to the finite diameter will modify the pulse shape. Since the full rod theory, (5), is nonlinear, it is to be expected that wave speeds will be modified as well.

Although transform methods cannot be used on the full equations, it is possible, by suitable scaling and choice of independent variables, to find asymptotic results for the various wave orders directly from the equations themselves rather than from representations for exact solutions. The method of relatively undistorted waves, introduced by Varley and Cumberbatch [20], has been used recently by Seymour and Parker [10] to examine the leading fast wave in (5). The same approach, also called the method of modulated simple waves, may be used to investigate the structure of the main pulse.

To simplify the analysis, it will be assumed that the stored energy in (3) has the decomposition

$$W(w', u, u') = W_1(w', u) + W_2(u'). \tag{42}$$

Let the dependent variables be u, $v = \dot{w}$, and $\varepsilon = w'$. Equations (5) are rewritten as

$$\begin{aligned} S' &= \rho_1 \dot{v}, \\ \dot{Q}' - \dot{P} &= \rho_2 \ddot{u}, \\ v' &= \dot{\varepsilon}. \end{aligned} \tag{43}$$

First define a new set of independent variables

$$\zeta = \delta^{1/2} Z, \qquad \tau = \delta^{1/2} t, \tag{44}$$

where δ is a small parameter. This scaling implies that fixed values of ζ, τ will correspond to long times and large distances as δ tends to zero. Next define a second set of independent variables in terms of ζ, τ:

$$\begin{aligned} \alpha &= \alpha(\zeta, \tau), \\ \mathfrak{z} &= \delta \zeta. \end{aligned} \tag{45}$$

In terms of these new variables, derivatives with respect to Z, t are replaced as follows:

$$
\begin{aligned}
\frac{\partial}{\partial Z} &= \delta^{1/2}\alpha_\zeta \frac{\partial}{\partial \alpha} + \delta^{3/2}\frac{\partial}{\partial \mathfrak{z}}, \\
\frac{\partial}{\partial t} &= \delta^{1/2}\alpha_\tau \frac{\partial}{\partial \alpha}.
\end{aligned} \tag{46}
$$

In (45) α is a "fast" and $\mathfrak{z}$ a "slow" variable. The reason for the odd combination of scaling in (44) and (45) will become apparent later. It turns out that α has properties like a characteristic variable. Accordingly, let ω and κ be defined as

$$
\omega = \alpha_\tau, \qquad \kappa = \alpha_\zeta. \tag{47}
$$

α is constant along curves with slope

$$
\frac{d\zeta}{d\tau} = c = -\frac{\omega}{\kappa}. \tag{48}
$$

It is convenient, instead of ω and κ to use the slowness, $s = 1/c$, and the incremental arrival time $\ell = 1/\omega$. In terms of these variables, the compatibility condition $\omega_\zeta = \kappa_\tau$ becomes

$$
s_\alpha = \delta \ell_{\mathfrak{z}}, \tag{49}
$$

and, taking (42) into account, equation (43) becomes

$$
\begin{aligned}
sS_u u_\alpha + \rho_1 v_\alpha + sS_{w'}\varepsilon_\alpha &= \delta\ell\{S_u u_{\mathfrak{z}} + S_{w'}\varepsilon_{\mathfrak{z}}\}, \\
P_u u_\alpha - sP_{w'} v_\alpha &= \delta\ell\{-P_{w'} v_{\mathfrak{z}} + Q_{u'} u_{\tau\zeta\zeta} - \rho_2 u_{\tau\tau\tau}\}, \\
sv_\alpha + \varepsilon_\alpha &= \delta\ell v_{\mathfrak{z}},
\end{aligned} \tag{50}
$$

where derivatives in τ and ζ are retained on the right-hand side of $(50)_2$ for compactness.

To lowest order the right-hand sides of (50) may be set equal to zero. This requires that the determinant of coefficients of u_α, v_α, ε_α must vanish giving

$$
\frac{1}{s^2} = c_b^2 = \frac{1}{\rho_1}\left(S_{w'} - \frac{P_{w'}}{P_u} S_u\right), \tag{51}
$$

and u_α, v_α, ε_α must occur in the ratio

$$
u_\alpha : v_\alpha : \varepsilon_\alpha = -\nu : -c_b : 1. \tag{52}
$$

With the energy partition of (42) neither S nor P depends on u'. Thus (52) may be integrated to find two of u, v, ε in terms of the third. For example

$$
\begin{aligned}
v &= \tilde{v}(u), \\
\varepsilon &= \tilde{\varepsilon}(u).
\end{aligned} \tag{53}
$$

The right-hand side of (50) must satisfy a compatibility condition. Thus multiplication of (50) by the left proper vector

$$(-1, s\nu, sS_{w'}) \tag{54}$$

gives the transport equation

$$\begin{aligned} S_{\mathfrak{z}} - sEv_{\mathfrak{z}} = s\nu\rho_2 \Big\{ c_2^2 \frac{1}{\ell}\frac{\partial}{\partial\alpha}\left[\frac{s}{\ell}\frac{\partial}{\partial\alpha}\left(\frac{s}{\ell}u_\alpha\right)\right] \\ - \frac{1}{\ell}\frac{\partial}{\partial\alpha}\left[\frac{1}{\ell}\frac{\partial}{\partial\alpha}\left(\frac{1}{\ell}u_\alpha\right)\right]\Big\}, \end{aligned} \tag{55}$$

where $\rho_2 c_2^2 = Q_{u'}$. Since S, s, E and v all depend on u through (53), the left-hand side of (55) may be written as $\Phi(u)u_{\mathfrak{z}}$ so that the dependent variable is u alone. However, the mapping $\alpha(\zeta, \tau)$ must be developed simultaneously with the solution $u(\alpha, \mathfrak{z})$ since the incremental arrival time ℓ is also unknown.

The properties of (55) will not be studied further here except for the following two remarks. First, the linearized version of (55) can be shown to have solutions like (39), and second, for small (but finite) amplitude waves, the transport equation reduces to the Korteweg–de Vries equation. This latter fact may also be deduced directly from (5) following the prescription of Leibovich and Seebass [21]. Many other authors have also noted that the Korteweg–de Vries equation describes longitudinal waves in thin rods, e.g. [22].

Steady waves

Equation (49) indicates that if $\ell_{\mathfrak{z}} = O(1)$, then s, and hence c_b, is almost constant. Therefore, some solutions of (55) should be nearly steady waves, perhaps approaching them asymptotically in time. Accordingly, the properties of steady solutions to (5) will be examined.

A steady wave is one in which the solution to (5) may be written

$$\begin{aligned} w &= w(Z - ct), \\ u &= u(Z - ct), \end{aligned} \tag{56}$$

where c is an unspecified, but constant, wave speed. The equations become

$$\begin{aligned} S' &= \rho_1 c^2 w'', \\ Q' - P &= \rho_2 c^2 u'', \end{aligned} \tag{57}$$

where the prime now denotes differentiation with respect to $Z - ct$.

There are two integrals of the motion:

$$S = \rho_1 c^2 w' + A,$$

(58)
$$Qu' + Aw' - W = \tfrac{1}{2}c^2(\rho_2 u'^2 - \rho_1 w'^2) + B,$$

where A and B are constants. The second integral is found by multiplying $(57)_1$ by w' and $(52)_2$ by u', adding the two equations, and noting that the result may be written as

$$\left(\frac{\partial W}{\partial u'} u' + \frac{\partial W}{\partial w'} w' - W\right)' = \tfrac{1}{2}c^2(\rho_1 w'^2 + \rho_2 u'^2)'.$$

After integration $(58)_1$ is substituted into this expression to give $(58)_2$.

If (58) can be solved for w' and u', they may be expressed as follows:

$$w' = F(u; A, B, c),$$

(59)
$$u'^2 = G(u; A, B, c).$$

Equation $(59)_2$ is written in terms of u'^2 rather than u' since both of (58) are even in u'. Now $(59)_2$ may be solved by quadrature and then $(59)_1$ gives the distribution in longitudinal strain parametrically through u.

To examine the qualitative nature of solutions and the role of the constants A, B, and c, first consider the case $A = B = 0$. For ease of analysis the strain energy will be assumed to be partitioned as in (42). For small but finite amplitude waves let W be expressed as a power series. First note that

(60)
$$W = \frac{1}{\pi a^2} \int \tilde{W} dA = \frac{2}{a^2} \int_0^a R\tilde{W} dA,$$

where $\tilde{W}$ is the three-dimensional strain energy per unit volume. With (12) used in (60), W may be expressed as

$$W = \tfrac{1}{2}(\lambda + 2\mu)w'^2 + 2\lambda\mu u w' + 2(\lambda + \mu)u^2 + \tfrac{1}{4}a^2\mu u'^2$$

(61)
$$+ C_1 u^3 + C_2 u^2 w' + C_3 u w'^2 + C_4 w'^3 + \cdots,$$

where the C's are combinations of higher order elastic moduli.

Equation $(58)_1$ with $A = 0$ gives

(62)
$$w' = -\frac{2\lambda}{\lambda + 2\mu - \rho c^2} u + O(u^2),$$

and $(58)_2$ with $A = B = 0$ and the use of (62) gives

(63)
$$u'^2 = Cu^2 + Du^3 + \cdots,$$

where

(64)
$$C = \frac{8(\lambda + \mu)(E - \rho c^2)}{a^2(\mu - \rho c^2)(\lambda + 2\mu - \rho c^2)},$$

and D may be either positive or negative, depending on the exact combination of higher order elastic moduli. C may be either positive or negative as well:

$$(65)\qquad \begin{aligned} C &> 0 \quad \text{if } c_b < c < c_1, \\ C &< 0 \quad \text{if } c_2 < c < c_b, \end{aligned}$$

where c_1, c_2, and c_b are the linear elastic wave speeds in this context.

Figure 4 shows sketches of u'^2 vs. u and u' vs. u for the various combinations of signs of C and D. The closed loops in (4e) and (4f) correspond to solitary waves that are symmetrical in Z about the point of maximum amplitude and extend from $Z = -\infty$ to $Z = +\infty$. The open curves in (4e)–(4h) correspond to waves in which the amplitude grows without limit. Clearly, to be useful, solutions of this kind must join onto unsteady solutions before the amplitudes become too extreme.

Now consider the case where $A = 0$ but $B \neq 0$. Instead of (63) the following holds:

$$(66)\qquad u'^2 = \hat{B} + Cu^2 + Du^3.$$

The constant $\hat{B}$ is proportional to B, but has opposite sign. $\hat{B}$ merely shifts

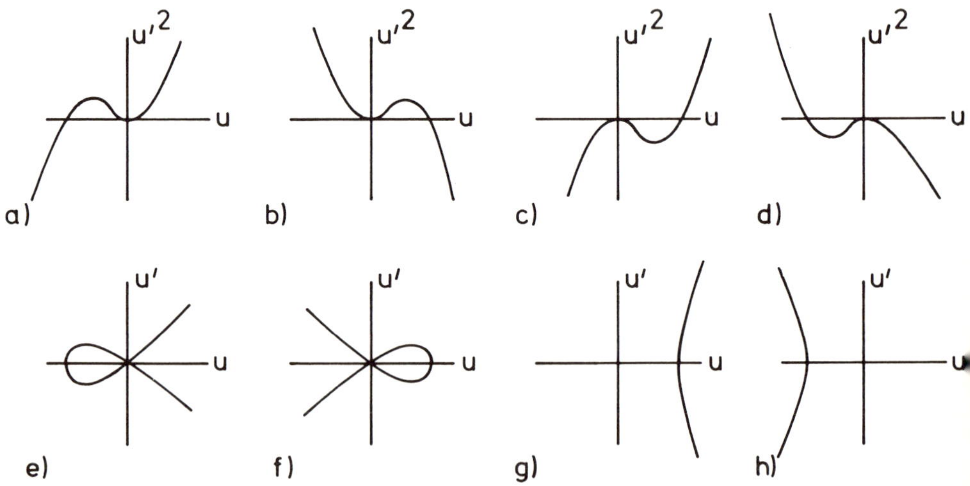

Figure 4. Phase plane sketches for steady waves, $A = B = 0$ in every case: (a) and (e) both $C > 0, D > 0$; (b) and (f) $C > 0, D < 0$; (c) and (g) $C < 0, D > 0$; (d) and (h) both $C < 0, D < 0$.

the curves in (4a)–(4d) vertically. Figure 5 shows two of the possibilities. The closed loops in (5c) and (5d) correspond to periodic solutions.

Next consider the case when both $A \neq 0$ and $B \neq 0$. The integrals of the motion may be rewritten as follows:

$$S - S_0 = \rho_1 c^2(w' - w'_0),$$

(67) $$Qu' + S_0(w' - w'_0) - (W - W_0) = \tfrac{1}{2}c^2\{\rho_2 u'^2 - \rho_1(w' - w'_0)^2\} + B,$$

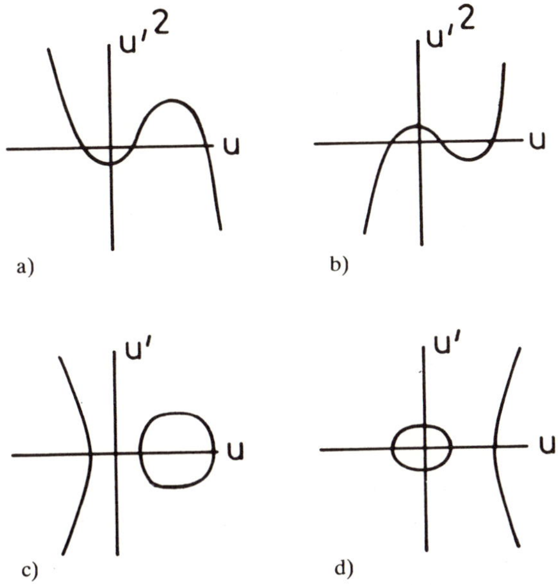

Figure 5. Phase plane sketches for steady waves, $A = 0$, but $B \neq 0$: (a) and (c) $B > 0$ ($\hat{B} < 0$), otherwise like Figure 4b; (b) and (d) $B < 0$ ($\hat{B} > 0$), otherwise like Figure 4c.

where $W_0 = W(w'_0, u_0, 0)$, etc. These integrals are the same as before, but now they are centered on the point $(w'_0, u_0, 0)$ and S_0 plays the role of A. The character of the solutions is exactly the same as in Figures 4 and 5, but now the curves are shifted to the right ($S_0 < 0$) or left ($S_0 > 0$).

Finally suppose that the elastic constants are such that $D = 0$. Instead of (66), equation (68) must be used:

(68) $$u'^2 = \hat{B} + Cu^2 + Hu^4.$$

This should occur when the expansion is centered at an inflexion point of the static stress–strain curve. C is the same as before and H is determined by higher order elastic moduli. Some of the possibilities are shown in Figure 6. Either positive or negative solitary waves are possible according to (6c), and (6d) is a periodic solution of unusual shape.

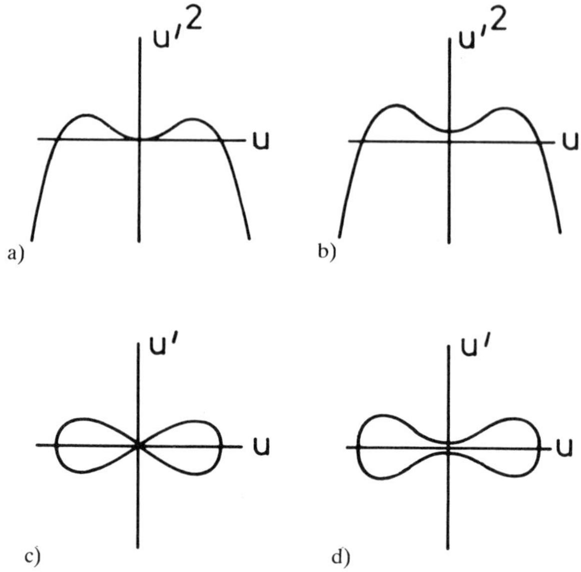

Figure 6. Phase plane sketches for steady waves, $D = 0$: (a) and (c) $A = B = 0$; (b) and (d) $A = 0$, but $B < 0$ ($\hat{B} > 0$).

The dynamic trajectories, $\mathcal{D}$, of the nonlinear steady waves may deviate considerably from the static curve $P(w', u, 0) = 0$ in the $w' - u$ plane. To see this, again suppose that the strain energy is partitioned as in (42) and consider the case $A = 0$. Figure 7a shows a contour plot of W_1 on the $w' - u$ plane with the curve $P = W_{1u} = 0$ and several lines of $u =$ constant drawn in. In Figure 7b several curves of $S = W_{1w'}$ vs. w' with $u =$ constant are shown. Shown as a dashed curve on (7b) is the static stress–strain curve, which has somewhat smaller slope than any of the curves for $u =$ constant, according to (17). From reference to $(58)_1$ it can be observed that points of the nonlinear trajectories lie on the intersection of the surfaces $S_1 = S(w', u)$ and $S_2 = \rho_1 c^2 w'$. In the projection of Figure 7b S_2 is a straight line through the origin with slope $\rho_1 c^2$, for example the line $p_1 p_2$ (for c less than c_b, the bar speed at $w' = 0$). By comparing the line $p_1 p_2$ with the curves for S at $u =$ constant and the static stress–strain curve, it can be seen that between points p_1 and p_2 the dynamic trajectory deviates to values of u less than those on $P = 0$, and at points outside the segment $p_1 p_2$ the trajectory deviates to greater values of u than those on $P = 0$. Thus, the trajectory, $\mathcal{D}$, is qualitatively like the dashed curve in Figure 7c.

Maximum or minimum amplitudes of u will occur where $u' = 0$. With $u' = 0$, equation $(58)_2$ generates curves, $\mathcal{C}$, in the $w' - u$ plane that correspond

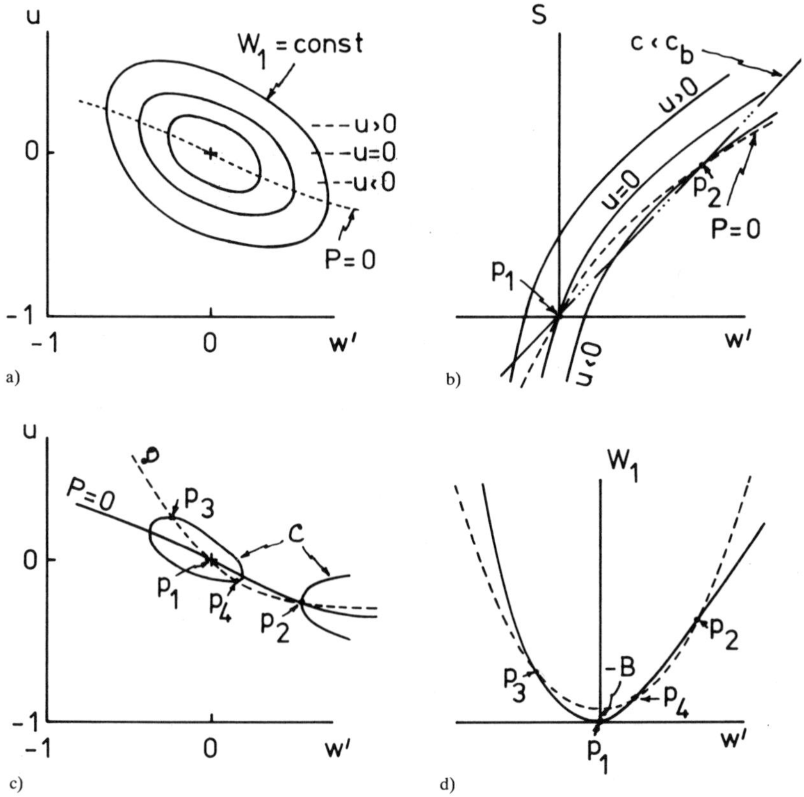

Figure 7. Comparison of steady dynamic trajectory with static trajectory: (a) contours of W_1 = constant and static curve $P = 0$; (b) intersection of surfaces $S(w', u)$ and $\rho_1 c^2 w'$; (c) dynamic trajectory deduced from Figure 4b; (d) intersection of cylinder $(\frac{1}{2}\rho_1)c^2 w'^2 - B$ with widest projection of $W_1(w', u)$.

to the intersection of the surfaces $\hat{W}_1 = W_1(w', u)$ and $\hat{W}_2 = \frac{1}{2}\rho_1 c^2 w'^2 - B$. The intersection of curves $\mathscr{C}$ and $\mathscr{D}$ marks the ends of the dynamic trajectory. By differentiating $(58)_2$ with respect to w', it may be seen that the points on $\mathscr{C}$ satisfy

$$P\frac{du}{dw'} + S = \rho_1 c^2 w'. \tag{69}$$

At the intersection of $\mathscr{C}$ and $\mathscr{D}$, $(58)_1$ holds as well so that at those points

$$P\frac{du}{dw'} = 0. \tag{70}$$

That is, either $P = 0$ or the slope of $\mathscr{C}$ is horizontal.

Now consider further the intersection of surfaces $\hat{W}_1$ and $\hat{W}_2$ with $B < 0$ ($\hat{B} > 0$). Figure 7d shows the projection parallel to the u-axis of the widest points of $\hat{W}_1 = W_1(w', u)$. These points correspond to the curve $P = 0$ in Figure 7a, and, in fact, the projected potential well in Figure 7d generates the static stress–strain curve. The surface $\hat{W}_2 = \frac{1}{2}\rho_1 c^2 w'^2 - B$ is a parabolic cylinder with generators parallel to the u-axis. An end view of the cylinder is shown as the dashed parabola in Figure 7d. Since the level curves of $\hat{W}_1$ are closed, as shown in Figure 7a, and since the level curves of $\hat{W}_2$ are straight lines parallel to the u-axis, the curves $\mathscr{C}$ are qualitatively similar to those sketched in Figure 7c. Thus, for the case considered, which corresponds to Figure 5d, the motion lies along $\mathscr{D}$ and either oscillates between points p_3 and p_4 or begins at p_2 and moves to the right until interrupted by an unsteady motion. In either case, note the divergence of $\mathscr{D}$ from the curve $P = 0$.

Other cases for different shapes of the potential well $W_1(w', u)$ may be considered in a similar manner. However, from the preceding qualitative discussion it is clear that the cubic expansion in equation (63) leads to the qualitatively correct curves in the phase plane as shown in Figures 4 and 5, provided that the function

$$K(u'^2) = Q(u')u' - W_2(u') - \tfrac{1}{2}\rho_2 c^2 u'^2$$

is monotonic and hence single-valued in u'^2. If this is not the case, then the curves of u'^2 vs. u in Figures 4, 5 and 6 will have bifurcation points. The analysis would then be somewhat more complicated and will not be attempted here.

References

1. J. F. Bell, *The Experimental Foundations of Solid Mechanics*, Handbuch der Physik, Vol. VIa/1, Springer-Verlag, New York, 1973.
2. G. E. Hauver, *Penetration with instrumented rods*, Int. J. Eng. Sci. **16** (1978), 871–877.
3. R. Skalak, *Longitudinal impact of a semi-infinite circular elastic bar*, J. Appl. Mech. **24** (1957), 59–64.
4. W. A. Green, *Dispersion relations for elastic waves in bars*, in *Progress in Solid Mechanics*, Vol. I (I. N. Sneddon and R. Hill, eds.), North-Holland, Amsterdam, 1960.
5. J. H. Shea, *Propagation of plastic strain pulses in cylindrical lead bars*, J. Appl. Phys. **39** (1968), 4004–4011.
6. G. P. DeVault, *The effect of lateral inertia on the propagation of plastic strain in a cylindrical rod*, J. Mech. Phys. Sol. **13** (1965), 55–68.

7. S. S. Antman, *The Theory of Rods*, Handbuch der Physik, Vol. VIa/2, Springer-Verlag, New York, 1972.

8. J. W. Nunziato and E. K. Walsh, *On the influence of void compaction and material non-uniformity on the propagation of one-dimensional acceleration waves in granular materials*, Arch. Rat. Mech. Anal. **64** (1977), 299–316 and Addendum, Arch. Rat. Mech. Anal. **67** (1977), 395–397.

9. J. W. Nunziato and E. K. Walsh, *One-dimensional shock waves in uniformly distributed granular materials*, Int. J. Solids and Structures **14** (1978), 681–689.

10. D. F. Parker and B. R. Seymour, *Finite amplitude one-dimensional pulses in an inhomogeneous granular material*, Arch. Rat. Mech. Anal. **72** (1980), 265–284.

11. M. F. McCarthy, private communication.

12. S. S. Antman, *Qualitative theory of the ordinary differential equations of nonlinear elasticity*, in *Mechanics Today*, Vol. 1 (S. Nemat-Nasser, ed.), Pergamon, New York, 1972.

13. R. D. Mindlin and G. Herrmann, *A one-dimensional theory of compressional waves in an elastic rod*, Proceedings of the First U.S. National Congress of Applied Mechanics, 1950, pp. 187–191.

14. G. B. Whitham, *Some comments on wave propagation and shock wave structure with applications to magnetohydrodynamics*, Comm. on Pure and Appl. Math. **12** (1959), 113–158.

15. G. B. Whitham, *Linear and Nonlinear Waves*, John Wiley, New York, 1974.

16. T. T. Wu, *A note on the stability condition for certain wave propagation problems*, Comm. on Pure and Appl. Math. **14** (1961), 745–747.

17. F. Bowman, *Introduction to Bessel Functions*, Dover, New York, 1958.

18. M. Abromowitz and I. A. Stegun, *Handbook of Mathematical Functions*, Dover, New York, 1965.

19. J. W. Nunziato and E. K. Walsh, *Small-amplitude wave behavior in one-dimensional granular solids*, J. Appl. Mech. **44** (1977), 559–564.

20. E. Varley and E. Cumberbatch, *Non-linear high frequency sound waves*, J. Inst. Maths. Applics. **2** (1966), 133–143.

21. S. Leibovich and A. R. Seebass, *Examples of dissipative and dispersive systems leading to the Burgers and the Korteweg–de Vries equations*, in *Nonlinear Waves* (S. Leibovich and A. R. Seebass, eds.), Cornell University Press, Ithaca, N.Y., 1974.

22. G. A. Nariboli and A. Sedov, *Burger's–Korteweg–De Vries equation for viscoelastic rods and plates*, J. Math. Anal. and Appl. **32** (1970), 661–667.

THE STRAIN-ENERGY FUNCTION FOR RUBBER-LIKE MATERLIALS

L. J. ZAPAS

National Bureau of Standards, Center for Materials Science, Polymer Science and Standards Division, Washington, D.C. 20234, USA

(Received October 17, 1980)

ABSTRACT

Some results are presented on theories which attempt to describe the relation between force and deformation of rubbers where the linear laws of the classical theories of elasticity are not applicable. A few experimental results are presented in order to show the present status of the form of the strain-energy function of rubber-like materials.

1. Introduction

The development of the theory of rubber elasticity from its very early stages up until the present makes a fascinating story which is beyond the scope of this review. Rather, some specific results have been chosen, which are intended to lead the reader to a better understanding of the recent status of the strain-energy function for rubber-like materials.

Vulcanized rubbers and other cross-linked non-crystalline synthetic polymers are solids and at ambient temperatures can be subjected to very high deformations with relatively small forces. Their low rigidity and high extensibility attracted the interest of many scientists in the development of theories describing the relation between force and deformation of rubbers where the linear laws of the classical theories of elasticity are not applicable. One approach in the description of the non-linear behavior of rubbers was based on a microscopic or molecular model which started with the work of Meyer, von Susich and Valco [1] in 1932, who predicted that the tension in a specimen of vulcanized rubber subjected to a constant stretch will increase with a rising temperature. The difficulty with this approach is that even if the molecular structure is well understood, the passage to the constitutive equation,

Proceedings of the IUTAM Symposium on Finite Elasticity, Lehigh University, August 10–15, 1980. Invited paper.

Carlson, D.E. and Shield, R.T. (eds.)
Proceedings of the IUTAM Symposium on Finite Elasticity
ISBN 90 247 2629 8

expressed in phenomenological terms, is usually very difficult and cannot be made at all without so many idealizations of both the model and the mathematics as to leave the significance of the result in serious question. Some results are presented in Section 2a. The other approach, the phenomenological, stems largely from the work of Rivlin [2] starting in 1947 on the finite elasticity of isotropic materials. Though the fundamental equations in the theory of finite elasticity existed at the end of the last century, it was not till thirty-two years ago that serious progress was made in the solution of problems. Rivlin [3–5] considered the class of elastic materials which are isotropic and incompressible, such as rubbers. With these restrictions only, he was able to solve a number of problems: simple extension of a rod, biaxial deformation of a sheet, simple shear, simple torsion and so on, which collectively provide a means for the experimental check of the theory. In 1951 Rivlin and Saunders [6] published a classic paper, in which they characterized some rubbers by comparing experimental measurements with the results of the calculation and in fact checked the validity of this theory. That was the starting point for more work on various rubber-like materials, some results of which are presented in Section 2b.

2. The strain-energy function

a. *Kinetic theory*

About forty-five years ago polymer physicists sought to describe through molecular theories the behavior of vulcanized rubbers in simple extension. In 1934, Guth and Mark [7] and Kuhn [8] described and treated in a quantitative manner the statistical properties of a single long-chain molecule which led to mathematical expressions for the entropy of the chain as a function of the distance between its ends, and hence to a more complete description of the elasticity of the molecule. This was followed in 1942–43 by the treatment of a network of such long-chain molecules by statistical methods whereby the tensile force on a vulcanized rubber can be calculated as a function of elongation [9–12] given by the following equation

$$F = ANkT(\lambda - 1/\lambda^2) \tag{2.1}$$

where A is the area in the undistorted state, N is the number of chains between cross-links per unit volume, k is Boltzman's constant, T is the absolute temperature and λ is the stretch. (If l_0 was the initial length and l was the stretched length then $\lambda = l/l_0$.) The main assumptions are that the chains are Gaussian, the chain ends are displaced affinely, there are no intermolecular interactions and that rubbers are incompressible. In Figure 1 a schematic representation of the behavior of cross-linked rubbers in extension is shown, where $F/(\lambda - 1/\lambda^2)$, the reduced force, is plotted versus $1/\lambda$. From $\lambda = 1$ to $1/\lambda = \alpha$ the reduced force is monotonically decreasing. The reasons behind the rise for values of λ bigger than $1/\alpha$ remains a controversial question. Many scientists believe that this occurrence is due to crystallization of the rubber. Equation (2.1) suggests that the reduced force obtained from isothermal experiments is a constant. A line of zero slope drawn through the data, shown in Figure 1 as a dotted line, deviates $\pm 20\%$. To polymer

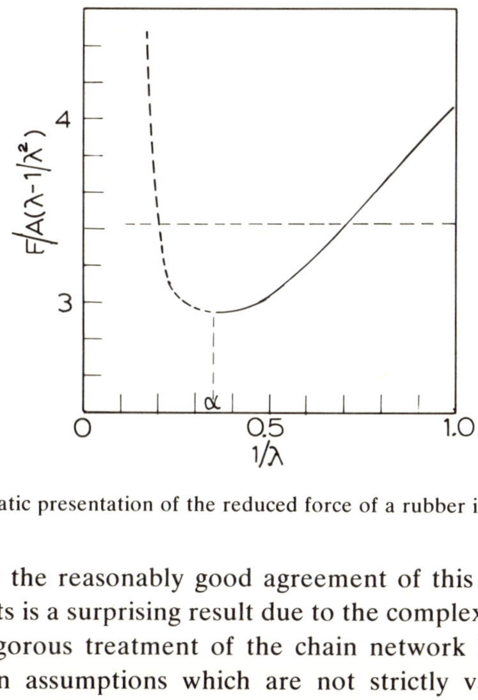

Figure 1. A schematic presentation of the reduced force of a rubber in simple extension versus $1/\lambda$.

physicists today the reasonably good agreement of this early kinetic theory with experiments is a surprising result due to the complexity that the problem presents if a rigorous treatment of the chain network is to be considered. Later, based on assumptions which are not strictly valid, solutions were obtained for a non-Gaussian network model. In 1952 Wang and Guth [13] showed that for the biaxial deformation of a sheet with the principal extension ratios λ_1, λ_2, and λ_3 the difference of the principal stresses is given by the following expression

$$\sigma_{11} - \sigma_{22} = \frac{NkT}{3} n^{1/2} \left\{ \lambda_1 \mathscr{L}^{-1}\left(\frac{\lambda_1}{n^{1/2}}\right) - \lambda_2 \mathscr{L}^{-1}\left(\frac{\lambda_2}{n^{1/2}}\right) \right\}, \tag{2.2}$$

where n is the number of equivalent random links ($\sim$ inversely proportional to N) and $\mathscr{L}^{-1}(x)$ is the inverse Langevin function. For values for which $\mathscr{L}^{-1}(x) \simeq 3x$ this reduces to neo-Hookean behavior. For the case of simple extension it predicts reasonably well the upturn of the reduced force versus $1/\lambda$ for high values of λ, but for values of $1/\lambda$ bigger than α it fails to show the behavior shown in Figure 1, since it yields a monotonically increasing reduced force as $1/\lambda$ goes to zero.

The first theory to approximate the sigmoidal behavior of Figure 1 was by DiMarzio [14] who in 1962 calculated the contribution to the configurational entropy of a rubber arising from the competition of chain segments for space, but as he points out the magnitude of the correction term is from 1/10 to 1/2 the magnitude of the observed difference. More recently Flory [15] considered the effect of the restriction on the fluctuations of network junctions imposed by neighboring chains. In this theory, the network junctions do not transform affinely but their positions depend on the strain. As he points out, "the model entails arbitrary physical approximations, the representation of the constraints affecting the junctions is intuitive and may involve appreciable error, thus precise conformity with experiment should not necessarily be expected". This last statement reflects the great difficulties encountered in formulating tractable molecular models.

b. *Phenomenological method*

The most highly developed part of the subject of non-linear continuum mechanics is concerned with the mechanics of elastic materials which undergo deformations sufficiently large so that classical elasticity theory is not applicable. As mentioned in the introduction, Rivlin in a series of papers [2–5] using only the restrictions of isotropy and incompressibility solved a number of problems in finite elasticity such as simple extension of a rod, biaxial deformation of a sheet, simple shear, simple torsion and so on. The solution for the case of a biaxial deformation of a sheet provided the way for the experimental determination of the strain-energy function for rubber-like materials which can be considered isotropic and incompressible. If W is the strain-energy function and λ_1, λ_2, λ_3 are the three principal stretches in a rectangular Cartesian coordinate system, then for an incompressible elastic material the strain energy is given by

$$W = W(I_1, I_2), \tag{2.3}$$

where

$$I_2 = \frac{1}{\lambda_1^2} + \frac{1}{\lambda_2^2} + \lambda_1^2\lambda_2^2. \tag{2.5}$$

and

$$I_1 = \lambda_1^2 + \lambda_2^2 + \frac{1}{\lambda_1^2\lambda_2^2}, \tag{2.4}$$

The third invariant $I_3 = \lambda_1^2\lambda_2^2\lambda_3^2$ is equal to unity, due to the incompressibility. The quantities I_1, I_2, I_3 are the invariants of the Cauchy–Green deformation tensor. The partial derivatives of W with respect to I_1 and I_2 can be obtained from experiments in biaxial deformation of a sheet and are given by the following expressions

$$2\left(\frac{\partial W}{\partial I_1}\right) = \left(\frac{\lambda_1^2\sigma_1}{\lambda_1^2\lambda_3^2} - \frac{\lambda_2^2\sigma_2}{\lambda_1^2-\lambda_3^2}\right)(\lambda_1^2-\lambda_2^2)^{-1}, \tag{2.6}$$

$$2\left(\frac{\partial W}{\partial I_2}\right) = l\left(\frac{\sigma_1}{\lambda_1^2-\lambda_3^2} - \frac{\sigma_2}{\lambda_2^2-\lambda_3^2}\right)(\lambda_1^2-\lambda_2^2)^{-1}, \tag{2.7}$$

where $\sigma_\kappa = \sigma_{\kappa\kappa} - \sigma_{33}$ and $\sigma_{\kappa\kappa}$ $(\kappa = 1,2,3)$ is the component of the stress tensor in the principal direction κ. For a uniform rod subjected to a tensile force F, the force F is given in terms of the extension ratio λ by the relation

$$F = 2A\left(\lambda - \frac{1}{\lambda^2}\right)\left(\frac{\partial W}{\partial I_1} + \frac{1}{\lambda}\frac{\partial W}{\partial I_2}\right), \tag{2.8}$$

where A is the cross-sectional area of the rod in its undeformed state and I_1 and I_2 are given by

$$I_1 = \lambda^2 + 2/\lambda \quad \text{and} \quad I_2 = 2\lambda + \frac{1}{\lambda^2}. \tag{2.9}$$

A few years before Rivlin's work, Mooney's publication [16] in 1940 was the *first* significant attempt to describe phenomenologically the behavior of rubbers under large deformations. He presented his theory in two forms: a general theory and a special theory. In the special theory he assumed that the rubber is incompressible and isotopic in the unstrained state and that Hooke's law is obeyed in simple shear or in simple shear superimposed in a plane transverse to a prior uniaxial extension or compression. (The most general theory was based on an unspecified non-linear stress–strain relation in shear.) Based on these assumptions and symmetry arguments he obtained for the strain-energy function, W, a function, which can be written as

$$W = C_1(I_1-3) + C_2(I_2-3), \tag{2.10}$$

where C_1 and C_2 are constants, and I_1, I_2 are the invariants given by (2.4) and (2.5).

In 1948 Rivlin [3] derived a general form of the strain-energy function for an incompressible isotropic elastic material, which can be presented as the sum of a series of terms involving powers of (I_1-3) and (I_2-3) or

$$W = \sum_{i,j=0}^{\infty} C_{ij}(I_1-3)^i(I_2-3)^j, \tag{2.11}$$

where C_{ij} are constants and $C_{00} = 0$.

Equation (2.10) is then the first order term of (2.11) in I_1 and I_2 and today is known as the Mooney–Rivlin strain-energy function. Its success in representing the behavior in simple extension over a reasonable range of elongations made such a deep impression that even today some polymer physicists are trying to give an explanation for the C_2 term, ignoring completely later developments and measurements which clearly show the inadequacy of this form for the strain-energy function.

In 1951 Rivlin and Saunders [6] presented the results obtained from the biaxial deformation of rubber sheets, from which they determined $\partial W/\partial I_1$ and $\partial W/\partial I_2$ as a function of I_1 and I_2. In Figure 2 and Figure 3 their results are

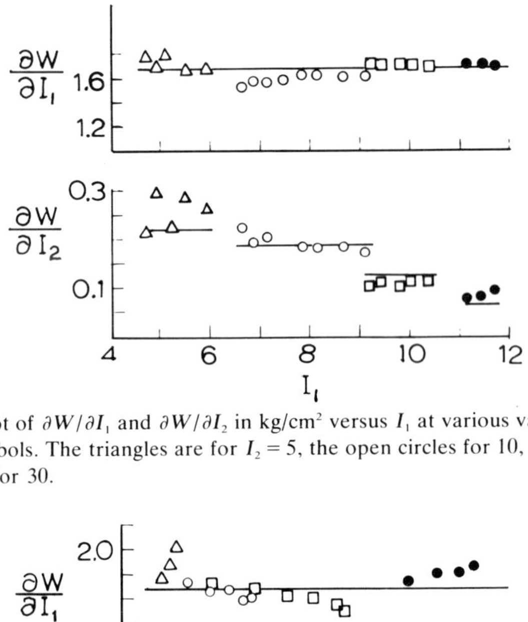

Figure 2. Plot of $\partial W/\partial I_1$ and $\partial W/\partial I_2$ in kg/cm^2 versus I_1 at various values of I_2, represented by different symbols. The triangles are for $I_2 = 5$, the open circles for 10, the squares for 20 and the solid circles for 30.

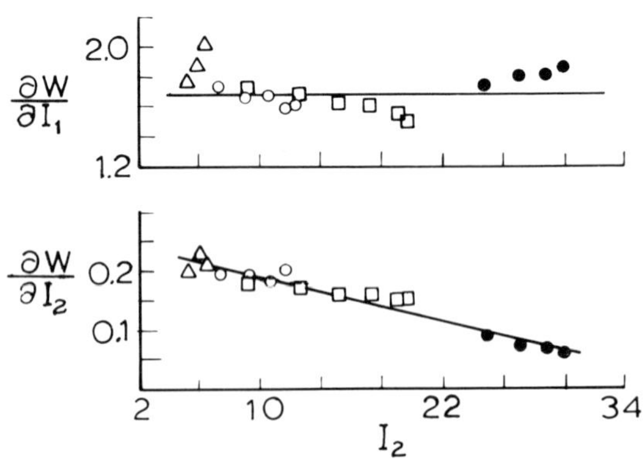

Figure 3. Plot of $\partial W/\partial I_1$ and $\partial W/\partial I_2$ in kg/cm^2 versus I_2 at various values of I_1 represented by different symbols. The triangles are for $I_1 = 5$, the open circles for 7, the squares for 9 and the solid circles for 11.

plotted. Within the range of their data, $\partial W/\partial I_1$ is essentially constant and $\partial W/\partial I_2$ is a function of I_2 alone.

From experiments performed on similar rubbers for other types of deformations, they were able to compare the data with results calculated using the above results. In all cases they found the agreement to be very good. In Figure 4, we show the data that they obtained in simple extension and compression. These data showed emphatically the inadequacy of the Mooney strain-energy function, which for the compression predicts a continuous rise of the reduced stress. Indeed, this work of Rivlin and Saunders is not only the first work in the proper determination of the strain-energy of rubber-like materials, but it became also the guide for some more work on various rubbers, quite a few years later.

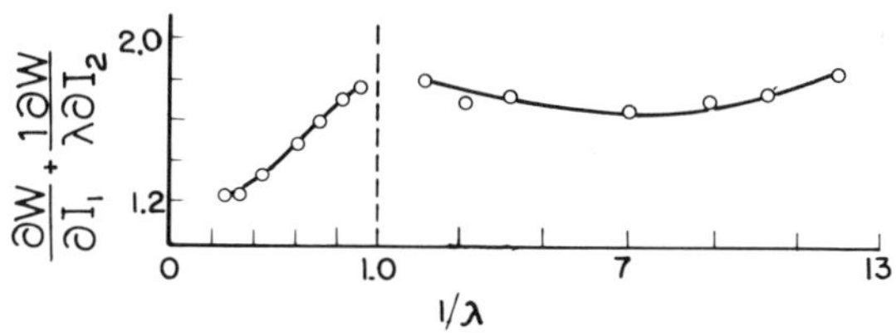

Figure 4. Plot of $[(\partial W/\partial I_1)+(1/\lambda)(\partial W/\partial I_2)]$ in kg/cm^2 versus $1/\lambda$ from experiments on simple extension and simple compression.

From experiments on the torsion of a tube which is prevented from contracting in diameter by the application of an internal pressure, in 1953 Gent and Rivlin [17] found that $\partial W/\partial I_2$ is time-dependent and is the main contributor to the hysteresis observed in their experiments. This interesting result as well as experimental results obtained by others show that most of the vulcanized rubbers (natural or synthetic) are really viscoelastic solids. It also shows that the mechanisms on the molecular level, which determine the behavior of rubbers, are different for $\partial W/\partial I_1$ and $\partial W/\partial I_2$.

Rivlin [18] later showed a one to one correspondence between the behavior of a stress relaxing incompressible elastic material and that of an incompressible ideal elastic material, for single step stress relaxation experiments (i.e., the material in the undistorted state is suddenly deformed and the deformation is kept constant).

If biaxial single step stress-relaxation experiments are performed and isochronal data are obtained after the introduction of the step, one can determine values of $\partial \hat{W}(\ ,t)/\partial I_1$ and $\partial \hat{W}(\ ,t)/\partial I_2$, where $\hat{W}(I_1, I_2, t)$ is not a strain-energy function, but can be used to predict the isochronal behavior in single step stress relaxation for various other deformations by treating it as a strain-energy function in elasticity theory. For viscoelastic solids, where after very long times the stress relaxes to a constant value, we may consider $\hat{W}$ as a strain-energy function. Researchers in the 1960's, who essentially repeated Rivlin and Saunders' experiments on a variety of other rubbers, obtained $\hat{W}$ either at times where the material was still relaxing or at very long times approximating equilibrium conditions.

In the course of experiments on the biaxial deformation of various vulcanizates of butyl rubber, Zapas [19] reported negative values of $\partial\hat{W}/\partial I_2$ at very small extensions, while at higher extension $\partial\hat{W}/\partial I_2$ was positive. Even for the most highly crosslinked vulcanizate, after 164 hours there were still no indications of reaching a plateau as shown in Figure 5. In the same work Zapas concluded from 24 hour isochronal data that $\partial\hat{W}/\partial I_1$ and $\partial\hat{W}/\partial I_2$ each depended on both I_1 and I_2. In 1967 Becker [20] reported biaxial single step stress relaxation data on a natural rubber vulcanizate (with 10 minute isochrones) for which it was shown that at small deformation $\partial\hat{W}/\partial I_2$ was negative. Extensive data obtained in biaxial measurements by the group of Kawabata and Kawai [21] show that $\partial\hat{W}/\partial I_1$ and $\partial\hat{W}/\partial I_2$ are both functions of I_1 and I_2. In the range of deformations where I_1 and I_2 are greater than 5 the data could be represented reasonably well with the strain-energy function proposed by Rivlin and Saunders. (But this is also the range for which Rivlin and Saunders obtained data.) In all this work the implicit assumption was made that rubbers are "simple materials" for which the stress at a particle depends on the deformation gradient at the particle (and not on their spatial

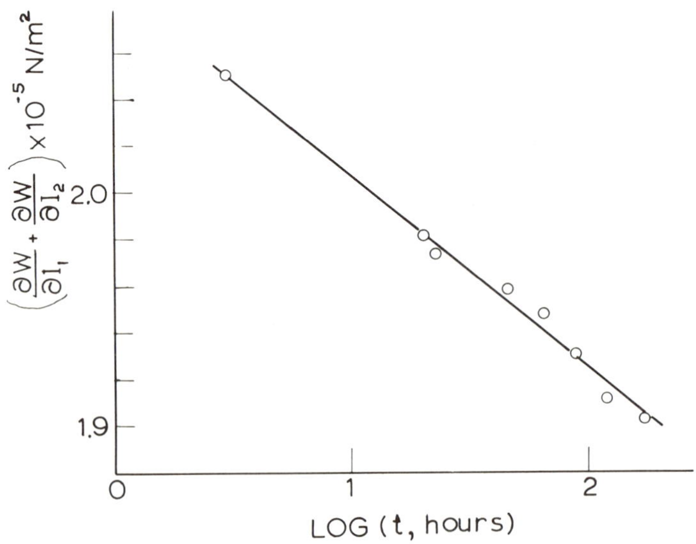

Figure 5. Plot of $((\partial W/\partial I_1)+(\partial\hat{W}/\partial I_2))$ versus $\log t$.

derivatives). Rivlin and Saunders [6] presented experimental evidence supporting this assumption through their study of the torsion of a cylinder where they measured simultaneously both the total couple and the total axial thrust required to keep the axial length constant as a function of amount of torsion. They also obtained data on combined torsion and extension. The hysteresis effects in their rubber samples together with the scatter in the data led Penn and Kearsley [22] to reexamine the assumption of simple materials, with further experimental work on a rubber highly cross-linked with peroxide which shows almost no hysteresis effects and minimal time dependence in stress relaxation experiments. The torque and total axial thrust were measured as a function of twist. The measurements were repeated on small cylinders cut from the larger cylinders to insure that the tests were on identical material. Plots of reduced torque and of the reduced normal force versus reduced twist for the different sized cylinders were in coincidence to within one percent, showing that rubbers are indeed "simple materials". They were also able to obtain values of $\partial W/\partial I_1$ and $\partial W/\partial I_2$ at different values of $I_1 = I_2$. The behavior resembled very much the behavior observed by Zapas [19]. At very small deformations ($I_1, I_2 \sim 3.1$), $\partial W/\partial I_1$ tended to increase and $\partial W/\partial I_2$ decrease with decreasing strain in a manner such that their sum remains roughly constant.

Up to this point this section has been involved with the general strain-energy function as derived by Rivlin for isotropic and incompressible materials. In 1967 Valanis and Landel [23] put forward the hypothesis that the strain-energy function could be represented as

$$W = f(\lambda_1) + f(\lambda_2) + f(\lambda_3), \tag{2.12}$$

where λ_1, λ_2, λ_3 represent the three principal extension ratios. This form is very similar to the general theory of Mooney already mentioned, as was pointed out by Rivlin and Sawyers [24]. The general validity of this strain-energy function has not been shown. To check the validity of the Valanis–Landel hypothesis, experiments are needed where the second derivatives of the strain-energy function are involved. It is not surprising that a reasonable agreement has been obtained with data on biaxial deformations using this form of the strain-energy. Rivlin and Sawyers [24] have shown that a strain energy W where $\partial W/\partial I_1$ is a constant and $\partial^2 W/\partial I^2$ is a constant is consistent with equation (2.12). From the experimental results shown in Figure 2 and Figure 3 it can be seen that one can approximate $\partial W/\partial I_2$ with a straight line. Parenthetically it can be mentioned that all the aforementioned molecular theories use the Valanis–Landel hypothesis. Lately Treloar [25] has shown that a non-Gaussian theory for rubber elasticity that he had proposed earlier does not yield a strain energy of the form given by (2.12), but his calculations reveal that the deviation from a Valanis–Landel form is very small.

An interesting result has been shown by Kearsley and Zapas [26] for elastic materials whose strain-energy function obeys (2.12). If the curve $f_1(\lambda)$ plotted versus λ is flat in the neighborhood of $\lambda = 1$ ($f_1(1)$ is conventionally taken to be zero) then $\partial W/\partial I_2$ must be negative for values of I_1 and I_2 very close to 3. Viewed in this way, the sudden change of $\partial W/\partial I_2$ close to the undeformed state, the experimental observation mentioned earlier, does not seem unreasonable.

At this point it is worth mentioning a restriction of a different character on the strain-energy function of rubbers. The incompressibility of rubber-like materials is usually invoked in the determination of the strain-energy function. But in reality rubbers are slightly compressible. In 1970 Penn [27] measured the volume changes accompanying the extension of a rubber. In analyzing these data he introduced a set of invariants, two of which are independent of pure volume changes, and he showed that the strain energy can not be represented as the sum of a part depending on shear alone and a part depending on volume alone.

From all those selected examples, it may be concluded that the strain-energy function of rubber-like materials is a much more complicated function of the invariants I_1 and I_2 than initially thought. The subject is by no means closed. It should be possible to use the present knowledge of synthetic rubbers and sophisticated means of cross-linking to produce a material which has minimal time dependence in stress relaxation and which does not crystallize even at very high elongations. With such a material it should be possible to obtain data on deformations testing ideas such as the finite extensibility of the chains and in general providing measurements for a better description of the strain-energy function.

References

1. K. H. Meyer, G. Von Susich and E. Valco, Kolloidzschr. **59** (1932), 208.
2. R. S. Rivlin, J. Appl. Phys. **18** (1947), 444, 837.
3. R. S. Rivlin, Phil. Trans. Roy. Soc. **A241** (1948), 379.
4. R. S. Rivlin, Proc. Roy. Soc. **A195** (1949), 463.
5. R. S. Rivlin, Phil. Trans. Roy. Soc. **A242** (1949), 173.
6. R. S. Rivlin and D. W. Saunders, Phil. Trans. Roy. Soc. **A243** (1951), 251.
7. E. Guth and H. Mark, Monatsh. Chem. **65** (1934), 93.
8. W. Kuhn, Kolloidzschr. **101** (1934), 248.
9. F. T. Wall, J. Chem. Phys. **10** (1942), 485.
10. P. J. Flory and J. Rehner, J. Chem. Phys. **11** (1943), 512.
11. H. M. James and E. Guth, J. Chem. Phys. **11** (1943), 455.

12. L. R. G. Treloar, Trans. Faraday Soc. **39** (1943), 36.
13. M. C. Wang and E. Guth, J. Chem. Phys. **20** (1952), 1144.
14. E. A. DiMarzio, J. Chem. Phys. **36** (1962) 1563.
15. P. J. Flory, J. Chem. Phys. **66** (1977), 5720.
16. M. Mooney, J. Appl. Phys. **11** (1940), 582.
17. A. N. Gent and R. S. Rivlin, Proc. Phys. Soc. **B65** (1952), 645.
18. R. S. Rivlin, Q. Appl. Math. **14** (1956), 83.
19. L. J. Zapas, J. Res. Nat'l. Bur. Std. **8A** (1966), 525.
20. G. W. Becker, J. Polymer Sci. **C16** (1967), 2893.
21. S. Kawabata and H. Kawai, *Advances in Polymer Science*, Vol. 24, *Molecular Properties*, Springer-Verlag, New York, 1977.
22. R. W. Penn and E. A. Kearsley, Trans. Soc. of Rheology **20** (1976), 227.
23. K. G. Valanis and R. F. Landel, J. App. Phys. **38** (1967), 2997.
24. R. S. Rivlin and K. N. Sawyers, Trans. Soc. of Rheology **20** (1976), 545.
25. L. R. G. Treloar and G. Riding, Proc. R. Soc. Lond. **A369** (1980), 261.
26. E. A. Kearsley and L. J. Zapas, J. of Rheology **24**(4) (1980), 483.
27. R. Penn, Trans. Soc. of Rheology **14** (1970), 509.

SPEAKERS

Dr. J. M. Ball, Department of Mathematics, Heriot-Watt University, Riccarton, Currie, Edinburgh EH14 4AS, Scotland

Professor P. G. Ciarlet, Analyse Numérique, Tour 55–65, 5e étage, Université Pierre et Marie Curie, 4 Place Jussieu, 75230 Paris Cedex 05, France

Professor G. Duvaut, Université de Paris VI, Mécanique Théorique, Tour 66–4, Place Jussieu, Paris 75230 Cedex 05, France

Professor J. L. Ericksen, Dept. of Theoretical Mechanics, Latrobe 122, The Johns Hopkins University, Baltimore, MD 21218, USA

Professor A. N. Gent, Dean of Graduate Studies and Research, The University of Akron, Akron, OH 44325, USA

Professor G. Grioli, Universita di Padova, Seminario Matematico, Via Belzoni 3, 35100 Padova, Italy

Professor M. E. Gurtin, Department of Mathematics, Carnegie-Mellon University, Schenley Park, Pittsburgh, PA 15213, USA

Professor M. Hayes, Department of Mathematical Physics, University College, Belfield, Dublin 4, Ireland

Dr. A. Golobiewska Herrmann, Dept. of Applied Mechanics, Stanford University, Stanford, CA 94305, USA

Dr. J. M. Hill, Department of Mathematics, The University of Wollongong, P. O. Box 1144, Wollongong, N.S.W.-2500, Australia

Professor J. W. Hutchinson, Division of Applied Sciences, Pierce Hall, Harvard University, Cambridge, MA 02138, USA

Professor F. John, Courant Institute of Mathematical Sciences, 251 Mercer Street, New York, NY 10012, USA

Professor J. K. Knowles, Division of Engineering and Applied Science, California Institute of Technology, Pasadena, CA 91125, USA

Prof. Dr. Ir. W. T. Koiter, Delft University of Technology, Department of Mechanical Engineering, Mekelweg 2, 2600 GA Delft, The Netherlands

Professor Ingo Müller, Hermann-Föttinger Institut für Thermo- und Fluiddynamik, Technische Universität Berlin, Strasse des 17 Juni 135, 1000 Berling 12, German Federal Republic

Professor P. M. Naghdi, Department of Mechanical Engineering, University of California, Berkeley, Berkeley, CA 94720, USA

Professor J. T. Oden, Department of Aerospace Engineering and Engineering Mechanics, The University of Texas at Austin, Austin, TX 78712, USA

Dr. R. W. Ogden, School of Mathematics, University of Bath, Claverton Down, Bath BA2 7AY, Somerset, England

Professor A. C. Pipkin, Division of Applied Mathematics, Brown University, Providence, RI 02912, USA

Professor R. S. Rivlin, Center for the Application of Mathematics, Lehigh University, 203 E. Packer Ave., Bethlehem, PA 18015, USA

Professor K. N. Sawyers, Center for the Application of Mathematics, Lehigh University, 203 E. Packer Ave., Bethlehem, PA 18015, USA

Professor R. T. Shield, Dept. of Theoretical and Applied Mechanics, 212 Talbot Laboratory, University of Illinois, Urbana, IL 61801, USA

Professor R. Skalak, Dept. of Civil Engineering and Engineering Mechanics, Seeley W. Mudd Building, Columbia University, New York, NY 10027, USA

Professor A. J. M. Spencer, Department of Theoretical Mechanics, The University of Nottingham, University Park, Nottingham, NG7 2RD, England

Professor Dr.-Ing. E. Stein, Lehrstuhl für Baumechanik, Technische Universität Hannover, Callinstrasse 32, D-3000 Hannover 1, German Federal Republic

Dr. A. G. Thomas, The Malysian Rubber Producers' Research Association, Brickendonbury, Hertford, SC13 8NL, England

Professor T. Valent, Istituto di Analisi e Meccanica, Via Belzoni 7, 35100 Padova, Italy

Dr. T. W. Wright, U.S. Army Armament Research & Development Command, U.S. Army Ballistic Research Laboratory, Aberdeen Proving Ground, MD 21005, USA

Dr. L. J. Zapas, National Bureau of Standards, Gaithersburg, MD 20760, USA

PARTICIPANTS OTHER THAN SPEAKERS, AND SPONSORS

Dr. R. Abeyaratne, Division of Engineering, Brown University, Providence, RI 02912, USA

Dr. C. Astill, Solid Mechanics Program, Division of Engineering, National Science Foundation, Washington, DC 20550, USA

Dr. N. Basdekas, Office of Naval Research, Department of the Navy, Arlington, VA 22217, USA

Professor M. F. Beatty, Department of Engineering Mechanics, University of Kentucky, Lexington, KY 40506, USA

Dr. A. Cardon, Dienst voor Toegepaste Mechanica van het Continuum, Pleinlaan 2, B-1050, Brussels, Belgium

Professor M. M. Carroll, Department of Mechanics, School of Engineering, University of California, Berkeley, Berkeley, CA 94720, USA

Dr. J. Chandra, Mathematics Division, U.S. Army Research Office, Research Triangle Park, NC 27709, USA

Dr. P. K. Currie, Shell Research B.V. Rijswijk, Den Haag, The Netherlands

Professor D. G. B. Edelen, Center for the Application of Mathematics, Lehigh University, 4 West 4th Street, Bethlehem, PA 18015, USA

Professor F. Erdogan, Department of Mechanical Engineering and Mechanics, Lehigh University, 550 Packard Lab., Bldg. #19, Bethlehem, PA 18015, USA

Professor J. D. Eshelby, University of Sheffield, Faculty of Materials, Sheffield S10 2TN, England

Professor G. Fichera, Via Pietro Mascagni 7, 00199 Rome, Italy

Dr. J. E. Fitzgerald, Director, School of Civil Engineering, Georgia Institute of Technology, Atlanta, GA 30332, USA

Professor R. Fosdick, Dept. of Aerospace Engineering and Mechanics, University of Minnesota, Minneapolis, MN 55455, USA

Dr. W. A. Green, Department of Theoretical Mechanics, University of Nottingham, University Park, Nottingham NG7 2RD, England

Dr. J. M. Greenberg, Mathematical Sciences Division, National Science Foundation, Washington, DC 20550, USA

Dr. Guo Zhong-Heng, Lehrstuhl für Mechanik 1, Rurh-Universität Bochum, Postfach 102148, D-4630 Bochum, German Federal Republic

Professor George Herrmann, Department of Applied Mechanics, Stanford University, Stanford, CA 94305, USA

Professor C. O. Horgan, Dept. of Metallurgy, Mechanics and Materials Science, Michigan State University, East Lansing, MI 48824, USA

Dr. R. D. James, Division of Engineering, Brown University, Providence, RI 02912, USA

Dr. E. A. Kearsley, National Bureau of Standards, Gaithersburg, MD 20760, USA

Professor N. Laws, Department of Mathematics, Cranfield Institute of Technology, Cranfield, Bedford, England

Professor E. H. Lee, Department of Applied Mechanics, Stanford University, Stanford, CA 94305, USA

Professor M. F. McCarthy, Department of Mathematical Physics, University College, Galway, Ireland

Dr. K. W. Neale, Department of Civil Engineering, Université de Sherbrooke, Sherbrooke, Quebec, Canada J1K 2R1

Dr. D. F. Parker, Department of Theoretical Mechanics, University of Nottingham, University Park, Nottingham, NG7 2RD, England

Dr. G. P. Parry, Department of Mathematics, University of Bath, Bath, Somerset, England

Dr. N. Perrone, Director, Structural Mechanics Program, Department of the Navy, Office of Naval Research, Arlington, VA 22217, USA

Dr. P. Podio-Guidugli, Department of Mathematics, Carnegie-Mellon University, Schenley Park, Pittsburgh, PA 15213, USA

Dr. T. G. Rogers, Department of Theoretical Mechanics, The University, University Park, Nottingham, NG7 2RD, England

Dr. E. Saibel, Engineering Sciences Division, Department of the Army, US Army Research Office, Box 12211, Research Triangle Park, NC 27709, USA

Professor George C. M. Sih, Director, Institute of Fracture & Solid Mechanics, Lehigh University, Packard Laboratory, Room 451A, Building #19, Bethlehem, PA 18015, USA

Professor Gerald F. Smith, Director, Center for the Application of Mathematics, Lehigh University, 203 E. Packer Ave., Bethlehem, PA 18015, USA

Professor A. S. Wineman, Department of Applied Mechanics and Engineering Science, The University of Michigan, Ann Arbor, MI 48109, USA